MATHEMATICAL METHODS IN ELECTROMAGNETISM

Linear Theory and Applications

Series on Advances in Mathematics for Applied Sciences – Vol. 41

MATHEMATICAL METHODS IN ELECTROMAGNETISM

Linear Theory and Applications

Michel Cessenat

CEA/DAM
Centre d'Etudes de Bruyères-le-Châtel
France

World Scientific
Singapore • New Jersey • London • Hong Kong

Published by

World Scientific Publishing Co. Pte. Ltd.

5 Toh Tuck Link, Singapore 596224

USA office: 27 Warren Street, Suite 401-402, Hackensack, NJ 07601

UK office: 57 Shelton Street, Covent Garden, London WC2H 9HE

Library of Congress Cataloging-in-Publication Data

Cessenat, Michel.

Mathematical methods in electromagnetism : linear theory and applications / Michel Cessenat.

396 p. 22.5 cm. -- (Series on advances in mathematics for applied sciences ; vol. 41)

Includes bibliographical references and index.

ISBN-13 978-981-02-2467-7 -- ISBN-10 981-02-2467-2

1. Electromagnetism -- Mathematics. I. Title. II. Series.

QC760.C43 1996

537'.01'51--dc20 96-11628

CIP

British Library Cataloguing-in-Publication Data

A catalogue record for this book is available from the British Library.

Acknowledgements

I wish to thank Professor R. Dautray for drawing my attention to certain problems of electromagnetism. I also wish to thank Professor J.L. Lions for suggesting to me, one day at Orly airport, to write this book. Applying his ideas on asymptotic analysis to electromagnetism is an exciting challenge! I wish to express my gratitude to Professor A. Bossavit for constructive remarks on ways to improve this book. I am grateful to Professors M. Artola, P. Benilan, R. Petit and also to Dr. A. Gervat for useful comments. Dr. G. Zerah must be thanked for fruitful discussions on ferromagnetism.

I am also gratefully indebted to Dr. J.-M. Clarisse, and O. Cessenat, and also to C. Averseng and C. Mares for their help.

Preface

Electromagnetism has numerous applications whether in energy transport or signal transmission in devices as diverse as antennas, waveguides, optical fibers, gratings, electrical circuits, light in lasers and plasmas. Thus, much research has been devoted to problems in electromagnetism essentially from a physical point of view. One aim of this book is to present a more global analysis encompassing both the physical and mathematical aspects of electromagnetism. In particular, powerful modern mathematical methods are emphasized, which lead to numerical applications.

Phenomena under consideration are modelled as "distributed system" by the "electromagnetic field" which must satisfy Maxwell's equations. In general, media are described using the "continuum assumption" by introducing specific constants and thus, without entering their fine structures. This description leads to constitutive relations. Various conditions must also be treated such as: initial, boundary or transmission conditions, conditions at infinity, finite (or locally finite) energy conditions.

In the absence of any coupling with other phenomena such as fluid motion, determining the electromagnetic field which satisfies all the necessary conditions is already a difficult problem.

The student or the engineer may feel quite uneasy in face of these problems and think that usual mathematical books are helpless. Indeed, problems in electromagnetism can be very diverse, and differ from classical problems in partial differential equations in the vectorial nature of the electromagnetic field. Thus, differential geometry is present at all stages whether when modelling, choosing a mathematical framework, or computing theoretical or numerical solutions. Another peculiarity is inherent to the constitutive relations and to the specific "constants" of the media. These constants may be complex or take the form of matrices for usual dissipative and anisotropic media, or be positive real numbers for free space (or ideal conservative media). Therefore, different methods must be applied depending on the media. It must also be the case that these specific constants strongly depend on frequency, thus leading to evolution problem with time delay, or depend on the electromagnetic field itself, giving rise to nonlinear effects.

Last but not least, electromagnetism problems are wave propagation problems, therefore presenting classical features of such problems. Hence stationary problems in the frequency domain are often ill-posed, and become well-posed only by imposing radiation conditions at infinity or on the direction of propagation. This results in difficulties when treating scattering by infinite obstacles.

One aim of this book is to provide the reader with the basic tools and the functional analysis concepts corresponding to the usual physical hypotheses of the modelling stage. We also present in the Appendix, certain properties of differential geometry, the corner stone of electromagnetism. As examples of application of these tools and concepts, we treat several fundamental problems of electromagnetism, e.g.: scattering of an incident electromagnetic wave by a bounded obstacle, scattering by a grating (periodically infinite obstacle), wave propagation in a waveguide or in an optical fiber.

Several recent approaches suitable for solving electromagnetism problems and related to numerical methods are presented, in particular:

Integral methods, resulting in solving an integral equation on a given surface (typically the surface of the obstacle in a stationary scattering problem) thus sparing one spacial dimension. However, the resulting matrices are full.

Semigroup methods, allowing to treat evolution problems, and also stationary problems both with planar or cylindrical geometries, and a given direction of wave propagation.

Variational methods (particularly suited to numerical applications), in the case of dissipative media.

Spectral methods, when spectral decompositions (of normal operators) are possible. However, these approaches, well known in physics as *modal decompositions*, and which are the core of many textbooks on electromagnetism are often used beyond their scope.

In many cases, hybrid approaches are preferable which combine the respective advantage of different methods. For instance, when solving a stationary scattering problem, a finite element method is applied to an inhomogeneous obstacle and coupled to an integral method for the outer domain. The determination of constants for a sample in a waveguide is another typical example: a finite element method is used in the sample and coupled to a spectral method for the rest of the homogeneous waveguide.

One of the main ideas developed in this book is the use of Calderon projectors and operators (also called impedance or admittance surface operators). These operators are surface operators containing all the information relevant to the electromagnetic properties of the domain bounded by the surfaces where they are defined. Such operators are especially useful when tackling asymptotic problems in electromagnetism: e.g. the scattering by an obstacle of high conductivity or with fine periodic inclusions.

We hope that this book is useful for students or engineers having to solve problems in electromagnetism. We also hope that mathematical notions will not conceal physical concepts from the reader, but rather allow a better understanding of modelization in electromagnetism, and emphasize the essential features related to the geometry and nature of materials.

Some prerequisites in functional analysis may be useful as for example Dautray-Lions [1].

Contents

CHAPTER 1

MATHEMATICAL MODELLING OF THE ELECTROMAGNETIC FIELD IN CONTINUOUS MEDIA

MAXWELL EQUATIONS AND CONSTITUTIVE RELATIONS

1. EVOLUTION MAXWELL EQUATIONS

Electromagnetic phenomena in a domain $\Omega \subseteq R^3$ occupied by a medium and for all time t of R (or only for positive t or for a finite interval of time) are described with the help of four functions D, E, B, H of $(x,t) \in \Omega \times R$ with values in R^3. These are fairly generally called (but there is no universal agreement on these names):

D: *electric induction*, E: *electric field*
B: *magnetic induction*, H: *magnetic field*.

These functions are related to two functions (or distributions) defined on $\Omega \times R$, for all t (or only some interval), with values in R and R^3 (resp.), called *charge density* ρ, and *current density* J (resp.), by the *Maxwell equations*:

$$(1)\quad \left|\begin{array}{lll} \text{i)} & -\dfrac{\partial D}{\partial t} + \operatorname{curl} H = J, & \textit{Maxwell Ampere law} \\ \text{ii)} & \dfrac{\partial B}{\partial t} + \operatorname{curl} E = 0, & \textit{Maxwell Faraday law} \\ \text{iii)} & \operatorname{div} D = \rho, & \textit{Gauss electrical law} \\ \text{iv)} & \operatorname{div} B = 0, & \textit{Gauss magnetic law} \end{array}\right.$$

with the following usual notations, in a Cartesian system of coordinates, for $E = (E_1, E_2, E_3)$, $x = (x_1, x_2, x_3)$:

(2) $\quad \operatorname{div} E = \sum_i \frac{\partial E_i}{\partial x_i}, \quad \operatorname{curl} E = (\frac{\partial E_3}{\partial x_2} - \frac{\partial E_2}{\partial x_3}, \frac{\partial E_1}{\partial x_3} - \frac{\partial E_3}{\partial x_1}, \frac{\partial E_2}{\partial x_1} - \frac{\partial E_1}{\partial x_2}).$

The densities of current and of charge satisfy the continuity relation or *charge conservation*:

(3) $\quad \frac{\partial \rho}{\partial t} + \operatorname{div} J = 0.$

The system of units employed here is the "système international" formerly called "MKSA". The units for the quantities ρ, J, D, E, B, H, and their relations with the fundamental units M (mass), L (length), T (time), Q (charge) are:

$[\rho]$ = Coulomb/cubic meter = $L^{-3}Q$, $[J]$ = Ampere/square meter = $L^{-2}T^{-1}Q$
$[D]$ = Coulomb/square meter = $L^{-2}Q$, $[E]$ = Volt/meter = $MLT^{-2}Q^{-1}$,
$[B]$ = Weber/square meter = $MT^{-1}Q^{-1}$, $[H]$ = Ampere/meter = $L^{-1}T^{-1}Q$.

From the identity, for all vector fields u: div curl u = 0, we see that equations (1)iii) and iv) are partially redundant: applying the divergence to (1)i) and ii), then taking account of (3), we obtain

(4) $\quad \frac{\partial}{\partial t}(\operatorname{div} D - \rho) = 0, \quad \frac{\partial}{\partial t}(\operatorname{div} B) = 0,$

which implies the relations (1)iii) and iv) for all time t if these relations are satisfied at some initial instant. Besides there are additional relations, which are seen below.

To some extent, it is possible to act on ρ and J to produce a required electromagnetic field as in transmitting antenna. A typical example is a thin wire modelled by a line, with ρ and J concentrated on this line. Besides, they may also be partially known and subject to a random process.

Free space case. For free space in a domain Ω (with only given charges and currents in it), the quantities D, E, B, H are linked by the following "*constitutive relations*"

(5) $\quad D = \varepsilon_0 E, \quad B = \mu_0 H,$

with ε_0 and μ_0 called respectively *permittivity* and *permeability* of the free space

(6) $\quad \varepsilon_0 = \frac{1}{36\pi} 10^{-9}$ Farad/meter, $\mu_0 = 4\pi.10^{-7}$ Henry/meter, $\varepsilon_0 \mu_0 c^2 = 1,$

with $[\varepsilon_0] = M^{-1}L^{-3}T^2Q^2$, $[\mu_0] = MLQ^{-2}$, c being the *velocity of light in free space*. The electromagnetic field, reduced to (E, B), satisfies Maxwell equations

$$
(1)' \quad \left|
\begin{array}{l}
\text{i)}\ -\dfrac{1}{c^2}\dfrac{\partial E}{\partial t} + \text{curl}\, B = \mu_0 J \\
\text{ii)}\ \dfrac{\partial B}{\partial t} + \text{curl}\, E = 0 \\
\text{iii)}\ \text{div}\, E = \dfrac{1}{\varepsilon_0}\rho \\
\text{iv)}\ \text{div}\, B = 0,
\end{array}
\right.
$$

which are also the equations of electromagnetism for the *microscopic scale*, naturally posed in all the space R^3. By homogenization or by any averaging process, these are believed to give Maxwell equations (1) in continuous media for the *macroscopic scale*. When there is no charge and no current, using the relation curl curl u = – Δu + grad div u (for all vector fields u), we deduce from (1)' that E and B satisfy the wave equation (with velocity c) which shows the *hyperbolic* nature of Maxwell equations:

$$-\frac{1}{c^2}\frac{\partial^2 u}{\partial t^2} + \Delta u = 0.$$

⊗

Energy Balance in free space (without charges or currents). We define the *electromagnetic energy* W(t) in a domain Ω of free space by

$$(7) \qquad W(t) = \frac{1}{2}\int_\Omega (D.E + B.H)\, dx,$$

with the density of energy:

$$(8) \qquad w = \frac{1}{2}(D.E + B.H) = \frac{1}{2}(\varepsilon_0 E^2 + \mu_0 H^2).$$

The *Poynting vector* is then defined with the vector product ∧ by

$$
(9) \quad \left|
\begin{array}{l}
\qquad\qquad S = E \wedge H, \\
\text{i.e., } S_1 = E_2H_3 - E_3H_2,\ S_2 = E_3H_1 - E_1H_3,\ S_3 = E_1H_2 - E_2H_1.
\end{array}
\right.
$$

Using the formula:

$$(10) \qquad \text{div}\,(E \wedge H) = \text{curl}\, E.H - E.\text{curl}\, H,$$

and Maxwell equations, we have (for J = 0):

$$(11) \qquad \frac{\partial w}{\partial t} + \text{div}\, S = 0.$$

Integrating on Ω gives:

$$(12) \qquad \frac{\partial W}{\partial t}(t) = -\int_\Gamma n.S\, d\Gamma,$$

where n is the exterior normal to the boundary Γ of Ω. Thus *the time derivative of the energy in Ω (i.e. the power P(t)) is equal to the opposite of the flux of the Poynting vector through the boundary Γ of Ω.*

These relations are a priori formal and must be justified but we already know that E and H must be *square integrable* on Ω, in order to have a finite electromagnetic energy which gives the *natural mathematical framework* for the study of the electromagnetic field.
When there are particles (at rest) in the free space, which are modelled by point charges, the electromagnetic energy is not finite. This is generally viewed as a defect of the modelling.
In the SI unit system, the unit for energy is Joule and the unit for power is Watt, with $[W] = ML^2T^{-2}$, $[P] = ML^2T^{-3}$. ⊗

Remark 1. When the medium that fills Ω is not the free space, formula (7) is not representative of the electromagnetic energy but it is still possible to balance the power with the expression:

$$(13) \qquad P(t) = \int_\Omega \left(\frac{\partial B}{\partial t}.H + \frac{\partial D}{\partial t}.E\right) dx$$

in the framework of square integrable fields. The problem of defining the energy of the electromagnetic field by some formula with respect to D, E, B, H, comes from the difficulty of separating it from the energy of matter.

In any case, writing that D, E, B, H are square integrable fields on every compact set is believed to impose that the electromagnetic energy be locally finite. This property often ensures uniqueness of the solution, notably for problems with wedges.

⊗

2. STATIONARY MAXWELL EQUATIONS

If we assume that the electromagnetic field is defined and satisfies (1) for all t in R (which is not so obvious in comparison with diffusion phenomena), it is possible to apply a Laplace transformation (see Dautray-Lions [1] Chap. 16). To be specific, let $\mathcal{S}'(R)$ denotes the set of tempered distributions, we define, for each distribution f on R, the following set

$$(14) \qquad I_f = \{\xi \in R, \quad e^{-\xi t} f \in \mathcal{S}'(R)\}.$$

It is an interval of R. The *bilateral Laplace transform* of f is defined by

$$(14)' \qquad \mathcal{L}f(p) = \mathcal{F}(e^{-\xi t} f)(\eta), \quad p = \xi + i\eta, \quad \xi \in I_f, \quad \eta \in R,$$

where $\mathcal{F}g$ denotes the usual *Fourier transform* of the tempered distribution g.

We write formally: $\mathcal{F}g(\eta) = \int_{-\infty}^{+\infty} e^{-i\eta t} g(t)\,dt$, $\mathcal{L}f(p) = \int_{-\infty}^{+\infty} e^{-pt} f(t)\,dt$.

In electromagnetism it is more common to use the *Fourier-Laplace* transform:

$$\hat{f}(\omega) = \mathcal{L}f(-i\omega) = \int_{-\infty}^{+\infty} e^{it\omega} f(t)\,dt, \quad \omega = ip = -\eta + i\xi. \tag{15}$$

Let z' and z" be the real and imaginary parts of every complex number z = z' + iz". Thus for $\omega = \omega' + i\omega''$, (15) is defined for $\omega'' = \xi \in I_f$. Restricted to real numbers ω, $\hat{f}$ is the usual Fourier transform of f (up to a change of sign): $\hat{f}(\omega) = \mathcal{F}f(-\omega)$. Then using the following Fourier-Laplace property:

$$\left(\frac{df}{dt}\right)\hat{}\,(\omega) = -i\omega\hat{f}(\omega), \quad \omega \in \overset{\circ}{I}_f \text{(the interior of } I_f), \tag{16}$$

the Fourier-Laplace transform of (1) is:

$$\left|\begin{array}{l} \text{i) } i\omega\hat{D}(\omega) + \text{curl } \hat{H}(\omega) = \hat{J}(\omega), \\ \text{ii) } -i\omega\hat{B}(\omega) + \text{curl } \hat{E}(\omega) = 0, \\ \text{iii) } \text{div } \hat{D}(\omega) = \hat{\rho}(\omega), \\ \text{iv) } \text{div } \hat{B}(\omega) = 0, \end{array}\right. \tag{17}$$

with:

$$-i\omega\hat{\rho}(\omega) + \text{div } \hat{J}(\omega) = 0. \tag{18}$$

These equations are called the ***stationary (or steady-state) Maxwell equations***. Note that, using (15), the real fields D, E, B, H are transformed into complex fields satisfying

$$\overline{\hat{D}(\omega)} = \hat{D}(-\bar{\omega}), \quad \overline{\hat{E}(\omega)} = \hat{E}(-\bar{\omega}), \quad \omega \in C \text{ with } \omega'' \in I_D, I_E, \ldots \tag{19}$$

since (15) implies

$$\overline{\hat{f}(\omega)} = \int_{-\infty}^{\infty} e^{-i\bar{\omega}t} f(t)\,dt = \hat{f}(-\bar{\omega}). \tag{19'}$$

Thus the knowledge of $\hat{D}(\omega), \ldots$ for positive numbers ω also implies that of $\hat{D}(-\omega)$, ... when Im $\omega = 0$ is allowed, i.e., when D, E... are tempered distributions in time. The variable $\omega > 0$ is the ***(angular) frequency (or pulsation)***. We keep this terminology for a complex ω. The variable $\nu = \omega/2\pi$ is the ***frequency***. Its unit is Herz (Hz). Frequencies are often given in MHz (10^6 Hz) or GHz (10^9 Hz). In free space, the wavelength λ is define by $\lambda = c/\nu$; then 1 GHz corresponds to 30 cm. The visible domain corresponds to 10^{15} Hz ($\lambda = 0.4$ to $0.8\ \mu$m, $1\mu\text{m} = 10^{-6}$ m). Frequencies of γ-rays, X-rays are higher than 10^{15} Hz; hyperfrequencies or microwaves frequencies are about 10^{11} or 10^{10} Hz, λ is about millimeters or centimeters. Frequencies for radio, TV, radar are less than 10^{10} Hz.

For electromagnetic fields defined for positive times only, we can use a *unilateral Laplace transformation* defined by: $L_u f(p) = \int_0^\infty e^{-pt} f(t)\, dt$.

Then the initial values of the field D(0), B(0) would appear in (17). But this would be inconsistent with the usual constitutive relations (see (22) below).

⊗

Remark 2. We could also consider the Maxwell equations (1) for complex fields and notably for vector functions of the form:

(20) $D(t) = D_0\, e^{-i\omega t}, \quad E(t) = E_0\, e^{-i\omega t}, \quad \text{with } D_0, E_0 \in C^3,$

then equations (1) and (3) give equations (17), (18) for D_0, E_0 directly. It must not be forgotten that the modelling of electromagnetic phenomena imposes real fields. In fact, *stationary fields* correspond to the real part of (20).

⊗

Remark 3. The modelling of physical situations prevents us from having any distribution for ρ and J: they are nothing worse than first-order distributions.

⊗

Remark 4. *Some properties of Fourier-Laplace transforms*

We recall the analyticity property of some Fourier-Laplace transforms:

i) If f is a distribution with compact support, then $\hat{f}(\omega)$ is a holomorphic (or analytic) function on all the complex plane C.

ii) If f is a distribution on R (with $I_f \neq \varnothing$) with its support in $[\alpha, \infty[$, then its Fourier-Laplace transform is holomorphic in a half-plane $\omega'' > \xi_0$, and satisfies:

$$|\hat{f}(\omega)| < e^{-\omega''\alpha}\, \mathrm{Pol}(|\omega|), \ \omega = \omega' + i\omega'', \ \omega'' > \xi_0 ,$$

where Pol is a polynomial with positive coefficients (see Schwartz [1] p. 310). Converse statements are *Paley-Wiener* theorems.

If S and f are distributions with support bounded to the left, with: $\overset{\circ}{I}_S$ and $\overset{\circ}{I}_f \neq \varnothing$, then $I_{S*f} \supseteq I_S \cap I_f$ and: $\widehat{S*f}(\omega) = \hat{S}(\omega).\hat{f}(\omega)$, for $\omega'' \in \overset{\circ}{I}_S \cap \overset{\circ}{I}_f$.

Some useful examples of Fourier-Laplace transforms

Let Y be the Heaviside function: Y(t) = 1 for t > 0, 0 for t < 0; we will also use the sgn function: sgn t = 1 for t > 0, − 1 for t < 0, and the constant function 1, 1(t) = 1 for all t. Then their intervals (14) are respectively given by:

$$I_Y = [0, +\infty), \quad I_{1-Y} = (-\infty, 0], \quad I_{sgn} = \{0\}, \quad I_1 = \{0\}.$$

Denote pv the Cauchy principal value, δ the Dirac distribution, their Fourier-Laplace transforms are

i) for f = Y, $\hat{f}(\omega) = i/\omega$, $\omega'' > 0$; for the boundary value $\omega'' = 0$, we have:

$$\hat{f}(\omega) = i\, \mathrm{pv}\frac{1}{\omega'} + \pi\delta(\omega'),$$

ii) $f = 1 - Y$, $\hat{f}(\omega) = -i/\omega$, $\omega'' < 0$, and for $\omega'' = 0$: $\hat{f}(\omega) = -i\,\mathrm{pv}\frac{1}{\omega'} + \pi\delta(\omega')$,

iii) $f = \mathrm{sgn}$, $\hat{f}(\omega) = 2i\,\mathrm{pv}\frac{1}{\omega'}$, for $\omega'' = 0$,

iv) $f = 1$, $\hat{f}(\omega) = 2\pi\,\delta(\omega')$, for $\omega'' = 0$.

⊗

3. CONSTITUTIVE RELATIONS

In a given medium occupying the domain Ω, we define the vector fields P and M the (electric) *polarization* and the *magnetization* respectively by:

(21) $$P + \varepsilon_0 E = D, \quad M + \mu_0 H = B.$$

For certain media, P and M depend only on E and H, this gives relations by (21), called *constitutive relations* expressing D and B by means of E and H. They may be of very different types: linear or not, local or not. So we give only some of them. We will assume here that the medium is at rest. Moving media will be considered in the Appendix.

These relations may come either from a microscopic theory and (or) from homogenization, or from results of experiments (by solving some inverse problems or by mathematical modelling as in hysteresis phenomena). They enable us to partially avoid the description of the electromagnetic field interaction with matter. This is certainly a weak point of the modelling. The mere distributional framework which was sufficient for Maxwell equations (1), does not allow us to give a meaning to the constitutive relations even in very simple situations. The framework of (locally) square integrable functions (with respect to x) will be useful to give a meaning to such situations.

3.1. Linear isotropic dielectric media

3.1.1. Conditions obtained from invariance by time translation and causality

In a linear dielectric medium, the constitutive relations have to be:

i) *linear* (and local, free of x-derivatives) between D and E on the one hand, and between B and H on the other hand,

ii) *invariant by time translation*, therefore expressed by means of a convolution in time (here we assume that the electromagnetic field exists for all time):

(22) $$D = \varepsilon \underset{t}{*} E, \quad B = \mu \underset{t}{*} H, \text{ for certain distributions in t, } \varepsilon \text{ and } \mu,$$

iii) *causal*, i.e., $E(x,t')$ and $H(x,t')$ cannot influence $P(x,t)$ and $M(x,t)$ for $t' > t$.

These conditions imply (for *isotropic* media) that ε and μ are distributions with respect to t with support in R^+. The relations (22) are formally written as:

(22)' $$D(t) = \int_{-\infty}^{t} \varepsilon(t-t')E(t')dt', \quad B(t) = \int_{-\infty}^{t} \mu(t-t')H(t')dt'.$$

We furthermore assume that the relations are local in x, and that $\varepsilon(t)$ and $\mu(t)$ are (bounded) functions of x. If the medium is homogeneous and isotropic, the real valued distributions ε and μ are independent of x in Ω; ε and μ correspond to the permittivity and the permeability of the medium.
The Fourier-Laplace transforms of relations (22) are:

$$(23) \qquad \hat{D}(\omega) = \hat{\varepsilon}(\omega)\,\hat{E}(\omega), \quad \hat{B}(\omega) = \hat{\mu}(\omega)\,\hat{H}(\omega).$$

The quantities $\hat{\varepsilon}(\omega)$ and $\hat{\mu}(\omega)$ are called *permittivity and permeability* of the medium at the frequency ω. Their units are the Farad/meter and the Henry/meter. Then we define, δ being the time Dirac distribution,

$$(24) \qquad \kappa = \varepsilon - \varepsilon_0\delta, \quad \chi = \mu - \mu_0\delta,$$

and thus:

$$(24)' \qquad \hat{\kappa} = \hat{\varepsilon} - \varepsilon_0, \quad \hat{\chi} = \hat{\mu} - \mu_0.$$

Comparing (24) with (21), we see that P and M are related to E, H by:

$$(25) \qquad \left| \begin{array}{ll} \text{i) } P = \kappa \underset{t}{*} E, & \text{i.e., } \hat{P}(\omega) = \hat{\kappa}(\omega)\,\hat{E}(\omega), \\ \text{ii) } M = \chi \underset{t}{*} H, & \text{i.e., } \hat{M}(\omega) = \hat{\chi}(\omega)\,\hat{H}(\omega); \end{array} \right.$$

$\hat{\kappa}(\omega)$ and $\hat{\chi}(\omega)$ are called *electric and magnetic susceptibilities* at ω. We often replace them by $\varepsilon_0\hat{\kappa}_r(\omega)$ and $\mu_0\hat{\chi}_r(\omega)$, hence: $\varepsilon = \varepsilon_0\hat{\varepsilon}_r = \varepsilon_0(1+\hat{\kappa}_r)$, $\mu = \mu_0\hat{\mu}_r = \mu_0(1+\hat{\chi}_r)$:

$$(25)' \qquad \hat{P}(\omega) = \varepsilon_0\hat{\kappa}_r(\omega)\,\hat{E}(\omega), \quad \hat{M}(\omega) = \mu_0\hat{\chi}_r(\omega)\,\hat{H}(\omega);$$

$\hat{\kappa}_r(\omega)$, $\hat{\chi}_r(\omega)$ are called *relative electric and magnetic susceptibilities* at ω; ε_r, μ_r are the relative permittivity and permeability of the medium.

In order to give a sense to the convolution product in time (22) we have to make various hypotheses on (E, H), and on (ε, μ).

Hypotheses on (E, H). We will study two possibilities:

H1) E, H are distributions whose support in time is bounded below:

supp E and supp H $\subset [t_0, +\infty)$ for some t_0; we write E and H $\in \mathcal{D}'_+(\mathbb{R})^3$.

In this way the convolution product (22) has a meaning since the supports of ε and μ are also bounded from the left. Before the time t_0, $E(t) = H(t) = 0$, and we assume also $J(t) = 0$, $\rho(t) = 0$, $t < t_0$; *the history of the medium begins at time* t_0. We often take t_0 as the initial moment.

H2) There is p, $1 \le p \le \infty$, such that E, H $\in L^p(\mathbb{R}, L^2(\Omega)^3)$.
We will make additional hypotheses on ε and μ (see for example (27) below) which will give a meaning to (22): in this case, *all the past of the electromagnetic field intervenes in the domain* Ω.

The hypotheses H1) and H2) exclude electromagnetic fields of the form:

$$E(t) = \mathrm{Re}\,(E_0 e^{i\omega_0 t}), \quad H(t) = \mathrm{Re}\,(H_0 e^{i\omega_0 t}), \quad \forall t \in R,\ \omega_0'' > 0,$$

(E_0, H_0 independent of t), although their Fourier-Laplace transforms exist!

Hypotheses on (ε, μ), or on (κ, χ). We will study four possibilities. In all of these, ε and μ are, at least, tempered distributions of t, thus I_ε and $I_\mu \supseteq R^+$.

H3) $\hat{\varepsilon}(\omega)$ and $\hat{\mu}(\omega)$ are continuous functions of ω on R ($\omega'' = 0$), with:

(26) $$\lim_{|\omega|\to\infty} \hat{\varepsilon}(\omega) = \varepsilon_0, \quad \lim_{|\omega|\to\infty} \hat{\mu}(\omega) = \mu_0.$$

A hypothesis, a priori a little stronger, giving the continuity of the mappings: $E \to \varepsilon * E$, and $H \to \mu * H$ in the spaces $L^p(R^+, L^q(\Omega)^3)$, $\forall\, p,q \geq 1$, is:

H4) κ and $\chi \in L^1(R_t)$, with: supp κ and supp $\chi \subset R^+ = [0, \infty)$.

Let $C_0(R)$ be the space of continuous functions on R, tending to 0 at infinity. The Fourier-Laplace transforms of ε and μ with H4) are:

(27) $\hat{\varepsilon} = \varepsilon_0 + \hat{\kappa}$, $\hat{\mu} = \mu_0 + \hat{\chi}$, defined for $\omega'' \geq 0$, with: $\hat{\kappa}(\omega), \hat{\chi}(\omega) \in C_0(R_{\omega'})$, $\forall\, \omega'' \geq 0$.

Many other hypotheses, even stronger than (26) can be made, for example

H5) $\kappa, \chi, \frac{\partial \kappa}{\partial t}, \frac{\partial \chi}{\partial t} \in L^1(R_t)$ with support in $[0, \infty)$.

This hypothesis implies: $\hat{\kappa}(\omega), \hat{\chi}(\omega), \omega\hat{\kappa}(\omega), \omega\hat{\chi}(\omega) \in C_0(R_{\omega'})$, for all $\omega'' \geq 0$, then

(28) $i\omega\hat{\kappa}(\omega)$ and $i\omega\hat{\chi}(\omega) \to 0$ for $|\omega| \to \infty$, $\omega \in R$ or C,

thus:

$$\hat{\varepsilon} - \varepsilon_0 = o\left(\frac{1}{|\omega|}\right), \quad \hat{\mu}(\omega) - \mu_0 = o\left(\frac{1}{|\omega|}\right).$$

This hypothesis H5) also implies that if $E, H \in L^p(R, L^2(\Omega)^3)$, then:

$$D, \frac{\partial D}{\partial t}, B, \frac{\partial B}{\partial t} \in L^p(R, L^2(\Omega)^3).$$

H6) ε and μ have a compact support, ***that is the medium keeps the memory of its past only for times*** t such as: $-T < t < 0$.

In this case no hypothesis on (E,H) need to be made since in this case the convolution product is always defined. From Remark 4, we have:

$\hat{\varepsilon}(\omega)$, $\hat{\mu}(\omega)$ for $\omega \in R$ are restrictions to the real axis of analytical functions with exponential type, which is important for the determination of ε, μ by sampling: the question is to determine a function $\hat{f}$ from the family of numbers $(\hat{f}(j\delta))$, $j \in Z$, the results of experiments for angular frequencies with the same spacing $\delta > 0$.

The answer is given by the "first sampling theorem" (or Shannon theorem), in fact a dual form of it; see for example, Meyer [1] p. 10:

Let $T_m = \inf\{T,\ \operatorname{supp} f \subseteq [-T,+T]\}$; if $0 < \delta < \pi/T_m$ (or $T_m < \omega_e/2 = \pi/\delta$), then $\hat{f}$ (and f) is determined by the family $(\hat{f}(j\delta))$, $j \in Z$.
On the contrary, if $\delta > \pi/T_m$, then the family $(\hat{f}(j\delta))$, $j \in Z$, does not determine f; the tighter the spacing of the sampling, the greater the memory T_m of the material is.

It is moreover often implicitly assumed (see Landau-Lifschitz [1]) that:
H7) κ and χ are positive functions,
which implies:
H7)' $\hat{\kappa}(\omega)$ and $\hat{\chi}(\omega)$ are positive decreasing functions on the imaginary axis (note that $\hat{\kappa}$ and $\hat{\chi}$ are real valued on this axis).

⊗

3.1.2. Consequences of the hypothesis H4). Kramers-Kronig relations

The hypothesis on the support of ε and μ, therefore of κ and χ, being in R^+, along with H4) implies:

(29) $\qquad (1 - \operatorname{sgn} t)\,\kappa = 0, \quad (1 - \operatorname{sgn} t)\,\chi = 0.$

The Fourier-Laplace transform of (29) is (the product is changed into a convolution, and the Fourier-Laplace transform of sgn is given in Remark 4):

(30) $\qquad (1 + iH)\,\hat{\kappa} = 0, \quad (1 + iH)\,\hat{\chi} = 0,$

with H the Hilbert transformation on R, defined by the Cauchy principal value:

(31) $$Hf(\omega) = \frac{1}{\pi}\,\mathrm{pv}\int_{-\infty}^{+\infty} \frac{f(\tilde{\omega})}{\tilde{\omega} - \omega}\, d\tilde{\omega} \overset{\mathrm{def}}{=} \frac{1}{\pi} \lim_{\eta \to 0} \int_{|\tilde{\omega} - \omega| > \eta} \frac{f(\tilde{\omega})}{\tilde{\omega} - \omega}\, d\tilde{\omega}.$$

Thus the functions $\hat{\kappa}$ and $\hat{\chi}$ (and thus $\hat{\varepsilon}$ and $\hat{\mu}$) are holomorphic functions of ω in the complex half-plane:

(32) $\qquad C^+ = \{\omega \in C,\ \omega'' = \operatorname{Im}\omega > 0\},$

with $\hat{\kappa}$ and $\hat{\chi} \in L^\infty(C^+)$, and are given thanks to their restrictions to the real axis by

(33) $$\hat{\kappa}(\omega) = \frac{1}{2\pi i}\int_{-\infty}^{+\infty} \frac{\hat{\kappa}(\tilde{\omega})}{\tilde{\omega} - \omega}\, d\tilde{\omega} \quad \text{(similarly for } \hat{\chi}\text{)}, \quad \omega \in C^+.$$

By decomposition into real and imaginary parts (with usual notations):

(34) $\qquad \hat{\varepsilon}(\omega) = \hat{\varepsilon}'(\omega) + i\hat{\varepsilon}''(\omega), \quad \hat{\mu}(\omega) = \hat{\mu}'(\omega) + i\hat{\mu}''(\omega), \quad \omega \in C^+,$

the relations (30) give, for all real ω:

$$(35)\quad \left|\begin{array}{l} \text{i) } \hat{\varepsilon}'(\omega) - \varepsilon_0 = \frac{1}{\pi}\,\mathrm{pv}\int_{-\infty}^{+\infty} \frac{\hat{\varepsilon}''(\tilde{\omega})}{\tilde{\omega} - \omega}\,d\tilde{\omega} = H\hat{\varepsilon}''(\omega), \\ \text{ii) } \hat{\varepsilon}''(\omega) = -\frac{1}{\pi}\,\mathrm{pv}\int_{-\infty}^{+\infty} \frac{\hat{\varepsilon}'(\tilde{\omega}) - \varepsilon_0}{\tilde{\omega} - \omega}\,d\tilde{\omega} = -H(\hat{\varepsilon}' - \varepsilon_0)(\omega), \end{array}\right.$$

and similarly for $\hat{\mu}'(\omega) - \mu_0$, and $\hat{\mu}''(\omega)$.

These relations are called *Kramers-Kronig* relations: the imaginary parts of $\hat{\varepsilon}$ and $\hat{\mu}$ are fixed by giving the real parts, and the opposite is true.

Furthermore as ε and μ are real, $\hat{\varepsilon}$ and $\hat{\mu}$ satisfy (19), i.e.,

$$(36)\qquad \hat{\varepsilon}(-\bar{\omega}) = \overline{\hat{\varepsilon}(\omega)},\quad \hat{\mu}(-\bar{\omega}) = \overline{\hat{\mu}(\omega)},\quad \omega \in C,\ \omega'' \in I_\varepsilon, I_\mu, \ldots$$

thus:

$$(36)'\qquad \hat{\varepsilon}(-\omega) = \overline{\hat{\varepsilon}(\omega)},\quad \hat{\mu}(-\omega) = \overline{\hat{\mu}(\omega)},\quad \forall\,\omega \in R,$$

therefore:

$$(36)''\quad \left|\begin{array}{l} \text{i) } \hat{\varepsilon}'(-\omega) = \hat{\varepsilon}'(\omega),\quad \hat{\mu}'(-\omega) = \hat{\mu}'(\omega), \\ \text{ii) } \hat{\varepsilon}''(-\omega) = -\hat{\varepsilon}''(\omega),\quad \hat{\mu}''(-\omega) = -\hat{\mu}''(\omega),\quad \forall\,\omega \in R. \end{array}\right.$$

Substituting (36)" into (35), we obtain:

$$(35)'\quad \left|\begin{array}{l} \text{i) } \hat{\varepsilon}'(\omega) - \varepsilon_0 = \frac{1}{\pi}\,\mathrm{pv}\int_0^{\infty} \hat{\varepsilon}''(\tilde{\omega})\frac{2\tilde{\omega}}{\tilde{\omega}^2 - \omega^2}\,d\tilde{\omega}, \\ \text{ii) } \hat{\varepsilon}''(\omega) = -\frac{1}{\pi}\,\mathrm{pv}\int_0^{\infty} (\hat{\varepsilon}'(\tilde{\omega}) - \varepsilon_0)\frac{2\tilde{\omega}}{\tilde{\omega}^2 - \omega^2}\,d\tilde{\omega}, \end{array}\right.$$

and similarly for $\hat{\mu}'$, $\hat{\mu}''$ which gives the imaginary (resp. real) parts of $\hat{\varepsilon}$, $\hat{\mu}$ from the real (resp. imaginary) parts of $\hat{\varepsilon}$ and $\hat{\mu}$ for $\omega > 0$ only.

A medium with $\hat{\varepsilon}''(\omega) = 0$, for all real ω, i.e. $\hat{\varepsilon}$ real will be such that $\hat{\varepsilon}'(\omega)$ is constant, independent of ω. This is similar for μ.

Such a medium is called *perfect*. In fact only the free space is perfect.

Other consequences of H4) Notice the following property due to the continuity of the Hilbert transformation in $FL^1(R) \subset C_0(R)$.

If $\|\varepsilon' - \varepsilon_1'\|_{L^1} \le \eta$, which implies $|\hat{\varepsilon}'(\omega) - \hat{\varepsilon}_1'(\omega)| \le \eta$, for all real ω, then:

$$|\hat{\varepsilon}''(\omega) - \hat{\varepsilon}_1''(\omega)| \le \|\varepsilon'' - \varepsilon_1''\|_{L^1} \le \eta,\quad \forall\,\omega \in R,$$

or also if $\kappa \in L^1(R^+)$, then:

$$\sup_{\omega \in R} |\hat{\kappa}(\omega)| = \|\hat{\kappa}\|_{C_0(R)} \le \int_0^{\infty} |\kappa(t)|\,dt = \|\kappa\|_{L^1(R^+)}. \qquad \otimes$$

3.1.3. Constitutive relation between J and E: Ohm's law

In general, electromagnetic field produces currents and charges in media: these currents are called *eddy currents*. In certain materials (called *ohmic conductors*) we can add to (22) a relation constitutive between J and E, (of the same type, for the same reason):

(37) $$J = \sigma \underset{t}{*} E \quad (\text{or } J = \sigma \underset{t}{*} E + J_e, \; J_e \text{ a given exterior current})$$

with σ a real function (or distribution) of t (for an isotropic medium) with support in R^+ also for causality reasons; σ corresponds to the *conductivity* of the medium. The Fourier-Laplace transform of (37) is:

(38) $$\hat{J} = \hat{\sigma}\,\hat{E} \quad \textit{Ohm's law};$$

$\hat{\sigma}(\omega)$ is called the *conductivity* of the medium (at ω); its unit is ohm^{-1}/meter.
It seems fairly natural to have hypotheses on σ similar to that on ε, μ, notably to H4):

H8) $\sigma = \sigma_1 \delta + \sigma_2$, σ_1 is a positive constant, with: supp $\sigma_2 \subseteq R^+$, $\sigma_2 \in L^1(R)$,

giving a sense to (37), with the hypothesis H1) or H2) on (E, H), and the properties: supp $J \subseteq$ supp E with H1) when supp $E \subset [t_1, +\infty[$), or $J \in L^p(R, L^2(\Omega)^3)$ with H2).

Modelling an energy loss by Joule effect (see below) leads one to assume:

H9) $\operatorname{Re}\hat{\sigma} = \operatorname{Re}(\sigma_1 + \hat{\sigma}_2) \geq 0$.

3.1.4. Stationary Maxwell equations with Contitutive relations

From (17) with (23) and (38), we obtain the *stationary Maxwell equations* in the form:

(39) $$\left|\begin{array}{l} \text{i) } (i\omega\hat{\varepsilon}(\omega) - \hat{\sigma}(\omega))\hat{E}(\omega) + \operatorname{curl}\hat{H}(\omega) = 0, \\ \text{ii) } - i\omega\hat{\mu}(\omega)\hat{H}(\omega) + \operatorname{curl}\hat{E}(\omega) = 0, \\ \text{iii) } \operatorname{div}(\hat{\varepsilon}(\omega)\,\hat{E}(\omega)) = \hat{\rho}(\omega), \\ \text{iv) } \operatorname{div}(\hat{\mu}(\omega)\,\hat{H}(\omega)) = 0, \end{array}\right.$$

with:

(40) $$- i\omega\hat{\rho}(\omega) + \operatorname{div}(\hat{\sigma}(\omega)\hat{E}(\omega)) = 0.$$

Then we can change D and ε into D_c and ε_c defined (formally) by:

(41) $$\left|\begin{array}{l} D_c(t) = D(t) + \int_{-\infty}^{t} J(s)\,ds = \varepsilon \underset{t}{*} E + \int_{-\infty}^{t} \sigma \underset{t}{*} E\,ds, \\ \\ D_c = D + Y * J = \varepsilon * E + Y * (\sigma * E) = (\varepsilon + Y * \sigma) * E. \end{array}\right.$$

Let:

(42) $\Sigma(t) = \int_{-\infty}^{t} \sigma(s)\, ds = \int_{0}^{t} \sigma(s)\, ds = (Y * \sigma)(t), \quad \text{supp}\, \Sigma \subseteq R^+,$

and

(43) $\varepsilon_c = \varepsilon + \Sigma = \varepsilon + Y * \sigma.$

The Fourier-Laplace transforms of (43) and (41) are:

(44) $\hat{\varepsilon}_c(\omega) = \hat{\varepsilon}(\omega) + i\frac{\hat{\sigma}(\omega)}{\omega}, \quad \hat{D}_c(\omega) = \hat{\varepsilon}_c(\omega)\hat{E}(\omega), \quad \omega'' > 0.$

The stationary Maxwell equations (39) without exterior currents and charges may be written, according to (44):

(45) $$\left| \begin{array}{l} \text{i) } i\omega\hat{\varepsilon}_c\hat{E} + \text{curl}\, \hat{H} = 0, \\ \text{ii) } -i\omega\hat{\mu}\hat{H} + \text{curl}\, \hat{E} = 0, \\ \text{iii) } \text{div}\, (\hat{\varepsilon}_c\hat{E}) = 0, \\ \text{iv) } \text{div}\, (\hat{\mu}\hat{H}) = 0. \end{array} \right.$$

Then in a homogeneous medium ($\hat{\varepsilon}_c$ and $\hat{\mu}$ x-independent) the components of $\hat{E}$ and $\hat{H}$ satisfy the *Helmholtz equation*:

$$\Delta u + k^2 u = 0, \quad \text{with } k^2 = \omega^2\hat{\varepsilon}_c\hat{\mu}.$$

3.1.5. General constitutive relations with Ohm's law

With the hypothesis H3) on ε and H8) on σ, we see that:

(46) $\varepsilon_c = \varepsilon_0\delta + \sigma_1 Y + \kappa + Y * \sigma_2$

looks like ε but does not satisfy (27): for instance $Y * \sigma_2$ is a bounded measurable function, but in the general case it is not integrable:

(47) $\kappa_c \overset{\text{def}}{=} \kappa + \sigma_1 Y + Y * \sigma_2 \in L^1(R) + L^\infty(R), \text{ but } \kappa_c \notin L^1(R).$

However the Fourier-Laplace transform of κ_c is such that:

(48) $\hat{\kappa}_c(\omega) = \hat{\kappa}(\omega) + i\,(\hat{\sigma}(\omega)/\omega) \longrightarrow 0 \text{ for } |\omega| \longrightarrow \infty, \quad \omega'' > 0.$

The hypotheses on ε and σ are not at the same level, since $d\varepsilon_c/dt = d\varepsilon/dt + \sigma$. Thus we have to make one of the following hypotheses:

H10) $\sigma_1 = 0$, $\sigma = \sigma_2$ is the time derivative of a function θ, $\theta \in L^1(R)$, $\text{supp}\,\theta \subseteq R^+$.

Then $\Sigma = Y * \sigma = Y * \theta' = Y' * \theta = \delta * \theta = \theta$, the prime denoting the derivative. Thus:

$$\varepsilon_c = \varepsilon + \theta = \varepsilon_0\delta + \kappa_c, \text{ with } \kappa_c = \kappa + \theta \in L^1(R), \text{ thus } \hat{\varepsilon}_c = \varepsilon_0 + \hat{\kappa}_c, \quad \hat{\kappa}_c = \hat{\kappa} + \hat{\theta} \in C_0(R_{\omega'}).$$

With this hypothesis we see that ε_c has exactly the same properties as ε, there is no singularity of $\hat{\varepsilon}_c$ at 0, and therefore it is not valuable for all media.

H11) $\sigma_1 \neq 0$, and σ_2 is the time derivative of a function θ_2 with $\theta_2 \in L^1(R)$, supp $\theta_2 \subseteq R^+$, with the hypothesis H8),

then we have, as in (46) and (47):

(49) $$\varepsilon_c = \varepsilon_0 \delta + \sigma_1 Y + \nu_c, \quad \nu_c = \kappa + \theta_2 \in L^1(R), \quad \text{supp}\, \nu_c \subseteq R^+,$$

and thus $\varepsilon_0 + \hat{\nu}_c$ also satisfies the Kramers-Kronig relations (32).
Note that the hypotheses H5) on (κ, χ), and H8) on σ also imply that $d\kappa_c/dt$, hence (ν_c, χ) also satisfies the hypothesis H5).
The notions of conductor or of insulator are generally defined with respect to their microscopic properties (the possibility of having "free electrons" or not). At the macroscopic level it seems that a conductor has σ_1 different from 0, whereas $\sigma_1 = 0$ for an insulator.

Besides, a good conductor (at the frequency ω) has a high value for $\hat{\varepsilon}''_c(\omega)$, and a perfect conductor would correspond to the limit $\hat{\varepsilon}''_c(\omega) \longrightarrow \infty$ (or $\sigma_1 \longrightarrow \infty$).
These terminologies may be somewhat confusing; sometimes the word "dielectric" concerns only insulator.

3.1.6. Conditions on ε and μ for a dissipative medium in a bounded domain
Usually the "energy (power) balance" in a bounded domain Ω is obtained in the following way. We consider the expression:

(50) $$P(t) = \int_\Omega \left(\frac{\partial D}{\partial t} . E + \frac{\partial B}{\partial t} . H\right)(x,t)\, dx.$$

Using the Maxwell equations (1) and relation (10) (similar to (11) (12) for free space), we obtain:

(51) $$\begin{aligned} P(t) &= \int_\Omega (\text{curl}\, H.E - H.\text{curl}\, E)\, dx - \int_\Omega J.E\, dx \\ &= - \int_\Omega \text{div}\,(E \wedge H)\, dx - \int_\Omega J.E\, dx. \end{aligned}$$

With the Poynting vector S (see (9)), we define:

(52) $$P_S(t) = \int_\Omega \text{div}\,(E \wedge H)\, dx = \int_\Gamma n.E \wedge H\, d\Gamma = \int_\Gamma n.S\, d\Gamma, \quad P_J(t) = \int_\Omega J.E\, dx.$$

Thus (51) with (52) gives the "power balance":

(53) $$- P(t) = P_S(t) + P_J(t).$$

The usual interpretation of (53) is as follows (Jones [1] p.51): the left-hand side of (53) is the *rate at which electromagnetic energy decreases* in Ω; $P_S(t)$ is the *rate at which electromagnetic energy crosses* Γ; $P_J(t)$ is the *rate at which electromagnetic energy is converted into heat* in Ω.

We also define:

(54) $\quad P_c(t) = \int_\Omega (\frac{\partial D_c}{\partial t}.E + \frac{\partial B}{\partial t}.H)\,dx = P(t) + P_J(t),$

thus

(54)' $\quad P_c(t) = \int_\Omega (\varepsilon_0 \frac{\partial E}{\partial t}.E + \mu_0 \frac{\partial H}{\partial t}.H)\,dx + \int_\Omega (\frac{\partial P}{\partial t}.E + \frac{\partial M}{\partial t}.H + J.E)\,dx.$

Then with $W_0(t) = \frac{1}{2}\int_\Omega (\varepsilon_0 E^2 + \mu_0 H^2)\,dx$, we define:

(55) $\quad P_0(t) = \int_\Omega (\varepsilon_0 \frac{\partial E}{\partial t}.E + \mu_0 \frac{\partial H}{\partial t}.H)\,dx = \frac{\partial W_0}{\partial t}(t),$

(56) $\quad P_m(t) = \int_\Omega (\frac{\partial P}{\partial t}.E + \frac{\partial M}{\partial t}.H + J.E)\,dx,$

and thus

(57) $\quad P_c(t) = P_0(t) + P_m(t).$

At this point, we note that:

i) $P_0(t)$ is the derivative of an energy, which looks like the energy of the electromagnetic field in free space. This is natural when comparing to the microscopic situation,

ii) $P_m(t)$ corresponds to the interaction of the electromagnetic field with matter. As $P_J(t)$ is a part of $P_m(t)$, it seems fairly natural that $P_m(t)$ correspond to a transfer of electromagnetic energy into matter, by Joule effect, absorption,..., that is, by electric or magnetic loss.

Generalizing the case where $J = \sigma_1 E$ with $\sigma_1 > 0$, thus giving $P_J(t) > 0$, we can think that the modelling of a *closed system* (without exterior currents) must satisfy

H12) $P_m(t) \geq 0, \quad \forall t \in R,$

with $P_m(t)$ defined by (56), and thus with (27) and (37):

(58) $\quad P_m(t) = \int_\Omega [((\frac{\partial \kappa}{\partial t} + \sigma) \underset{t}{*} E).E + (\frac{\partial \chi}{\partial t} \underset{t}{*} H).H]\,dx.$

Recalling that by (54), (53), (57):

(59) $\quad - P_c(t) = - P_0(t) - P_m(t) = P_S(t),$

then H12) gives:

(60) $\quad - P_0(t) \geq P_S(t).$

Remark 5. If electromagnetic energy flows out of Ω by crossing Γ, the flux of the Poynting vector through Γ is positive and we have:

(61) $\mathcal{P}_S(t) \geq 0,$

thus by (60), $\mathcal{P}_o(t)$ is negative and the "electromagnetic energy" $W_o(t)$ is decreasing in Ω using (55).

⊗

Related to H12), but not equivalent to H12), the following hypothesis may be made:

H13) $\hat{\varepsilon}_c''(\omega) = \operatorname{Im} \hat{\varepsilon}_c(\omega) \geq 0, \quad \hat{\mu}''(\omega) = \operatorname{Im} \hat{\mu}(\omega) \geq 0, \quad \forall \omega \geq 0,$

that is also (with (35)'):

H13)' $\operatorname{Re}(i\omega\hat{\varepsilon}_c(\omega)) \leq 0, \quad \operatorname{Re}(i\omega\hat{\mu}(\omega)) \leq 0, \quad \forall \omega \in R.$

The link between H12) and H13) may be viewed in one of the following ways:
i) Assume that:

(62) $$E, H \in L^2(R, L^2(\Omega)^3), \text{ and } \frac{\partial B}{\partial t}, \frac{\partial D_c}{\partial t} \in L^2(R, L^2(\Omega)^3),$$

then $\mathcal{P}_m(t) \in L^1(R)$ and using the fact that the Fourier transformation in $L^2(R)$ is an isometry (up to a constant factor) we have:

(63) $$\left| \begin{aligned} \int_R \mathcal{P}_m(t)\,dt &= \frac{1}{2\pi} \operatorname{Re} \int_R (i\omega\hat{\varepsilon}_c(\omega)|\hat{E}(\omega)|^2 + i\omega\hat{\mu}(\omega)\,|\hat{H}(\omega)|^2)\,d\omega \\ &= \frac{1}{2\pi} \operatorname{Re} \int_R [(i\omega\hat{\kappa}(\omega) + \hat{\sigma}(\omega))|\hat{E}(\omega)|^2 + i\omega\hat{\chi}(\omega)|\hat{H}(\omega)|^2]\,d\omega. \end{aligned} \right.$$

We see that hypothesis H13) gives, when (62) is satisfied:

(64) $$\int_R \mathcal{P}_m(t)\,dt \geq 0.$$

ii) Assume that the electromagnetic field is stationary, that is, D, E,... are the real parts of (20), for real ω. Then we have:

(65) $$\mathcal{P}_o(t) = \frac{i\omega\varepsilon_0}{4} \int_\Omega [E_0^2 e^{2i\omega t} - \bar{E}_0^2 e^{-2i\omega t}]\,dx + \frac{i\omega\mu_0}{4} \int_\Omega [H_0^2 e^{2i\omega t} - \bar{H}_0^2 e^{-2i\omega t}]\,dx.$$

Thus the mean value of $\mathcal{P}_o(t)$ over a period $T = 2\pi/\omega$ is zero. Then P and M defined by (25), are:

(66) $$\left| \begin{aligned} P(t) &= \kappa * E(t) = \int_0^\infty \kappa(s)E(t-s)ds = \operatorname{Re}(E_0 \int_0^\infty \kappa(s)\, e^{i\omega(t-s)}ds) \\ &= \operatorname{Re}(E_0\,\hat{\kappa}(-\omega)\,e^{i\omega t}) = \operatorname{Re}(\bar{E}_0\hat{\kappa}(\omega)\,e^{-i\omega t}), \end{aligned} \right.$$

(67) $$M(t) = \chi * H(t) = \operatorname{Re}(H_0\hat{\chi}(-\omega)\,e^{i\omega t}) = \operatorname{Re}(\bar{H}_0\hat{\chi}(\omega)\,e^{-i\omega t}).$$

Similarly:

(68) $$J(t) = \sigma * E(t) = \operatorname{Re}(E_0\hat{\sigma}(-\omega)\,e^{i\omega t}) = \operatorname{Re}(\bar{E}_0\hat{\sigma}(\omega)\,e^{-i\omega t}),$$

and thus $P_m(t)$ is given (according to (56)) by:

$$(69)\quad \left| \begin{aligned} & P_m(t) = \frac{1}{2} \mathrm{Re}\,[\, e^{-2i\omega t}(- i\omega \hat{\kappa}(\omega) + \hat{\sigma}(\omega)) \int_\Omega \bar{E}_0^2\, dx - i\omega\, \hat{\chi}(\omega) \int_\Omega \bar{H}_0^2\, dx \,] \\ & \qquad + \mathrm{Re}\,[\, (\omega \hat{\kappa}''(\omega) + \hat{\sigma}'(\omega)) \int_\Omega |E_0|^2 dx + (\omega \hat{\chi}''(\omega)) \int_\Omega |H_0|^2 dx \,]. \end{aligned} \right.$$

For *real* ω, we have from (44), (27):

$$(70)\qquad \mathrm{Re}\,(- i\omega\, \hat{\varepsilon}_c(\omega)) = \omega \hat{\kappa}''(\omega) + \hat{\sigma}'(\omega), \qquad \mathrm{Re}\,(- i\omega\, \hat{\mu}(\omega)) = \omega \hat{\chi}''(\omega).$$

Thus hypothesis H13) implies that the mean value of $P_m(t)$ over a period is always positive. It is not true that $P_m(t)$ is positive for any t, but it is true with H13). It seems that the discrepancy between H12) and H13) may come either from time delay due to convolution or from matter giving (during a short interval of time) energy to the electromagnetic field.
We define:

$$(71)\qquad C_+ = \{ z \in C,\ 0 \le \arg z < \pi,\ z \ne 0 \}; \qquad C^+ = \{ z \in C,\ z'' = \mathrm{Im}\, z > 0 \}.$$

We generally adopt inequalities a little more precise than H13):

H14) $\omega \hat{\varepsilon}_c(\omega) \in C_+$, $\omega \hat{\mu}(\omega) \in C_+$, $\forall\, \omega \in C_+$.

Thus for real positive ω:

H14)' $\hat{\varepsilon}''_c(\omega) \ge 0$ with $\hat{\varepsilon}'_c(\omega) > 0$ for $\hat{\varepsilon}''_c(\omega) = 0$, similarly for $\hat{\mu}(\omega)$, $\forall\, \omega > 0$.

Remember that we have, thanks to (36) and (36)':

$$(72)\qquad \hat{\varepsilon}''_c(-\omega) = -\, \hat{\varepsilon}''_c(\omega) \le 0, \qquad \hat{\mu}''(-\omega) = -\, \hat{\mu}(\omega) \le 0 \quad \text{for } \omega > 0.$$

If we suppose that H5) and H7) are satisfied, the functions

$$(73)\qquad \phi(\omega) = \mathrm{Im}\,(\omega \hat{\kappa}_c(\omega)), \quad \psi(\omega) = \mathrm{Im}\,(\omega\, \hat{\chi}(\omega))$$

are imaginary parts of holomorphic functions of ω in C^+, which tend to zero when $|\omega| \longrightarrow \infty$. Thus ϕ and ψ are harmonic functions on C^+ (i.e., for $\omega'' > 0$).

Since their boundary values are non-negative, then from the (strong) maximum principle (see Dautray-Lions [1] chap. 2), we have

$$(74)\qquad \phi(\omega) > 0,\ \psi(\omega) > 0, \quad \forall\, \omega \in C^+ \ \ (\omega'' > 0),$$

and thus we deduce H14) from H14)'.

Remark 6. With above notations (see H8) it is often separately assumed that:

H15) $\left| \begin{array}{l} \hat{\varepsilon}''(\omega) = \hat{\kappa}''(\omega) \geq 0, \hat{\mu}''(\omega) \geq 0, \\ \hat{\sigma}'(\omega) \geq 0, \hat{\sigma}'_2 \geq 0, \quad \forall \omega \geq 0, \end{array} \right.$

giving:

(75) $\quad \hat{\varepsilon}''_c(\omega) = \hat{\varepsilon}''(\omega) + (\hat{\sigma}'(\omega)/\omega) \geq 0, \quad \text{for } \omega \geq 0.$

With hypotheses H4) and H8), we can apply the strong maximum principle in:
i) the first quadrant of the complex plane:

$$C^{++} = \{\omega = \omega' + i\omega'', \ \omega' > 0, \ \omega'' > 0\},$$

to $\hat{\kappa}''$ and $\hat{\mu}''$. Recall that $\hat{\kappa}''(\omega) = 0$ for $\omega = i\omega''$ from (36), thus:

(76) $\quad \hat{\kappa}''(\omega) = \hat{\varepsilon}''(\omega) > 0, \quad \hat{\mu}''(\omega) > 0 \quad \text{for } \omega \in C^{++},$

and then:

(76)' $\quad \hat{\varepsilon}'' < 0$ and $\hat{\mu}''(\omega) < 0$ for ω in the quarter of plane $\omega' < 0, \omega'' > 0$,

ii) the half-plane C^+ ($\omega'' > 0$), to $\hat{\sigma}'_2$, thus

(77) $\quad \hat{\sigma}'_2(\omega) > 0, \quad \forall \omega \in C^+.$

Note that although $\hat{\varepsilon}_c(\omega)$ is singular at 0, we can apply the maximum principle to $\hat{\varepsilon}''_c(\omega)$ in C^{++}, because as $\omega \longrightarrow 0$,

(78) $\left| \begin{array}{l} i\hat{\sigma}(\omega)/\omega \approx i\hat{\sigma}(0)/\omega = i\hat{\sigma}'(0)/\omega, \text{ because } \hat{\sigma}''(0) = 0, \\ \text{with } \operatorname{Im}(i\hat{\sigma}'(0)/\omega) = \omega' (\hat{\sigma}'(0)/|\omega|^2) > 0, \text{ and } \hat{\sigma}'(0) \neq 0. \end{array} \right.$

On the imaginary axis with $\omega'' > 0$, $\hat{\sigma}(\omega)$ is real, $\hat{\varepsilon}''_c(\omega) = \hat{\varepsilon}''(\omega) = 0$. Thus:

(79) $\quad \hat{\varepsilon}''_c(\omega) > 0 \ \text{ in } C^{++}.$

Note that H14) implies the inequality:

(80) $\quad \omega'\hat{\varepsilon}''_c(\omega) + \omega''\hat{\varepsilon}'_c(\omega) \geq 0 \quad \text{for } \omega'' > 0,$

similarly for $\hat{\mu}$ and thus:

(80)' $\quad \hat{\varepsilon}'_c(\omega) \geq -\frac{\omega'}{\omega''}\hat{\varepsilon}''_c(\omega)$ (or $-\arg\omega \leq \arg\hat{\varepsilon} \leq \pi - \arg\omega$) for $\omega'' > 0$.

With the hypothesis (see H7)) that κ (or κ_c) and χ are positive functions of t, $\hat{\kappa}_c$ and $\hat{\chi}$ are decreasing functions of ω along the imaginary axis, and thus $\hat{\varepsilon}_c$ and $\hat{\mu}$ have the same properties; since $\hat{\varepsilon}_c$ and $\hat{\mu}$ tend to ε_0 and μ_0 (resp.) as $|\omega| \longrightarrow \infty$, we have

(81) $\quad \hat{\varepsilon}_c(\omega) \geq \varepsilon_0, \ \hat{\mu}(\omega) \geq \mu_0, \quad \forall \omega = i\omega'', \ \omega'' > 0.$

This implies, thanks to H14)' that $\hat{\varepsilon}_c$ and $\hat{\mu}$ never vanish in $\bar{C}_+$, as asserted by H14).

Often a stronger hypothesis than H14)' is made (see Landau-Lifschitz [1]))

H15) $\hat{\varepsilon}_c''(\omega) > 0,\ \hat{\mu}''(\omega) > 0,\quad \forall \omega > 0.$

With H7)', this implies that $\hat{\varepsilon}_c$ and $\hat{\mu}$ never vanish in $\bar{C}_+$. Thus they have inverses which tend to ε_0^{-1} and μ_0^{-1} at infinity (with H4) and H11)). Hence we can inverse the causality relations (22)' and (41)! ⊗

3.1.7. Some other definitions

From the permittivity and the permeability of a medium at the frequency ω, we usually define (with hypotheses H14):

i) the ***wavenumber*** $k = k(\omega)$ by:

(82) $\quad k^2 = \omega^2\, \hat{\varepsilon}_c\, \hat{\mu},\quad 0 \le \arg k < \pi,$

thus, with (71), $k \in C_+$ for $\omega \in C_+$, and also:

$$k = \sqrt{\omega^2\, \hat{\varepsilon}_c \hat{\mu}},\quad \text{with } \arg k = \frac{1}{2}(\arg \omega\hat{\varepsilon}_c + \arg \omega\hat{\mu});$$

ii) the ***wavelength*** $\lambda = 2\pi/|k'|$, $k' = \mathrm{Re}\, k$, but generally defined only for real k;

iii) the ***complex index*** of the medium $n = n(\omega)$ by $n = ck/\omega = \pm\sqrt{\hat{\varepsilon}_c(\omega)\hat{\mu}(\omega)/\varepsilon_0\mu_0}$,

(with the choice $n = c\sqrt{\hat{\varepsilon}_c\hat{\mu}}$, with $0 \le \arg n < \pi$, we would have $ck = \pm\, \omega n$).

For real positive ω, we can take $n = c\sqrt{\hat{\varepsilon}_c\hat{\mu}}$ and $ck = n\omega$;

iv) the ***impedance*** of the medium $Z = Z(\omega)$ by:

(83) $$Z = \frac{\omega\hat{\mu}}{k} = \frac{k}{\omega\hat{\varepsilon}_c} = \sqrt{\hat{\mu}/\hat{\varepsilon}_c}\quad \text{with } -\pi/2 < \arg Z < \pi/2,$$

since $\arg Z = \arg(\omega\hat{\mu}) - \arg k = \frac{1}{2}(\arg(\omega\hat{\mu}) - \arg(\omega\hat{\varepsilon}_c))$, thus $Z' = \mathrm{Re}\, Z > 0$.

Note that Z exists for all ω in C_+ (with $Z \neq 0$), since $\hat{\varepsilon}_c(\omega)$ and $\hat{\mu}(\omega)$ never vanish according to H14). We have $\hat{\varepsilon}_c(\omega)$, $\hat{\mu}(\omega)$ from k, n, Z by:

(84) $\quad \omega\hat{\varepsilon}_c = k/Z,\quad \omega\hat{\mu} = kZ,\quad \hat{\varepsilon}_c = n/cZ,\quad \hat{\mu} = nZ/c,$

but we cannot choose arbitrarily k and Z, with $0 \le \arg k < \pi$, $-\pi/2 < \arg Z < \pi/2$, to have $\hat{\varepsilon}_c$ and $\hat{\mu}$ in C_+. Note that the impedance unit is $[Z] = \mathrm{Ohm} = ML^2T^{-1}Q^{-2}$.

The ***admittance*** of the medium is defined by $Y = 1/Z$.

Remark 7. Using (36), the quantities k, n, Z as functions of ω satisfy:

$$(f(-\bar{\omega}))^2 = \overline{(f(\omega))^2},\ \text{for } f = k, n, Z,\quad \text{thus: } f(-\bar{\omega}) = \pm\, \overline{f(\omega)}.$$

With the previous definitions of k, n, Z, we have:

(85) $k(-\bar{\omega}) = -\overline{k(\omega)}, \quad Z(-\bar{\omega}) = \overline{Z(\omega)}, \quad n(-\bar{\omega}) = \overline{n(\omega)}.$

Thus Z *and* n *satisfy* (36) and are the Fourier-Laplace transforms of some real functions or distributions. Besides:

i) when $|\omega| \longrightarrow \infty$, $k(\omega) \approx \omega/\sqrt{\varepsilon_0\mu_0} = \omega\, c$, $\lim_{\omega \to \infty} Z(\omega) = Z_0$, and $\lim_{\omega \to \infty} n(\omega) = 1$,

with $Z_0 = \sqrt{\mu_0/\varepsilon_0}$ the impedance of the vacuum, $Z_0 = 120\,\pi$;

ii) when $\omega \to 0$, and $\sigma_1 \neq 0$ (see H8)), we have: $Z(\omega) \to 0$, $k(\omega) \to 0$, $n(\omega)$ is singular at 0.

⊗

3.1.8. Summary of the main properties for a linear isotropic dielectric medium

The electromagnetic behavior of a linear isotropic dielectric medium is given by the "constants" of the constitutive laws, the permittivity ε or ε_c (linked by (42)) and the permeability μ, with the following "natural" properties (see H4), H11))

(86) $\varepsilon_c = \varepsilon_0\delta + \sigma_1 Y + \nu_c, \qquad \mu = \mu_0\delta + \chi,$

Y the Heaviside function, σ_1 constant ≥ 0, κ_c, and $\chi \in L^1(R)$, real, with support in R^+. Real and imaginary parts of the Fourier-Laplace transforms of ν_c and χ have to satisfy the Kramers-Kronig relations (with H the Hilbert transformation, see (31)):

(87) $\hat{\nu}'_c = H\hat{\nu}''_c, \quad \hat{\nu}''_c = -H\hat{\nu}'_c,$ similarly for χ.

Finally the energy loss in the medium is expressed by relations H14) (or H14)'). Apart from vacuum and perfect media, we have to assume that:

(88) $\hat{\varepsilon}''_c(\omega) > 0, \quad \hat{\mu}''(\omega) > 0, \quad \forall\, \omega > 0.$

There is no natural hypothesis of positivity for real parts of $\hat{\varepsilon}_c$ or $\hat{\kappa}_c$ and $\hat{\mu}$ or $\hat{\chi}$. H14) requires positivity only when the imaginary part is null. Note that these real parts are Hilbert transforms of positive functions, in the "regular" case.

3.2. Linear anisotropic dielectric media

Certain media called *anisotropic media* have their physical properties which depend on the direction taken at each point. In the linear anisotropic dielectric media, the constitutive relations are given by (22), with ε and μ real matrices, functions (or distributions) of time with support contained in R^+ (from causality). Their Fourier-Laplace transforms will be complex valued matrices, with properties similar to those of isotropic media, under similar hypotheses, see H3): we only have to replace ε_0, μ_0 by $\varepsilon_0 I$, $\mu_0 I$, I being the unit 3x3 matrix, in (26), (27),... We have as in (36), (26):

(89) $\hat{\varepsilon}(-\bar{\omega}) = \overline{\hat{\varepsilon}(\omega)}$, $\hat{\mu}(-\bar{\omega}) = \overline{\hat{\mu}(\omega)}$, $\omega \in \mathbb{C}$, $\omega'' \in I_\varepsilon, I_\mu, \ldots$

(thus $\hat{\varepsilon}(\omega)$ and $\hat{\mu}(\omega)$ are real matrices on the imaginary axis) and

(90) $\lim_{|\omega|\to\infty} \hat{\varepsilon}(\omega) = \varepsilon_0 I$, $\lim_{|\omega|\to\infty} \hat{\mu}(\omega) = \mu_0 I$.

Decomposing $\hat{\varepsilon}(\omega)$ and $\hat{\mu}(\omega)$ into real and imaginary parts:

$$\hat{\varepsilon}(\omega) = \hat{\varepsilon}'(\omega) + i\hat{\varepsilon}''(\omega), \qquad \hat{\mu}(\omega) = \hat{\mu}'(\omega) + i\hat{\mu}''(\omega), \quad \omega \in \mathbb{C}^+,$$

we again obtain the Kramers-Kronig relations as in (35) or (35)' between $\hat{\varepsilon}' - \varepsilon_0 I$ and $\hat{\varepsilon}''$, or $\hat{\mu}' - \mu_0 I$ and $\hat{\mu}''$, and also for the components of these matrices:

(91) $\hat{\varepsilon}'_{ij}(\omega) - \varepsilon_0\delta_{ij} = H\hat{\varepsilon}''_{ij}(\omega)$, $\hat{\varepsilon}''_{ij}(\omega) = -H(\hat{\varepsilon}'_{ij} - \varepsilon_0\delta_{ij})(\omega)$, $\forall\, i,j$.

Similarly for "ohmic anisotropic conductors", an electric field creates interior currents as in (37), but with σ a real matrix. The Fourier-Laplace transform of (37) is Ohm's law with $\hat{\sigma}(\omega)$ a complex matrix also satisfying (89). We can thus define quantities D_c and ε_c as in (41) and (43), giving the stationary "homogeneous" Maxwell equations (45) for anisotropic media, with matrices $\hat{\varepsilon}_c$, $\hat{\mu}$.
The only interesting new point is the positivity relations according to dissipation and energy losses in the medium, corresponding to H12) or H13) for all $\omega > 0$:

H16) $\mathrm{Re}\,(-i\omega\sum\hat{\varepsilon}_{cij}\xi_i\bar{\xi}_j) \geq 0$, $\mathrm{Re}\,(-i\omega\sum\hat{\mu}_{ij}\xi_i\bar{\xi}_j) \geq 0$, $\forall\, \xi = (\xi_1,\xi_2,\xi_3) \in \mathbb{C}^3$,

or stronger (see H15)): there is $\alpha = \alpha(\omega) > 0$, for all $\omega > 0$, such that:

H17) $\mathrm{Re}\,(-i\omega\,\hat{\varepsilon}_c(\omega)\xi.\bar{\xi}) \geq \alpha\,|\xi|^2$, $\mathrm{Re}\,(-i\omega\,\hat{\mu}(\omega)\xi.\bar{\xi}) \geq \alpha\,|\xi|^2$, $\forall\, \xi \in \mathbb{C}^3$,

that is: for all $\omega > 0$, the following matrices are definite positive:

$$\hat{\varepsilon}_c^{\,I}(\omega) = \frac{1}{2i}(\hat{\varepsilon}_c - \hat{\varepsilon}_c^{\,*})(\omega), \quad \hat{\mu}^{\,I}(\omega) = \frac{1}{2i}(\hat{\mu} - \hat{\mu}^{\,*})(\omega).$$

But note that for $\omega \to +\infty$, $\hat{\varepsilon}_c(\omega) \to \varepsilon_0 I$, $\hat{\mu}(\omega) \to \mu_0 I$, thus necessarily $\alpha(\omega) \to 0$.
For $\omega \to 0$, $\hat{\mu}(\omega) \to \hat{\mu}'(0)$, a real matrix, $-i\omega\hat{\varepsilon}_c = -i\omega\hat{\varepsilon} + \hat{\sigma} \to \hat{\sigma}'(0)$, thus $\alpha(\omega) \to 0$.
From the strong maximum principle, we can assert that H17) implies that these relations are also satisfied, with hypotheses at infinity, for all $\omega \in \mathbb{C}^{++}$.

Example. When the medium is a ferrite (with an applied magnetic field parallel to the z-axis), the permeability matrix is, with complex μ_j, j = 1, 2, 3

(92) $$\hat{\mu} = \begin{vmatrix} \mu_1 & 0 & 0 \\ 0 & \mu_2 & i\mu_0 \\ 0 & -i\mu_0 & \mu_3 \end{vmatrix}.$$

The matrix $\mu^I = \frac{1}{2i}(\hat{\mu} - (\hat{\mu})^*)$ is positive definite if and only if:

(93) $\mu''_i > 0$, $i = 1,2,3$ and $(\mu''_0)^2 < \mu''_2\mu''_3$.

⊗

Remark 8. If the matrices $\hat{\varepsilon}_c(\omega)$ and $\hat{\mu}(\omega)$ satisfy H17), they are invertible, and their inverses also satisfy: there is α_0 depending on ω so that for all $\omega > 0$:

(94) $\operatorname{Re}(-(i\omega\hat{\varepsilon}_c(\omega))^{-1}\xi.\bar{\xi}) \geq \alpha_0\, |\xi|^2, \quad \operatorname{Re}(-(i\omega\hat{\mu}(\omega))^{-1}\xi.\bar{\xi}) \geq \alpha_0\, |\xi|^2, \quad \forall \xi \in C^3.$

⊗

The importance of such linear anisotropic media comes notably from the fact that they are stable by homogenization: mixing periodically two linear homogeneous dielectric media (isotropic or not) with H15) or H17), we obtain a linear composite anisotropic medium with H17).

3.3. Linear chiral media

A medium whose constitutive relations are given by:

(95) $\qquad D = \varepsilon * E + \tilde{\varepsilon} * H, \quad B = \tilde{\mu} * E + \mu * H,$

with real functions (or distributions) of time ε, $\tilde{\varepsilon}$, $\tilde{\mu}$, μ with support in R^+, is called a *chiral medium*. The Fourier-Laplace transform of (95) is:

(96) $\qquad \hat{D}(\omega) = \hat{\varepsilon}(\omega)\hat{E}(\omega) + \hat{\tilde{\varepsilon}}(\omega)\hat{H}(\omega), \quad \hat{B}(\omega) = \hat{\tilde{\mu}}(\omega)\hat{E}(\omega) + \hat{\mu}(\omega)\hat{H}(\omega),$

with $\hat{\varepsilon}$, $\hat{\tilde{\varepsilon}}$, $\hat{\tilde{\mu}}$, $\hat{\mu}$ also satisfying (36), and with the usual analytic properties (see 3.1). Note that if Ohm's law is also satisfied, we can replace D and ε by D_c and ε_c resp., as in (41),(44)), in relations (96). Furthermore we generally assume that $\tilde{\varepsilon} = -\tilde{\mu}$. Modelling dissipation and loss of energy in a chiral medium leads to the following hypothesis, similar to H12),H13),H15) or H17): for all $\omega > 0$,

H18) $\left| \begin{array}{l} \operatorname{Re}(-i\omega \sum \hat{\theta}_{\alpha\beta}\, \xi_\alpha\, \bar{\xi}_\beta) \geq 0, \quad \forall \xi = (\xi_1,\xi_2),\ \xi_\alpha = (\xi_{\alpha j}) \in C^3,\ \alpha = 1, 2, \\ \text{with } \hat{\theta}_{11} = \hat{\varepsilon}\ (\text{or } \hat{\varepsilon}_c),\ \hat{\theta}_{12} = \hat{\tilde{\varepsilon}},\ \hat{\theta}_{21} = \hat{\tilde{\mu}},\ \hat{\theta}_{22} = \hat{\mu}, \end{array}\right.$

or a stronger one: there is $a = a(\omega) > 0$, for all $\omega > 0$ so that

H19) $\operatorname{Re}(-i\omega \sum \hat{\theta}_{\alpha\beta}\xi_{\alpha j}\bar{\xi}_{\beta k}) \geq a\, |\xi|^2 = a \sum |\xi_{\alpha j}|^2, \quad \forall \xi \in C^3 \times C^3,$

giving therefore:

(97) $\hat{\varepsilon}''(\omega) > 0,\ \hat{\mu}''(\omega) > 0, \text{ and: } |\lambda(\omega)|^2 < \hat{\varepsilon}''(\omega)\,\hat{\mu}''(\omega), \quad \lambda = \frac{1}{2i}(\hat{\tilde{\varepsilon}} - \bar{\hat{\tilde{\mu}}}), \quad \forall \omega > 0.$

More generally ε, $\tilde{\varepsilon}$, μ, $\tilde{\mu}$ may be real matrices corresponding to a chiral anisotropic medium; then each $(\hat{\theta}_{\alpha\beta})$ is a 3x3 matrix, satisfying (36), with some usual holomorphic properties, and with a hypothesis generalizing H19): there is $a > 0$, depending on ω, so that, for all $\omega > 0$:

H20) $\quad \operatorname{Re}(-i\omega \sum \hat{\theta}_{\alpha j\beta k}\,\xi_{\alpha j}\bar{\xi}_{\beta k}) \geq a \sum |\xi_{\alpha j}|^2, \quad \forall \xi \in C^3.$

Note also that this family of media is stable by homogenization: by mixing periodically a linear chiral medium with another linear dielectric medium, chiral or not, isotropic or not, we obtain a linear anisotropic chiral medium (with H17) for dissipative media).

3.4. Nonlinear constitutive relations

The coefficients of the constitutive relations in general depend on the state of the medium, notably on the temperature, on the strain (in an elastic medium), and also on the electromagnetic field itself, thus giving nonlinear constitutive laws. If the medium is near equilibrium and if the electromagnetic field is not too strong, we can often linearize these relations. We give some examples where nonlinearity appears. In the first examples, nonlinearity implies anisotropy:

i)"birefringence" in an electric field; the permittivity is given by:

$$\hat{\varepsilon}_{ik} = \hat{\varepsilon}^{o}\,\delta_{ik} + \alpha\;\hat{E}_i\hat{E}_k, \quad \alpha \text{ constant;}$$

ii)"magneto optic effect"; the permittivity is a (non-symmetric) tensor function of H but satisfies:

$$\varepsilon_{ik}(H) = \varepsilon_{ik}(-H);$$

iii)"Hall effect": the conductivity is a (non-symmetric) tensor of H:

$$\sigma_{ik}(H) = \sigma_{ik}(-H).$$

iv) *ferroelectric media* and *ferromagnetic media* have fairly similar properties, with respect to the electric field or to the magnetic field; here we develop only the question of ferromagnetic media.

A *ferromagnetic material* may be defined as one that possesses a *spontaneous magnetization*, and its magnitude M_s depends only on the temperature of the material and not on the presence or absence of an applied magnetic field. Induced magnetization may be neglected. The magnetization vector M(x) is:

(98) $$M(x) = M_s U(x), \text{ with } |U(x)| = 1,\ x \in \Omega.$$

In a static case (that is without evolution) the magnetization field is deduced from H thanks to a minimization principle, as follows.

We consider a ferromagnetic medium occupying a domain Ω in R^3; we assume that the exterior domain Ω' of Ω is occupied by free space. An "exterior" magnetic field H_o (a constant vector) is applied onto this medium. Then the magnetic field and the magnetic induction must satisfy:

(99) $$\operatorname{div} B = 0, \quad \operatorname{curl} H = 0 \quad \text{in } R^3,$$

with the following conditions:

i) in $\Omega' = R^3 \setminus \bar{\Omega}$:

(100) $H = H_o + H_r$, $B = \mu_o H = \mu_o (H_o + H_r)$, with $H_r(x) \to 0$ for $|x| \to \infty$.

ii) in Ω, we set $H = H_o + H'$ and we have:

(101) $B = \mu_o H + M$, with $M = M_s U$, $U \in S^2$.

From (99) and (100), H_r is given from a potential ϕ with:

(102) $H_r = \text{grad}\,\phi$ in Ω', with $\Delta\phi = 0$, $\phi \to 0$ at infinity.

From equations (99), (since we search local finite energy solutions) we have the following transmission conditions across the boundary Γ of Ω:

(103) $[n \wedge H]_\Gamma = 0$, $[n.B]_\Gamma = 0$,

where $[u]_\Gamma$ is the jump of u across Γ. We assume that the domain Ω is simply connected. Then from (99) there is a potential v (in the Sobolev space $H^1(\Omega)$):

(104) $H' = \text{grad}\, v$ in Ω,
so that, from (99) and (101):

(105) $\text{div}\,(\mu_o\, \text{grad}\, v + M_s U) = 0$ in Ω,

with the boundary conditions on Γ, from (102), (103) and (104):

(106) $v|_\Gamma = \phi|_\Gamma$, $\mu_o \frac{\partial v}{\partial n} + n.M|_\Gamma = \mu_o \frac{\partial \phi}{\partial n}|_\Gamma$.

We define the exterior capacity operator C^e (see Dautray-Lions [1] chap. 2) by:

(107) $C^e(\phi|_\Gamma) = \frac{\partial \phi}{\partial n}|_\Gamma$, ϕ such that $\Delta\phi = 0$ in Ω', $\phi(x) \to 0$ for $|x| \to \infty$.

Then v must satisfy the following boundary condition:

(108) $\mu_o \frac{\partial v}{\partial n}|_\Gamma - \mu_o C^e(v|_\Gamma) = -\, n.M|_\Gamma = -\, M_s\, n.U|_\Gamma$.

The magnetization vector M is determined from a *minimization principle* in the following way: U has to minimize the *Gibbs function* G (or *free energy*)

(109) $G = E_e + E_a + E_d + E_Z + E_{as}$, with $E_e = \int_\Omega w_e\, dx$, $E_a = \int_\Omega w_a dx$,

E_e the *exchange energy*, E_a the *anisotropy energy*, E_d the *demagnetizing energy*, E_Z the *Zeeman energy*, E_{as} the *surface anisotropy energy*.
They are defined for a cubic crystal (for example) by:

(110) $$w_e = \frac{1}{2} C |\text{grad } U|^2 = \frac{1}{2} C \sum |\text{grad } U_i|^2, \quad C > 0, \ U = (U_1, U_2, U_3),$$

(111) $$w_a = K_1(U_1^2 U_2^2 + U_2^2 U_3^2 + U_3^2 U_1^2) + K_2(U_1^2 U_2^2 U_3^2),$$

(112) $$\left| \begin{array}{l} E_d = -\frac{1}{2} M_s \int_\Omega U.H' dx \text{ with (104)}, \quad E_Z = - M_s \int_\Omega U.H_o dx, \\ E_{as} = \frac{1}{2} K_s \int_{\partial\Omega} (n.U)^2 d\Gamma, \ K_s \text{ constant.} \end{array} \right.$$

In the case of a hexagonal crystal, w_e and w_a are given by

(110)' $$w_e = \frac{1}{2} \sum_{i=1,2,3} \{C_1 (\sum_{j=1,2} |\frac{\partial U_i}{\partial x_j}|^2) + C_2 |\frac{\partial U_i}{\partial x_3}|^2\}, \quad \text{with } C_1, C_2 > 0$$

(111)' $$w_a = K_1(1 - U_3^2) + K_2(1 - U_3^2)^2, \ K_j \text{ constant}, j = 1,2.$$

We have a problem with two unknowns U and v, with (105), (108). Plotting the component of the volume average magnetization in the direction of H_o against H_o gives the magnetization curve which is typical of a hysteresis phenomenon. For other situations in ferromagnetic media, see Brown [1].

These problems are not very different (at least from a mathematical point of view) from problems relative to liquid crystals where the main unknown is a vector field n of unit modulus, the "*director*" which gives the orientation of the "*anisotropic axis*". In a liquid crystal, a magnetic field H and an electric field E induce respectively a magnetization M and a polarization P

$$M = \chi_1 H + \chi_2(n.H)n, \quad P = (\varepsilon_1 - 1)E + \varepsilon_2(n.E)n,$$

with constants χ_1, χ_2, ε_1, ε_2 corresponding to the magnetic susceptibilities and the dielectric permittivity with the field parallel and perpendicular to the director. For the modelling of such situations, see Ericksen [1] and Leslie [1].

In the evolution case, we have to write thermomechanical balance laws; observed phenomena are typical of *hysteresis phenomena* which correspond to a certain nonlinearity with a certain memory of the past. We refer to Kranosel'skii-Pokrovskii [1], and also to Mayergoyz [1], for Preisach's hysteresis modelling.

Many physical phenomena are based on a coupling of continuum mechanics (elastic solids or fluids) with electromagnetism; see Eringen-Maugin [1].

CONCLUSION

The use of constitutive laws generally avoids taking into account the interaction of electromagnetic field with matter: this macroscopic modelling cannot represent all the fine situations in physics, notably when matter is not in (local) thermodynamic equilibrium. Then we must use finer modelling, coupling Maxwell equations with other equations (Vlasov equation, or Schrödinger equation for quantum effects). Such situations lead to nonlinear equations and thus are outside the scope of this study.

CHAPTER 2

MATHEMATICAL FRAMEWORK FOR ELECTROMAGNETISM

INTRODUCTION

An electromagnetic field given by the electric field E and the magnetic field H, in a bounded free space domain Ω, has finite energy if E and H are square integrable, which immediately implies, at least in the stationary case, that their curls are also square integrable. Thus the L^2-functional framework is a natural one for electromagnetism.

Owing to the repeated use in electromagnetism of operators grad, curl and div, we are naturally led to define functional spaces based on these operators. For the grad operator the natural functional spaces are the Sobolev spaces. At first we briefly recall the main properties of these spaces, referring to Dautray-Lions [1], Brezis [1], Adams [1], Necas [1] for developments. Then we give functional spaces based on the operators div and curl.

Sobolev Spaces

Let Ω be an open set in R^n. We briefly recall the definition and the main properties of the Sobolev spaces on Ω; at first order

(1) $H^1(\Omega) = \{u \in L^2(\Omega),\ \frac{\partial u}{\partial x_i} \in L^2(\Omega),\ i = 1 \text{ to } n\} = \{u \in L^2(\Omega),\ \ \text{grad}\, u \in L^2(\Omega)^n\}$,

with $\text{grad}\, u = (\frac{\partial u}{\partial x_i})_{i=1 \text{ to } n}$ and more generally using multi-index notations

$\alpha = (\alpha_1,\ldots,\alpha_n)$, $|\alpha| = \alpha_1 + \ldots + \alpha_n$ with $\alpha_i \in \mathrm{N}$, $D^\alpha u = \dfrac{\partial^{|\alpha|}}{\partial^{\alpha_1} x_1 \ldots \partial^{\alpha_n} x_n}$,

$H^k(\Omega) = \{u \in L^2(\Omega),\ D^\alpha u \in L^2(\Omega),\ \forall\, \alpha,\ |\alpha| \le k\},\ k \in \mathrm{N}.$

They are complex or real Hilbert spaces for the norm

$$\|u\|_k = \Big(\sum_{|\alpha|\le k} \int_\Omega |D^\alpha u|^2\, dx \Big)^{1/2}.$$

We denote by $D(\Omega)$ the space of C^∞ functions in Ω with compact support in Ω; let $H_0^k(\Omega)$ be the closure of $D(\Omega)$ in $H^k(\Omega)$. Then with

(2) $\quad D(\bar{\Omega}) = \{v,\ \ \exists\, u \in D(\mathrm{R}^n),\ v = u|_\Omega\}$,

and supposing that Ω is "regular" (locally on one side of its boundary Γ, with suitable regularity), we have:

Lemma 1. *The space $D(\bar{\Omega})$ is dense in $H^k(\Omega)$, $\forall\, k \in \mathrm{N}$.*

For the proof, see Lions-Magenes [1], Necas [1], Grisvard [1]. Recall also the result under the hypothesis Γ Lipschitzian:

the space of Lipschitz functions on $\bar{\Omega}$, $C^{0,1}(\bar{\Omega})$ is dense in $H^1(\Omega)$.

The Sobolev spaces on Ω for real k (and not only positive integer) are also defined by duality and interpolation, thus allowing one to define Sobolev spaces on the boundary Γ. We now give the first trace theorem in Sobolev spaces.

Theorem 1. *Let Ω be a bounded open set in R^n of class $C^{k-1,1}$* (see App. 1). Let n *be a unit normal to Γ. The trace mapping $u \rightarrow (u|_\Gamma, \frac{\partial u}{\partial n}|_\Gamma, \ldots, \frac{\partial^{k-1} u}{\partial n^{k-1}}|_\Gamma)$ defined on $D(\bar{\Omega})$, extends continuously from $H^k(\Omega)$ into $H^{k-1/2}(\Gamma) \times \ldots \times H^{1/2}(\Gamma)$.*

The kernel of this mapping is the space $H_0^k(\Omega)$.

We give here two interesting trace results (see Lions [2]); we define the spaces

(3) $\quad H^1(\Delta,\Omega) \overset{\text{def}}{=} \{u \in H^1(\Omega),\ \Delta u \in L^2(\Omega)\},\quad H(\Delta,\Omega) \overset{\text{def}}{=} \{u \in L^2(\Omega),\ \Delta u \in L^2(\Omega)\}.$

Proposition 1. *Let $u \in H^1(\Delta,\Omega)$, then $u|_\Gamma \in H^{1/2}(\Gamma)$, $\frac{\partial u}{\partial n}|_\Gamma \in H^{-1/2}(\Gamma)$.*

Let $u \in H(\Delta,\Omega)$, then $u|_\Gamma \in H^{-1/2}(\Gamma)$, $\frac{\partial u}{\partial n}|_\Gamma \in H^{-3/2}(\Gamma)$.

1. SPACES FOR CURL AND DIV; TRACE THEOREMS

Let Ω be an open set in R^n. For v in $L^2(\Omega)$ or $L^2(\Omega)^n$, let

(4) $$\|v\| = (\int_\Omega |v(x)|^2 dx)^{1/2}$$

be the norm of v in these spaces, and we denote by (u,v) the corresponding scalar product of u and v. We define the following Hilbert spaces with their "natural" norms:

(5) $$\begin{cases} H(\mathrm{div},\Omega) = \{v \in L^2(\Omega)^n,\ \mathrm{div}\, v \in L^2(\Omega)\}, \\ \|v\|_{H(\mathrm{div},\Omega)} = (\|v\|^2 + \|\mathrm{div}\, v\|^2)^{1/2}; \end{cases}$$

(6) $H_0(\mathrm{div},\Omega)$ = the closure of $D(\Omega)^n$ in $H(\mathrm{div},\Omega)$.

In a similar manner, we define for n = 3:

(7) $$\begin{cases} H(\mathrm{curl},\Omega) = \{v \in L^2(\Omega)^3,\ \mathrm{curl}\, v \in L^2(\Omega)^3\}, \\ \|v\|_{H(\mathrm{curl},\Omega)} = (\|v\|^2 + \|\mathrm{curl}\, v\|^2)^{1/2}; \end{cases}$$

(8) $H_0(\mathrm{curl},\Omega)$ = the closure of $D(\Omega)^3$ in $H(\mathrm{curl},\Omega)$.

In definitions (5) and (7), div v and curl v are taken in the sense of distributions. We also have to consider open sets in R^2, and curl operators in this case. We define two "curl" operators as linear differential operators, one (scalar) from $D'(\Omega)^2$ into $D'(\Omega)$, and the second (vectorial) from $D'(\Omega)$ into $D'(\Omega)^2$ by:

(9) $$\mathrm{curl}\, v = \frac{\partial v_2}{\partial x_1} - \frac{\partial v_1}{\partial x_2},\ \forall v = (v_1, v_2) \in D'(\Omega)^2,$$

(10) $$\overrightarrow{\mathrm{curl}}\, \phi = (\frac{\partial \phi}{\partial x_2}, -\frac{\partial \phi}{\partial x_1}),\ \forall \phi \in D'(\Omega).$$

Note that for all $\phi \in D(\Omega)$ and $v \in D'(\Omega)^2$, we have:

(11) $$\langle \overrightarrow{\mathrm{curl}}\, \phi, v\rangle = \langle \frac{\partial \phi}{\partial x_2}, v_1\rangle + \langle -\frac{\partial \phi}{\partial x_1}, v_2\rangle = \langle \phi, -\frac{\partial v_1}{\partial x_2} + \frac{\partial v_2}{\partial x_1}\rangle = \langle \phi, \mathrm{curl}\, v\rangle.$$

Then we can define the spaces $H(\mathrm{curl},\Omega)$, $H_0(\mathrm{curl},\Omega)$ for Ω in R^2 as in (7).
We make the following regularity hypothesis:

H1) The open set Ω in R^n is locally on one side of its boundary Γ, with Γ bounded and Lipschitzian (i.e., defined by Lipschitzian local charts).
We define:

(12) $C_0^{0,1}(\bar{\Omega})$ the space of (restrictions to Ω of) Lipschitzian functions with compact support in R^n.

Then we have a first property of spaces H(div,Ω) and H(curl,Ω):

Lemma 2. *Spaces* $C_0^{0,1}(\bar{\Omega})^n$ and $D(\bar{\Omega})^n$ *are dense in* H(div,Ω) *and* H(curl,Ω), with $n = 3$ or 2.

PROOF. Using Lipschitzian charts we first prove the Lemma for the half-space, then we operate by truncation and regularization by convolution, as usually done in Sobolev spaces (see Duvaut-Lions [1], for Γ of class C^1). We can also prove that the orthogonal complement of the space of the regular functions with compact support (in R^n) in H(div,Ω) and H(curl,Ω) reduces to {0} (see Girault-Raviart [1], Dautray-Lions [1], chap. 9).

⊗

We also prove extension properties using similar arguments:

Definition 1. *An extension operator from a functional space* X(Ω) *into* $X(R^n)$ *is a continuous operator* P *from* X(Ω) *into* $X(R^n)$ *such that*:

$$Pu|_{\Omega} = u \,, \quad \forall u \in X(\Omega).$$

Lemma 3. *With hypothesis H1), there are continuous extension operators from* H(div,Ω) *into* $H(div,R^n)$ *(for any* n*) and from* H(curl,Ω) *into* $H(curl,R^n)$ $n = 2, 3$.

The following trace theorems are basic in functional analysis for electromagnetism and fluid mechanics (see Duvaut-Lions [1]).

Theorem 2. *Trace in* H(div,Ω). *With hypothesis H1) the trace mapping*:

$\gamma_n : v \rightarrow n.v|_{\Gamma}$, *defined on* $D(\bar{\Omega})^n$, *has a continuous extension, also denoted by* γ_n *from* H(div,Ω) onto $H^{-1/2}(\Gamma)$. *The kernel of this mapping*, $\ker \gamma_n$, *is* $H_0(div,\Omega)$.

Let us denote by $n \wedge v$ the vector product of the unit normal (outgoing or incoming) with vector v; let $\pi_\Gamma v$ be the projection on the tangent plane to Γ of the restriction $v|_{\Gamma}$ of any element $v \in D(\bar{\Omega})$ to Γ (in the Appendix we will note t_Γ for π_Γ). We have:

$$(13) \qquad \pi_\Gamma v = - \, n \wedge (n \wedge v|_{\Gamma}), \quad \forall v \in D(\bar{\Omega}) \,.$$

Theorem 3. *Trace in* H(curl,Ω). *With hypothesis H1), the trace mapping*:

$\gamma_\tau : v \rightarrow n \wedge v|_{\Gamma}$ (resp. π_Γ) *defined on* $D(\bar{\Omega})^n$ *has a continuous extension also denoted* by γ_τ (resp. π_Γ), from H(curl,Ω) into $H^{-1/2}(\Gamma)^m$ with $m = 3$ for $n = 3$, $m = 1$ for $n = 2$. *The kernel of this mapping*, $\ker \gamma_\tau$ (resp. $\ker \pi_\Gamma$) *is the space* $H_0(curl,\Omega)$.

Remark 1. *On the equivalence of the trace theorem for* γ_τ and π_Γ.
i) With only the hypothesis H1), it is not obvious that we have
$n \wedge u \in H^{-1/2}(\Gamma)^3$ for all u in $H^{-1/2}(\Gamma)^3$, because n is in $L^\infty(\Gamma)^3$ only.

ii) If Γ has $C^{1,\eta}$ regularity ($\eta > 1/2$), then the normal n is $C^{0,\eta}$ (i.e. there exists $C > 0$, so that $|n(x) - n(y)| \le C\,|x - y|^\eta$), the mapping $\phi \longrightarrow n_i\phi$ is continuous in $H^s(\Gamma)$ for $|s| \le 1/2$ (see Grisvard [1] p. 21), so that *the trace theorems for* γ_τ *and* π_Γ are then equivalent.

iii) Note that the decomposition of any vector function u into $u = \pi_\Gamma u + (n.u)n$, gives the orthogonal decomposition of $L^2(\Gamma)^n$ into $L^2(\Gamma)^n = L^2_t(\Gamma) \oplus L^2_n(\Gamma)$, but does not a priori (with only hypothesis H1)) decomposes the space $H^{1/2}(\Gamma)^n$ into $H^{1/2}(\Gamma)^n = H^{1/2}_t(\Gamma) \oplus H^{1/2}_n(\Gamma)$. But obviously this is true for the $C^{1,\eta}$ ($\eta > 1/2$) regularity of the boundary, and then by duality the $H^{-1/2}(\Gamma)^n$ space is also decomposed.

⊗

PROOF of Theorem 2. i) We start from the Green formula

(14) $\quad (v, \text{grad}\,\phi) + (\text{div}\, v, \phi) = \int_\Gamma n.v\,\phi\, d\Gamma,$

for all v in $C_0^{0,1}(\bar{\Omega})^n$, ϕ in $C_0^{0,1}(\bar{\Omega})$ (or in $D(\bar{\Omega})^n$ and in $D(\bar{\Omega})$), and by density of $C_0^{0,1}(\bar{\Omega})$ in $H^1(\Omega)$, for all $\phi \in H^1(\Omega)$. Thus we have:

(15) $\quad \left|\int_\Gamma n.v\,\phi\, d\Gamma\right| \le \|v\|_{H(\text{div},\Omega)} \cdot \|\phi\|_{H^1(\Omega)}, \quad \forall v \in D(\bar{\Omega}),\ \phi \in H^1(\Omega).$

The left hand side (l.h.s.) depends on ϕ only through its trace $\mu = \phi|_\Gamma$ on Γ. Since

(16) $\quad \|\mu\|_{H^{1/2}(\Gamma)} = \inf_{\phi \in H^1(\Omega),\, \phi|_\Gamma = \mu} \|\phi\|_{H^1(\Omega)},$

we get

(17) $\quad \left|\int_\Gamma n.v\,\mu\, d\Gamma\right| \le \|v\|_{H(\text{div},\Omega)} \cdot \|\mu\|_{H^{1/2}(\Gamma)}.$

Thus the mapping $\gamma_n : v \longrightarrow n.v|_\Gamma$ defined on $D(\bar{\Omega})$ equipped with the H(div,Ω) norm, is continuous into the space $H^{-1/2}(\Gamma)$ and (from Lemma 2) has a natural continuous extension also denoted by γ_n from H(div,Ω) into $H^{-1/2}(\Gamma)$.

ii) We have to prove that γ_n is onto. Let μ be in $H^{-1/2}(\Gamma)$. Then there exists u in $H^1(\Omega)$ solution of the Neumann problem:

(18) $\quad -\Delta u + u = 0 \ \text{ in } \Omega, \quad \frac{\partial u}{\partial n}\Big|_\Gamma = \mu \quad \text{on } \Gamma.$

Then v = grad u is in H(div,Ω) and satisfies $\gamma_n v = \mu$, which proves that γ_n is onto.
Remark that the Green formula (14) is also true for v in H(div,Ω) and ϕ in $H^1(\Omega)$.

iii) Proof about the kernel of γ_n.
Let w be in ker γ_n orthogonal to $\mathcal{D}(\Omega)^n$ in H(div,Ω). Then w satisfies:

(19) $$((v,w)) = (v,w) + (\operatorname{div} v, \operatorname{div} w) = 0, \quad \forall\, v \in \mathcal{D}(\Omega)^n.$$

Let $w_0 = \operatorname{div} w$. Then w_0 is in $L^2(\Omega)$ and satisfies:

$$(v, w) + (\operatorname{div} v, w_0) = 0, \quad \forall\, v \text{ in } \mathcal{D}(\Omega)^n.$$

Therefore $w = \operatorname{grad} w_0$, thus $w_0 \in H^1(\Omega)$. Applying the Green formula with $\phi = w_0$, and $v = w$, we obtain $(v, \operatorname{grad} w_0) + (\operatorname{div} v, w_0) = (w, w) + (w_0, w_0) = 0$, thus $w = 0$.

⊗

PROOF of Theorem 3.
i) We start from the Green formula:

(20) $$(\operatorname{curl} v, \phi) - (v, \operatorname{curl} \phi) = \int_\Gamma n \wedge v . \phi \, d\Gamma,$$

for all v and ϕ in $\mathcal{D}(\bar{\Omega})^3$ or also v and ϕ in $H^1(\Omega)^3$. Thus

(21) $$\left|\int_\Gamma n \wedge v . \phi \, d\Gamma\right| \le \|v\|_{H(\operatorname{curl},\Omega)} . \|\phi\|_{H^1(\Omega)^3}, \quad \forall\, v \in \mathcal{D}(\bar{\Omega})^3,\ \phi \in H^1(\Omega)^3.$$

The left hand side of (21) depends on ϕ only through its trace $\mu = \phi|_\Gamma$ on Γ; thus

(21)' $$\left|\int_\Gamma n \wedge v . \mu \, d\Gamma\right| \le \|v\|_{H(\operatorname{curl},\Omega)} . \|\mu\|_{H^{1/2}(\Gamma)^3}, \quad \forall\, v \in \mathcal{D}(\bar{\Omega})^3,\ \mu \in H^{1/2}(\Gamma)^3,$$

hence

(22) $$\|n \wedge v\|_{H^{-1/2}(\Gamma)^3} \le \|v\|_{H(\operatorname{curl},\Omega)}, \quad \forall\, v \text{ in } \mathcal{D}(\bar{\Omega})^3.$$

Thus the mapping γ_τ: $v \longrightarrow n \wedge v|_\Gamma$ defined on $\mathcal{D}(\bar{\Omega})^3$ equipped with the H(curl,Ω) norm, is continuous into the space $H^{-1/2}(\Gamma)^3$ and (from Lemma 2) has a natural continuous extension, also denoted by γ_τ, from H(curl,Ω) into $H^{-1/2}(\Gamma)$.
Note that the Green formula (20) is also true for all v in H(curl,Ω) and ϕ in $H^1(\Omega)$, with the duality bracket of $H^{1/2}(\Gamma)^3$ and $H^{-1/2}(\Gamma)^3$ in the r.h.s. instead of the integral.
ii) Proof about the kernel of γ_τ.
Let u be in ker γ_τ, orthogonal to $\mathcal{D}(\Omega)^3$ in H(curl,Ω); then u satisfies

(23) $$\begin{cases} (\operatorname{curl} u, \operatorname{curl} \phi) + (u, \phi) = 0, & \forall\, \phi \in \mathcal{D}(\Omega)^3, \\ n \wedge u|_\Gamma = 0 . \end{cases}$$

Let v be in $L^2(\Omega)^3$, with v = curl u; v satisfies:

$$(v, \text{curl } \phi) + (u,\phi) = 0, \quad \forall \phi \in D(\Omega)^3,$$

thus curl v + u = 0 , therefore v ∈ H(curl,Ω). Then the Green formula gives:

$$(u, u) + (v, v) = -(u, \text{curl } v) + (\text{curl } u, v) = 0;$$

hence we have u = v = 0.

⊗

Remark 2. We emphasize that the mapping γ_τ is not *onto* $H^{-1/2}(\Gamma)^3$. A more precise trace theorem is seen later on. ⊗

2. JUMP FORMULAS ACROSS A BOUNDED HYPERSURFACE Γ IN R^n

Let Ω be a bounded open set in R^n, with a regular (Lipschitzian) boundary Γ; let φ and u be "regular" functions on R^n, (resp. with values in C and C^n), that is of C^1-regularity on each side of Γ, having limits up to the boundary. Then the jump of a function φ (or u) across Γ is denoted by:

$$(24) \qquad [\phi]_\Gamma \overset{\text{def}}{=} \phi|_{\Gamma-} - \phi|_{\Gamma+},$$

with $\phi|_{\Gamma-}$, $\phi|_{\Gamma+}$ the boundary value of φ on each side of Γ, Γ_- being the interior face. The normal n to Γ is oriented from Γ_- to Γ_+.

We denote by $\partial_i\phi$ the derivative in the sense of distributions in R^n of φ with respect to x_i and by $(\partial_i\phi)$ the derivative in the classical sense of φ. Then denoting by δ_Γ the Dirac distribution on Γ, we have:

Proposition 2. *With the above regularity hypotheses on Ω, φ, u, we have:*

$$(25) \qquad \left| \begin{array}{l} \partial_i\phi = (\partial_i\phi) - [\phi]_\Gamma n_i\delta_\Gamma, \\ \text{grad } \phi = (\text{grad } \phi) - [\phi]_\Gamma n\, \delta_\Gamma, \qquad \text{div } u = (\text{div } u) - [n.u]_\Gamma\, \delta_\Gamma \\ \text{curl } u = (\text{curl } u) - [n \wedge u]_\Gamma\, \delta_\Gamma \text{ for } n = 3. \end{array} \right.$$

These formulas are also true for φ and u satisfying:

$$(26) \qquad \left| \begin{array}{l} \phi|_\Omega \in H^1(\Omega),\ \phi|_{\Omega'} \in H^1_{loc}(\bar\Omega'),\ \Omega' = R^n \backslash \bar\Omega, \\ u|_\Omega \in H(\text{div},\Omega),\ u|_{\Omega'} \in H_{loc}(\text{div},\bar\Omega') \\ u|_\Omega \in H(\text{curl},\Omega),\ u|_{\Omega'} \in H_{loc}(\text{curl},\bar\Omega') \text{ for } n = 3, \end{array} \right.$$

with the notation for X = H^1, H(div,.), H(curl,.)

$$(27) \qquad X_{loc}(\bar\Omega') \overset{\text{def}}{=} \{ \varphi \in D'(\Omega') \text{ (or } \in D'(\Omega')^n),\ \zeta\varphi \in X(\Omega'),\ \forall \zeta \in D(R^n)\}.$$

PROOF. Formulas (25) are straightforward from the Green formulas. As an example, we have with $\phi_\Omega \overset{\text{def}}{=} \phi|_\Omega$, $\phi_{\Omega'} \overset{\text{def}}{=} \phi|_{\Omega'}$:

$$<\partial_i\phi, \zeta> \overset{\text{def}}{=} <\phi, -\partial_i\zeta> = \int_\Omega \phi(-\partial_i\zeta)dx + \int_{\Omega'} \phi(-\partial_i\zeta)dx$$

$$= \int_\Omega \partial_i\phi\, \zeta\, dx + \int_{\Omega'} \partial_i\phi\, \zeta\, dx - \int_\Gamma n_i\phi_\Omega\, \zeta\, d\Gamma + \int_\Gamma n_i\phi_{\Omega'}\, \zeta\, d\Gamma$$

$$= <(\partial_i\phi), \zeta> + \int_\Gamma n_i(\phi_{\Omega'} - \phi_\Omega)\, \zeta\, d\Gamma.$$

The generalizations to spaces (26) follow naturally from Theorems 2 and 3.

Remark that in the r.h.s. of (25), $(\partial_i\phi)$ denotes the function in $L^2_{loc}(R^n)$ whose restriction to Ω or Ω' is equal to the derivative in the sense of distributions in Ω or Ω' of ϕ_Ω or $\phi_{\Omega'}$.

⊗

3. DIFFERENTIAL OPERATORS ON A "REGULAR" SURFACE Γ

We define here some differential operators on a surface Γ in R^n, with fairly weak regularity hypotheses, that is Γ Lipschitzian. But we often have to suppose more regularity, for example that Γ is $C^{1,\alpha}$, for $0 < \alpha < 1$ at least on pieces Γ_j whose union is Γ (see App.1 for these notations).
The ideas on differential operators are in fact more natural from the point of view of differential geometry, so we refer to the appendix at the end of the book. But we generally have to assume some more regularity hypotheses ($C^{1,1}$). In order to be self-contained we define here all what we need for "usual" electromagnetism.
Let $d\Gamma$ be the surface measure on Γ, $\pi_\Gamma = \pi_{\Gamma,x}$ be the orthogonal projection on the tangent plane at the point x of Γ, which exists almost everywhere for $d\Gamma$.
It is possible to define differential operators on Γ either by restriction or by extension: for every "regular" function ϕ (or vector function u) on the closure of Ω, with trace ϕ_0 on Γ (with projection on the tangent plane u_0), we have:

Definition 2. *The operators* grad_Γ, div_Γ *and* curl_Γ *are defined by*

(28)
i) $\text{grad}_\Gamma \phi_0 = \pi_\Gamma \text{grad}\, \phi$ on Γ, *and for* n = 3: $\text{curl}_\Gamma u_0 = n.(\text{curl}\, u)|_\Gamma$,

ii) *then by duality*: $<\text{div}_\Gamma v, \zeta> = - <v, \text{grad}_\Gamma \zeta>$, $\forall \zeta$ *regular function,*

and for n = 3, $<\overrightarrow{\text{curl}}_\Gamma \phi, v> = - <\phi, \text{curl}_\Gamma v>$, $\forall$ v *regular vector field.*

We can also operate by extension: for every "regular" function ϕ_0 on Γ, we assume that there is a "regular" extension ϕ of ϕ_0 in Ω such that $\phi = \phi_0$ on Γ. Then we have to verify that the l.h.s. of (28) does not depend on the chosen particular extension; but this is a direct consequence of the Green formula.

These definitions are somewhat formal; we can give a precise sense in Sobolev spaces in the case of Lipschitz Γ; let

(29) $H_t^{-1/2}(\Gamma)$ be the closure in $H^{-1/2}(\Gamma)^n$ of $L_t^2(\Gamma) \overset{\text{def}}{=} \{u \in L^2(\Gamma)^n,\ n.u = 0\}$.

The mapping $\phi_0 \longrightarrow \mathrm{grad}_\Gamma\, \phi_0$ *has a continuous extension from* $H^{1/2}(\Gamma)$ into $H_t^{-1/2}(\Gamma)$.

PROOF. Using the continuous mapping $\phi \longrightarrow \mathrm{grad}\, \phi$, from $H^1(\Omega)$ into $H(\mathrm{curl},\Omega)$, and the continuous mapping π_Γ from $H(\mathrm{curl},\Omega)$ into $H_t^{-1/2}(\Gamma)$ of theorem 3, we obtain the property.

Then by duality we see that div_Γ is a continuous mapping from $H_t^{1/2}(\Gamma)$ (the dual space of $H_t^{-1/2}(\Gamma)$) into $H^{-1/2}(\Gamma)$. The operator curl_Γ is continuous from $H_t^{1/2}(\Gamma)$ into $H^{-1/2}(\Gamma)$, and then $\overrightarrow{\mathrm{curl}}_\Gamma$ is continuous from $H^{1/2}(\Gamma)$ into $H_t^{-1/2}(\Gamma)$.

⊗

We can give immediately some properties of these operators. At first we have:

Lemma 4. *For regular* Γ *(for example* $C^{1,1}$*), and* $n = 3$, *we have for all* u *in* $H(\mathrm{curl},\Omega)$ *and* ϕ *in* $H^1(\Omega)$ *(or with* (26)*)*:

$$(30)\quad \left|\begin{array}{l} \mathrm{curl}_\Gamma\, \pi_\Gamma u = -\,\mathrm{div}_\Gamma\,(n \wedge u|_\Gamma)\,, \qquad \mathrm{curl}_\Gamma\,(n \wedge u|_\Gamma) = \mathrm{div}_\Gamma\,(\pi_\Gamma u), \\ \overrightarrow{\mathrm{curl}}_\Gamma\, \phi|_\Gamma = n \wedge (\mathrm{grad}\, \phi)|_\Gamma = n \wedge \mathrm{grad}_\Gamma\,(\phi|_\Gamma) \end{array}\right.$$

in $H^{-1/2}(\Gamma)$, *and thus*:

$$(30)'\qquad \overrightarrow{\mathrm{curl}}_\Gamma\, \phi_0 = n \wedge \mathrm{grad}_\Gamma\, \phi_0\,, \qquad \forall\, \phi_0 \in H^{1/2}(\Gamma).$$

PROOF. i) We first apply the div operator to the curl formula of (25). Remember that $v = (\mathrm{curl}\ u)$ is in $H(\mathrm{div},\Omega)$ and in $H_{loc}(\mathrm{div},\bar{\Omega}')$, since $\mathrm{div}\ \mathrm{curl}\ u = 0$ in Ω and Ω', hence $[n.v]_\Gamma \in H^{-1/2}(\Gamma)$; then

$$(31)\qquad -\,[n.\mathrm{curl}\ u]_\Gamma\, \delta_\Gamma - \mathrm{div}\,([n \wedge u]_\Gamma \delta_\Gamma) = 0.$$

For any ϕ in $D(R^3)$, we have:

$$\langle \mathrm{div}\,([n \wedge u]_\Gamma\, \delta_\Gamma),\ \phi\rangle = -\,\langle [n \wedge u]_\Gamma\, \delta_\Gamma\,,\ \mathrm{grad}\, \phi\rangle$$
$$= -\,\langle [n \wedge u]_\Gamma,\ \mathrm{grad}_\Gamma\, \phi_0\rangle_\Gamma = \langle \mathrm{div}_\Gamma\,([n \wedge u]_\Gamma),\ \phi_0\rangle_\Gamma\,,$$

by duality, thus:

$$\mathrm{div}\,([n \wedge u]_\Gamma\, \delta_\Gamma) = \mathrm{div}_\Gamma\,([n \wedge u]_\Gamma)\, \delta_\Gamma.$$

Then (31) gives:

$$(32)\qquad [n.\mathrm{curl}\ u]_\Gamma = -\,\mathrm{div}_\Gamma\,([n \wedge u]_\Gamma), \quad \text{in } H^{-1/2}(\Gamma).$$

With definition (29)i), we obtain the first formula in (30).

ii) We apply the curl operator to the grad formula of (25) (recall that: $v = (\text{grad}\,\phi)$ is in $H(\text{curl},\Omega)$ and $H_{loc}(\text{curl},\bar{\Omega}')$, then $n \wedge \text{grad}\,\phi\,|_\Gamma \in H^{-1/2}(\Gamma)^3$, from Theorem 3). Thus:

(33) $$[n \wedge \text{grad}\,\phi]_\Gamma\,\delta_\Gamma + \text{curl}\,(n\,[\phi]_\Gamma\,\delta_\Gamma) = 0.$$

Now for any ζ in $D(R^3)$, we have:

$$<\text{curl}\,(n[\phi]_\Gamma\delta_\Gamma), \zeta> = <n[\phi]_\Gamma\delta_\Gamma, \text{curl}\,\zeta> = <[\phi]_\Gamma, n.\text{curl}\,\zeta>_\Gamma$$

$$= <[\phi]_\Gamma, \text{curl}_\Gamma\,\pi_\Gamma\zeta> = -<\overrightarrow{\text{curl}}_\Gamma\,[\phi]_\Gamma, \zeta>,$$

therefore:

(34) $$\text{curl}\,(n[\phi]_\Gamma\delta_\Gamma) = -(\overrightarrow{\text{curl}}_\Gamma[\phi]_\Gamma)\,\delta_\Gamma.$$

Then using formula (33), we obtain the last formula in (30).

Note that we can easily verify that $\text{curl}_\Gamma\,\text{grad}_\Gamma\,\phi = 0$, $\quad \text{div}_\Gamma\,\overrightarrow{\text{curl}}_\Gamma\,\phi = 0$.

⊗

Then we define the *Laplace-Beltrami* operator Δ_Γ by:

(35) $$\Delta_\Gamma\phi = \text{div}_\Gamma\,\text{grad}_\Gamma\,\phi = -\,\text{div}_\Gamma\,(n \wedge \overrightarrow{\text{curl}}_\Gamma\,\phi) = \text{curl}_\Gamma\,\overrightarrow{\text{curl}}_\Gamma\,\phi.$$

4. THE SPACES $H^{-1/2}(\text{div},\Gamma)$, $H^{-1/2}(\text{curl},\Gamma)$. TRACES FOR $H(\text{curl},\Omega)$

Let Ω be a regular open set, with boundary Γ (for example $C^{1,1}$), then we define

(36) $$\left| \begin{aligned} &H^{-1/2}(\text{div},\Gamma) = \{ v \in H^{-1/2}(\Gamma)^3,\ \ n.v = 0,\ \ \text{div}_\Gamma\,v \in H^{-1/2}(\Gamma) \},\\ &H^{-1/2}(\text{curl},\Gamma) = \{ v \in H^{-1/2}(\Gamma)^3,\ \ n.v = 0,\ \ \text{curl}_\Gamma\,v \in H^{-1/2}(\Gamma) \}. \end{aligned} \right.$$

Theorem 4. *The trace mappings* $\gamma_\tau: v \to n \wedge v|_\Gamma$, and $\pi_\Gamma: v \to \pi_\Gamma v$ *are continuous from the space* $H(\text{curl},\Omega)$ *onto* $H^{-1/2}(\text{div},\Gamma)$ and $H^{-1/2}(\text{curl},\Gamma)$ *respectively*.

Remark 3. It follows easily from Theorem 4 that for the exterior domain Ω', the above trace mappings defined on:

(37) $$H_{loc}(\text{curl},\bar{\Omega}') = \{ v \in L^2_{loc}(\Omega', C^3),\ \text{with } v\zeta \in H(\text{curl},\Omega'),\ \forall\,\zeta \in D(R^3)\}$$

are also continuous and *onto*. Theorem 4 implies continuous extension properties, like Lemma 3. ⊗

Remark 4. Regularity $C^{1,1}$ of Γ does not seem to be necessary (see Remark 1 for some regularity conditions), but optimal regularity conditions are not known. ⊗

PROOF of Theorem 4. From Lemma 4, we know that for u in $H(\mathrm{curl},\Omega)$, we have

(38) $\qquad \mathrm{div}_\Gamma(n\wedge u|_\Gamma)$ and $\mathrm{curl}_\Gamma(\pi_\Gamma u)\in H^{-1/2}(\Gamma)$.

Thus we only have to prove that the trace mappings are onto, hence to make a lifting of the trace. We use a proof due to Tartar, for the mapping π_Γ only. We note that we can localize and that the property is preserved by Lipschitz charts with Lipschitz inverse. Thus we are reduced to the case of the half-space $\Omega = R^3_+$. Let $u=(u_1,u_2)$ on R^2 (for $x_3=0$) with:

(39) $\qquad u_1,\ u_2,\ \dfrac{\partial u_1}{\partial x_2}-\dfrac{\partial u_2}{\partial x_1}\in H^{-1/2}(\Gamma)$.

Let $\hat{w}$ be the Fourier transform of w, in R^3 or R^2; $\hat{w}(\xi)=\int_{R^n} e^{-ix.\xi}w(x)dx$, with $\xi=(\xi',\xi_3)=(\xi_1,\xi_2,\xi_3)$ the dual variables to $x=(x_1,x_2,x_3)$. We define $\hat{v}$ by:

$$(40)\qquad \left|\begin{array}{l}\hat{v}_j(\xi',\xi_3)=\hat{u}_j(\xi')\dfrac{1}{(1+|\xi'|^2)^{1/2}}\,\phi\Big(\dfrac{\xi_3}{(1+|\xi'|^2)^{1/2}}\Big),\ j=1,2,\\[2ex] \hat{v}_3(\xi',\xi_3)=\dfrac{\xi_3}{1+|\xi'|^2}(\xi_1\hat{v}_1+\xi_2\hat{v}_2),\end{array}\right.$$

with $\phi\in\mathcal{D}(R)$, and $\int_R\phi(\lambda)\,d\lambda=1$. Then we have, with $K=\int_R\phi^2(\lambda)\,d\lambda$:

$$\int_{R^3}|\hat{v}_j|^2d\xi=K\int_{R^2}\frac{1}{(1+|\xi'|^2)^{1/2}}|\hat{u}_j|^2\,d\xi'<+\infty,\ |\hat{v}_3|\le\frac{\xi_3}{(1+|\xi'|^2)^{1/2}}(|\hat{v}_1|+|\hat{v}_2|).$$

Thus the inverse Fourier transform v of $\hat{v}$ satisfies: $v=(v_1,v_2,v_3)\in L^2(R^3)^3$.

We also have

$$\xi_1\hat{v}_2-\xi_2\hat{v}_1=(\xi_1\hat{u}_2-\xi_2\hat{u}_1)\frac{1}{(1+|\xi'|^2)^{1/2}}\,\phi\Big(\frac{\xi_3}{(1+|\xi'|^2)^{1/2}}\Big)\in L^2(R^3),$$

thus $\dfrac{\partial v_1}{\partial x_2}-\dfrac{\partial v_2}{\partial x_1}\in L^2(R^3)$; furthermore

$$|\xi_1\hat{v}_3-\xi_3\hat{v}_1|=\frac{|\xi_3|}{1+|\xi'|^2}\,|(-\hat{v}_1-\xi_2^2\hat{v}_1+\xi_1\xi_2\hat{v}_2)|\le\frac{|\xi_3|}{(1+|\xi'|^2)^{1/2}}(|\hat{v}_1|+|\xi_2\hat{v}_1-\xi_1\hat{v}_2|).$$

Thus $(\xi_1\hat{v}_3-\xi_3\hat{v}_1)$ is in $L^2(R^3)$, whence $\Big(\dfrac{\partial v_3}{\partial x_1}-\dfrac{\partial v_1}{\partial x_3}\Big)$ is in $L^2(R^3)$; similarly we have $\Big(\dfrac{\partial v_3}{\partial x_2}-\dfrac{\partial v_2}{\partial x_3}\Big)\in L^2(R^3)$, and thus $\mathrm{curl}\,v\in L^2(R^3)^3$.

We only have to verify that the trace of v_j on the plane $x_3=0$ is u_j for $j=1,2$, that is $\int_R\hat{v}_j(\xi',\xi_3)\,d\xi_3=\hat{u}_j(\xi')$, which is a consequence of $\int_R\phi(\lambda)d\lambda=1$.

⊗

Remark 5. *Regularity result.* For a regular Γ (for example $C^{1,1}$) we easily see that the trace mappings γ_τ and π_Γ are also continuous from the space

(41) $$H^1(\text{curl},\Omega) \stackrel{\text{def}}{=} \{u \in H^1(\Omega)^3, \ \text{curl}\, u \in H^1(\Omega)^3\},$$

onto the spaces $H^{1/2}(\text{div},\Gamma)$ and $H^{1/2}(\text{curl},\Gamma)$ respectively, with for $\alpha = 1/2$:

(42) $$\begin{cases} H^\alpha(\text{div},\Gamma) \stackrel{\text{def}}{=} \{v \in H^\alpha(\Gamma)^3, \ \ n.v = 0, \ \ \text{div}_\Gamma v \in H^\alpha(\Gamma)\}, \\ H^\alpha(\text{curl},\Gamma) \stackrel{\text{def}}{=} \{v \in H^\alpha(\Gamma)^3, \ \ n.v = 0, \ \ \text{curl}_\Gamma v \in H^\alpha(\Gamma)\}. \end{cases}$$

⊗

Corollary 1. *The trace mappings* γ_τ *and* π_Γ *are also continuous from the space*:

(43) $$H(\text{curl, div}\, 0,\Omega) \stackrel{\text{def}}{=} \{u \in H(\text{curl},\Omega), \ \text{div}\, u = 0\},$$

onto the spaces $H^{-1/2}(\text{div},\Gamma)$ and $H^{-1/2}(\text{curl},\Gamma)$ *respectively*.

PROOF. We know from Theorem 3 that the kernel of the trace mappings is the space $H_0(\text{curl},\Omega)$. Therefore the trace mappings are isomorphisms from the orthogonal space $H_0^{\perp}(\text{curl},\Omega)$ to $H_0(\text{curl},\Omega)$ in $H(\text{curl},\Omega)$ onto their ranges; $H_0^{\perp}(\text{curl},\Omega)$ is the set of u in $H(\text{curl},\Omega)$ satisfying, with κ a real constant:

(44) $$\int_\Omega (\text{curl}\, u.\ \text{curl}\,\zeta + \kappa^2 u.\ \zeta)\, dx = 0\,, \ \ \forall \zeta \in D(\Omega)^3,$$

that is:

(44)' $$\text{curl}\,\text{curl}\, u + \kappa^2 u = 0 \text{ in } D'(\Omega).$$

Hence this is equivalent to:

(44)" $$\text{div}\, u = 0, \quad -\Delta u + \kappa^2 u = 0,$$

thanks to:

(45) $$\text{curl}\,\text{curl}\, u = -\Delta u + \text{grad}\,\text{div}\, u.$$

Therefore $H_0^{\perp}(\text{curl},\Omega)$ is also the set of u in $H(\text{curl},\Omega)$ which satisfies (44)", and thus is contained in the space $H(\text{curl, div}\, 0,\Omega)$. ⊗

Remark 6. From the corollary, the following problem: find u in $H(\text{curl},\Omega)$ such that

(46) $$\begin{cases} \text{curl}\,\text{curl}\, u + \kappa^2 u = 0 \ \ \text{in } D'(\Omega) \\ n \wedge u|_\Gamma = j, \ \ j \text{ given in } H^{-1/2}(\text{div},\Gamma), \end{cases}$$

has a unique solution. It is also possible to prove this result using a lifting of the boundary condition (which exists from Theorem 4), then using a variational method, with the Lax-Milgram lemma. ⊗

Corollary 2. *The mapping* $v \longrightarrow n \wedge v$ *is an isomorphism from* $H^{-1/2}(\mathrm{div},\Gamma)$ onto $H^{-1/2}(\mathrm{curl},\Gamma)$, *with inverse* $w \longrightarrow -n \wedge w$, *and we have*

(47) $$\mathrm{curl}_\Gamma v = -\mathrm{div}_\Gamma (n \wedge v), \quad \mathrm{div}_\Gamma w = \mathrm{curl}_\Gamma (n \wedge w),$$

for v in $H^{-1/2}(\mathrm{curl},\Gamma)$, and w in $H^{-1/2}(\mathrm{div},\Gamma)$.

PROOF. The isomorphism property is straightforward on regular functions, then by closure. Formula (47) follows directly from (30) and Theorem 4.

⊗

Proposition 3. *The space* $H^{-1/2}(\mathrm{div},\Gamma)$ *is naturally identified with the dual space of* $H^{-1/2}(\mathrm{curl},\Gamma)$, *when we use* $L^2_t(\Gamma)$ *as "pivot" space.*

PROOF. We apply Green formula (which results from (20)):

(48) $$\int_\Omega (\mathrm{curl}\, u \,.v - u \,.\mathrm{curl}\, v)\, dx = \int_\Gamma n \wedge u \,.v\, d\Gamma,$$

whose l.h.s. is defined for all u and v in H(curl,Ω). Thus in the r.h.s. the integral is in fact a duality <,>, and we have:

(49) $$|\langle n \wedge u, v\rangle| \leq \| u \|_{H(\mathrm{curl},\Omega)} \| v \|_{H(\mathrm{curl},\Omega)}.$$

Then since the l.h.s. of (48) depends only on traces $\gamma_\tau u = n \wedge u|_\Gamma$ and $\pi_\Gamma v$ of u and

(49)' $$|\langle n \wedge u, v\rangle| \leq \| n \wedge u \|_{H^{-1/2}(\mathrm{div},\Gamma)} \| v \|_{H^{-1/2}(\mathrm{curl},\Gamma)}.$$

This implies that the duality is continuous.

⊗

Note also that from Corollary 2 and Proposition 3, we have the duality property

(50) $$\langle n \wedge u \,, v\rangle = -\langle u \,, n \wedge v\rangle, \qquad \forall u, v \in H^{-1/2}(\mathrm{curl},\Gamma),$$

as a consequence of the Green formula (48).

⊗

The duality of Prop. 3 may seem fairly surprising, if we refer to the usual dualities for Sobolev spaces. We can verify this duality from two points of view.
i) At first we use local charts to work on space R^2, then we use a Fourier transform method as in Theorem 4.

We have $u = (u_1, u_2)$ in $H^{-1/2}(\mathrm{curl},R^2)$ if and only if its Fourier transform satisfies:

(51) $$\int_{R^2} \frac{1}{(1+\xi^2)^{1/2}} [|\hat{u}_1|^2 + |\hat{u}_2|^2 + |\xi_2 \hat{u}_1 - \xi_1 \hat{u}_2|^2]\, d\xi < +\infty.$$

Let M be the matrix defined by:

(52) $$M = \frac{1}{(1+\xi^2)^{1/2}} \begin{pmatrix} 1+\xi_2^2 & -\xi_1\xi_2 \\ -\xi_1\xi_2 & 1+\xi_1^2 \end{pmatrix},$$

we see that u is in $H^{-1/2}(\mathrm{curl}, R^2)$ if and only if:

$$(M\begin{pmatrix} \hat{u}_1 \\ \hat{u}_2 \end{pmatrix}, \begin{pmatrix} \hat{u}_1 \\ \hat{u}_2 \end{pmatrix}) < +\infty .$$

The dual space is thus composed of elements $v = (v_1, v_2)$ whose Fourier transform satisfies:

$$(\begin{pmatrix} \hat{v}_1 \\ \hat{v}_2 \end{pmatrix}, M^{-1}\begin{pmatrix} \hat{v}_1 \\ \hat{v}_2 \end{pmatrix}) < +\infty,$$

with

(53) $$M^{-1} = \frac{1}{(1+\xi^2)^{1/2}} \begin{pmatrix} 1+\xi_2^2 & \xi_1\xi_2 \\ \xi_1\xi_2 & 1+\xi_1^2 \end{pmatrix}.$$

Thus $v = (v_1, v_2)$ has to satisfy:

(54) $$\int_{R^2} \frac{1}{(1+\xi^2)^{1/2}} [|\hat{v}_1|^2 + |\hat{v}_2|^2 + |\xi_1 \hat{v}_1 + \xi_2 \hat{v}_2|^2] \, d\xi < +\infty,$$

that is $v \in H^{-1/2}(\mathrm{div}, R^2)$.

ii) We use a "Hodge decomposition" of the space $H^{-1/2}(\mathrm{div}, \Gamma)$ as follows:

Lemma 5. *With the $C^{1,1}$ regularity of the boundary Γ, every element* u *in* $H^{-1/2}(\mathrm{div}, \Gamma)$ (*resp.* v *in* $H^{-1/2}(\mathrm{curl}, \Gamma)$) *has a decomposition of the form*:

(55)
$$\begin{aligned} &u = \mathrm{grad}_\Gamma \phi + \overrightarrow{\mathrm{curl}}_\Gamma \psi + a, \\ &\text{with } \phi \in H^{3/2}(\Gamma),\ \psi \in H^{1/2}(\Gamma),\ \mathrm{div}_\Gamma a = 0,\ \mathrm{curl}_\Gamma a = 0 \\ &(\textit{resp.}\ v = \mathrm{grad}_\Gamma \tilde{\phi} + \overrightarrow{\mathrm{curl}}_\Gamma \tilde{\psi} + \tilde{a}, \\ &\text{with } \tilde{\phi} \in H^{1/2}(\Gamma),\ \tilde{\psi} \in H^{3/2}(\Gamma),\ \mathrm{div}_\Gamma \tilde{a} = 0,\ \mathrm{curl}_\Gamma \tilde{a} = 0). \end{aligned}$$

PROOF. The two decompositions are equivalent from Corollary 2. Thus we will only prove decomposition (55) of $H^{-1/2}(\mathrm{div}, \Gamma)$. Applying successively div_Γ, curl_Γ to (55), we obtain:

(56) $$\Delta_\Gamma \phi = \mathrm{div}_\Gamma u \in H^{-1/2}(\Gamma), \qquad \Delta_\Gamma \psi = \mathrm{curl}_\Gamma u \in H^{-3/2}(\Gamma).$$

This gives (up to a constant on each component of Γ) ϕ in $H^{3/2}(\Gamma)$ and φ in $H^{1/2}(\Gamma)$, (thanks to a variational method for example), and then development (55).

⊗

From the two decompositions (55), the scalar product of u and v is given by:

$$(57) \qquad \langle u, v\rangle = \langle \mathrm{grad}_\Gamma\, \phi, \mathrm{grad}_\Gamma\, \tilde{\phi}\rangle + \langle \overrightarrow{\mathrm{curl}}_\Gamma\, \varphi, \overrightarrow{\mathrm{curl}}_\Gamma\, \tilde{\varphi}\rangle + \langle a, \tilde{a}\rangle,$$

and thus, using the duality between $H^{1/2}(\Gamma)$ and $H^{-1/2}(\Gamma)$:

$$(57)' \qquad \langle u, v\rangle = -\langle \Delta_\Gamma\, \phi, \tilde{\phi}\rangle - \langle \varphi, \Delta_\Gamma\, \tilde{\varphi}\rangle + \langle a, \tilde{a}\rangle.$$

⊗

5. TRACES FOR $W^1(\mathrm{div},\Omega)$ (FOR THE POYNTING VECTOR)

Let u and v be regular fields in (the closure of) Ω. Then we can apply the Green formula in the following form:

$$(58) \quad \int_\Omega (\mathrm{curl}\, u.v - u.\mathrm{curl}\, v)\, dx = \int_\Gamma n \wedge u.v\, d\Gamma = -\int_\Gamma u.n \wedge v\, d\Gamma = \int_\Gamma n.u \wedge v\, d\Gamma.$$

When u and v are in space $H(\mathrm{curl},\Omega)$, the last formula must make sense. Note that we have:

$$(59) \qquad \mathrm{div}\,(u \wedge v) = \mathrm{curl}\, u.v - u.\mathrm{curl}\, v.$$

Therefore if we define the space:

$$(60) \qquad W^1(\mathrm{div}, \Omega) \overset{\mathrm{def}}{=} \{ w \in L^1(\Omega)^n\, ,\ \mathrm{div}\, w \in L^1(\Omega) \},$$

we have $w = u \wedge v \in W^1(\mathrm{div}, \Omega)$, and then we have to justify a trace formula for w.

Proposition 4. *Let Ω be a bounded open set in R^n, with Lipschitzian boundary Γ. Then the trace mapping $w \longrightarrow n.w|_\Gamma$ is continuous from $W^1(\mathrm{div}, \Omega)$ into the dual space of Lipschitzian functions on Γ.*

PROOF. It is based on the Green formula:

$$(61) \qquad \int_\Omega (u.\mathrm{grad}\, \phi + \mathrm{div}\, u\, .\phi)\, dx = \int_\Gamma n\, .u\, \phi\, d\Gamma\, ,$$

for all u in $W^1(\mathrm{div},\Omega)$ and all bounded ϕ with bounded gradient, hence $\phi \in W^{1,\infty}(\Omega)$. We have $\phi|_\Gamma \in C^{0,1}(\Gamma)$, and $n.u|_\Gamma$ has a sense from (61) by duality with Lipschitzian functions on Γ.

⊗

Application to electromagnetism. Let (E, H) be an electromagnetic field in Ω with finite energy, then E and H are in $H(\mathrm{curl},\Omega)$ and the Poynting vector $S = E \wedge H$ (in the real case) is in $W^1(\mathrm{div},\Omega)$, and therefore has a trace n.S on Γ, so that

(62) $\int_\Omega \operatorname{div} S \, dx = \int_\Gamma n.S \, d\Gamma = \int_\Gamma n .E \wedge H \, d\Gamma = \int_\Gamma n \wedge E .H \, d\Gamma = - \int_\Gamma E.\, n \wedge H \, d\Gamma.$

⊗

6. SOME COMPLEMENTARY RESULTS. POLAR SETS

In order to have a better understanding of finite energy for an electromagnetic field, we give the definition (see Lions [2], Dautray-Lions [1] chap. 2 for some applications):

Definition 3. *Let Λ be a closed set in* R^n. Let $H^{-1}_\Lambda(R^n)$ *be the space of distributions in* $H^{-1}(R^n)$ *with support in* Λ. *Then* Λ *is a polar set if:*

(63) $H^{-1}_\Lambda(R^n) = \{0\}.$

Lines and points are polar sets in R^3, but surfaces are not polar sets.
One of the main properties of polar sets is:

$H^1_0(R^n \backslash \Lambda) = H^1(R^n)$ *if and only if* Λ *is a polar set.*

We can prove a similar result in the framework of spaces on curl and div:

Proposition 5. *Let Ω be a bounded open set in* R^3, *Λ be a bounded polar set in* R^3, *contained in Ω. Then*:

(64) $$\begin{cases} H(\mathrm{curl}, \Omega\backslash\Lambda) = H(\mathrm{curl}, \Omega), \ H_0(\mathrm{curl}, \Omega\backslash\Lambda) = H_0(\mathrm{curl}, \Omega) \\ H(\mathrm{div}, \Omega\backslash\Lambda) = H(\mathrm{div}, \Omega), \ H_0(\mathrm{div}, \Omega\backslash\Lambda) = H_0(\mathrm{div}, \Omega). \end{cases}$$

PROOF. i) Let u be in $H(\mathrm{curl}, \Omega\backslash\Lambda)$. Since Λ is a polar set, it is a set of Lebesgue measure zero, and thus u is in $L^2(\Omega)$. Let $G = \mathrm{curl}_\Lambda u$, with $\mathrm{curl}_\Lambda u$ in the sense of distributions in $\Omega\backslash\Lambda$, that is

$$<u, \mathrm{curl}\, \phi> = <G, \phi>, \quad \forall \phi \in D(\Omega\backslash\Lambda)^3.$$

Then $G \in L^2(\Omega)^3$ and curl u (in the sense of distributions in Ω) has the following form

$$\mathrm{curl}\, u = G + \mu_\Lambda, \text{ with } \mu_\Lambda \text{ a distribution with support in } \Lambda.$$

We have $\operatorname{div} G = \nu_\Lambda \overset{\mathrm{def}}{=} - \operatorname{div} \mu_\Lambda$ in $D'(\Omega)$, and $\nu_\Lambda \in H^{-1}(\Omega)$, with support in Λ.

Thus $\nu_\Lambda = 0$, and then (see Theorem 8 in sec. 9) G is a curl (up to a regular vector function if Ω is not *simply connected*): $G = \mathrm{curl}\, \tilde{u}$ with $\tilde{u} \in L^2(\Omega)^3$, hence: $\mu_\Lambda = \mathrm{curl}\, v$, $v = (u - \tilde{u}) \in L^2(\Omega)^3$. Therefore μ_Λ is in $H^{-1}(\Omega)^3$, with support in Λ, thus $\mu_\Lambda = 0$, and we have proved the first part of Proposition 5.

ii) Let $u \in H(\mathrm{div}, \Omega\backslash\Lambda)$; thus using similar notations, we have $u \in L^2(\Omega)^3$, and $\operatorname{div}_\Lambda u \in L^2(\Omega)$. Thus we have in $D'(\Omega)$:

$$\operatorname{div} u = \operatorname{div}_\Lambda u + \nu_\Lambda, \quad \text{with} \quad \operatorname{supp} \nu_\Lambda \subseteq \Lambda;$$

but $\operatorname{div} u \in H^{-1}(\Omega)$, thus $\nu_\Lambda \in H^{-1}(\Omega)$ with $\operatorname{supp} \nu_\Lambda \subseteq \Lambda$, therefore $\nu_\Lambda = 0$, and we have proved the last part of Proposition 5.

⊗

Application to electromagnetism.
Let j_Λ be a current concentrated on a line Λ (or a point), at an angular frequency ω, then the electromagnetic field (E, H) created by this current in a domain Ω being a solution of the Maxwell equations:

$$(65) \qquad \left| \begin{array}{l} \operatorname{curl} H + i\omega\varepsilon E = j_\Lambda \\ -\operatorname{curl} E + i\omega\mu H = 0 \end{array} \right. \quad \text{in } \Omega,$$

cannot be of finite energy in Ω.
In electrostatics, if we have some constant electric charge density ρ concentrated on a bounded line Λ, the electric field E is the opposite of the gradient of the electric potential u which is the solution going to 0 at infinity, of

$$(66) \qquad \Delta u = -\rho\, \delta_\Lambda \quad \text{in } R^3;$$

u is the "*Newtonian potential*" (in electrostatics it is called the "*Coulomb potential*"). Since $\delta_\Lambda \in H^{-1-\nu}(R^3)$, for all $\nu > 0$, then $u \in H^{1-\nu}_{loc}(R^3)$ and therefore $E \in H^{-\nu}_{loc}(R^3)$.

A very simple example (but for an unbounded line Λ, the z axis) is the case of a constant charge density (taken equal to 1) on Λ; then we have:

$$\delta_\Lambda = \delta(r),\ r = (x^2 + y^2)^{1/2},\ u = \frac{1}{2\pi} \operatorname{Log} r,\ E = -\operatorname{grad} u = -\frac{1}{2\pi r}\left(\frac{x}{r}, \frac{y}{r}, 0\right) \notin L^2_{loc}(R^3)^3.$$

⊗

7. TRACES ON A SHEET

We often have to consider domains whose boundary is an irregular surface, such as a cube, a tetrahedron (and it is therefore convenient to decompose these surfaces into regular parts) as well as very thin obstacles that we have to model as sheets (that is surfaces with a boundary). So we study some trace theorems for these surfaces.
Let Γ_1 and Γ_2 be two regular sheets ($C^{1,1}$ up to their boundary) in R^n, with the same boundary Λ. We assume that:
H) There is a neighborhood V_δ of Λ in Γ_1 and Γ_2 and a symmetry operator R_o exchanging the neighborhoods in Γ_1 and Γ_2, and preserving distances on the surfaces

$$R_o(V_\delta \cap \Gamma_2) = V_\delta \cap \Gamma_1,\ R_o(V_\delta \cap \Gamma_1) = V_\delta \cap \Gamma_2.$$

Thus, we assume that we have local charts so that:

$$R_0(x', x_{n-1}) = (x', -\tilde{x}_{n-1}), \quad x' \in R^{n-2}, \quad x_{n-1}, \quad \tilde{x}_{n-1} \in R.$$

Let ϕ_2 be a function on $V_\delta \cap \Gamma_2$; we denote by $R_0^* \phi_2$ the function on $V_\delta \cap \Gamma_1$ defined by $R_0^* \phi_2 = \phi_2 o R_0^{-1}$.

Thus the use of the operator R_0 reduces the problems of matching two functions on the two different sheets Γ_1 and Γ_2 to problems on the same sheet Γ_0, but with two faces Γ_{0+} and Γ_{0-}. At first we study matching problems in scalar cases.

7.1. Trace problems on sheets (the scalar case)

Since functions in the usual Sobolev space $H^1(\Omega)$ for a bounded domain Ω with boundary Γ have traces in the space $H^{1/2}(\Gamma)$, we study matching problems in such a framework. At first, we give some definitions.

We denote by ρ (a function equivalent to) the distance of the point x to the boundary Λ of the sheet Γ_1 (a regular open set in the boundary Γ of Ω), in a neighborhood of Λ. We define, as usual (see for example Lions-Magenes [1]):

$$(67) \qquad H_{0\,0}^{1/2}(\Gamma_1) \overset{def}{=} \{ u \in H^{1/2}(\Gamma_1), \rho^{-1/2} u \in L^2(\Gamma_1) \} = \{ u \in H^{1/2}(\Gamma_1), \tilde{u} \in H^{1/2}(\Gamma) \},$$

where $\tilde{u}$ is the extension of u by 0 on the outside of Γ_1. We denote by:

(68) $H^{-1/2}(\Gamma_1)$ the dual space of $H^{1/2}(\Gamma_1)$, $(H_{0\,0}^{1/2}(\Gamma_1))'$ the dual space of $H_{0\,0}^{1/2}(\Gamma_1)$, that is also:

$$(68)' \qquad (H_{0\,0}^{1/2}(\Gamma_1))' = \{ f = f_0 + f_1, \ f_0 \in H^{-1/2}(\Gamma_1), \quad \rho^{1/2} f_1 \in L^2(\Gamma_1) \}.$$

Let $\Gamma_2 = \Gamma \backslash \bar{\Gamma}_1$ be the (regular) complementary sheet in Γ. For every function v on Γ, we denote by v_i the restriction of v to Γ_i for $i = 1, 2$.

Lemma 6. *We have the following equivalences*:

$$(69) \quad \left| \begin{array}{l} \text{i) } w \in H^{1/2}(\Gamma) \Leftrightarrow w_i \in H^{1/2}(\Gamma_i), \ i = 1, 2, \text{ and } w_1 - R_0^* w_2 \in H_{0\,0}^{1/2}(\Gamma_1), \\ \text{ii) } f \in H^{-1/2}(\Gamma) \Leftrightarrow f_i \in (H_{0\,0}^{1/2}(\Gamma_i))', \ i = 1, 2, \text{ and } R_0 f_1 + f_2 \in H^{-1/2}(\Gamma_2), \\ \qquad\qquad\qquad\qquad\qquad\qquad (\text{or } f_1 + \tilde{R}_0 f_2 \in H^{-1/2}(\Gamma_1)), \end{array} \right.$$

with $R_0 = {}^t R_0^*$, $\tilde{R}_0 = R_0^{-1}$.

PROOF. We first reduce to a neighborhood V_δ of Λ, thanks to a regular function ϕ_δ such that: $\phi_\delta(x) = 1$ if $d(x, \Lambda) \geq \delta$, $\quad \phi_\delta(x) = 0$ if $d(x, \Lambda) < \delta/2$.

Then for every f in $H^{-1/2}(\Gamma)$, v in $H^{1/2}(\Gamma)$, we have:

(70) $\quad <f, v> = <f, v\phi_\delta> + <f, v(1-\phi_\delta)>,$

and we have only to consider $w = v(1-\phi_\delta) \in H^{1/2}(\Gamma)$ with supp $w \subseteq V_\delta$.
Then i) is a consequence of properties of space $H^{1/2}_{0\,0}(\Gamma_1)$, see Lions-Magenes [1].
For part ii), we write (first at least formally for all $f \in H^{-1/2}(\Gamma)$, $w \in H^{1/2}(\Gamma)$):

$$(71)\quad \left|\begin{array}{l} <f, w> \stackrel{\text{def}}{=} <f, w>_\Gamma = <f_1, w_1>_{\Gamma_1} + <f_2, w_2>_{\Gamma_2} \\ \stackrel{\text{def}}{=} <f_1, w_1 - R_o^* w_2>_{\Gamma_1} + <{}^t R_0^* f_1 + f_2, w_2>_{\Gamma_2}, \end{array}\right.$$

with the transposed operator ${}^tR_0$ of R_0. From formula (71), we see that:

$$f \in (H^{1/2}(\Gamma))' = H^{-1/2}(\Gamma) \text{ if and only if } {}^tR_0^* f_1 + f_2 \in H^{-1/2}(\Gamma_2),$$

or also if and only if $f_1 + ({}^tR_0^*)^{-1} f_2 \in H^{-1/2}(\Gamma_1)$, that is (69)ii). ⊗

Remark 7. The difference in sign of matching formulas (69) follows from the change in orientation of the sheets Γ_1 and Γ_2 by R_0 (in the terminology of differential geometry, see the appendix, we can consider the elements of $H^{1/2}(\Gamma)$ as even 0-forms, and those of $H^{-1/2}(\Gamma)$ as odd $(n-1)$-currents). ⊗

Let Γ_0 be a regular bounded orientable sheet, $\tilde{\Gamma}_0$ its canonical oriented covering, that is identified to the union of the two faces $\Gamma_1 = \Gamma_{0+}$ and $\Gamma_2 = \Gamma_{0-}$ of the sheet Γ_0 (see Schwartz [1] pp.315, 316 and Dieudonné [1] T.3 p.145, for these notions). Let σ be the symmetry which exchange the two faces.
Then we define the following Sobolev spaces:

$$(72)\quad \left|\begin{array}{l} \text{i) } H^{1/2}(\tilde{\Gamma}_0) \stackrel{\text{def}}{=} \{w,\ w_i \in H^{1/2}(\Gamma_i),\ i=1,2 \text{ and } w_1 - w_2 \in H^{1/2}_{o\,o}(\Gamma_o)\} \\ \text{ii) } H^{-1/2}(\tilde{\Gamma}_0) \stackrel{\text{def}}{=} \{f,\ f_i \in (H^{1/2}_{0\,0}(\Gamma_i))',\ i=1,2 \text{ and } f_1 + f_2 \in H^{-1/2}(\Gamma_0)\}. \end{array}\right.$$

The two spaces $H^{1/2}(\tilde{\Gamma}_0)$ and $H^{-1/2}(\tilde{\Gamma}_0)$ are dual spaces by:

$$(73)\quad <f,w> = <f_1,w_1>_{\Gamma_1} + <f_2,w_2>_{\Gamma_2} \stackrel{\text{def}}{=} <f_1, w_1 - w_2>_{\Gamma_1} + <f_2 + f_1, w_2>_{\Gamma_2}.$$

Let Ω be a bounded open set in R^n, which contains Γ_0. Then we have the trace property (adapted to fissured materials):

Corollary 3. *The trace mapping* $\gamma_0: u \to u|_{\tilde{\Gamma}_0}$ *is continuous from* $H^1(\Omega\backslash\Gamma_0)$ *onto* $H^{1/2}(\tilde{\Gamma}_0)$. *The trace mapping* $\gamma_1: u \to \frac{\partial u}{\partial n}|_{\Gamma_0}$ *is continuous from*

$$(74)\quad H^1(\Delta, \Omega\backslash\Gamma_0) \stackrel{\text{def}}{=} \{u \in H^1(\Omega\backslash\Gamma_0),\ \Delta u \in L^2(\Omega\backslash\Gamma_0)\}$$

onto $H^{-1/2}(\tilde{\Gamma}_0)$.

PROOF. We only have to consider a regular open set Ω_1 contained in the closure of Ω, with Γ_0 contained in its boundary. Let $\Omega_2 = \Omega\backslash\bar{\Omega}_1$, Γ'_1 be such that $\partial\Omega_1 = \bar{\Gamma}_0 \cup \Gamma'_1$. Then $u \in H^1(\Omega\backslash\Gamma_0)$ implies $u|_{\Omega_i} \in H^1(\Omega_i)$, $i = 1, 2$. Therefore the traces:

$$u|_{\Omega_1}|_{\Gamma_0} = u|_{\Gamma_{0-}} \text{ (resp. } u|_{\Omega_2}|_{\Gamma_0} = u|_{\Gamma_{0+}}) \text{ and } u|_{\Omega_1}|_{\Gamma'_1} = u|_{\Omega_2}|_{\Gamma'_1}$$

must satisfy the matching conditions; the first part of the corollary follows. The same argument applies to the matching of the traces of $\gamma_1 u$ on Γ_0.

⊗

Remark 8. In the situation of Lemma 6, let R_1 be a lifting of $H^{1/2}(\Gamma_1)$ in $H^{1/2}(\Gamma)$, and let $\tilde{R}_1$ be a lifting of $(H^{1/2}_{o\,o}(\Gamma_1))'$ in $H^{-1/2}(\Gamma)$. Then we can write:

$$(75) \qquad \left|\begin{array}{l} u = R_1(u|_{\Gamma_1}) + (u - R_1(u|_{\Gamma_1})),\ \forall u \in H^{1/2}(\Gamma), \\ f = -\tilde{R}_1(f|_{\Gamma_1}) + (f + \tilde{R}_1(f|_{\Gamma_1})),\ \forall f \in H^{-1/2}(\Gamma), \end{array}\right.$$

thus we obtain the decomposition of spaces $H^{1/2}(\Gamma)$ and $H^{-1/2}(\Gamma)$ into the direct sum

$$(76) \qquad H^{1/2}(\Gamma) = H^{1/2}(\Gamma_1) \oplus H^{1/2}_{o\,o}(\Gamma_2), \quad H^{-1/2}(\Gamma) = (H^{1/2}_{o\,o}(\Gamma_1))' \oplus H^{-1/2}(\Gamma_2),$$

and we can identify the orthogonal (or polar) space of $H^{1/2}(\Gamma_1)$ to $(H^{1/2}_{o\,o}(\Gamma_2))'$, and the orthogonal space of $H^{1/2}_{o\,o}(\Gamma_2)$ to the space $H^{-1/2}(\Gamma_1)$ in the duality $H^{1/2}(\Gamma)$, $H^{-1/2}(\Gamma)$, thanks to the formula, for all w in $H^{1/2}(\Gamma)$, f in $H^{-1/2}(\Gamma)$:

$$(77) \quad \langle w,f\rangle = \langle R_1(w|_{\Gamma_1}),f\rangle + \langle w - R_1(w|_{\Gamma_1}),f\rangle = \langle w|_{\Gamma_1}, {}^tR_1 f\rangle + \langle w - R_1(w|_{\Gamma_1}), f|_{\Gamma_2}\rangle.$$

⊗

Remark 9. In the case of the orientable sheet of Corollary 3, we have the natural mappings:

$$S: \left|\begin{array}{l} w \in H^{1/2}(\tilde{\Gamma}_0) \rightarrow \frac{1}{2}(w|_{\Gamma_{0+}} + w|_{\Gamma_{0-}}) \in H^{1/2}(\Gamma_0), \\ f \in H^{-1/2}(\tilde{\Gamma}_0) \rightarrow \frac{1}{2}(f|_{\Gamma_{0+}} + f|_{\Gamma_{0-}}) \in (H^{1/2}_{o\,o}(\Gamma_o))', \end{array}\right.$$

and:

$$D: \left|\begin{array}{l} w \in H^{1/2}(\tilde{\Gamma}_0) \rightarrow \frac{1}{2}(w|_{\Gamma_{0+}} - w|_{\Gamma_{0-}}) \in H^{1/2}_{o\,o}(\Gamma_o), \\ f \in H^{1/2}(\tilde{\Gamma}_0) \rightarrow \frac{1}{2}(f|_{\Gamma_{0+}} - f|_{\Gamma_{0-}}) \in H^{-1/2}(\Gamma_0). \end{array}\right.$$

The mapping $w \rightarrow (Sw, Dw)$ is then an isomorphism from space $H^{1/2}(\tilde{\Gamma}_0)$ onto $H^{1/2}(\Gamma_0) \times H^{1/2}_{o\,o}(\Gamma_o)$ and from $H^{-1/2}(\tilde{\Gamma}_0)$ onto $(H^{1/2}_{o\,o}(\Gamma_o))' \times H^{-1/2}(\Gamma_o)$. This allows us to identify these spaces.

⊗

7.2. Trace problems on sheets (the vectorial case)
Here we assume, in order to avoid technical details, that the sheets are very regular (that is, they are infinitely differentiable manifolds with boundary). Then we have the chain of spaces (for i = 1, 2):

$$(78)\quad D(\Gamma_i) \to H^{1/2}_{o\,o}(\Gamma_i) \to H^{1/2}(\Gamma_i) \to L^2(\Gamma_i) \to H^{-1/2}(\Gamma_i) \to (H^{1/2}_{o\,o}(\Gamma_i))' \to D'(\Gamma_i).$$

Thus we can naturally use surface operators in the sense of distributions on Γ_i (or in the sense of generalized differential forms or currents, see the Appendix).

Let Ω be a regular bounded open set in R^n, with boundary Γ. We define:

$$(79)\quad H^{-s}(\mathrm{div},\Omega) \stackrel{\mathrm{def}}{=} \{u \in H^{-s}(\Omega)^n,\ \mathrm{div}\, u \in H^{-s}(\Omega)\},\quad s \in R.$$

Then we have:

Proposition 6. *For every* s *with* $0 \le s \le 1/2$, *and every* u in $H^{-s}(\mathrm{div},\Omega)$, n.u *has a trace on the boundary* Γ of Ω *such that:*

$$(80)\quad n.u \in H^{-s-1/2}(\Gamma).$$

PROOF. We use the Green formula (with n the exterior normal to Γ):

$$(81)\quad (\mathrm{div}\, u, \phi) + (u, \mathrm{grad}\, \phi) = \int_\Gamma n.u\, \phi\, d\Gamma.$$

Let ϕ be in $H^{s+1}(\Omega)$, then by the usual trace theorems, we have $\phi|_\Gamma \in H^{s+1/2}(\Gamma)$. Then for u in $H^{-s}(\mathrm{div}, \Omega)$, the l.h.s. of (81) has a sense and defines a continuous mapping on $H^{s+1}(\Omega)$. Thus from (81), $n.u|_\Gamma$ is a continuous form on $H^{s+1/2}(\Gamma)$.

⊗

Replacing the domain Ω by the sheet Γ_1 into definition (79), we have:

$$(82)\quad H^{-s}(\mathrm{div},\Gamma_1) \stackrel{\mathrm{def}}{=} \{u \in H^{-s}(\Gamma_1)^n,\ n.u = 0,\ \mathrm{div}_\Gamma\, u \in H^{-s}(\Gamma_1)\},$$

and then we obtain a proposition similar to the preceding one, with ν the exterior normal to the boundary Λ of Γ_1:

Proposition 6'. *For every* s *with* $0 \le s \le 1/2$, *and every* u in $H^{-s}(\mathrm{div},\Gamma_1)$, then $\nu.u$ *has a trace on the boundary* Λ of Γ_1 *so that*: $\nu.u \in H^{-s-1/2}(\Lambda)$.

PROOF. We only have to replace the Green formula (81) by:

$$(83)\quad (\mathrm{div}_\Gamma\, u, \phi) + (u, \mathrm{grad}_\Gamma\, \phi) = \int_\Lambda \nu.u\, \phi\, d\Lambda.$$

⊗

Remark 10. For $n = 3$, we also have the following trace result: the space

(84) $$H^{-s}(\mathrm{curl},\Gamma_1) \stackrel{\mathrm{def}}{=} \{u \in H^{-s}(\Gamma_1)^n,\ \ n.u = 0,\ \ \mathrm{curl}_\Gamma u \in H^{-s}(\Gamma_1)\},$$

is such that for every $s \in [0, 1/2]$, $v \in H^{-s}(\mathrm{curl},\Gamma_1)$, then (with τ a unit tangent vector to the boundary Λ of Γ_1) $\tau.v$ has a trace on Λ such that $\tau.v \in H^{-s-1/2}(\Lambda)$.

⊗

Proposition 6' and Remark 10 allow us to define the following spaces:

(85) $$\left|\begin{array}{l} H_o^{-1/2}(\mathrm{div},\Gamma_1) \stackrel{\mathrm{def}}{=} \{u \in H^{-1/2}(\mathrm{div},\Gamma_1),\ \ \nu.u|_\Lambda = 0\} \\ H_o^{-1/2}(\mathrm{curl},\Gamma_1) \stackrel{\mathrm{def}}{=} \{u \in H^{-1/2}(\mathrm{curl},\Gamma_1),\ \ \tau.u|_\Lambda = 0\}. \end{array}\right.$$

For Γ the boundary of a regular open set in R^n, let j be in $H^{-1/2}(\mathrm{div},\Gamma)$. Then by restriction to Γ_i we obtain that j_i for $i = 1, 2$ is in the following space:

(86) $$X(\mathrm{div},\Gamma_i) \stackrel{\mathrm{def}}{=} \{f \in (H_{o\,o}^{1/2}(\Gamma_i)^n)',\ \ n.f = 0,\ \ \mathrm{div}_\Gamma f \in (H_{o\,o}^{1/2}(\Gamma_i))'\}$$

with $\mathrm{div}_\Gamma f$ in the sense of distributions in Γ_i. For $n = 3$, we also define:

(87) $$X(\mathrm{curl},\Gamma_i) \stackrel{\mathrm{def}}{=} \{f \in (H_{o\,o}^{1/2}(\Gamma_i)^3)',\ \ n.f = 0,\ \ \mathrm{curl}_\Gamma f \in (H_{o\,o}^{1/2}(\Gamma_i))'\}.$$

Contrary to the elements of $H^{-1/2}(\mathrm{div},\Gamma_i)$ and $H^{-1/2}(\mathrm{curl},\Gamma_i)$, the elements of $X(\mathrm{div},\Gamma_i)$, or of $X(\mathrm{curl},\Gamma_i)$ have no trace on the boundary of Γ_i. This may be seen from Green formula (83).
For tangent vector fields (or 1-forms) we have matching properties similar to the scalar case in the framework of $H^{1/2}$ spaces, as in Lemma 6; this may be verified on each component of the field.
From now on, we have to specify the behavior of "fields" under orientation change, so using the language of differential geometry (see the Appendix) we consider the elements of space $H^{-1/2}(\mathrm{div},\Gamma)$ as odd 1-currents (or also as generalized odd 1-forms), and the elements of $H^{-1/2}(\mathrm{curl},\Gamma)$ as (generalized) even 1-forms.
We only recall that the pullback $R^*_o w$ of the 1-form w on Γ_1(resp. the pushforward $R_o f$ of the 1-field f) by the symmetry operator R_o is defined by:

(88) $$R_o^* w_x(v) = w_{R_o x}(TR_o v), \qquad \langle R_o f, w\rangle = \langle f, R_o^* w\rangle,$$

for all tangent vector v to Γ_1; TR_o corresponds to the Jacobian of the transformation R_o (see App. (326)). In the situation of Lemma 6, we have

Theorem 5. *(Matching property in* $H^{-1/2}(\mathrm{div},\Gamma)$ *and* $H^{-1/2}(\mathrm{curl},\Gamma)$*).*
We have the following equivalences:

(89) $\quad f \in H^{-1/2}(\mathrm{div},\Gamma) \Leftrightarrow f_i \in X(\mathrm{div},\Gamma_i),\ i=1,2 \text{ and } R_o f_1 + f_2 \in H_0^{-1/2}(\mathrm{div},\Gamma_2),$
and for $n = 3$,

(90) $\quad w \in H^{-1/2}(\mathrm{curl},\Gamma) \Leftrightarrow w_i \in X(\mathrm{curl},\Gamma_i),\ i=1,2 \text{ and } R_o^* w_2 - w_1 \in H_0^{-1/2}(\mathrm{curl},\Gamma_1).$

PROOF. We only have to show for (89) that (f_1,f_2) as in the r.h.s. of (89) defines an element f in $H^{-1/2}(\mathrm{div},\Gamma)$.

i) At first the last condition of (89) implies that: $R_o f_1 + f_2 \in H_t^{-1/2}(\Gamma_2)$, and thus f defined by $f|_{\Gamma_i} = f_i$ satisfies $f \in H_t^{-1/2}(\Gamma)$ from Lemma 6.

ii) The element g defined on Γ by $g|_{\Gamma_i} = g_i = \mathrm{div}_\Gamma(f|_{\Gamma_i})$, i = 1, 2 is also in $H^{-1/2}(\Gamma)$ from Lemma 6, and from commutation: $\mathrm{div}_\Gamma R_o f_1 = R_o \mathrm{div}_\Gamma f_1$.

iii) We have in the sense of distributions (or currents) on Γ:

(91) $$\mathrm{div}_\Gamma f = (\mathrm{div}_\Gamma f) + [\nu.f]_\Lambda \delta_\Lambda,$$

with $(\mathrm{div}_\Gamma f)$ the surface divergence in the sense of $D'(\Gamma_i)$ i = 1,2 and $[\nu.f]_\Lambda$ the jump of ν.f across Λ, which is zero from hypothesis.

Therefore we have proved that $\mathrm{div}_\Gamma f = g$, and (89). Equivalence (90) is similar, and can also be deduced from (89).

⊗

In the situation of Corollary 3, with a sheet Γ_o, we have:

Corollary 4. *The mappings* $u \to n \wedge u|_{\tilde{\Gamma}_o}$ and $u \to \pi_{\tilde{\Gamma}_o} u$ *are continuous from* $H(\mathrm{curl},\Omega\backslash\Gamma_o)$ onto $H^{-1/2}(\mathrm{div},\tilde{\Gamma}_o)$ and onto $H^{-1/2}(\mathrm{curl},\tilde{\Gamma}_o)$ *respectively, with*:

(92) $$\left| \begin{array}{l} H^{-1/2}(\mathrm{div},\tilde{\Gamma}_o) \overset{\mathrm{def}}{=} \{f,\ f_i = f|_{\Gamma_i} \in X(\mathrm{div},\Gamma_i),\ i=1,2 \text{ and } f_1 + f_2 \in H_0^{-1/2}(\mathrm{div},\Gamma_o)\}, \\ H^{-1/2}(\mathrm{curl},\tilde{\Gamma}_o) \overset{\mathrm{def}}{=} \{f,\ f_i = f|_{\Gamma_i} \in X(\mathrm{curl},\Gamma_i),\ i=1,2 \text{ and } f_1 - f_2 \in H_0^{-1/2}(\mathrm{curl},\Gamma_o)\}. \end{array} \right.$$

Remark 11. Using a lifting R_1 of $X(\mathrm{div},\Gamma_1)$ into $H^{-1/2}(\mathrm{div},\Gamma)$, we can write:

(93) $$f = R_1(f|_{\Gamma_1}) + (f - R_1(f|_{\Gamma_1})).$$

Thus we have the following decomposition into direct sum (like in Remark 8):

(94) $$H^{-1/2}(\mathrm{div},\Gamma) = X(\mathrm{div},\Gamma_1) \oplus H_0^{-1/2}(\mathrm{div},\Gamma_2),$$

and similarly, for the curl:

(95) $$H^{-1/2}(\mathrm{curl},\Gamma) = X(\mathrm{curl},\Gamma_1) \oplus H_0^{-1/2}(\mathrm{curl},\Gamma_2).$$

We verify that in the duality $H^{-1/2}(\mathrm{div},\Gamma)$, $H^{-1/2}(\mathrm{curl},\Gamma)$, the dual space of $X(\mathrm{div},\Gamma_1)$ is $H_o^{-1/2}(\mathrm{curl},\Gamma_1)$, and the dual space of $H_o^{-1/2}(\mathrm{div},\Gamma_2)$ is $X(\mathrm{curl},\Gamma_2)$.
Therefore the orthogonal (polar) space of $X(\mathrm{div},\Gamma_1)$ (resp. $H_o^{-1/2}(\mathrm{curl},\Gamma_1)$) is the space $X(\mathrm{curl},\Gamma_2)$ (resp. $H_o^{-1/2}(\mathrm{div},\Gamma_2)$).

⊗

Remark 12. We note that grad_Γ and $\overrightarrow{\mathrm{curl}}_\Gamma$ are continuous operators from $H_{o\,o}^{1/2}(\Gamma_i)$ into $(H_{oot}^{1/2}(\Gamma_i))'$ and by duality div_Γ and curl_Γ are continuous operators from $H_{oot}^{1/2}(\Gamma_i)$ into $(H_{o\,o}^{1/2}(\Gamma_i))'$, where as usual the index t refers to tangent fields.

⊗

Application to electromagnetism

Let j_Σ be some current concentrated on a (regular, bounded) sheet $\Sigma = \Gamma_o$ in a free space domain Ω, with angular frequency ω. Then the electromagnetic field (E, H) created by this current, which is solution of the Maxwell equations:

$$(96)\qquad \left|\begin{array}{l} \mathrm{curl}\,H + i\omega\varepsilon_o\,E = j_\Sigma \\ -\,\mathrm{curl}\,E + i\omega\mu_o\,H = 0 \end{array}\right. \quad \text{in } \Omega,$$

is of finite energy in $\Omega\backslash\Sigma$, i.e., E and H are in $H(\mathrm{curl},\Omega\backslash\Sigma)$, if and only if j_Σ satisfies

$$(97)\qquad j_\Sigma \in H_o^{-1/2}(\mathrm{div},\Sigma).$$

Therefore in the scattering of an incoming electromagnetic field (E_I, H_I) by a thin obstacle that we model by a sheet Σ, the jump of the scattered field across this sheet necessarily satisfies (97).

⊗

8. SOME REGULARITY RESULTS

It is often interesting to give up the $H(\mathrm{curl},\Omega)$ or the $H(\mathrm{div},\Omega)$ space for the $H^k(\Omega)$ space, $(k > 0)$, at first to use regularity results in Sobolev spaces and better to use the compacity of the natural injection of Sobolev spaces in $L^2(\Omega)$ for (regular) bounded sets Ω. We will use the following lemma extensively; for its proof, we only refer to Peetre [1], Tartar [1], Lions-Magenes [1].

Lemma 7 (Peetre). *Let* E_o, E_1, E_2 *be Banach spaces,* A_1, A_2 *be two continuous linear mappings from* E_o *into* E_1 *and* E_2 *respectively, with:*
i) A_2 *is a compact mapping;*
ii) *there is a constant* $c > 0$ *such that:*

$$(98)\qquad \|v\|_{E_o} \le c\,\{\|A_1 v\|_{E_1} + \|A_2 v\|_{E_2}\},\quad \forall v \in E_o.$$

Then:

i) $\ker A_1$ *has finite dimension, and* $\operatorname{Im} A_1$ *is closed*
ii) *there is a constant* $c_0 > 0$ *such that*:

(99) $$\inf_{w \in \ker A_1} \|v + w\|_{E_0} \le c_0 \|A_1 v\|_{E_1}, \quad \forall v \in E_0.$$

Theorem 6. *Let* Ω *be a regular* $(C^{1,1})$ *bounded open set in* $\mathbb{R}^3$. *Let* u *be in* $L^2(\Omega)^3$ *with* $\operatorname{curl} u \in L^2(\Omega)^3$, $\operatorname{div} u \in L^2(\Omega)$, *and* $n.u|_\Gamma \in H^{1/2}(\Gamma)$ (resp. $n \wedge u|_\Gamma \in H_t^{1/2}(\Gamma)$). *Then* $u \in H^1(\Omega)^3$.

PROOF. i) We first reduce to "homogeneous" boundary conditions with a lifting: there exists ϕ in $H^2(\Omega)$ with $\frac{\partial\phi}{\partial n}|_\Gamma = n.\operatorname{grad}\phi|_\Gamma = n.u|_\Gamma$. Then $v = u - \operatorname{grad}\phi$ satisfies: $v \in L^2(\Omega)^3$, $\operatorname{curl} v \in L^2(\Omega)^3$, $\operatorname{div} v \in L^2(\Omega)$, and $n.v|_\Gamma = 0$;
i)'(resp. there exists φ in $H^1(\Omega)^3$ so that $\varphi|_\Gamma = n \wedge u|_\Gamma \in H_t^{1/2}(\Gamma)$, and then $v = u - \varphi$ satisfies: $v \in L^2(\Omega)^3$, $\operatorname{curl} v \in L^2(\Omega)^3$, $\operatorname{div} v \in L^2(\Omega)$, and $n \wedge v|_\Gamma = 0$).

ii) Now let v be a regular vector field on Ω. Using the Green formulas (14), (20), and (45) we obtain:

(100) $$\begin{aligned}\int_\Omega |\operatorname{grad} v|^2\,dx &= \operatorname{Re}\int_\Omega -\Delta v.\bar v\,dx + \operatorname{Re}\int_\Gamma \frac{\partial v}{\partial n}.\bar v\,d\Gamma \\ &= \int_\Omega [|\operatorname{curl} v|^2 + |\operatorname{div} v|^2]\,dx + I_\Gamma\end{aligned}$$

where I_Γ is given by:

(100)' $$I_\Gamma = \operatorname{Re}\int_\Gamma \{-\operatorname{curl} v.\, n \wedge \bar v - \operatorname{div} v\,(n.\bar v) + \frac{\partial v}{\partial n}.\bar v\}\,d\Gamma.$$

We can obtain the last term in many ways, for instance using the *covariant derivative* or *covariant differentiation* along n (for this notion, see Gilkey [1] p.104, Dieudonné [1]) and then the formulas (310) of the Appendix on div and curl. With the notations of the App. 6, Remark 20, with R_m the mean curvature, v_Γ and v^n the projections of v on Γ and n (and using (47) or (30)), we have:

(100)" $$\left|\begin{aligned} I_\Gamma &= \operatorname{Re}\int_\Gamma [-(\operatorname{div}_\Gamma v_\Gamma)\,\bar v^{\,n} + (\overrightarrow{\operatorname{curl}}_\Gamma v^n).\, n \wedge \bar v]\,d\Gamma + \int_\Gamma (\tilde\Gamma v, v)\,d\Gamma \\ &= \operatorname{Re}\int_\Gamma [\operatorname{grad}_\Gamma v^n.\,\bar v - (\operatorname{curl}_\Gamma (n \wedge v))\,\bar v^{\,n}]\,d\Gamma + \int_\Gamma (\tilde\Gamma v, v)\,d\Gamma,\end{aligned}\right.$$

with $(\tilde\Gamma v, v) = -\int_\Gamma (\frac{1}{2}\sum_i \alpha_i |v^i|^2 + 2R_m |v_n|^2)\,d\Gamma$, $\alpha_i = g_i^{-1}\frac{\partial g_i}{\partial n}$, and $v = \sum v^i e_i + v^n n$ in an orthonormal coordinate system (e_1, e_2, n), $e_i = g_i^{-1/2}\partial_i$.

Thus for all v in $H^1(\Omega)^3$ with $n \wedge v|_\Gamma = 0$, or with $n.v|_\Gamma = 0$, there exists $C > 0$ so that

(101) $$\int_\Omega |\operatorname{grad} v|^2\,dx \le \int_\Omega [|\operatorname{curl} v|^2 + |\operatorname{div} v|^2]\,dx + C\int_\Gamma |v(x)|^2\,d\Gamma.$$

iii) We only have to apply Peetre Lemma, with E_o equal to the space:

(102) $H^1_{to}(\Omega) \overset{\text{def}}{=} \{u \in H^1(\Omega)^3,\ n \wedge u|_\Gamma = 0\}$ (resp. $H^1_{no}(\Omega) \overset{\text{def}}{=} \{u \in H^1(\Omega)^3,\ n.u|_\Gamma = 0\}$),

and $E_1 = L^2(\Omega)^3 \times L^2(\Omega)^3 \times L^2(\Omega)^3$, $E_2 = L^2(\Gamma)^3$, and the mappings:

$$A_1 : v \in E_o \to (\text{curl } v, \text{div } v, v) \in E_1,\ \ A_2 : v \in E_o \to v|_\Gamma \in L^2(\Gamma)^3.$$

(the compactness of A_2 is due to the compactness of the mapping $H^{1/2}(\Gamma) \to L^2(\Gamma)$).

⊗

From Peetre Lemma again, we have proved: the norm of $H^1(\Omega)^3$ on $H^1_{to}(\Omega)$ and on $H^1_{no}(\Omega)$ is equivalent to the norm:

$$\|v\| \overset{\text{def}}{=} \{\int_\Omega [|\text{curl } v|^2 + |\text{div } v|^2 + |v|^2] dx\}^{1/2}.$$

From Theorem 6, we obviously have:

Corollary 5. *With hypotheses of Theorem 6, we have the following equivalences:*

(103) $u \in H^1(\Omega)^3 \Leftrightarrow u \in L^2(\Omega)^3$, $\text{curl } u \in L^2(\Omega)^3$, $\text{div } u \in L^2(\Omega)$, $n.u|_\Gamma \in H^{1/2}(\Gamma)$ (resp. $n \wedge u|_\Gamma \in H^{1/2}_t(\Gamma)$),

with equivalence of the norms: $\|u\|_{H^1(\Omega)^3}$, *and:*

$[\|\text{curl } u\| + \|\text{div } u\| + \|u\| + \|n.u\|_{H^{1/2}(\Gamma)}]$ *and* $[\|\text{curl } u\| + \|\text{div } u\| + \|u\| + \|n \wedge u\|_{H^{1/2}(\Gamma)^3}]$.

Then we define very useful spaces (see section 10 below and the Appendix)

Definition 4. *We call cohomology spaces the following spaces:*

(104)
$$\begin{cases} H_1(\Omega) \overset{\text{def}}{=} \{u \in L^2(\Omega)^3,\ \text{curl } u = 0,\ \text{div } u = 0,\ n.u|_\Gamma = 0\}, \\ H_2(\Omega) \overset{\text{def}}{=} \{u \in L^2(\Omega)^3,\ \text{curl } u = 0,\ \text{div } u = 0,\ n \wedge u|_\Gamma = 0\}. \end{cases}$$

From Peetre Lemma, we have:

Corollary 6. *The cohomology spaces are finite dimensional.*

We have also further regularity results for n = 3 (or also for n = 2). For the proof, see Dautray-Lions [1] chap. 9:

Proposition 7. *Let Ω be a regular ($C^{k+1,1}$) bounded open set in* R^3. *Then:*

$u \in L^2(\Omega)^3$, $\text{curl } u \in H^k(\Omega)^3$, $\text{div } u \in H^k(\Omega)$, $n.u|_\Gamma \in H^{k+1/2}(\Gamma)$ or $n \wedge u|_\Gamma \in H^{k+1/2}_t(\Gamma)$ *satisfies* $u \in H^{k+1}(\Omega)^3$.

9. THE HODGE DECOMPOSITION

It is often very useful in electromagnetism to know (that is to characterize) the range and the kernel of the operators grad, curl and div in the space of square integrable fields in a domain Ω. This is part of differential geometry, known as the Hodge decomposition. Here we will give only main results, referring to the Appendix for developments, and to Dautray-Lions [1] chap. 9. Proofs are often an easy application of Petree Lemma.

Theorem 7. *Let Ω be an open bounded set in* R^3, *with boundary* Γ. *The following sequences are so that the range of one operator is contained in the kernel of the following in the sequence, with finite codimension*:

$$(105)\qquad \begin{array}{l} H^1(\Omega) \xrightarrow{\text{grad}} H(\text{curl},\Omega) \xrightarrow{\text{curl}} H(\text{div},\Omega) \xrightarrow{\text{div}} L^2(\Omega) \\ H^1_o(\Omega) \xrightarrow{\text{grad}} H_0(\text{curl},\Omega) \xrightarrow{\text{curl}} H_0(\text{div},\Omega) \xrightarrow{\text{div}} L^2(\Omega), \end{array}$$

and (therefore) the ranges of the operators grad, curl, div *(defined on their natural domains* $H^1(\Omega)$, $H(\text{curl},\Omega)$, $H(\text{div},\Omega)$, *or also with vanishing boundary conditions) are closed in the corresponding* $L^2(\Omega)$ *spaces*.

Remark 13. Let Γ_0 be an open set in Γ; we define

$$(106)\qquad \left| \begin{array}{l} H^1_{\Gamma_o}(\Omega) \overset{\text{def}}{=} \{u \in H^1(\Omega),\ u = 0 \text{ on } \Gamma_0\}, \\ H_{\Gamma_0}(\text{curl},\Omega) \overset{\text{def}}{=} \{u \in H(\text{curl},\Omega),\ n \wedge u = 0 \text{ on } \Gamma_0\}, \\ H_{\Gamma_0}(\text{div},\Omega) \overset{\text{def}}{=} \{u \in H(\text{div},\Omega),\ n.u = 0 \text{ on } \Gamma_o\}. \end{array} \right.$$

Then we can prove that *the following sequence has the same properties as* (105):

$$H^1_{\Gamma_o}(\Omega) \xrightarrow{\text{grad}} H_{\Gamma_0}(\text{curl},\Omega) \xrightarrow{\text{curl}} H_{\Gamma_0}(\text{div},\Omega) \xrightarrow{\text{div}} L^2(\Omega). \qquad \otimes$$

A *simply connected domain* Ω is such that grad $H^1(\Omega)$ = ker curl. More generally in the domain Ω let Σ_i, $i = 1$ to N, be cuttings of Ω so that the domain $\dot{\Omega} = \Omega \setminus \cup \Sigma_i$ be simply connected. Let Γ_j, $j = 0$ to m, be the components of Γ. Then we have

Proposition 8. *Characterization of the cohomology spaces. The cohomology spaces* (104) *are also given by*

$$\begin{array}{l} \mathcal{H}_1(\Omega) = \{u \in L^2(\Omega)^n,\ u = (\text{grad } q) \text{ in } \dot{\Omega},\ q \in H^1(\dot{\Omega}), \text{ with} \\ \qquad \Delta q = 0 \text{ in } \dot{\Omega},\ \frac{\partial q}{\partial n}\Big|_\Gamma = 0,\ [q]_{\Sigma_i} = C_i \text{ constant},\ [\frac{\partial q}{\partial n}]_{\Sigma_i} = 0\} \\ \mathcal{H}_2(\Omega) = \{u \in L^2(\Omega)^n,\ u = \text{grad } \phi,\ \phi \in H^1(\Omega),\ \Delta\phi = 0,\ \phi|_{\Gamma_j} = C_j \ \textit{constant}\} \end{array}$$

with

$$\left| \begin{array}{l} \dim \mathcal{H}_1(\Omega) = N \ (\textit{the "cutting" number of } \Omega), \\ \dim \mathcal{H}_2(\Omega) = m \text{ if } m+1 \textit{ is the number of components of } \Gamma. \end{array} \right.$$

Remark 14. Then we define:

$$(107)\quad \left| \begin{array}{l} \ker_0 \operatorname{curl} \overset{\text{def}}{=} \{v \in L^2(\Omega)^3, \quad \operatorname{curl} v = 0, \quad n \wedge v|_\Gamma = 0\} \\ \ker_0 \operatorname{div} \overset{\text{def}}{=} \{v \in L^2(\Omega)^3, \quad \operatorname{div} v = 0, \quad n.v|_\Gamma = 0\}. \end{array} \right.$$

We have:

$$\left| \begin{array}{l} H_1(\Omega) = \ker \operatorname{curl} \cap \ker_0 \operatorname{div} \subseteq \operatorname{curl} H^1(\Omega)^3, \text{ but is not in Im grad}, \\ H_2(\Omega) = \ker_0 \operatorname{curl} \cap \ker \operatorname{div} \subseteq \operatorname{grad} H^1(\Omega), \text{ but is not in Im curl}. \end{array} \right.$$

⊗

Characterization of the intersection of the kernels. We have the decomposition:

$$(108)\quad \ker \operatorname{curl} \cap \ker \operatorname{div} = H_1(\Omega) \oplus \operatorname{Im} \operatorname{grad} \cap \operatorname{Im} \operatorname{curl} \oplus H_2(\Omega),$$

with:

$$(109)\ \operatorname{Im} \operatorname{grad} \cap \operatorname{Im} \operatorname{curl} = \{u \in L^2(\Omega)^3, \ \exists\, \phi \in H^1(\Omega), v \in H^1(\Omega)^3, u = \operatorname{grad} \phi = \operatorname{curl} v\}.$$

Thus ϕ (resp. v) is a harmonic (resp. vector) function on Ω. We can choose v with:

$$\operatorname{div} v = 0, \quad n.v|_\Gamma = 0, \quad \int_{\Sigma_i} n.v\, d\Sigma = 0, \quad i = 1 \text{ to } N,$$

so that ϕ (up to a constant) and v are unique; they are conjugated harmonic functions in Ω.

⊗

Theorem 8. *Characterization of kernels for curl and div (with Ω in R^3).*

With the hypotheses of Theorem 7, let u *be in* $L^2(\Omega)^3$ *with*

i) $\operatorname{curl} u = 0$; *then there exist* $\phi \in H^1(\Omega)$, $h_1 \in H_1(\Omega)$ such that: $u = \operatorname{grad} \phi + h_1$.
Moreover, ϕ *is unique up to constant on each connected component of* Ω;

ii) $\operatorname{curl} u = 0$, $n \wedge u|_\Gamma = 0$; *then there exist unique* $\phi_0 \in H^1_o(\Omega)$, $h_2 \in H_2(\Omega)$ such that

$$u = \operatorname{grad} \phi_0 + h_2\,;$$

iii) $\operatorname{div} u = 0$; *then there exist* $v \in H^1(\Omega)^3$, $h_2 \in H_2(\Omega)$ such that $u = \operatorname{curl} v + h_2$;
v *is unique under the supplementary conditions*:

$$\operatorname{div} v = 0, \quad n.v|_\Gamma = 0, \quad \int_{\Sigma_i} n.v\, d\Sigma = 0, \quad i = 1 \text{ to } N.$$

iv) $\operatorname{div} u = 0$, $n.u|_\Gamma = 0$; *then there exist* $v_0 \in H^1_{to}(\Omega)$, $h_1 \in H_1(\Omega)$ such that:

$$u = \operatorname{curl} v_0 + h_1;$$

v_0 *is unique with the supplementary conditions*: $\operatorname{div} v_0 = 0$, $\int_{\Gamma_j} n.v_0\, d\Gamma = 0$, $j = 0$ to m.

Theorem 8'. *The Hodge decompositions of the kernels of curl and div. The spaces* ker curl, ker div, $\ker_0$ curl, $\ker_0$ div *have the orthogonal decompositions in* $L^2(\Omega)^3$:

$$(110)\quad \left| \begin{array}{ll} \ker \operatorname{curl} = \operatorname{Im} \operatorname{grad} \oplus H_1(\Omega), & \ker \operatorname{div} = \operatorname{Im} \operatorname{curl} \oplus H_2(\Omega), \\ \ker_0 \operatorname{curl} = \operatorname{Im} \operatorname{grad}_0 \oplus H_2(\Omega), & \ker_0 \operatorname{div} = \operatorname{Im} \operatorname{curl}_0 \oplus H_1(\Omega), \end{array} \right.$$

with: $\text{Im grad} = \text{grad } H^1(\Omega)$, $\text{Im grad}_0 = \text{grad } H^1_o(\Omega)$,
$\text{Im curl} = \text{curl } H^1(\Omega)^3 = \text{curl } H(\text{curl},\Omega)$, $\text{Im curl}_0 = \text{curl } H^1_o(\Omega)^3 = \text{curl } H_0(\text{curl},\Omega)$.

Definition 5. *Let* $J(\Omega)$, $\tilde{J}(\Omega)$, $V = J^1_v(\Omega)$, $\tilde{V} = J^1_\tau(\Omega)$ *be the following spaces:*

$$(111)\quad \begin{cases} J(\Omega) = \{ u \in L^2(\Omega)^3,\ \text{div}\, u = 0,\ \int_{\Gamma_j} n.u\, d\Gamma = 0,\ j = 0 \text{ to } m\}, \\ \tilde{J}(\Omega) = \{ u \in L^2(\Omega)^3,\ \text{div}\, u = 0,\ \int_{\Sigma_i} n.u\, d\Sigma = 0,\ i = 1 \text{ to } N \} \\ \tilde{V} = J^1_\tau(\Omega) = \{ u \in H(\text{curl},\Omega),\ \text{div}\, u = 0,\ n \wedge u|_\Gamma = 0,\ \int_{\Gamma_j} n.u\, d\Gamma = 0,\ j = 0 \text{ to } m\} \\ V = J^1_v(\Omega) = \{ u \in H(\text{curl},\Omega),\ \text{div}\, u = 0,\ n.u|_\Gamma = 0,\ \int_{\Sigma_i} n.u\, d\Sigma = 0,\ i = 1 \text{ to } N \}. \end{cases}$$

Theorem 9. *The ranges of the curl operator are given by*:

$$(112)\quad \text{curl } H^1(\Omega)^3 = \text{curl } V = J(\Omega), \qquad \text{curl } H^1_o(\Omega)^3 = \text{curl } \tilde{V} = \tilde{J}(\Omega);$$

the curl operator is an isomorphism from space V *onto* $J(\Omega)$ *and from* $\tilde{V}$ *onto* $\tilde{J}(\Omega)$.

Then from theorems 7 and 8 we obtain the following decomposition theorem:

Theorem 10. *Every* u *in* $L^2(\Omega)^3$ *has the decompositions*:

$$(113)\quad \begin{cases} u = \text{grad}\, p_0 + h_2 + \text{curl}\, v, & \text{with } p_0 \in H^1_o(\Omega),\ h_2 \in \mathcal{H}_2(\Omega),\ v \in V, \\ u = \text{grad}\, p + h_1 + \text{curl}\, v_0, & \text{with } p \in H^1(\Omega),\ h_1 \in \mathcal{H}_1(\Omega),\ v_0 \in \tilde{V}. \end{cases}$$

Moreover with these conditions, p_0, h_2, v, p (*up to constants on the components of* Ω), h_1, v_0 *are unique*.

Theorem 10'. *The Hodge decompositions of* $L^2(\Omega)^n$. *The space* $L^2(\Omega)^n$ *has the orthogonal decomposition*:

$$(114)\quad \begin{cases} L^2(\Omega)^n = \text{grad } H^1_o(\Omega) \oplus \ker \text{div} = \text{grad } H^1(\Omega) \oplus \ker_0 \text{div}, \\ L^2(\Omega)^3 = \text{Im curl}_0 \oplus \ker \text{curl} = \text{Im curl} \oplus \ker_0 \text{curl}, \quad \text{for } n = 3, \end{cases}$$

and thus:

$$(114)'\quad \begin{cases} L^2(\Omega)^3 = \text{grad } H^1_o(\Omega) \oplus \mathcal{H}_2 \oplus \text{curl } H^1(\Omega)^3 \\ L^2(\Omega)^3 = \text{grad } H^1(\Omega) \oplus \mathcal{H}_1 \oplus \text{curl } H^1_o(\Omega)^3. \end{cases}$$

For $n = 2$, *we have similarly* (*but with* $\dim \mathcal{H}_1(\Omega) = \dim \mathcal{H}_2(\Omega)$):

$$(114)''\quad L^2(\Omega)^2 = \text{grad } H^1_o(\Omega) \oplus \mathcal{H}_2 \oplus \overrightarrow{\text{curl}}\, H^1(\Omega) = \text{grad } H^1(\Omega) \oplus \mathcal{H}_1 \oplus \overrightarrow{\text{curl}}\, H^1_o(\Omega).$$

We can determine p_0, h_2, v, p ,h_1, v_0 in (113) by a variational method:

i) *for* p_0 and p:

(115)
$$\begin{cases} (\text{grad } p_0, \text{grad } \phi_0) = (u, \text{grad } \phi_o), & \forall \phi_o \in H^1_0(\Omega), \\ (\text{grad } p, \text{ grad } \phi) = (u, \text{ grad } \phi), & \forall \phi \in H^1(\Omega); \end{cases}$$

ii) *for* v_0:

(116) $$(\text{curl } v_0, \text{curl } \phi_0) = (u, \text{curl } \phi_0), \quad \forall \phi_0 \in \tilde{V}.$$

The sesquilinear form $\tilde{a}(v_0, \phi_0) \stackrel{\text{def}}{=} (\text{curl } v_0, \text{ curl } \phi_0)$ is coercive on $\tilde{V}$ from Peetre Lemma (with $E_0 = \tilde{V}$, $E_1 = E_2 = L^2(\Omega)^3$, $A_1 = \text{curl}$, $A_2 = I$);

iii) *for* v:

(117) $$(\text{curl } v, \text{curl } \phi) = (u, \text{curl } \phi), \quad \forall \phi \in V.$$

The sesquilinear form $a(v, \phi) \stackrel{\text{def}}{=} (\text{curl } v, \text{curl } \phi)$ is coercive on V from Peetre Lemma (with $E_0 = V$, $E_1 = E_2 = L^2(\Omega)^3$, $A_1 = \text{curl}$, $A_2 = I$);

iv) *for* h_2, we first use a basis of the space $H_2(\Omega)$, from the elements:

(118) $$v_j = \text{grad } \chi_j, \quad \text{with } \chi_j \in H^1(\Omega), \ \Delta\chi_j = 0 \text{ in } \Omega, \ \ \chi_j|_{\Gamma_i} = \delta_{ij}, \ j = 0 \text{ to } m.$$

Note that $\sum v_j = 0$, and the rank of the basis is only m.
Then we define the *capacitance matrix*:

(119) $$C_{ij} = \int_\Omega \text{grad } \chi_i . \text{grad } \chi_j \, dx = \int_{\Gamma_i} \frac{\partial \chi_j}{\partial n} d\Gamma.$$

This matrix is symmetric positive definite, of rank m (see for example Dautray-Lions [1] chap. 2). Then h_2 is given by:

(120)
$$\begin{cases} h_2 = \text{grad } p_2 = \sum_{j=0}^{m} y_j \text{ grad } \chi_j, \quad \text{with } y_j = p_2|_{\Gamma_j}, \text{ and} \\ \int_\Omega u . \text{grad } \chi_i dx = \sum_{j=0}^{m} C_{ij} y_j . \end{cases}$$

Since we can determine p_2 only up to a constant, we can choose $y_0 = 0$, and then the system (121) has one solution.

v) *for* h_1 we use the following basis of $H_1(\Omega)$:

(121)
$$\begin{cases} v_j = \text{grad } q_j, \quad \text{with } q_j \in H^1(\dot{\Omega}) \text{ satisfying:} \\ \Delta q_j = 0 \text{ in } \dot{\Omega}, \quad \frac{\partial q_j}{\partial n}|_\Gamma = 0, \quad [q_j]_{\Sigma_i} = \delta_{ij}, \quad [\frac{\partial q_j}{\partial n}]_{\Sigma_i} = 0. \end{cases}$$

Then we define the *inductance matrix*:

(122) $$L_{ij} = \int_\Omega \text{grad } q_j . \text{ grad } q_i \, dx = \int_{\Sigma_i} \frac{\partial \theta_j}{\partial n} d\Sigma,$$

which is a symmetric positive definite matrix. Then h_1 is given from:

(123) $$\left| \begin{array}{l} h_1 = (\text{grad } q), \text{ in } \dot{\Omega}, \ q = \sum_{j=1}^{N} y_j q_j, \text{ with } y_j \text{ given by:} \\ \int_\Omega u.\text{grad } q_i \, dx = \sum_{j=1}^{N} L_{ij} y_j . \end{array} \right.$$

⊗

Remark 15. For "static" electromagnetism, that is for frequency equal to zero, the above matrices very naturally occur:
i) in electrostatics: in a cavity, that is, in a free space bounded domain with a perfectly conducting medium on its boundary Γ, the electric field E satisfies

$$\text{curl } E = 0, \ \text{div } E = 0, \ n \wedge E|_\Gamma = 0, \text{ that is } E \in H_2(\Omega);$$

E is obtained by (120) from the values of the potential p_2 on the surfaces Γ_i,
ii) in magnetostatics we have to find the magnetic field H (or the magnetic induction B) solution of:

$$\text{curl } H = 0, \ \text{div } H = 0, \ n.H|_\Gamma = 0, \ \text{that is } H \in H_1(\Omega),$$

and then H is obtained by formula (123), with the N values y_j to be determined.

⊗

Remark 16. *The Hodge projectors.* Let us define the following operators G^o, G, R^o, R, P^1, P^2 by:

(124) $$\left| \begin{array}{l} G^o u = p_o \text{ given by (115)i), } Gu = p \text{ given by (115)ii),} \\ Ru = v \text{ by (117), } R^o u = v_o \text{ by (116),} \\ P^1 u = h_1, \ P^2 u = h_2 . \end{array} \right.$$

Note that we have $G^o = (-\Delta_D)^{-1}\text{div}$, but we cannot give such a formula for the other operators, since the space $(D(\Omega))$ is not dense in $H^1(\Omega)$, and the same for V and $\tilde{V}$. Then the operators:

(125) $$P_g^o \stackrel{\text{def}}{=} \text{grad } G^o, \ P_r \stackrel{\text{def}}{=} \text{curl } R, \ P^2 \ (\text{resp. } P_g \stackrel{\text{def}}{=} \text{grad } G, \ P_r^o \stackrel{\text{def}}{=} \text{curl } R^o, \ P^1)$$

are orthogonal and complementary projectors in $L^2(\Omega)^3$; they are called *the Hodge projectors associated with the Hodge decomposition* (113)i) (resp. (113)ii); they are also (the Hodge) projectors in $H^1(\Omega)^3$; thus:

(125)' $$P_g^o + P_r + P^2 = I \ \ (\text{resp } P_g + P_r^o + P^1 = I).$$

⊗

Remark 17. *The Hodge decomposition for a "closed" surface.* We recall that the Hodge decomposition for a surface Γ (the boundary of a regular bounded domain Ω) have been given by Lemma 5 for the trace space $H^{-1/2}(\mathrm{div},\Gamma)$; see also the Appendix. Here we recall only the decompositions:

$$(126)\quad \left|\begin{array}{l} L_t^2(\Gamma) = \mathrm{grad}_\Gamma H^1(\Gamma) \oplus \mathcal{H}(\Gamma) \oplus \overrightarrow{\mathrm{curl}}_\Gamma H^1(\Gamma)\,, \\ H_t^{-1/2}(\Gamma) = \mathrm{grad}_\Gamma H^{1/2}(\Gamma) \oplus \mathcal{H}(\Gamma) \oplus \overrightarrow{\mathrm{curl}}_\Gamma H^{1/2}(\Gamma)\,, \\ H^{-1/2}(\mathrm{div},\Gamma) = \mathrm{grad}_\Gamma H^{3/2}(\Gamma) \oplus \mathcal{H}(\Gamma) \oplus \overrightarrow{\mathrm{curl}}_\Gamma H^{1/2}(\Gamma)\,, \\ H^{-1/2}(\mathrm{curl},\Gamma) = \mathrm{grad}_\Gamma H^{1/2}(\Gamma) \oplus \mathcal{H}(\Gamma) \oplus \overrightarrow{\mathrm{curl}}_\Gamma H^{3/2}(\Gamma). \end{array}\right.$$

These decompositions are associated with the Hodge projectors (orthogonal in L^2):

$$(127)\qquad P_g = \mathrm{grad}_\Gamma \Delta_\Gamma^{-1} \mathrm{div}_\Gamma\,, \quad P_r = \overrightarrow{\mathrm{curl}}_\Gamma \Delta_\Gamma^{-1} \mathrm{curl}_\Gamma\,, \quad P_c = P_{\mathcal{H}(\Gamma)},$$

with:

$$(127)'\qquad P_g + P_r + P_c = I.$$

⊗

Remark 18. For an unbounded domain, with bounded complement, we have similar results, but the Sobolev spaces must be replaced by the Beppo Levi spaces; we refer to the Appendix and to Dautray-Lions [1] chap. 9. ⊗

10. INTERPOLATION RESULTS

We often have interesting results by the interpolation theory, see for example Lions-Magenes [1], Triebel [1]. We give here some results useful in electromagnetism, for the trace and regularity properties. With the usual notation $[X,Y]_\theta$ the interpolated space of θ-order between X and Y, we have:

Proposition 9. *Let Ω be a bounded regular open set in* $\mathbb{R}^3$, *and θ a real number with* $0<\theta<1$. *Then with usual notations (see* (79), (42), (84)):

$$(128)\qquad H^s(\mathrm{curl},\Omega) = \{u \in H^s(\Omega)^3,\ \mathrm{curl}\, u \in H^s(\Omega)^3\}, \quad s \in \mathbb{R},$$

we have, with $s = 1-\theta$:

(129)

$$\left|\begin{array}{l} [H^1(\mathrm{curl},\Omega), H(\mathrm{curl},\Omega)]_\theta = H^s(\mathrm{curl},\Omega)\,, \quad [H_0^1(\mathrm{curl},\Omega), H_0(\mathrm{curl},\Omega)]_\theta = H_0^s(\mathrm{curl},\Omega), \\ [H^1(\mathrm{div},\Omega), H(\mathrm{div},\Omega)]_\theta = H^s(\mathrm{div},\Omega)\,, \quad [H_0^1(\mathrm{div},\Omega), H_0(\mathrm{div},\Omega)]_\theta = H_0^s(\mathrm{div},\Omega), \end{array}\right.$$

and for the boundary Γ:

$$(130)\quad [H^1(\mathrm{curl},\Gamma), H(\mathrm{curl},\Gamma)]_\theta = H^s(\mathrm{curl},\Gamma)\,, \ [H^1(\mathrm{div},\Gamma), H(\mathrm{div},\Gamma)]_\theta = H^s(\mathrm{div},\Gamma).$$

PROOF. i) Results on the domain Ω can be obtained by extension to the whole space R^3, thanks to the trace theorem 4: every u in H(curl,Ω) has an extension Ru in H(curl,R^3). Then we have to prove the results for $\Omega = R^3$; this is easily obtained by Fourier transformation. Taking the restriction to Ω gives the first result in (129). The other results easily follow.
ii) We can also use the Hodge projectors (see (125)): since they are the same in $L^2(\Omega)^3$ and in $H^1(\Omega)^3$, they operate also in the interpolating spaces, so we have (129), as a straightforward application to retractions and coretractions properties with respect to interpolation (see for example Triebel [1] p. 22, and 118). This gives also results (130), using the Hodge projectors for the surface Γ.⊗

Prop. 9 implies that the interpolation results (129) are also true for every real positive θ, and in the case of (130) for every real θ. Then we obtain a generalization of the trace result of Prop. 6

Proposition 10. *Interpolation trace results. Let Ω be a bounded regular open set in R^3, and s a real positive number. Then the following trace mappings are continuous and surjective mappings*:

$$(131)\quad \left|\begin{array}{l} u \in H^s(\mathrm{curl},\Omega) \longrightarrow n \wedge u|_\Gamma \in H^{s-1/2}(\mathrm{div},\Gamma), \\ u \in H^s(\mathrm{curl},\Omega) \longrightarrow \pi_\Gamma u \in H^{s-1/2}(\mathrm{curl},\Gamma), \\ u \in H^s(\mathrm{div},\Omega) \longrightarrow n.u|_\Gamma \in H^{s-1/2}(\Gamma). \end{array}\right.$$

11. SOME VARIATIONAL FRAMEWORKS

Here we give variational frameworks useful in electromagnetism, which are particular cases of more general frameworks in differential geometry. These are given in Appendix, sections 3.1 and 3.2 (see Morrey [1]). There are essentially two types of variational frameworks in electromagnetism: either based on the H(curl,Ω) space (or some subspace) with the curl curl operator, or based on the Sobolev space $H^1(\Omega)^3$ (or some subspaces) with the Laplacian. We refer to Dautray-Lions [1] chap. 7 (for instance) for the definition of a variational framework, and for many examples with usual functions spaces.

11.1. Variational frameworks based on H(curl,Ω) spaces

11.1.1. A variational framework based on H_0(curl,Ω)

Let Ω be Lipschitz open set in R^3. Then let $V = H_0(\mathrm{curl},\Omega)$, and a(u,v) be the sesquilinear form on V:

$$(132)\qquad a(u,v) = \int_\Omega (\mathrm{curl}\, u.\ \mathrm{curl}\, \bar{v} + u.\ \bar{v})\, dx\,.$$

Then (V, $H = L^2(\Omega)^3$, a(u,v)) is a variational framework which defines the unbounded (selfadjoint and positive) operator A in H by:

$$(133)\quad \left|\begin{array}{l} D(A) \stackrel{\mathrm{def}}{=} \{u \in V, \text{ such that } v \longrightarrow a(u,v) \text{ is continuous on } V \\ \qquad\qquad\qquad \text{equipped with the topology of } H\}, \\ (Au,v) = a(u,v), \qquad \forall\, u \in D(A),\ v \in V. \end{array}\right.$$

A sesquilinear form a(u,v) is *coercive on* V if there exists C > 0 such that

$$\mathrm{Re}\, a(u,u) \geq C \|u\|_V^2, \quad \forall u \in V.$$

Then the *Lax-Milgram lemma* implies that the "standard" problem: given f in H or in V', find u in V so that

$$(134) \qquad a(u,v) = (f,v), \quad \forall v \in V,$$

has a unique solution in V.
From (133), and the Green formula, we characterize the operator A for (132) by:

$$(135) \quad \left| \begin{array}{l} Au = \mathrm{curl\,curl}\, u + u, \ \ u \in D(A) \\ D(A) = \{u \in V = H_0(\mathrm{curl},\Omega), \ \ \mathrm{curl\,curl}\, u \in L^2(\Omega)^3\}. \end{array} \right.$$

Since V' is a space of distributions, (134) with (132) is equivalent to the problem:

$$(136) \quad \left| \begin{array}{l} \mathrm{curl\,curl}\, u + u = f \ \text{ in } \Omega, \\ n \wedge u|_\Gamma = 0. \end{array} \right.$$

Since the sesquilinear form a(u,v) given by (132) is *coercive on* V (by definition), there is a unique solution u in V of (135) thanks to *Lax-Milgram lemma*.

11.1.2. A variational framework based on H(curl,Ω)

We can also consider the same framework (V,H,a(u,v)) but with V = H(curl,Ω). This defines a new operator A with

$$(137) \quad \left| \begin{array}{l} \text{i) } Au = \mathrm{curl\,curl}\, u + u, \\ \text{ii) } D(A) = \{u \in V, \ \mathrm{curl\,curl}\, u \in L^2(\Omega)^3, \ \ n \wedge \mathrm{curl}\, u|_\Gamma = 0\}. \end{array} \right.$$

This is easily verified from (133) using the Green formula. Then we can solve other problems similar to (136), but with the boundary condition $n \wedge \mathrm{curl}\, u|_\Gamma = 0$.
More generally, V may be any closed space with $H_0(\mathrm{curl},\Omega) \subseteq V \subseteq H(\mathrm{curl},\Omega)$.

For example if Γ_0 is a subset of Γ, $V = \{u \in H(\mathrm{curl},\Omega),\ n \wedge u|_{\Gamma_0} = 0\}$.

Example in electromagnetism. We consider a less "standard" problem in electromagnetism: let Ω be a domain bounded by a perfect conductor and Ω is occupied by a "dissipative" dielectric with permittivity ε and permeability μ, with given electric and magnetic current densities J, M, with J and M in $L^2(\Omega)^3$. We have to find the electromagnetic field (E,H) with finite energy in Ω, so that:

$$(138) \quad \left| \begin{array}{ll} \text{i) } -\mathrm{curl}\, E + i\omega\mu H = M, & \text{ii) } \mathrm{curl}\, H + i\omega\varepsilon E = J, \\ \text{iii) } n \wedge E|_\Gamma = 0. & \end{array} \right.$$

Then we can define a variational framework on the electric field E, with the following sesquilinear form in $V = H_0(\text{curl},\Omega)$:

(139) $$a(E,\tilde{E}) = \int_\Omega \left(-\frac{1}{i\omega\mu}\,\text{curl}\,E \,.\, \text{curl}\,\bar{\tilde{E}} - i\omega\varepsilon\, E \,.\, \bar{\tilde{E}}\right) dx, \quad \forall E, \tilde{E} \in V.$$

We see that problem (138) is equivalent to finding E in V solution of:

(140) $$a(E,\tilde{E}) = \int_\Omega \left(\frac{1}{i\omega\mu}\, M \,.\, \text{curl}\,\bar{\tilde{E}} - J \,.\, \bar{\tilde{E}}\right) dx, \quad \forall \tilde{E} \in V.$$

We define H by (138)i). Then (140) is equivalent to:

(141) $$\int_\Omega \left(-H \,.\, \text{curl}\,\bar{\tilde{E}} - i\omega\varepsilon\, E \,.\, \bar{\tilde{E}}\right) dx = \int_\Omega \left(-J \,.\, \bar{\tilde{E}}\right) dx, \quad \forall \tilde{E} \in V,$$

which is equivalent to (138)ii) thanks to the Green formula. Then from the usual hypotheses on a dissipative medium (see H14 in chap.1), we have $z'' = \text{Im}\, z > 0$ for $z = \varepsilon$ and μ with $\omega > 0$. Thus there is a positive number α so that:

(142) $$\text{Re}\, a(E,E) = \int_\Omega \left(\frac{\omega\mu''}{|\omega\mu|^2}\,|\text{curl}\,E|^2 + \omega\varepsilon''|E|^2\right) dx \geq \alpha\, |E|^2_{H(\text{curl},\Omega)},$$

that is, $a(E,\tilde{E})$ is a coercive (but not symmetric) sesquilinear form. Thanks to the Lax-Milgram lemma, this implies that problem (138) has a unique solution.

⊗

11.2. Variational frameworks based on $H^1(\Omega)^3$ and the Laplacian

Let Ω be a (regular, i.e., $C^{1,1}$) domain in R^3, either bounded or the complement to a bounded set. We define the spaces:

(143) $$\begin{cases} V_1^+ = H_t^1(\Omega) = \{u \in H^1(\Omega)^3, \;\; n.u|_\Gamma = 0\} \\ V_1^- = H_n^1(\Omega) = \{u \in H^1(\Omega)^3, \;\; n \wedge u|_\Gamma = 0 \;\; (\text{or } \pi_\Gamma u = 0)\}, \end{cases}$$

(with π_Γ the projection on the tangent plane) and the sesquilinear form:

(144) $$a(u,v) = (\text{curl}\,u,\, \text{curl}\,v) + (\text{div}\,u,\, \text{div}\,v) + (u,\, v).$$

Theorem 11. *The sesquilinear form* $a(u,v)$ *defined by* (144) *is coercive on* V_1^+ *and* V_1^-.

PROOF. When Ω is a bounded domain, this is equivalent to Corollary 5 (see (103)). When Ω is the complement of a bounded domain, this is also easily deduced from Peetre lemma (lemma 7), and from the inequality (101).
Theorem 11 is generalized by Theorems 4 and 15 of the Appendix, in differential geometry. ⊗

The variational frameworks $(V_1^+, L^2(\Omega)^3, a(u,v))$ and $(V_1^-, L^2(\Omega)^3, a(u,v))$ define selfadjoint operators A^+ and A^- by

(145) $$D(A^\pm) = \{u \in V^\pm, \;\; (A^\pm u, v) = a(u,v), \quad \forall v \in V_1^\pm\},$$

with:

$$(146)\quad \left|\begin{array}{l} \text{i) } A^{\pm}u = -\Delta u + u \text{ in } \Omega, \\ \text{ii) } D(A^+) = \{u \in H^1(\Omega)^3,\ \text{curl } u \in H^1(\Omega)^3,\ \text{div } u \in H^1(\Omega), \\ \qquad\qquad n.u|_{\Gamma} = 0,\ \ n \wedge \text{curl } u|_{\Gamma} = 0\}, \\ \text{ii)' } D(A^-) = \{u \in H^1(\Omega)^3,\ \text{curl } u \in H^1(\Omega)^3,\ \text{div } u \in H^1(\Omega), \\ \qquad\qquad n \wedge u|_{\Gamma} = 0,\ \ (\text{div } u)|_{\Gamma} = 0\}. \end{array}\right.$$

PROOF of (146) for A^- only. Taking v in $\mathcal{D}(\Omega)^3$ in (145), we obtain (146)i). Taking v in V_1^- with div v = 0 we have this: for u in $D(A^-)$, $v \to$ (curl curl u + u, v) is continuous on the subspace $\{v \in L^2(\Omega)^3,\ \text{div } v = 0\}$. But it is also continuous on grad $H_0^1(\Omega)$ (since (curl curl u, grad ϕ) = 0 for ϕ in $\mathcal{D}(\Omega)$. Therefore we obtain that curl curl u is in $L^2(\Omega)^3$. Thus v = curl u is in $L^2(\Omega)^3$ with curl v in $L^2(\Omega)^3$, div v = 0, and $n.v|_{\Gamma} = \text{div}_{\Gamma}(n \wedge u|_{\Gamma}) = 0$. Thus v is in $H^1(\Omega)^3$. Since Δu is in $L^2(\Omega)^3$, we obtain from (45) that grad div u is in $L^2(\Omega)^3$. Then we have div u = 0 on Γ thanks to:

$$(147)\qquad (\text{curl } u, \text{curl } v) + (\text{div } u, \text{div } v) + (u, v) = (-\Delta u, v) + \int_{\Gamma} \text{div } u\, (n.v)\, d\Gamma.$$

Proposition 7 implies that for regular Ω, $D(A^+)$ and $D(A^-)$ are in $H^2(\Omega)^3$. A more general result is given in the Appendix (see (156)"). ⊗

In order to solve the problem (138) in $H^1(\Omega)^3$, with suitable regularity of M and J, we use the following sesquilinear form:

$$(148)\qquad a_0(E,\tilde{E}) = \int_{\Omega} [-\frac{1}{i\omega\mu}(\text{curl } E \,.\, \text{curl } \bar{\tilde{E}} + \text{div } \varepsilon E.\ \text{div } \bar{\varepsilon}\bar{\tilde{E}}) - i\omega\varepsilon\, E.\bar{\tilde{E}}]\, dx.$$

For a dissipative medium, $a_0(E,\tilde{E})$ is coercive on $V_1 = \{E \in H^1(\Omega)^3,\ n \wedge E|_{\Gamma} = 0\}$. Then, instead of (140), we consider the problem: find E in V_1 such that

$$(148)'\qquad a_0(E,\tilde{E}) = L_{M,J}(\tilde{E}),\qquad \forall \tilde{E} \in V_1,$$

with $L_{M,J}(\tilde{E}) \stackrel{\text{def}}{=} \int_{\Omega} (\frac{1}{i\omega\mu} M \,.\, \text{curl } \bar{\tilde{E}} + \frac{1}{\omega^2\mu} \text{div } J.\ \text{div } \bar{\varepsilon}\bar{\tilde{E}} - J \,.\, \bar{\tilde{E}})\, dx.$

If $M \in H(\text{curl},\Omega)$, $J \in H(\text{div},\Omega)$, with $\phi_0 = \text{div } J \in H_0^1(\Omega)$, then $L_{M,J}$ is continuous on $L^2(\Omega)^3$. Thus the problem (148)' has a unique solution E in $D(A^-)$ (see (146)) and it is equivalent to (138).

Remark 19. *An obstacle to compactness.* One of the main differences between the spaces $H^1(\Omega)^3$ and $H(\text{curl},\Omega)$ for regular bounded Ω is: the imbeddings into $L^2(\Omega)^3$ of $H(\text{curl},\Omega)$, $H(\text{div},\Omega)$, $H(\text{curl},\text{div},\Omega)$, and their intersections with ker curl or ker div are not compact since all these spaces contain the space

(149) $\mathcal{H}^o(\Omega) = \ker \mathrm{curl} \cap \ker \mathrm{div} = \{u \in L^2(\Omega)^3,\ \mathrm{curl}\, u = 0,\ \mathrm{div}\, u = 0\}$,

which is a closed subspace of $L^2(\Omega)^3$ with infinite dimension. Thus it prevents all greater spaces to have a compact imbedding in $L^2(\Omega)^3$. Using the (Hodge) decomposition of $L^2(\Omega)^3$ in the form (with (143)):

$$L^2(\Omega)^3 = \mathrm{grad}\, H^1_o(\Omega) \oplus \mathcal{H}^o(\Omega) \oplus \mathrm{curl}\, H^1_n(\Omega)^3$$

we easily obtain the decomposition of the space

$$H(\mathrm{curl}, \mathrm{div}, \Omega) = \mathrm{grad}\, D(A_D) \oplus \mathcal{H}^o(\Omega) \oplus \mathrm{curl}\, D(A^-),$$

where $D(A_D)$ is the domain of the Dirichlet Laplacian, $D(A_D) = H^2(\Omega) \cap H^1_0(\Omega)$, for a regular domain Ω, and with $D(A^-)$ given by (146) which is also contained in $H^2(\Omega)^3$. Thus the spaces grad $D(A_D)$ and curl $D(A^-)$ are contained in $H^1(\Omega)^3$, with compact imbedding in $L^2(\Omega)^3$.

Now we easily obtain the result when Ω is a regular bounded open set in R^3 (in fact it is true with a Lipschitz-continuous boundary, see Costabel [2]).

Proposition 11. *The imbedding of each of the following spaces into* $L^2(\Omega)^3$ *is compact*

$$\{u \in H(\mathrm{curl},\mathrm{div},\Omega),\ n \wedge u|_\Gamma \in L^2(\Gamma)^3\}, \quad \{u \in H(\mathrm{curl},\mathrm{div},\Omega),\ n.u|_\Gamma \in L^2(\Gamma)\},$$

and more generally (see Weber [1], Hazard-Lenoir [1]) if $\zeta \in L^\infty(\Omega)$ *with* $\mathrm{Re}\,\zeta > \alpha > 0$

$$\{u \in H(\mathrm{curl},\Omega),\ \mathrm{div}\,\zeta u \in L^2(\Omega),\ n \wedge u|_\Gamma \in L^2(\Gamma)^3 \text{ (resp. } n.\zeta u|_\Gamma \in L^2(\Gamma)\}.$$

We refer to the Appendix, A.2 Remark 5 and to Theorem 9 A.3 for other compactness results, for example for the compensated compactness theorem. ⊗

12. FIRST PROBLEMS WITH INHOMOGENEOUS BOUNDARY CONDITIONS

12.1. Problems with $H^{-1/2}(\mathrm{div},\Gamma)$ boundary conditions

i) Previously (in section 4), we have solved some boundary problems in the $H(\mathrm{curl},\Omega)$ framework (with "regular" Ω, bounded or the complementary of a bounded set). Proving that the trace mapping γ_τ is onto (see Corollary 1), we solved (see (46)): find u in $H(\mathrm{curl},\Omega)$ satisfying, with m given in $H^{-1/2}(\mathrm{div},\Gamma)$

(150) $\mathrm{curl}\,\mathrm{curl}\, u + \kappa^2 u = 0 \ \text{in}\ \Omega, \quad n \wedge u|_\Gamma = m.$

This is equivalent to the system (of Maxwell type):

(151) $-\mathrm{curl}\, u + v = 0, \quad \mathrm{curl}\, v + \kappa^2 u = 0 \ \text{in}\ \Omega, \quad n \wedge u|_\Gamma = m,$

or also to:

(151)' $-\Delta u + \kappa^2 u = 0, \quad \mathrm{div}\, u = 0 \ \text{in}\ \Omega, \quad n \wedge u|_\Gamma = m.$

We denote by Cm and C^+m the boundary values (with $\kappa = 1$):

(152) $Cm = n \wedge \mathrm{curl}\, u|_\Gamma = n \wedge v|_\Gamma, \qquad C^+m = \pi_\Gamma \mathrm{curl}\, u,$

with u the solution of (151), and v in (151)'. Then we define:

Definition 6. *The Calderon or Capacity operators. The mapping* $C: m \rightarrow Cm$ and $C^+: m \rightarrow C^+m$ *are called the Calderon or the capacity operators.*

We verify that C is a continuous isomorphism in $H^{-1/2}(\text{div},\Gamma)$ with $C^2 = -I$.

ii) More generally, thanks to Lax-Milgram lemma, we know that there exists a unique solution to the following problem in variational form: find u in $H(\text{curl},\Omega)$ such that

(153) $\qquad (\text{curl } u, \text{curl } v) + (u, v) = \langle f, v\rangle$ for all v in $H(\text{curl},\Omega)$,

with f given in the dual space $(H(\text{curl},\Omega))'$ of $H(\text{curl},\Omega)$, which is not a space of distributions. So the only point is to recognize a boundary problem in (153). This problem may be understood in the following way: using the orthogonal decomposition

(154) $\qquad H(\text{curl},\Omega) = H_0(\text{curl},\Omega) \oplus H_0^{\perp}(\text{curl},\Omega)$,

we have by duality

(154)' $\qquad (H(\text{curl},\Omega))' = (H_0(\text{curl},\Omega))' \oplus (H_0^{\perp}(\text{curl},\Omega))'$,

and since $H_0^{\perp}(\text{curl},\Omega)$ is isomorphic to the trace space $H^{-1/2}(\text{div},\Gamma)$, see Corollary 1, we can identify dual spaces, that is, $(H_0^{\perp}(\text{curl},\Omega))'$ with $H^{-1/2}(\text{curl},\Gamma)$.

Thus we can identify f with the couple (f_0,g), with f_0 in the space of distributions $(H_0(\text{curl},\Omega))'$, and g in $H^{-1/2}(\text{curl},\Gamma)$, i.e., we can write:

(155) $\qquad \langle f, v\rangle = \langle f_0, v_0\rangle + \langle g, n \wedge v\rangle_\Gamma, \qquad \forall v \in H(\text{curl},\Omega)$,

v_0 being the projection of v onto $H_0(\text{curl},\Omega)$ and <,> the duality bracket in Ω and Γ. First taking v in $H_0(\text{curl},\Omega)$ in (153) we have:

(156) $\qquad \text{curl curl } u + u = f_0$.

Then taking v in $H_0^{\perp}(\text{curl},\Omega)$, (153) gives

(156)' $\qquad (\text{curl } u, \text{curl } v) + (u, v) = \langle g, n \wedge v\rangle_\Gamma$.

Using the Green formula (on the "unusual" side !), we have:

(157) $\qquad \langle n \wedge u, \text{curl } v\rangle_\Gamma = \langle g, n \wedge v\rangle_\Gamma$.

Using (152) we have:

(157)' $\qquad \langle n \wedge u, C^+(n \wedge v)\rangle_\Gamma = \langle g, n \wedge v\rangle_\Gamma$.

Thus by duality (see (173)' below), we have:

(158) $C^{+}(n \wedge u) = g, \quad \text{thus} \quad n \wedge \text{curl}\, u|_{\Gamma} = n \wedge g.$

Then we see that the variational problem (153) *is equivalent to* (156) *and* (158). Note that the dual space $(H_o(\text{curl},\Omega))'$ of $H_o(\text{curl},\Omega)$ is characterized by:

$$(H_o(\text{curl},\Omega))' = \{w = \text{curl}\, v + u,\ v \in L^2(\Omega)^3,\ u \in H_o(\text{curl},\Omega),\ \text{or } u = \text{grad}\, \phi,\ \phi \in H_o^1(\Omega)\},$$

since the operator curl curl + I is an isomorphism from $H_o(\text{curl},\Omega)$ onto $(H_o(\text{curl},\Omega))'$.

Remark 20. Using the same method we can interpret *the inhomogeneous Neumann problem for the Laplacian,* thanks to the corresponding capacity operator. More precisely for every u_o in $H^{1/2}(\Gamma)$ let Cu_o be defined by:

(159) $Cu_o = \frac{\partial u}{\partial n}|_{\Gamma},$

with u the solution in $H^1(\Omega)$ of the Dirichlet problem: $-\Delta u + u = 0$ in Ω, $u|_{\Gamma} = u_o$.

We call also C *the capacity operator* for the Laplacian, see Dautray-Lions [1] chap. 2; it is an isomorphism from $H^{1/2}(\Gamma)$ onto space $H^{-1/2}(\Gamma)$.
Then we consider the inhomogeneous Neumann problem:

(160) $\int_{\Omega} (\text{grad}\, u.\ \text{grad}\, \bar{v} + u\, \bar{v})\, dx = \langle f, v\rangle, \quad \forall v \text{ in } H^1(\Omega).$

with f given in $(H^1(\Omega))'$ which is not a space of distributions. We use the orthogonal decomposition:

(161) $H^1(\Omega) = H_o^1(\Omega) \oplus \mathcal{H}^1(\Omega), \quad \text{with } \mathcal{H}^1(\Omega) = \{u \in H^1(\Omega),\ -\Delta u + u = 0\},$

which is isomorphic to space $H^{1/2}(\Gamma)$. Then its dual space is identified to space $H^{-1/2}(\Gamma)$. Thus by duality we have:

(162) $(H^1(\Omega))' = H^{-1}(\Omega) \oplus (\mathcal{H}^1(\Omega))',$

and then f is identified to the pair (f_o,g) with f_o in $H^{-1}(\Omega)$, and g in $H^{-1/2}(\Gamma)$. Note that if f is in $L^2(\Omega)$, *we have to decompose* f into (f_o,g) also, i.e., we have to consider $L^2(\Omega)$ as a subspace of $H^1(\Omega)'$, not of $H^{-1}(\Omega)$. Thus we have:

(163) $\langle f, v\rangle = \langle f_o, v_o\rangle + \langle g, v\rangle_{\Gamma}, \quad \forall v \in H^1(\Omega).$

First taking v in $H_o^1(\Omega)$ in (160), we see that u satisfies:

(164) $-\Delta u + u = f_o \quad \text{in } \Omega.$

Then taking v in $H^1(\Omega)$, and applying Green formula to (160), we obtain:

(165) $$<u, \frac{\partial v}{\partial n}>_\Gamma = <g, v>_\Gamma,$$
that is:

(165)' $$<u, Cv_\Gamma> = <g, v_\Gamma>_\Gamma, \quad \forall v_\Gamma \in H^{1/2}(\Gamma),$$

using the capacity operator C given by (159). Thus by duality, since ${}^tC$ (or C^*) is identical to C:

(166) $$C(u|_\Gamma) = g.$$

Therefore the solution u of the inhomogeneous Neumann problem (160) is also the solution of (164),(166).
We note that in the two cases (problems (153) and (160)) there is no trace for curl u, with u in H(curl,Ω) or for the normal derivative of u on Γ for u in $H^1(\Omega)$.
⊗

Remark 21. A very interesting property of these capacity operators is given by their relations to the quotient norms. More precisely:
i) Norms in $H^{1/2}(\Gamma)$. Let ϕ be $H^{1/2}(\Gamma)$. Then its quotient norm is given by

(167) $$\|\phi\|^2_{H^{1/2}(\Gamma)} = \inf_{u \in H^1(\Omega),\, u|_\Gamma = \phi} \int_\Omega [|\text{grad}\, u|^2 + |u|^2]\, dx.$$

The infimum is reached at u solution of the following problem in $H^1(\Omega)$:

(168) $$-\Delta u + u = 0, \quad u|_\Gamma = \phi.$$

Thus thanks to the usual Green formula, we obtain (formally):

(169) $$\|\phi\|^2_{H^{1/2}(\Gamma)} = \int_\Gamma \frac{\partial u}{\partial n} \phi\, d\Gamma,$$

or using operator C, see (159), and the duality bracket <,> for $H^{1/2}(\Gamma)$, $H^{-1/2}(\Gamma)$

(170) $$\|\phi\|^2_{H^{1/2}(\Gamma)} = <C\phi, \phi>.$$

ii) Norm in $H^{-1/2}(\text{div},\Gamma)$. Let m be in $H^{-1/2}(\text{div},\Gamma)$. Its quotient norm is:

(171) $$\|m\|^2_{H^{-1/2}(\text{div},\Gamma)} = \inf_{u \in H(\text{curl},\Omega),\, n \wedge u|_\Gamma = m} \int_\Omega [|\text{curl}\, u|^2 + |u|^2]\, dx.$$

The infimum is reached at u solution of the problem in H(curl,Ω):

(172) $$\text{curl curl}\, u + u = 0 \text{ in } \Omega, \quad n \wedge u|_\Gamma = m.$$

Thus using Green formula in (171), we obtain:

(173) $\|m\|^2_{H^{-1/2}(\mathrm{div},\Gamma)} = \int_\Gamma \mathrm{curl}\, u \,.\, n \wedge \bar{u}\; d\Gamma = \langle C^+ m, m\rangle_\Gamma = \langle n \wedge Cm, m\rangle_\Gamma$,

and also:

(173)' $((m_u, m_v))_{H^{-1/2}(\mathrm{div},\Gamma)} = \langle m_u, C^+ m_v\rangle = \langle C^+ m_u, m_v\rangle$.

Thus we see that it is possible to evaluate norms in $H^{1/2}(\Gamma)$, or in $H^{-1/2}(\mathrm{div},\Gamma)$ thanks to the capacity operators, using (169), or (173), in correspondence with positivity properties of these operators. ⊗

12.2. Problems with $H^{1/2}(\Gamma)$ boundary conditions

Corresponding to the $H^1(\Omega)^3$ variational framework, we have the boundary problem: find u in $H^1(\Omega)^3$ satisfying, with given m and g

(174) $$\begin{cases} \text{i) } -\Delta u + u = 0 & \text{in } \Omega, \\ \text{ii) } n \wedge u|_\Gamma = m, & m \in H_t^{1/2}(\Gamma) \\ \text{iii) } (\mathrm{div}\, u)|_\Gamma = g, & g \in H^{-1/2}(\Gamma). \end{cases}$$

The uniqueness of the solution is a consequence of theorem (11). Then in order to solve (174), we proceed in two stages:

i) *Lifting of the boundary condition* (174)ii).

We note that $V_1^- = H_n^1(\Omega)$ is a closed subspace of H(curl, div, Ω). Let $V^{-\perp}$ be its orthogonal space. It is characterized by:

(175) $$V^{-\perp} = \{u \in H(\mathrm{curl},\mathrm{div},\Omega),\ -\Delta u + u = 0,\ (\mathrm{div}\, u)|_\Gamma = 0\}.$$

Recall that the trace mapping $\gamma: u \to n \wedge u|_\Gamma$ is continuous from H(curl, div, Ω) onto $H^{-1/2}(\mathrm{div},\Gamma)$ with kernel V_1^-; therefore γ is an isomorphism from $V^{-\perp}$ onto space $H^{-1/2}(\mathrm{div},\Gamma)$. Moreover, thanks to Theorem 6, we have $u \in H^1(\Omega)^3$ when $n \wedge u|_\Gamma$ is in $H_t^{1/2}(\Gamma)$. Thus γ is also an isomorphism from $V^{-\perp} \cap H^1(\Omega)^3$ onto $H_t^{1/2}(\Gamma)$.

Therefore for all m given in $H_t^{1/2}(\Gamma)$, there exists a unique U in $H^1(\Omega)^3$ such that:

(176) $$\begin{cases} \text{i) } -\Delta U + U = 0 \text{ in } \Omega, \\ \text{ii) } n \wedge U|_\Gamma = m, \quad \text{iii) } (\mathrm{div}\, U)|_\Gamma = 0. \end{cases}$$

ii) *Thus by difference* we have to find $w = u - U$, w in $H^1(\Omega)^3$ satisfying:

(177) $$\begin{cases} \text{i) } -\Delta w + w = 0 \text{ in } \Omega, \\ \text{ii) } n \wedge w|_\Gamma = 0, \quad \text{iii) } (\mathrm{div}\, w)|_\Gamma = g, \quad g \in H^{-1/2}(\Gamma). \end{cases}$$

This is solved by a variational method: using the sesquilinear form a(u,v) given by (144), w must satisfy

(178) $a(w,v) = \int_\Gamma g\, n.v\, d\Gamma, \quad \forall v \in V_I^-$.

Thanks to Lax-Milgram lemma we know that (178) has one solution.

Remark 22. The difference in tackling the two boundary conditions is very meaningful: the first one (174)ii) which is included in the Sobolev space $H^1(\Omega)^3$ is said to be stable, the second (174)iii) which is obtained by duality, by the weak formulation of the problem is said to be unstable (see Necas[1]). ⊗

Conclusion on the difference between the two types of boundary problems 12.1 and 12.2: in the first one, more general boundary conditions are allowed; the second gives a regularity result with respect to the first, and is based on the usual Sobolev space $H^1(\Omega)^3$ and thus compactness properties (for regular bounded Ω); also it seems easier to apply in numerical methods. We will study other more difficult boundary problems in chapter 3.

13. BOUNDARY PROBLEMS OF CAUCHY TYPE. UNIQUENESS THEOREMS

Here we give without proof (according to the case, proof may be quite obvious or quite difficult) some results which are fundamental to a deep understanding of boundary problems in the Helmholtz case and also in the Maxwell case. They concern only the question of uniqueness of the solution and not (a priori) the question of existence; furthermore there will often be no solution at all for such problems. Therefore they do not have (a priori) a physical flavor ! In all this section, Ω is a "regular" domain in R^n, (n = 3 for Maxwell equation) bounded or not, with boundary Γ. These uniqueness theorems are of Holmgren type (see for instance Dautray-Lions [1] sec. 2.1.5); see also Saut-Scheurer [1] and Müller [1] for other uniqueness results.

Theorem 12. *The Helmholtz case. The following problem: find u with*

(179)
$$\left|\begin{array}{l} \text{i) } \Delta u + k^2 u = 0 \ \text{ in } \Omega , \\ \text{ii) } u|_\Gamma = 0 , \quad \frac{\partial u}{\partial n}|_\Gamma = 0 , \textit{ or only on some regular part } \Gamma_0 \text{ of } \Gamma , \\ \text{iii) } u \in L^2_{loc}(\bar{\Omega}), \text{ i.e., } \zeta u \in L^2(\Omega) \text{ for all } \zeta \text{ in } \mathcal{D}(R^n), \end{array}\right.$$

(*or more generally with u in a functional space so that the trace condition* ii) *makes sense) has* 0 *as unique solution.*

When Γ_0 is *a part* of Γ we prove that u is null in a neighbourhood of Γ_0, then we use analyticity of the solution.

Corollary 7. *The following problem: find u in* $L^2_{loc}(\bar{\Omega})$ satisfying

(180)
$$\left|\begin{array}{l} \text{i) } \Delta u + k^2 u = f \ \text{ in } \Omega , \\ \text{ii) } u|_\Gamma = u_0 , \ \frac{\partial u}{\partial n}|_\Gamma = u_1 , \end{array}\right.$$

with given f in $L^2(\Omega)$, u_0 in $H^{1/2}(\Gamma)$, u_1 in $H^{-1/2}(\Gamma)$, *can have at most one solution.*

Theorem 13. *The Maxwell case. The following problem: find* (E,H) *with*

$$(181)\qquad \left|\begin{array}{l} \text{i) curl } H + i\omega\varepsilon E = 0\,, \quad \text{ii) } - \text{curl } E + i\omega\mu H = 0 \ \text{ in } \Omega, \\ \text{iii) } n \wedge E|_\Gamma = 0\,,\ n \wedge H|_\Gamma = 0 \ (\textit{or only on some part } \Gamma_0 \text{ of } \Gamma), \end{array}\right.$$

and with (ε,μ) *constant on* Ω *(and* (E,H) *in a functional space so that the trace condition* iii) *makes sense), has* (E,H) = (0,0) *as unique solution.*

Corollary 8. *The Maxwell problem: find* (E,H) *satisfying*

$$(182)\qquad \left|\begin{array}{l} \text{i) curl } H + i\omega\varepsilon E = J\,, \\ \text{ii) } - \text{curl } E + i\omega\mu H = M\,,\ \text{on } \Omega, \\ \text{iii) } n \wedge E|_\Gamma = M^o,\ n \wedge H|_\Gamma = J^o \ (\textit{or only on some part } \Gamma_0 \text{ of } \Gamma), \\ \text{iv) } (E,H) \ \textit{with finite energy}, \text{ i.e., } E \text{ and } H \text{ in } H_{loc}(\text{curl},\bar{\Omega}), \end{array}\right.$$

with given J, M in $L^2(\Omega)^3$, J^o, M^o in $H^{-1/2}(\text{div},\Gamma)$, *can have at most one solution.*

14. WHITNEY ELEMENTS; NUMERICAL TREATMENT

Let Ω be a bounded open set in R^3 where we have to solve an electromagnetic problem; we have to approximate vector fields, or n-forms, and to take into account the tangential (or the normal) continuity of the fields. Here we briefly describe finite elements, known as Whitney elements, adapted to these requirements. First we specify some notations; we essentially follow Bossavit [1]. Let (T_i) be a tesselation of Ω by a finite number t of tetrahedra, that is:

$$\bigcup_i \bar{T}_i = \bar{\Omega}\,,\quad T_i \cap T_j = \varnothing \quad \text{for } i \neq j,$$

with the usual properties of a finite element mesh. Note that tetrahedra at the boundary may be curved. Then we call *N, E, F, T* respectively the set of nodes (vertices), edges, facets and tetrahedra which constitute the mesh.

Let a_i, i = 1 to 4, be the nodes of a reference tetrahedron T_0, and let $(w_i(x))$, i = 1 to 4, be the barycentric coordinates of the point x; then:

$$(183)\qquad x = \sum_{i=1 \text{ to } 4} w_i(x)\, a_i, \text{ that is also } x_j = \sum_{i=1 \text{ to } 4} w_i(x)\, a_{ij},\quad 1 = \sum_{i=1 \text{ to } 4} w_i(x),$$

or in matrix form:

$$(183)'\qquad \begin{pmatrix} x_1 \\ x_2 \\ x_3 \\ 1 \end{pmatrix} = A_{T_0} \begin{pmatrix} w_1(x) \\ w_2(x) \\ w_3(x) \\ w_4(x) \end{pmatrix},\ \text{with } A_{T_0} = \begin{pmatrix} a_{11} & a_{12} & a_{13} & a_{14} \\ a_{21} & a_{22} & a_{23} & a_{24} \\ a_{31} & a_{32} & a_{33} & a_{34} \\ 1 & 1 & 1 & 1 \end{pmatrix};$$

then we can inverse relations (183)':

(184) $$\begin{pmatrix} w_1(x) \\ w_2(x) \\ w_3(x) \\ w_4(x) \end{pmatrix} = A_{T_o}^{-1} \begin{pmatrix} x_1 \\ x_2 \\ x_3 \\ 1 \end{pmatrix}, \quad w_i(x) = \sum_j (A_{T_o}^{-1})_{ij}\, x_j;$$

therefore we have for x in T_o:

(185) $$\frac{\partial w_i}{\partial x_j}(x) = A_{T_o ij}^{-1}, \quad i = 1 \text{ to } 4,\ j = 1 \text{ to } 3,\ \forall x \text{ in } T_o.$$

Since $w_i(x)$ is a linear function of x in T_o, its gradient is constant in T_o. Then in the mesh, for n in N, w_n is a "node-element". For every edge e = (i,j) in the mesh, we define "edge-elements":

(186) $$w_e = w_i \operatorname{grad} w_j - w_j \operatorname{grad} w_i;$$

then for evey facet f = (i, j, k), we define "face-elements":

(187) $$w_f = 2\,(w_i \operatorname{grad} w_j \wedge \operatorname{grad} w_k + w_j \operatorname{grad} w_k \wedge \operatorname{grad} w_i + w_k \operatorname{grad} w_i \wedge \operatorname{grad} w_j)$$

and for every tetrahedron T = v = (i,j,k,l), we define "volume-elements":

(188) $$w_v = 6 \operatorname{grad} w_i.(\operatorname{grad} w_j \wedge \operatorname{grad} w_k).$$

Note that these elements w depend a priori on the chosen orientation for the edge, facet, tetrahedron, and that "edge-elements" and "face-elements" are vector fields. We can prove fairly easily the following properties:

(189) $$\int_{e'} w_e . \vec{\nu}\, dl = \delta_{ee'},\ \int_{f'} w_f . \vec{\nu}\, dS = \delta_{ff'},\ \int_{v'} w_v\, dx = \delta_{vv'},\ (\text{with } w_n(n') = \delta_{nn'}),$$

with $\vec{\nu}$ a unit vector along e', or normal to f'.

Definition 7. *We call Whitney space of order* i, i = 0, 1, 2 , 3 *respectively and we denote it by* $W^i(\Omega)$, *the vector space generated by* (w_n), (w_e), (w_f), (w_v). *We also denote by* $W^i_{\ o}(\Omega)$ *the subspace of elements of* $W^i(\Omega)$, *vanishing at the boundary.*

The elements of Whitney space are linear combinations:

(190) $$u = \sum_{s \in S} u_s\, w_s, \text{ with } S = N, E, F, T,$$

where complex numbers u_s denote the degrees of freedom. From property (130) it follows that the dimension of space $W^i(\Omega)$ is n, e, f, t, that is the number of nodes, edges, facets, tetrahedra for i = 0, 1, 2, 3 respectively.

The elements of $W^i(\Omega)$, for i = 1 to 3, are certainly not continuous on Ω, but they have the following important properties:

$$\text{(191)}\qquad \begin{aligned} &\text{curl}\, w_e = 2\,\text{grad}\, w_i \wedge \text{grad}\, w_j, &&\text{for } e = (i, j),\\ &\text{div}\, w_f = 6\,\text{grad}\, w_i.(\text{grad}\, w_j \wedge \text{grad}\, w_k), &&\text{for } f = (i,j,k)\end{aligned}$$

in the sense of distributions in Ω, that is they are bounded on Ω: thus edge-elements have no jump of their tangential part across facets, and face-elements have no jump of their normal part across facets. Thus we have the inclusions:

$$\text{(192)}\quad W^0(\Omega)\subset H^1(\Omega),\ \ W^1(\Omega)\subset H(\text{curl},\Omega),\ \ W^2(\Omega)\subset H(\text{div},\Omega),\ \ W^3(\Omega)\subset L^2(\Omega).$$

Moreover, since we have from the definition of the barycentric coordinates, for example in T_0 (with a suitable choice of the edge orientations):

$$1 = \sum_{i=1 \text{ to } 4} w_i(x), \text{ then } \sum_{j\neq i} w_{(i,j)} = -\,\text{grad}\, w_i.$$

Therefore we also have the inclusions:

$$\text{(193)}\qquad \text{grad}\, W^0(\Omega)\subset W^1(\Omega),\quad \text{curl}\, W^1(\Omega)\subset W^2(\Omega),\quad \text{div}\, W^2(\Omega)\subset W^3(\Omega),$$

and we have the sequence, with the properties of (105):

$$\text{(194)}\qquad W^0(\Omega)\xrightarrow{\text{grad}} W^1(\Omega)\xrightarrow{\text{curl}} W^2(\Omega)\xrightarrow{\text{div}} W^3(\Omega).$$

Whitney elements are a generalization of mixed elements of Raviart-Thomas and of Nedelec elements, see Girault-Raviart [2] and Roberts-Thomas [1]. For some developments on these Whitney elements and their use in computation for electromagnetism, we refer to Bossavit [4].

CHAPTER 3

STATIONARY SCATTERING PROBLEMS WITH BOUNDED OBSTACLES

1. STATIONARY WAVES DUE TO SOURCES IN BOUNDED DOMAINS

First, we study in the whole space R^3 stationary waves, of wavenumber k, produced by a source f localized in a bounded domain. We have to specify conditions at infinity corresponding to outgoing or incoming waves; these conditions are known as the Sommerfeld radiation conditions.

Then we study electromagnetic fields produced by charges ρ and currents J localized in a bounded domain, at a given angular frequency ω or at a given wavenumber k, in free space. Conditions at infinity corresponding to outgoing waves or incoming waves are called the Silver-Müller conditions.

Then it will allow us to solve (by a surface integral method):
i) boundary problems, in a bounded domain or its complement,
ii) scattering problems for a bounded obstacle due an (exterior) incident wave.

1.1. The main properties for the Helmholtz equation in R^3

Here a *stationary wave* is a solution u of the Helmholtz equation in the whole space

(1) $$\Delta u + k^2 u = - f \quad \text{in } R^3,$$

with $k > 0$, and f a distribution with *compact support* (i.e., f in $\mathcal{E}'(R^3)$).

1.1.1. Local regularity properties

First it follows from the hypoelliptic property of the operator $(\Delta + k^2)$ that u is C^∞ in any open domain Ω where this is true for f. It is even an analytic function in Ω where this is true of f, since $(\Delta + k^2)$ is analytic-hypoelliptic (see Treves [1] p. 23); thus u is analytic outside the compact support of f. Using Sobolev spaces we have from the elliptic nature of $(\Delta + k^2)$, see Lions-Magenes [1]:

$$u \in H^{s+2}_{loc}(R^3) \text{ for } f \in H^s_{loc}(R^3), \ \forall s \in R.$$

1.1.2. Outgoing and incoming Sommerfeld conditions

The energy balance for the wave equation:

$$(2) \qquad \frac{1}{c^2}\frac{\partial^2 w}{\partial t^2} - \Delta w = g \ \text{ in } \ R_t^+ x R^3,$$

for a given "regular" function g, in a bounded domain Ω with boundary Γ, at time t, is:

$$(3) \qquad \frac{d}{dt}E_\Omega(w,t) + I_\Gamma(w,t) = \mathrm{Re}\,(\int_\Omega g\frac{\partial \bar{w}}{\partial t}\,dx),$$

with:

$$\left| \begin{array}{l} E_\Omega(w,t) = \frac{1}{2}\int_\Omega (|\mathrm{grad}\, w|^2 + \frac{1}{c^2}|\frac{\partial w}{\partial t}|^2)\,dx = \text{energy of } w \text{ in } \Omega, \\ I_\Gamma(w,t) = -\,\mathrm{Re}\,(\int_\Gamma \frac{\partial w}{\partial n}\frac{\partial \bar{w}}{\partial t}\,d\Gamma) = \text{flux of energy outgoing from } \Gamma; \end{array} \right.$$

the r.h.s. of (3) is the energy provided in Ω by the source g (see for example Dautray-Lions [1] chap. 2.8.7); n being *the exterior normal* to Ω.
For a source g of the form: $g = e^{-i\omega t} f$ (for f in $L^2(R^3)$, with compact support, and with angular frequency ω) and for a stationary wave $w = e^{-i\omega t}u$, u satisfies the Helmhotz equation with the wavenumber $k = \omega/c$:

$$(4) \qquad k^2 u + \Delta u = -\,f,$$

and the energy balance for the (bounded) domain Ω is

$$(5) \qquad I_\Gamma(u) \overset{\text{def}}{=} -\,\mathrm{Re}\,(\,i\omega\int_\Gamma \frac{\partial u}{\partial n}\bar{u}\,d\Gamma) = \mathrm{Re}\,(\,i\omega\int_\Omega f\bar{u}\,dx).$$

The condition: "f having a compact support K" implies that u is regular outside K and allows us to define:

Definition 1. *We call outgoing (resp. incoming) Sommerfeld condition the condition at infinity:*

$$(6) \qquad \frac{\partial u}{\partial n} - iku = o(\frac{1}{r}) \ \text{ with } \ u = O(\frac{1}{r}), \ \ r = |x| \to \infty,$$

((6)'resp.) $\frac{\partial u}{\partial n} + iku = o(\frac{1}{r})$ with $u = O(\frac{1}{r})$, $r = |x| \to \infty$,

with o *and* O *uniform with respect to* $\alpha = x/r$, *and we call outgoing (resp. incoming) wave a solution u of the Helmholtz equation that satisfies the outgoing (resp. incoming) Sommerfeld conditions.*

The flux of energy of an outgoing (resp. incoming) wave through every boundary S_r of a ball B_r containing the support of f is positive (resp. negative):

(7) $$I_{S_r}(u) = \mathrm{Re}\,(-i\omega \int_{S_r} \frac{\partial u}{\partial n} \bar{u}\, dS_r) = \mathrm{Re}\,(ck^2) \int_{S_r} |u|^2 dS_r + o(1) \geq 0 \quad (\text{resp.} \leq 0).$$

Note that Definition 1 is also valid for f a distribution with compact support.

1.1.3. Elementary outgoing (incoming) solution

The function $\Phi = \Phi^{out}$ defined on R^3 by:

(8) $$\Phi(x) = \Phi(r) = \frac{e^{ikr}}{4\pi r}, \qquad r = |x|,$$

which satisfies:

(9) $$\left| \begin{array}{l} (\Delta + k^2)\,\Phi = -\delta, \\ \frac{\partial \Phi}{\partial r} - ik\,\Phi = -\frac{e^{ikr}}{4\pi r^2} = o\,(\frac{1}{r}), \end{array} \right.$$

corresponds to an outgoing wave due to the (point) source δ. It is called *the elementary outgoing solution of the Helmholtz equation* (or Green function by physicists). Its conjugate is *the elementary incoming solution* Φ^{in}. Note that:

(10) $$\Phi \in L^{\alpha}_{loc}(R^3) \text{ for } 1 \leq \alpha < 3, \quad \text{grad}\,\Phi \in L^{\beta}_{loc}(R^3)^3 \text{ for } 1 \leq \beta < 3/2.$$

1.1.4. Fundamental properties

Let f be a distribution in R^3 with compact support. Then the convolution product:

(11) $$u = \Phi * f$$

satisfies the Helmholtz equation:

(12) $$(\Delta + k^2)\,u = (\Delta + k^2)\Phi * f = -\delta * f = -f\,.$$

We prove (see for example Dautray-Lions [1] chap. 2.8) that u given by (11) is an outgoing wave. Then we have:

Theorem 1. *The problem: find* u, *a distribution in* R^3, *satisfying*

$$(13)\quad \left|\begin{array}{l} \text{i) } \Delta u + k^2 u = -f, \quad f \text{ given in } E'(R^3),\ k>0, \\ \text{ii) } (\frac{\partial u}{\partial n} - ik\, u)(x) = o(\frac{1}{r}), \quad \text{with } u(x) = O(\frac{1}{r}) \text{ when } r = |x| \to \infty, \end{array}\right.$$

has a unique solution, given by (11).

We only need to prove uniqueness, which is a consequence of the following lemma:

Lemma 1 (Rellich). *Let* Ω' *be a connected domain which is the complement of a bounded domain, and let* u *be a wave satisfying*:

$$(14)\quad \left|\begin{array}{l} \text{i) } \Delta u + k^2 u = 0 \quad \text{in } \Omega', \\ \text{ii) } \int_{S^2} |u(r\alpha)|^2\, d\alpha = o(r^{-2}), \text{ i.e. } \int_{S_r} |u(x)|^2 d\Gamma_x = o(1) \text{ when } r \to \infty. \end{array}\right.$$

Then $u = 0$ *in* Ω'.

The "standard" proof of the Lemma uses a "multipole expansion", or "Rayleigh expansion" (see section 11). For such a proof, see Müller [1] p. 87. We also prove it in chap. 5.3.

PROOF *of the uniqueness in theorem 1.* We only have to prove that $f = 0$ implies $u = 0$. We use the energy balance (5) in the domain $\Omega = B_\rho$, with $c = 1$. Thus:
$Re(-ik \int_{S_r} \frac{\partial u}{\partial n} \bar{u}\, dS_r) = 0$, and thus $k^2 \int_{S_r} |u(x)|^2 dS_r = o(1)$ when $r \to \infty$, with the Sommerfeld condition (6). Therefore $u = 0$ from Rellich Lemma.

⊗

Proposition 1. *Behavior at infinity. The outgoing solution of the Helmholtz equation* (13) *has the following behavior at infinity* (*uniformly in* α):

$$(15)\qquad u(r\alpha) = \Phi(r)\, \hat{f}(k\alpha) + o(\frac{1}{r}), \quad r = |x|, \ \alpha = x/r,$$

$\hat{f}$ *being the Fourier transform of* f, *given by* $\hat{f}(\xi) = \int f(x)\, e^{-i\xi . x}\, dx$.

PROOF. There are many ways to prove this proposition. We give here a proof using the Green formula (see Müller [1] p. 114).
Let $\Phi_x(z) = \Phi(x - z)$ be the elementary outgoing solution centered at x. When r goes to infinity, Φ_x has the behavior:

$$(16)\qquad \frac{\partial \Phi_x}{\partial n_z}(z) = -ik\, \Phi(r)\, (n.\alpha)\, e^{-ik\alpha . z} + o(1), \quad \Phi_x(z) = \Phi(r)\, e^{-ik\alpha . z} + o(1),$$

with $r = |x|$, $\alpha = n = x/r$. Let Ω be a "spherical" domain $\Omega = B_{R_1} \backslash B_{R_2}$ with $R_1 > R_2$, with supp $f \subset B_{R_2}$. Then:

$$(17)\qquad \int_\Omega ((\Delta + k^2)\Phi_x \,.\, u - \Phi_x (\Delta + k^2)u)\, dz = \int_{\partial\Omega} (\frac{\partial \Phi_x}{\partial n} u - \Phi_x \frac{\partial u}{\partial n})\, d\Gamma_z,$$

therefore (since the boundary $\partial\Omega$ of Ω is $\partial\Omega = S_{R_1} \cup S_{R_2}$):

$$(18)\qquad -u(x) = -I(S_{R_2}) + I(S_{R_1}), \quad \text{with } I(S) = \int_S (\frac{\partial \Phi_x}{\partial n} u - \Phi_x \frac{\partial u}{\partial n})\, d\Gamma_z.$$

First let $R_1 \to \infty$; then:

$$(19)\qquad u(x) = I(S_R) \quad \text{for } R = R_2.$$

Then let $r = |x| \to \infty$; using (16), we obtain

$$(20)\qquad u(x) = u(r\alpha) = \Phi(r) \int_{S_R} e^{-ik\alpha.y} (ik(n.\alpha)\, u + \frac{\partial u}{\partial n})\, d\Gamma + o(1).$$

But we also have, using the hypothesis that f has a compact support, and the Green formula:

$$(21)\qquad \left|\begin{aligned} \hat{f}(k\alpha) &= \langle f, e^{-ik\alpha.x} \rangle_{B_R} = \int_{B_R} ((\Delta + k^2)u \,.\, e^{-ik\alpha.x} - u \,.\, (\Delta + k^2) e^{-ik\alpha.x})\, dx \\ &= \int_{S_R} (\frac{\partial u}{\partial n} e^{-ik\alpha.x} - u \frac{\partial}{\partial n}(e^{-ik\alpha.x}))\, dS_R = \int_{S_R} (\frac{\partial u}{\partial n} + ikn.\alpha\, u)\, e^{-ik\alpha.x}\, dS_R, \end{aligned}\right.$$

hence giving (15). ⊗

In scattering problems, it is very important to characterize the source f from $\hat{f}(k\alpha)$. The properties of $\hat{f}(k\alpha)$ will be specified at section 11. Here we only note that if $\hat{f}(k\alpha) = 0$ for all α in S^2, then $u(r\alpha) = o(\frac{1}{r})$ when $r \to \infty$.
From the Rellich Lemma, $u(x) = 0$ for $r > R$, outside the support of f, i.e., $u \in E'(R^3)$ which is possible only if $\hat{f}(\xi)/(\xi^2 - k^2)$ is an entire function, see Hörmander [1] p.78. Such sources are perfectly stealth, since they create waves which vanish outside a bounded domain.

1.2. The main properties for Maxwell equations in R^3
Here we study the following Maxwell problem in the whole space R^3: find (E, H) in $D'(R^3)^3 \times D'(R^3)^3$ satisfying

$$(22)\qquad \left|\begin{aligned} &\text{i) } \operatorname{curl} H + i\omega\varepsilon E = J, \\ &\text{ii) } -\operatorname{curl} E + i\omega\mu H = M \quad \text{in } R^3 \end{aligned}\right.$$

with ε and μ the permittivity and the permeability of free space ($\varepsilon > 0$, $\mu > 0$); J and M are electric and magnetic currents, and are distributions with compact support in R^3. Very generally, magnetic currents do not exist, but it is very useful to introduce them, so that Maxwell equations become more symmetric with respect to E and H.

1.2.1. Relation to the Helmholtz problem

We can eliminate in system (22) the electric field E or the magnetic field H, in order to obtain equations on H only (resp. E). Applying the curl operator to (22)i), or to (22)ii), we obtain:

$$(23)\qquad \left|\begin{array}{l} \text{i) curl curl } H - k^2 H = \text{curl } J + i\omega\varepsilon M \\ \text{ii) curl curl } E - k^2 E = -\text{ curl } M + i\omega\mu J \quad \text{in } R^3, \end{array}\right.$$

with $k^2 = \omega^2\varepsilon\mu$. Using the relation:

$$(24)\qquad \text{curl curl } u = -\Delta u + \text{grad div } u,$$

and applying the div operator to (23), we obtain:

$$(25)\qquad \left|\begin{array}{l} -k^2 \text{ div } H = i\omega\varepsilon \text{ div } M \\ -k^2 \text{ div } E = i\omega\mu \text{ div } J, \end{array}\right.$$

and thus (23) gives:

$$(26)\qquad \left|\begin{array}{l} -\Delta H - k^2 H = \text{curl } J + i\omega\varepsilon\,(M + k^{-2} \text{ grad div } M), \\ -\Delta E - k^2 E = -\text{ curl } M + i\omega\mu\,(J + k^{-2} \text{ grad div } J). \end{array}\right.$$

Let m and j be defined by:

$$(27)\qquad \left|\begin{array}{l} m \overset{\text{def}}{=} \text{curl } J + i\omega\varepsilon\,(M + k^{-2} \text{ grad div } M), \\ j \overset{\text{def}}{=} -\text{ curl } M + i\omega\mu\,(J + k^{-2} \text{ grad div } J). \end{array}\right.$$

We have obtained the (vector) Helmholtz equations:

$$(28)\qquad \Delta H + k^2 H = -m, \qquad \Delta E + k^2 E = -j,$$

where j and m $\in \mathcal{E}'(R^3)^3$, i.e., are given distributions with compact support.

As in the case of the Helmholtz equation, it is necessary to introduce conditions at infinity in order to obtain a well-posed problem. We make the hypothesis that E and H satisfy the outgoing Sommerfeld condition for $u = E_j$ or H_j. Then from Theorem 1, E and H are obtained by the convolution product:

(29) $E = \Phi * j, \quad H = \Phi * m.$

The behaviors of E and H at infinity are given by:

(30) $E(r\alpha) = \Phi(r)\, \hat{j}(k\alpha) + o(\frac{1}{r}), \quad H(r\alpha) = \Phi(r)\, \hat{m}(k\alpha) + o(\frac{1}{r}),$

with (from (27)):

(31) $$\left| \begin{array}{l} \hat{j}(k\alpha) = -\, ik\alpha \wedge \hat{M}(k\alpha) + i\omega\mu\, \Pi_\alpha \hat{J}(k\alpha), \\ \hat{m}(k\alpha) = ik\alpha \wedge \hat{J}(k\alpha) + i\omega\varepsilon\, \Pi_\alpha \hat{M}(k\alpha), \end{array} \right.$$

with Π_α the projection of the vector v on the tangent plane to the sphere S_k at $k\alpha$:

(32) $\Pi_\alpha v = \Pi v = v - \alpha\,(\alpha.v).$

We immediately verify the following properties:

(33) $$\left| \begin{array}{l} \text{i) } \alpha.\hat{j}(k\alpha) = 0, \quad \alpha.\hat{m}(k\alpha) = 0, \quad \forall \alpha \in S^2, \\ \text{ii) } -\alpha \wedge \hat{m}(k\alpha) = \frac{1}{Z}\hat{j}(k\alpha), \quad \alpha \wedge \hat{j}(k\alpha) = Z\,\hat{m}(k\alpha), \end{array} \right.$$

with Z the free space impedance:

(34) $Z = \frac{k}{\omega\varepsilon} = \frac{\omega\mu}{k}, \quad Z^2 = \frac{\mu}{\varepsilon}.$

We can also obtain these formulas in the following way. First applying the div operator to (7), we have:

(35) $v \overset{\text{def}}{=} \operatorname{div} E = \Phi * \operatorname{div} j,$

with the behavior at infinity:

(36) $v(r\alpha) = \Phi(r)\, \widehat{\operatorname{div} j}\,(k\alpha) + o\,(\frac{1}{r}) = \Phi(r)\, ik\alpha.\hat{j}(k\alpha) + o\,(\frac{1}{r}).$

Since div J has a compact support, we have from (25)

(37) $v = \operatorname{div} E = 0 \quad \text{for } r > R_o,$

hence $\alpha.\hat{j}(k\alpha) = 0$. Operating similarly for div H, we obtain (33)i).

Then applying the curl operator to (29), we have:

(38) $w \overset{\text{def}}{=} \operatorname{curl} E = \Phi * \operatorname{curl} j,$

whose behavior at infinity is given by:

(39) $w(r\alpha) = \Phi(r)\, \widehat{\operatorname{curl} j}\,(k\alpha) + o\,(\frac{1}{r}) = \Phi(r)\, ik\alpha \wedge \hat{j}(k\alpha) + o\,(\frac{1}{r}).$

Using the Maxwell equation (22)ii) outside the support of M, that is:

(40) $\quad - \text{curl}\, E + i\omega\mu\, H = 0,$

we get through (30) and (39)

(41) $\quad - ik\alpha \wedge \hat{j}(k\alpha) + i\omega\mu\, \hat{m}(k\alpha) = 0,$
that is (33)ii). ⊗

Conversely, condition (33)i) implies through (36) $\text{div}\, E = o(\frac{1}{r})$ when $r \to \infty$, thus from the Rellich Lemma $\text{div}\, E = 0$ on the outside of a ball B_a. Then if E satisfies the Helmholtz problem

$$(42)\quad \begin{cases} \text{i) } \Delta E + k^2 E = -j \quad \text{with } j \in E'(R^3)^3, \text{ and } \alpha.\hat{j}(k\alpha) = 0, \\ \text{ii) } \dfrac{\partial E}{\partial n} - ik\, E = o(\frac{1}{r}) \text{ when } r \to \infty, \end{cases}$$

we have $\text{div}\, E = 0$; if we define H on the outside of the support of j by $H \stackrel{\text{def}}{=} \frac{1}{i\omega\mu}\, \text{curl}\, E$ then (E, H) satisfies Maxwell equations, since

$$\text{curl}\, H = \frac{1}{i\omega\mu}\, \text{curl curl}\, E = -\frac{1}{i\omega\mu}\, \Delta E = \frac{k^2}{i\omega\mu}\, E = -\, i\omega\varepsilon\, E.$$

1.2.2. The Silver-Müller conditions

Conditions (33)ii) imply the following behavior at infinity:

$$(43)\quad \begin{cases} \omega\varepsilon\alpha \wedge E(r\alpha) - k\, H(r\alpha) = o(\frac{1}{r}) \text{ with } E(r\alpha) = O(\frac{1}{r}) \\ \omega\mu\alpha \wedge H(r\alpha) + k\, E(r\alpha) = o(\frac{1}{r}) \text{ with } H(r\alpha) = O(\frac{1}{r}), \end{cases} \quad \text{uniformly in } \alpha \in S^2.$$

These conditions are called *the Silver-Müller conditions* for the Maxwell equations (22). They are equivalent to the outgoing Sommerfeld conditions for the Helmholtz equation (if we only change i into $-i$ in (22), then (43) would correspond to the incoming condition). This is a consequence of the theorem:

Theorem 2. *The Maxwell problem: find the electromagnetic field* (E, H) *in the space of distributions in the whole space* R^3, *satisfying*
i) *Maxwell equations* (22) *with given distributions with compact support* (J, M),
ii) *the Silver-Müller conditions* (43),
has a unique solution, given from (29) *and* (27) *by*

$$(44)\quad \begin{cases} E = j * \Phi = [i\omega\mu\, (J + k^{-2}\, \text{grad div}\, J) - \text{curl}\, M] * \Phi \\ H = m * \Phi = [i\omega\varepsilon\, (M + k^{-2}\, \text{grad div}\, M) + \text{curl}\, J] * \Phi. \end{cases}$$

Its behavior at infinity is given by (30) *with* (31), (33).

Note that we can write (44) in many ways, since:

(45) $(\text{curl } M) * \Phi = - M \hat{*} \text{grad } \Phi, \quad \text{with } (M \hat{*} \text{grad } \Phi)_1 = M_2 * \frac{\partial \Phi}{\partial x_3} - M_3 * \frac{\partial \Phi}{\partial x_2}, \ldots$

The electric and magnetic charges ρ_e and ρ_m are obtained through J and M by

(46) $i\omega\rho_e = \text{div } J, \quad i\omega\rho_m = \text{div } M.$

Then (44) may also be written:

$$(47) \quad \begin{cases} E = i\omega\mu\, J * \Phi + M \hat{*} \text{grad } \Phi - \frac{1}{\varepsilon}\rho_e * \text{grad } \Phi, \\ H = i\omega\varepsilon\, M * \Phi - J \hat{*} \text{grad } \Phi - \frac{1}{\mu}\rho_m * \text{grad } \Phi. \end{cases}$$

These formulas are often called *Stratton-Chu* (or also *Kirchhoff*) formulas. When $M = 0$, if we define the vector potential A and the scalar potential ϕ by

$$A = \mu\, J * \Phi, \quad \phi = \frac{1}{\varepsilon}\rho * \Phi = \frac{1}{i\omega\varepsilon} \text{div } J * \Phi = \frac{c^2}{i\omega} \text{div } A, \quad \text{with } \rho = \rho_e,$$

then (E,H) is given by: $E = i\omega\, A - \text{grad } \phi, \quad H = \frac{1}{\mu} \text{curl } A.$

PROOF. We only have to prove uniqueness of the solution of the Maxwell problem with Silver-Müller conditions at infinity, for J and $M = 0$. This is obtained through the energy balance equation: let Γ be a surface which is the boundary of a regular bounded domain Ω; then *the flux of the Poynting vector* $S = \text{Re}\,(E \wedge \bar{H})$ *through* Γ *is zero*

(48) $P = \text{Re}\,(\int_\Gamma n.\, E \wedge \bar{H}\, d\Gamma) = \text{Re}\,(\int_\Omega \text{div}\,(E \wedge \bar{H})\, dx) = 0,$
since

(48)' $\text{div}\,(E \wedge \bar{H}) = \text{curl } E \,.\, \bar{H} - E.\, \text{curl } \bar{H} = i\omega\mu\, H.\bar{H} - i\omega\varepsilon\, E.\bar{E}.$

Then taking $\Gamma = S_r$, with high r and using (43), we have:

(49) $P = \text{Re}\,(\int_{S_r} n.\, E \wedge \bar{H}\, dS_r) = \text{Re}\,(\int_{S_r} n \wedge E.\, \bar{H}\, dS_r) = \frac{k}{\omega\varepsilon} \int_{S_r} |H|^2\, dS_r + o(1);$

therefore $\int_{S_r} |H|^2\, dS_r = o(1)$. From the Rellich Lemma, we have $H = 0$, and then $E = 0$ in the whole space.

⊗

Remark 1. The Silver-Müller conditions imply:

(50) $\alpha.E(r\alpha) = o(\frac{1}{r}), \quad \alpha.H(r\alpha) = o(\frac{1}{r}), \quad \text{i.e., } x.E(x) = o(1), \quad x.H(x) = o(1),$

(and also: $\alpha.\text{curl } E = o(\frac{1}{r})$, $\alpha.\text{curl } H = o(\frac{1}{r})$). Thus (x, E, H) is an orthogonal frame at infinity. If E_Σ, H_Σ are the projections of E and H onto the tangent plane to the sphere $r\Sigma$, then with the usual notations we have:

(51) $\quad \mathrm{curl}_\Sigma E_\Sigma = o(\frac{1}{r}), \quad \mathrm{curl}_\Sigma H_\Sigma = o(\frac{1}{r}).$

If we compare the Silver-Müller conditions with the outgoing Sommerfeld conditions, we also obtain:

(52) $\quad \overrightarrow{\mathrm{curl}}_\Sigma E_r = o(\frac{1}{r}), \quad \overrightarrow{\mathrm{curl}}_\Sigma H_r = o(\frac{1}{r}),$

which cannot be proved straightforwardly.

⊗

1.3. Transmission problems and surface integral operators

Here we study waves due to distributions of charges or currents located on a surface Γ which is the boundary of some bounded domain Ω, first in the Helmholtz case, then in the Maxwell case.

1.3.1. Helmholtz problems with charges on "regular" surfaces

First we recall (see Prop. 2, chap. 2) some jump formulas for a regular function u on each side of a (regular) surface Γ. Here by "regular" we mean that u is C^1 (up to the boundary). This is necessary so as to speak of the derivatives of u in a classical sense. Let δ_Γ be the Dirac distribution on Γ. We have

(53) $\quad \mathrm{grad}\, u = (\mathrm{grad}\, u) - [u]_\Gamma\, n\, \delta_\Gamma,$

and if u is a (regular) vector function of x in R^3, with values in C^3:

(54) $\quad \mathrm{curl}\, u = (\mathrm{curl}\, u) - [n \wedge u]_\Gamma \delta_\Gamma, \qquad \mathrm{div}\, u = (\mathrm{div}\, u) - [n.u]_\Gamma \delta_\Gamma.$

Expressions in brackets (grad u),... in (53),(54) correspond to derivatives in the classical sense, whereas the l.h.s. of these equalities are derivatives in the distribution sense; $[v]_\Gamma$ is the jump of the function v across the surface Γ, i.e., the normal n to Γ being oriented from Γ_i to Γ_e

(55) $\quad [v]_\Gamma = v|_{\Gamma_i} - v|_{\Gamma_e}.$

We can easily prove (53), (54) from Green formula. As an example for all $\phi \in D(R^3)^3$, we have:

$$<\mathrm{curl}\, u, \phi> \overset{\mathrm{def}}{=} <u, \mathrm{curl}\, \phi> = \int_{\Omega'} u.\mathrm{curl}\, \phi\, dx + \int_\Omega u.\mathrm{curl}\, \phi\, dx$$
$$= \int_{\Omega'} \mathrm{curl}\, u.\phi\, dx + \int_\Omega \mathrm{curl}\, u.\phi\, dx + \int_\Gamma n \wedge u|_{\Gamma_e} d\Gamma - \int_\Gamma n \wedge u|_{\Gamma_i} d\Gamma,$$

giving (54). These formulas (53), (54) are also valid in the Sobolev framework, with the following regularity conditions; let $X = X_\Gamma$ be the space defined by:

(56) $$\left| \begin{array}{l} X \overset{\mathrm{def}}{=} \{u \in L^2_{loc}(R^3),\ u|_\Omega \in H^1(\Omega),\ \Delta(u|_\Omega) \in L^2(\Omega), \\ \qquad u|_{\Omega'} \in H^1_{loc}(\bar\Omega'),\ \Delta(u|_{\Omega'}) \in L^2_{loc}(\bar\Omega')\}. \end{array} \right.$$

In these formulas, we recall that the Laplacian is taken in the sense of distributions in Ω and Ω', and that

$$(57)\quad \left|\begin{array}{l} H^1_{loc}(\bar{\Omega}') = \{u,\ \phi u|_{\Omega'} \in H^1(\Omega'),\quad \forall \phi \in D(R^3)\}, \\ L^2_{loc}(\bar{\Omega}') = \{u,\ \phi u|_{\Omega'} \in L^2(\Omega'),\quad \forall \phi \in D(R^3)\}. \end{array}\right.$$

When u is in X, then from the usual trace theorems (see Prop.1, Chap.2), u and its normal derivative have traces on each side of Γ, with

$$(57)' \quad u|_{\Gamma_{i,e}} \in H^{1/2}(\Gamma),\ \frac{\partial u}{\partial n}\Big|_{\Gamma_{i,e}} \in H^{-1/2}(\Gamma).$$

Then the Green formulas are valid for u in X and therefore also the jump formulas (53),(54). Let γ_i, γ_e be the trace mappings on X

$$(58) \quad \gamma_i(u) = \gamma_i u = (u|_{\Gamma_i}, \frac{\partial u}{\partial n}\Big|_{\Gamma_i}),\ \text{and } \gamma_e \text{ similar to } \gamma_i,$$

and let $\tilde{G}_i \stackrel{def}{=} \gamma_i X$, $\tilde{G}_e \stackrel{def}{=} \gamma_e X$. Then we have $\tilde{G}_i \neq \tilde{G}_e$, with $\tilde{G}_i$ and $\tilde{G}_e \neq Y$, with

$$(59) \quad Y \stackrel{def}{=} H^{1/2}(\Gamma) \times H^{-1/2}(\Gamma).$$

Thus γ_i and γ_e, which are continuous from X into Y, cannot be onto. Let π_1 (resp. π_2) be the projection on the first (resp. second) component. Then we can prove (from Lions [1]) that $\pi_1\gamma_i$ and $\pi_1\gamma_e$ (resp. $\pi_2\gamma_i$ and $\pi_2\gamma_e$) are surjective mappings onto $H^{1/2}(\Gamma)$ (resp. $H^{-1/2}(\Gamma)$).

The set $\tilde{G}_i \cap \tilde{G}_e$ corresponds to the traces of functions in X (which we can take with compact support) which are "continuous" with their normal derivatives across Γ; thus these functions are in $H^2(R^n)$: we can verify it by a Fourier transform method. Therefore we have (for regular Γ)

$$(59)' \quad \tilde{G}_i \cap \tilde{G}_e = H^{3/2}(\Gamma) \times H^{1/2}(\Gamma).$$

Also there is no extension mapping from $H^1(\Delta,\Omega)$ (nor from $H(\Delta,\Omega)$, see (3) chap. 2) into the space $H^1(\Delta,R^n)) = H(\Delta,R^n) = H^2(R^n)$.

Proposition 2. *The mapping* $\gamma = \gamma_- = \gamma_i - \gamma_e$ *is continuous from* X *onto* Y.

PROOF. In order to have a lifting of γ, we can use a variational method, with a lifting of the jump condition $[u]_\Gamma = \rho$, and with the bilinear form:

$$a(u,v) = \int_{R^n} (\text{grad } u . \text{grad } v + uv)\, dx$$

which is coercive on $H^1(R^n)$. Then we have a lifting u of $(0,u_1)$, as the solution of:

(60) $$a(u,v) = \int_\Gamma u_1 v\, d\Gamma, \quad \forall v \in H^1(R^n).$$

We obviously have the same property for $\gamma_+ = \gamma_i + \gamma_e$, *and then*: $\tilde{G}_i + \tilde{G}_e = Y$.

⊗

Let u be a (stationary) wave with locally finite energy in Ω and its complementary Ω' up to the boundary, which is discontinuous (with its normal derivative) across the surface Γ. Therefore, using formulas (53), (54), we see that u satisfies:

(60)' $$\Delta u + k^2 u = -([\tfrac{\partial u}{\partial n}]\,\delta_\Gamma + \text{div}([u]_\Gamma n \delta_\Gamma)).$$

Conversely, we have:

Proposition 3. *The following problem: find* u *satisfying*

(61)
| i) $\Delta u + k^2 u = -f$ in R^3, with $f = \rho' \delta_\Gamma + \text{div}(\rho n \delta_\Gamma)$,
| *with given* ρ, ρ' *resp. in* $H^{1/2}(\Gamma)$ and $H^{-1/2}(\Gamma)$;
| ii) *the outgoing Sommerfeld condition,*

has a unique solution u *in* X, *which is given by the Kirchhoff formula*:

(62) $$u = \Phi * f = \Phi * (\rho' \delta_\Gamma + \text{div}(\rho n \delta_\Gamma)),$$

and the jumps of u *and its normal derivative across* Γ *are given by*:

(63) $$[u]_\Gamma = \rho, \quad [\tfrac{\partial u}{\partial n}]_\Gamma = \rho'.$$

PROOF. Thanks to Theorem 1, we only have to prove that u given by (62) is in X. Let u_1 be a lifting of the jump conditions (59) (this comes from Proposition 2); then u_1 satisfies:

(64) $$\Delta u_1 + k^2 u_1 = -g - (\rho' \delta_\Gamma + \text{div}(\rho n \delta_\Gamma)), \quad \text{with } g \in L^2_{loc}(R^3).$$

We can choose u_1 with a compact support, so that g also has a compact support. Then $v = u - u_1$ satisfies:

(64)' $$\Delta v + k^2 v = -g,$$

as well as the outgoing Sommerfeld condition; thus v is given by $v = \Phi * g$ with g in L^2. Thus from the regularity properties (see 1.1.1) we have v in $H^2_{loc}(R^3)$, therefore v in X, and thus $u = v + u_1$ is in X.

⊗

Proposition 4. *The mappings* P_i *and* P_e *defined by:*

(65) $$P_i(\rho,\rho') = \gamma_i u, \quad \text{resp. } P_e(\rho,\rho') = -\gamma_e u,$$

with u *the solution of* (61), *are continuous complementary projectors in the space* Y *defined by* (59).

PROOF. The fact that these operators are continuous is a simple consequence of Propositions 2 and 3. They are obviously projectors. Furthermore the uniqueness of the solution in Proposition 3 implies that the intersection of their kernels (or of their ranges) is reduced to 0. From (63) we obtain:

(66) $$P_i + P_e = I.$$ ⊗

Definition 2. *The operators* P_i *and* P_e *are called interior and exterior Calderon projectors (relative to the Helmholtz problem).*

Then if we denote by Y_i and Y_e *their images in* Y, we have the decomposition of Y into the direct sum (but it is not an orthogonal decomposition, since the Calderon projectors are not Hermitian):

(66)' $$Y = Y_i \oplus Y_e.$$

Single and double layer potentials. We define the *single and double layer potentials* respectively by:

(67) $$\mathcal{L}\rho' = \Phi * (\rho'\delta_\Gamma), \qquad \mathcal{P}\rho = \Phi * \operatorname{div}(\rho n \delta_\Gamma),$$

or also, for x outside Γ, by:

(67)' $$\mathcal{L}\rho'(x) = \int_\Gamma \rho'(y)\Phi(x-y)\,d\Gamma_y, \qquad \mathcal{P}\rho(x) = -\int_\Gamma \rho(y)\frac{\partial\Phi}{\partial n_y}(x-y)\,d\Gamma_y.$$

Therefore the solution u of (61) is given by:

(68) $$u = u_s + u_d, \qquad \text{with } u_s = \mathcal{L}\rho',\ u_d = \mathcal{P}\rho.$$

Proposition 3 implies that u_s and u_d have the following properties:

(69) $$\left|\begin{array}{l} \text{i) } [u_s]_\Gamma = 0, \quad [\frac{\partial u_s}{\partial n}]_\Gamma = \rho', \\ \text{ii) } [u_d]_\Gamma = \rho, \quad [\frac{\partial u_d}{\partial n}]_\Gamma = 0, \end{array}\right.$$

since $\Delta u_s + k^2 u_s = -\rho'\delta_\Gamma$, and $\Delta u_d + k^2 u_d = -\operatorname{div}(\rho n \delta_\Gamma)$.

Note that $u_s \in H^1_{loc}(R^3)$ thanks to (69).

Then we define the following *surface integral operators* L, K, J, R:

$$(70)\qquad \left| \begin{array}{ll} L\rho' = \mathcal{L}\rho'|_{\Gamma_i} = \mathcal{L}\rho'|_{\Gamma_e}, & J\rho' = \frac{\partial}{\partial n}\mathcal{L}\rho'|_{\Gamma_i} + \frac{\partial}{\partial n}\mathcal{L}\rho'|_{\Gamma_e} \\ K\rho = \mathcal{P}\rho|_{\Gamma_i} + \mathcal{P}\rho|_{\Gamma_e}, & R\rho = \frac{\partial}{\partial n}\mathcal{P}\rho|_{\Gamma_i} = \frac{\partial}{\partial n}\mathcal{P}\rho|_{\Gamma_e}. \end{array} \right.$$

We also define the matrix operator S by:

$$(71)\qquad S = \begin{pmatrix} K & 2L \\ 2R & J \end{pmatrix}.$$

Then we obtain the relations between the Calderon projectors and the operator S:

$$(72)\qquad P_i = \frac{1}{2}(I+S),\ P_e = \frac{1}{2}(I-S),\ S = P_i - P_e,$$

or:

$$(73)\qquad P_- \overset{\text{def}}{=} P_i,\ P_+ \overset{\text{def}}{=} P_e,\quad P_\pm = \frac{1}{2}(I \mp S),\quad S = P_- - P_+;$$

therefore S is also defined (using Proposition 3) by:

$$(74)\qquad S\begin{pmatrix} \rho \\ \rho' \end{pmatrix} = \gamma_i u + \gamma_e u,$$

and since P_+ and P_- are projectors, S satisfies

$$(74)'\qquad S^2 = I.$$

Thus S is an isomorphism in Y. This implies the following relations and properties between the operators L, J, K and R:
i) the continuity properties are given by:

$$(75)\qquad \left| \begin{array}{ll} H^{-1/2}(\Gamma) \xrightarrow{L} H^{1/2}(\Gamma), & H^{1/2}(\Gamma) \xrightarrow{R} H^{-1/2}(\Gamma) \\ H^{-1/2}(\Gamma) \xrightarrow{J} H^{-1/2}(\Gamma), & H^{1/2}(\Gamma) \xrightarrow{K} H^{1/2}(\Gamma), \end{array} \right.$$

ii) the algebraic relations:

$$(76)\qquad \left| \begin{array}{ll} KL + LJ = 0, & RK + JR = 0, \\ K^2 + 4LR = I, & J^2 + 4RL = I. \end{array} \right.$$

Other properties, useful for the limit problems, will be given below. We only note that these operators are (at least formally) given by:

$$(77)\qquad \left| \begin{array}{ll} L\rho'(x) = \int_\Gamma \rho'(y)\, \Phi(x-y)\, d\Gamma_y, & R\rho(x) = -\int_\Gamma \rho(y) \frac{\partial^2\Phi}{\partial n_x \partial n_y}(x-y)\, d\Gamma_y \\ J\rho'(x) = 2\int_\Gamma \rho'(y) \frac{\partial\Phi}{\partial n_x}(x-y)\, d\Gamma_y, & K\rho(x) = -2\int_\Gamma \rho(y) \frac{\partial\Phi}{\partial n_y}(x-y)\, d\Gamma_y. \end{array} \right.$$

1.3.2. Maxwell problems with currents on "regular" surfaces
First we define the functional space:

(78) $\quad X_o = \{ u \in L^2_{loc}(R^3)^3,\ \text{curl}\,(u|_{\Omega}) \in L^2(\Omega)^3,\ \text{curl}\,(u|_{\Omega'}) \in L^2_{loc}(\bar{\Omega}')^3 \}$

with notations (57). Then, from the trace theorem (see chap. 2), if u is in X_o, it has traces on each side of Γ:

(79) $\quad n \wedge u|_{\Gamma_{i,e}} \in H^{-1/2}(\text{div},\Gamma)$, or $\pi_{\Gamma_{i,e}} u \in H^{-1/2}(\text{curl},\Gamma)$.

We also have trace properties, if we define the trace mappings γ_i, γ_e, $\tilde{\gamma}_i$, $\tilde{\gamma}_e$ by:

(80) $\quad \gamma_{i,e} u = n \wedge u|_{\Gamma_{i,e}}$ and $\tilde{\gamma}_{i,e} = \pi_{\Gamma_{i,e}} u$.

Proposition 5. *The mappings* γ_i, γ_e, $\gamma_i + \gamma_e$, $\gamma_i - \gamma_e$ *(resp. the same with tilde) are continuous from* X_o *onto* $H^{-1/2}(\text{div},\Gamma)$ (resp. $H^{-1/2}(\text{curl},\Gamma)$).

Then we will use the product spaces $X = X_o \times X_o$ and

(81) $\quad Y_{div} = H^{-1/2}(\text{div},\Gamma) \times H^{-1/2}(\text{div},\Gamma), \quad Y_{curl} = H^{-1/2}(\text{curl},\Gamma) \times H^{-1/2}(\text{curl},\Gamma).$

Let u = (E, H) be an electromagnetic wave with locally finite energy in Ω and its complement Ω' up to the boundary, u being discontinuous across the surface Γ. Therefore, thanks to formula (54), u = (E, H) is in X and satisfies:

(82) $$\left| \begin{array}{l} \text{i) curl } H + i\omega\varepsilon E = -[n \wedge H]_\Gamma \delta_\Gamma , \\ \text{ii) } - \text{curl } E + i\omega\mu H = [n \wedge E]_\Gamma \delta_\Gamma . \end{array} \right.$$

Remark 2. *Redundant matching conditions.* Note that we have from (54):

(83) $\quad \text{div}\, E = -[n.E]_\Gamma \delta_\Gamma, \quad \text{div}\, H = -[n.H]_\Gamma \delta_\Gamma.$

Applying the div operator to (82), we obtain that the jumps of n.E, and n.H are related to the other jumps through:

(84) $\quad i\omega\varepsilon [n.E]_\Gamma = \text{div}_\Gamma([n \wedge H]_\Gamma), \quad -i\omega\mu [n.H]_\Gamma = \text{div}_\Gamma([n \wedge E]_\Gamma),$
since:

(85) $$\langle \text{div}(v\,\delta_\Gamma), \phi \rangle = -\langle v\,\delta_\Gamma, \text{grad}\,\phi \rangle = -\int_\Gamma v.\,\text{grad}_\Gamma \phi \, d\Gamma$$
$$= \int_\Gamma \text{div}_\Gamma v \,.\, \phi \, d\Gamma = \langle \text{div}_\Gamma v\, \delta_\Gamma, \phi \rangle, \quad \forall \phi \in D(R^3),$$

for all (regular) tangent vector fields v. ⊗

Conversely, we have:

Theorem 3. *The following so-called "transmission" electromagnetic problem: find* (E, H) *with locally finite energy (i.e.,* E *and* H *in* $L^2_{loc}(R^3)^3$) *satisfying*

(86)
$$\begin{cases} \text{i) } \operatorname{curl} H + i\omega\varepsilon E = 0 \\ \text{ii) } -\operatorname{curl} E + i\omega\mu H = 0, \text{ in } \Omega \text{ and } \Omega', \\ \text{iii) } [n \wedge E]_\Gamma = M_\Gamma, \quad -[n \wedge H]_\Gamma = J_\Gamma, \quad \textit{with } J_\Gamma, M_\Gamma \in H^{-1/2}(\operatorname{div}, \Gamma), \\ \text{iv) } \textit{the Silver-Müller conditions at infinity,} \end{cases}$$

is equivalent to finding (E, H) *with locally finite energy satisfying*

(86)'
$$\begin{cases} \text{i) } \operatorname{curl} H + i\omega\varepsilon E = J_\Gamma \delta_\Gamma \\ \text{ii) } -\operatorname{curl} E + i\omega\mu H = M_\Gamma \delta_\Gamma \text{ in } R^3, \\ \text{iii) } \textit{the Silver-Müller conditions at infinity,} \end{cases}$$

and has a unique solution (E, H) *with locally finite energy up to* Γ, *that is* (E, H) *is in* X, *and it is given by the convolution product* (47), *which is written (formally)*:

(87)
$$\begin{cases} E(x) = \int_\Gamma [i\omega\mu\Phi(x-y)J_\Gamma(y) - \frac{1}{\varepsilon}\operatorname{grad}\Phi(x-y)\rho_e(y) - \operatorname{grad}\Phi(x-y) \wedge M_\Gamma(y)]\,d\Gamma_y \\ H(x) = \int_\Gamma [i\omega\varepsilon\Phi(x-y)M_\Gamma(y) - \frac{1}{\mu}\operatorname{grad}\Phi(x-y)\rho_m(y) + \operatorname{grad}\Phi(x-y) \wedge J_\Gamma(y)]\,d\Gamma_y \end{cases}$$

with:

(88) $$i\omega\rho_e \overset{def}{=} \operatorname{div}_\Gamma J_\Gamma, \qquad i\omega\rho_m \overset{def}{=} \operatorname{div}_\Gamma M_\Gamma.$$

The formulas (87) are also called *Stratton-Chu* formulas.

PROOF. From Proposition 5, there is (E_0, H_0) in X, with a compact support, so that

(89) $$[n \wedge E_0]_\Gamma = M_\Gamma, \qquad [n \wedge H_0]_\Gamma = -J_\Gamma,$$

thus there are J and M in $L^2(R^3)^3$ with compact support, so that:

(90)
$$\begin{cases} \operatorname{curl} H_0 + i\omega\varepsilon E_0 = J_\Gamma \delta_\Gamma - J, \\ -\operatorname{curl} E_0 + i\omega\mu H_0 = M_\Gamma \delta_\Gamma - M. \end{cases}$$

Let:

(91) $$E_1 = E - E_0, \quad H_1 = H - H_0.$$

Then (E_1, H_1) satisfies:

$$(92)\qquad \left|\begin{array}{l} \text{curl } H_1 + i\omega\varepsilon\, E_1 = J \\ -\text{curl } E_1 + i\omega\mu\, H_1 = M. \end{array}\right.$$

Thus from theorem 2, (E_1, H_1) is given by the convolution product (44), and we have (E_1, H_1) in X, thus from (91), (E,H) is in X. Then Theorem 3 follows easily. ⊗

We define the mappings P_i and P_e by:

$$(93)\qquad P_i\begin{pmatrix} M_\Gamma \\ J_\Gamma \end{pmatrix} = \begin{pmatrix} n\wedge E|_{\Gamma_i} \\ -n\wedge H|_{\Gamma_i} \end{pmatrix},\quad P_e\begin{pmatrix} M_\Gamma \\ J_\Gamma \end{pmatrix} = -\begin{pmatrix} n\wedge E|_{\Gamma_e} \\ -n\wedge H|_{\Gamma_e} \end{pmatrix},$$

with (E, H) the solution of (86). We have:

Proposition 6. *The mappings* P_i *and* P_e *are continuous complementary projectors in* Y_{div} *(see* (81)).

The proof is very easy, as in the case of the Helmholtz equation, see Proposition 4. The jump condition (86)iii) directly gives:

$$(94)\qquad P_i + P_e = I.$$

Definition 3. *The operators* P_i *and* P_e *are called interior and exterior Calderon projectors (for the Maxwell problem).*

Then if we denote by $G_i = Y_{div,\,i}$ and $G_e = Y_{div,\,e}$ their images in Y_{div}, we have the decomposition of Y_{div} (see (81)) into the direct sum (but not orthogonal, since the Calderon projectors are not hermitian):

$$(94)'\qquad Y_{div} = G_i \oplus G_e = Y_{div,\,i} \oplus Y_{div,\,e}.$$

Electric layer and magnetic layer. We define two new operators in a similar way to the Helmholtz problem (see (67)), associated with the *electric layer* and the *magnetic layer* respectively by:

$$(95)\qquad \left|\begin{array}{l} \mathcal{L}_m u_\Gamma = ik(u_\Gamma\delta_\Gamma + k^{-2}\text{grad div}(u_\Gamma\,\delta_\Gamma)) * \Phi = ik\,(\mathcal{L}\, u_\Gamma + k^{-2}\text{grad}\,\mathcal{L}\,\text{div}_\Gamma\, u_\Gamma) \\ \mathcal{P}_m u_\Gamma = -\text{curl}\,(u_\Gamma\,\delta_\Gamma * \Phi) = -\text{curl}\,(\mathcal{L}u_\Gamma), \end{array}\right.$$

for all u_Γ in $H^{-1/2}(\text{div},\Gamma)$, or also for x outside Γ, (formally) by:

$$(95)'\qquad \left|\begin{array}{l} \mathcal{L}_m u_\Gamma(x) = \int_\Gamma [\Phi(x-y)\, u_\Gamma(y) + k^{-2}\,\text{grad}\,\Phi\,(x-y)\,\text{div}_\Gamma\, u_\Gamma(y)]\, d\Gamma_y, \\ \mathcal{P}_m u_\Gamma(x) = -\int_\Gamma \text{grad}\,\Phi\,(x-y)\wedge u_\Gamma(y)\, d\Gamma_y. \end{array}\right.$$

Therefore the solution (E, H) of the Maxwell problem (86) is given by:

$$(96)\qquad \left|\begin{array}{l} E = P_m M_\Gamma + Z L_m J_\Gamma \\ H = \frac{1}{Z} L_m M_\Gamma - P_m J_\Gamma, \end{array}\right.$$

(with Z given by (34)), or in a matrix form:

$$(96)'\qquad \begin{pmatrix} E \\ H \end{pmatrix} = \begin{pmatrix} P_m & Z L_m \\ \frac{1}{Z} L_m & -P_m \end{pmatrix} \begin{pmatrix} M_\Gamma \\ J_\Gamma \end{pmatrix}.$$

Then from Theorem 3, we see that for u_Γ in $H^{-1/2}(\mathrm{div},\Gamma)$, we have $L_m u_\Gamma$ and $P_m u_\Gamma$ in X_o, and also:

Proposition 7. *The operators L_m and P_m have the following properties across Γ:*

$$(97)\qquad \left|\begin{array}{ll} [n \wedge L_m u_\Gamma]_\Gamma = 0, & [n.L_m u_\Gamma]_\Gamma = -\frac{1}{ik} \mathrm{div}_\Gamma u_\Gamma, \\ [n \wedge P_m u_\Gamma]_\Gamma = u_\Gamma, & [n.P_m u_\Gamma]_\Gamma = 0, \end{array}\right.$$

that is: the tangential component of $L_m u_\Gamma$ is continuous, whereas its normal component is discontinuous, and the tangential component of $P_m u_\Gamma$ is discontinuous, whereas its normal component is continuous.

PROOF. The tangential jump formulas are implied by the definition of the operators L_m and P_m, since for $u_\Gamma = J_\Gamma$ with $M_\Gamma = 0$, the jumps are obtained from the Maxwell equations (86)' and (86)iii). The normal jumps are due to formulas (84), with (34).

⊗

Remark 3. The tangential continuity of $L_m u_\Gamma$ implies that curl $L_m u_\Gamma$ is in $L^2_{loc}(R^3)^3$, and thus $L_m u_\Gamma$ is in $H_{loc}(\mathrm{curl}, R^3)^3$.

Note that the following relations between L_m and P_m are satisfied:

$$(98)\qquad \left|\begin{array}{l} \mathrm{curl}\, P_m u_\Gamma = ik L_m u_\Gamma - u_\Gamma \delta_\Gamma \\ \mathrm{curl}\, L_m u_\Gamma = -ik P_m u_\Gamma, \end{array}\right.$$

and that:

$$(99)\qquad \mathrm{div}\, P_m u_\Gamma = 0, \qquad \mathrm{div}\, L_m u_\Gamma = \frac{1}{ik} (\mathrm{div}_\Gamma u_\Gamma)\, \delta_\Gamma.$$

We can prove these relations by simply applying the curl operator to (96) or (95) and then the div operator to (96) or (95).

⊗

Then we define using (80):

(100) $\quad T\,u_\Gamma \overset{\text{def}}{=} -(\gamma_i+\gamma_e)\,P_m u_\Gamma, \qquad R\,u_\Gamma = 2\,n \wedge L_m u_\Gamma\,|_\Gamma.$

Thus (at least formally)

(100)'
$$\begin{cases} Tu_\Gamma(x) = 2\int_\Gamma n_x \wedge \left(u_\Gamma(y) \wedge \text{grad}_x \Phi(x-y)\right) d\Gamma_y, \\ Ru_\Gamma = 2ik\; n \wedge [Lu_\Gamma + \frac{1}{k^2}\,\text{grad}_\Gamma L\, \text{div}_\Gamma u_\Gamma]. \end{cases}$$

The traces of the electromagnetic field (E, H) solution of (86) satisfy (from (96)):

(101)
$$\begin{cases} (\gamma_i+\gamma_e)\,E = -\,T\,M_\Gamma + Z\,RJ_\Gamma, \\ -(\gamma_i+\gamma_e)\,H = -\frac{1}{Z}\,RM_\Gamma + T\,J_\Gamma. \end{cases}$$

If we define the matrix operator S by:

(102)
$$S = \begin{pmatrix} -T & ZR \\ -\frac{1}{Z}R & -T \end{pmatrix},$$

then we obtain:

(103)
$$\begin{pmatrix} (\gamma_i+\gamma_e)\,E \\ -(\gamma_i+\gamma_e)\,H \end{pmatrix} = S \begin{pmatrix} M_\Gamma \\ J_\Gamma \end{pmatrix};$$

we deduce the relations between the Calderon projectors and S:

(104) $\quad P_i = \frac{1}{2}(I+S), \quad P_e = \frac{1}{2}(I-S), \quad S = P_i - P_e.$

Since P_i and P_e are projectors, S satisfies:

(105) $\quad S^2 = I,$

and thus S is an isomorphism in Y_{div}. This implies that the operators T and R are continuous in $H^{-1/2}(\text{div},\Gamma)$, and also:

(106) $\quad T^2 - R^2 = I, \qquad TR + RT = 0.$

Remark 4. If we consider transmission problems across a line in the complex plane for the Cauchy-Riemann operator, we also obtain an operator S which is the Hilbert operator, and which is very similar to the operator S defined by (102) for the Maxwell equations. ⊗

1.3.3. Some regularity results

a) *The Helmholtz case*

We assume that Ω is a bounded domain with boundary Γ with convenient regularity. We define for real positive s the following spaces:

$$(107)\quad \left| \begin{array}{l} X^s = X^s_\Gamma \overset{\text{def}}{=} \{u \in D'(R^n),\ u|_\Omega \in H^s(\Omega),\ u|_{\Omega'} \in H^s_{loc}(\bar{\Omega}')\}, \\ \tilde{X}^s = X^s_\Gamma(\Delta) \overset{\text{def}}{=} \{u \in D'(R^n),\ u|_\Omega \in H^s(\Omega),\ \Delta u|_\Omega \in H^s(\Omega), \\ \qquad\qquad u|_{\Omega'} \in H^s_{loc}(\bar{\Omega}'),\ \ \Delta u|_{\Omega'} \in H^s_{loc}(\bar{\Omega}')\}. \end{array} \right.$$

If u is in X^s and satisfies the Helmholtz equation in Ω and Ω', then u is in $\tilde{X}^s$, and more generally, we have $\Delta^m u \in X^s$ and $\tilde{X}^s$, for all $m \in N$.

Proposition 8. *Let* u *be in* X^s *satisfying the Helmholtz equation in* Ω *and* Ω'; *then*:

$$(108)\quad \left| \begin{array}{ll} u|_{\Gamma_{i,e}} \in H^{s-1/2}(\Gamma), & \frac{\partial u}{\partial n}|_{\Gamma_{i,e}} \in H^{s-3/2}(\Gamma),\ldots \\ \Delta^m u|_{\Gamma_{i,e}} \in H^{s-1/2}(\Gamma), & \frac{\partial \Delta^m u}{\partial n}|_{\Gamma_{i,e}} \in H^{s-3/2}(\Gamma), \end{array} \right.$$

for interior traces and for exterior traces (and then for the jumps).

SKETCH OF PROOF. For s = 0 and 1, the proof is a consequence of the trace theorem in spaces $H^0(\Delta,\Omega) = H(\Delta,\Omega)$ and $H^1(\Delta,\Omega)$, see Prop. 1. chap. 2. For s with $0 \le s \le 1$, we have the result by interpolation, but we have to prove that:

$$(109)\qquad [H^1(\Delta,\Omega), H^0(\Delta,\Omega)]_\theta = H^s(\Delta,\Omega) = \{u \in H^s(\Omega),\ \Delta u \in L^2(\Omega)\}, \qquad s = 1 - \theta.$$

We identify $H^0(\Delta,\Omega)$ with the graph G_0 of the Laplacian in L^2, and $H^1(\Delta,\Omega)$ with the graph G_1 of the Laplacian in $H^1 \times L^2$. These graphs are closed subspaces in $L^2 \times L^2$ and $H^1 \times L^2$. The orthogonal space of G_0 in $L^2(\Omega) \times L^2(\Omega)$ is the space:

$$G_0^\perp = \{(f,g),\ f = -\Delta g,\ f \text{ and } g \in L^2(\Omega),\ g|_\Gamma = 0,\ \frac{\partial g}{\partial n}|_\Gamma = 0\} = H^2_0(\Omega).$$

Then we can prove that the orthogonal projector P_0 in $L^2(\Omega) \times L^2(\Omega)$ on G_0 satisfies: $P_0(H^1(\Omega) \times L^2(\Omega)) \subset G_1$. Now (109) is a consequence of theorem 1.2.4 of Triebel [1]. ⊗

Conversely we have:

Proposition 9. *Let* $\rho \in H^s(\Gamma)$, $\rho' \in H^{s-1}(\Gamma)$, *then* $P\rho \in \tilde{X}^{s+1/2}$, $L\rho' \in \tilde{X}^{s+1/2}$, *with* $s \ge -1/2$. *Then for* $(\rho,\rho') \in H^s(\Gamma) \times H^{s-1}(\Gamma)$, *we have* $u \overset{\text{def}}{=} P\rho + L\rho' \in \tilde{X}^{s+1/2}$.

PROOF. We define, for all real positive s:

(110) $X_s \stackrel{def}{=} \{ u \in D'(R^n),\ u|_\Omega \in H^s(\Omega),\ u|_{\Omega'} \in H^s(\Omega')\}.$

Then for all ϕ in $D(R^n)$, the mapping $\rho \longrightarrow \phi P\rho$ is continuous from $H^{1/2}(\Gamma)$ (resp. $H^{3/2}(\Gamma)$) into X_1 (resp. X_2). Thus by interpolation it is continuous from $H^s(\Gamma)$ into $[X_2, X_1]_\theta = X_\sigma$ with $s = \frac{3}{2}(1-\theta) + \frac{1}{2}\theta$, $\sigma = 2-\theta$. Thus $P\rho \in X^{s+1/2}$ for $1/2 < s < 3/2$. In a similar way we obtain the result for $-1/2 \le s \le 1/2$, then $L\rho' \in X^{s+1/2}$.

Proposition 9 for half integer s, $s \ge 3/2$, is due to regularity results.

⊗

Remark 5. The single layer and double layer potentials also have the jump properties for every integer j:

(111) $$\left| \begin{array}{ll} [\Delta^j P u_o]_\Gamma = (-k^2)^j u_o, & [\frac{\partial}{\partial n}\Delta^j P u_o]_\Gamma = 0, \\ [\Delta^j L u_1]_\Gamma = 0, & [\frac{\partial}{\partial n}\Delta^j L u_1]_\Gamma = (-k^2)^j u_1. \end{array} \right.$$

⊗

We define the space, for real s:

(112) $Y^s = Y^s(\Gamma) \stackrel{def}{=} H^{s+1/2}(\Gamma) \times H^{s-1/2}(\Gamma).$

Then Proposition 9 implies:

Proposition 10. *Let* (ρ,ρ') *be in* Y^s, s *a positive real. Then the solution* u *of the Helmholtz transmission problem* (61) *has its traces* $\gamma_i u$, $\gamma_e u$ *in* Y^s; *thus the Calderon projectors* P_i *and* P_e *(see* (65)) *act in* Y^s, *and the space* Y^s *has the decomposition*:

(113) $Y^s = Y^s_i \oplus Y^s_e$, with $Y^s_i = P_i Y^s$, $Y^s_e = P_e Y^s$.

Remark 6. By duality P_i and P_e (see (65)) also act in Y^s for negative real s, and then decomposition (113) is true for all real s.

⊗

Therefore S defined by (71) acts in Y^s for all real s and then we have the following properties for the operators L, R, J, K which is a generalization of (75)

(114) $$\left| \begin{array}{l} H^s(\Gamma) \xrightarrow{L} H^{s+1}(\Gamma),\ H^{s+1}(\Gamma) \xrightarrow{R} H^s(\Gamma) \\ H^s(\Gamma) \xrightarrow{J} H^s(\Gamma),\ H^s(\Gamma) \xrightarrow{K} H^s(\Gamma). \end{array} \right.$$

⊗

b) *The Maxwell case*
Regularity results for Maxwell equations will be obtained as a consequence of the above properties. We define the following spaces for s real, with (128) chap. 2:

$$(115)\quad \left| \begin{array}{l} X^s_{curl} \stackrel{def}{=} \{u \in (X^s)^3, \text{ with } (107), \text{ curl } u|_\Omega \in H^s(\Omega)^3, \text{ curl } u|_{\Omega'} \in H^s_{loc}(\bar{\Omega}')^3\}, \\ Y^s_{div} \stackrel{def}{=} H^s(\text{div},\Gamma) \times H^s(\text{div},\Gamma). \end{array} \right.$$

Then we have:

Proposition 11. *Let* $j \in H^s(\text{div},\Gamma)$, $s \geq -1/2$; then $L_m j$ and $P_m j \in (X^{s+1/2})^3$.

PROOF. From hypothesis we have $j \in H^s(\Gamma)^3$, $\text{div}_\Gamma j \in H^s(\Gamma)$.
Applying Proposition 9, we obtain $Lj \in (X^{s+3/2})^3$ and $L(\text{div}_\Gamma j) \in X^{s+3/2}$, thus: $\text{grad}\, L\, \text{div}_\Gamma j \in (X^{s+1/2})^3$, and from (95) we have $L_m j$ and $P_m j \in (X^{s+1/2})^3$. ⊗

The case s = 1/2 gives H^1 regularity. From Proposition 11, we have:

Proposition 12. *Let* $(J_\Gamma, M_\Gamma) \in Y^s_{div}$; *then the solution* $u = (E,H)$ *of the problem* (86) (or (86)') *satisfies*:

$$(116)\qquad u = (E,H) \in (X^{s+1/2})^3 \times (X^{s+1/2})^3.$$

Proposition 13. *The Calderon projectors* P_i *and* P_e *defined by* (93) (*relative to the Maxwell problem*) *operate in* $Y^s{}_{div}$ *for all real positive s, and by duality for all real s.*

This gives the decomposition:

$$(117)\qquad Y^s_{div} = Y^s_{div,\,i} \oplus Y^s_{div,\,e}, \quad \text{with } Y^s_{div,\,i} = P_i Y^s_{div},\; Y^s_{div,\,e} = P_e Y^s_{div}.$$

As a consequence we have:

Corollary 1. *The operator* S *defined by* (102) *operates in* $Y^s{}_{div}$ *for all real* s, *and the operators* T *and* R (*see* (100)) *operate in* $H^s(\text{div},\Gamma)$ *for all real* s.

1.3.4. Integral method for a sheet

a) *The Helmholtz case*
Let Γ_0 be a "regular" compact sheet, that is a regular part of the boundary Γ of an open set Ω in R^3.

We recall (see chap. 2) that the elements of $H^{1/2}_{o\,o}(\Gamma_o)$ and $H^{-1/2}(\Gamma_0)$ extended by 0 on $\Gamma\backslash\Gamma_0$ are elements of $H^{1/2}(\Gamma)$ and $H^{-1/2}(\Gamma)$ respectively. We define:

(118) $\quad X^1_{\Gamma_o} \overset{\text{def}}{=} \{u \in \mathcal{D}'(R^n),\ \zeta u \in H^1(R^n\backslash\Gamma_0),\ \zeta\Delta u \in H^1(R^n\backslash\Gamma_0),\ \ \forall\zeta \in \mathcal{D}(R^n)\}.$

If the tilde denotes the extension on Γ by 0 of an element defined on Γ_0, then Proposition 9 implies

Proposition 14. *Let* $\rho \in H^{1/2}_{o\,o}(\Gamma_o)$, $\rho' \in H^{-1/2}(\Gamma_0)$, *then* $P\rho \overset{\text{def}}{=} P\tilde{\rho}$ *and* $L\rho' \overset{\text{def}}{=} L\tilde{\rho}'$ *are in* $X^1_{\Gamma_o}$ *with* $L\rho' \in H^1_{loc}(R^n)$.

Corollary 2. *Let* $\rho \in H^{1/2}_{o\,o}(\Gamma_o)$, $\rho' \in H^{-1/2}(\Gamma_0)$, *then the problem: find* u *with locally finite energy in* $R^n\backslash\Gamma_0$, i.e., $u \in H^1_{loc}(\overline{R^n\backslash\Gamma_0})$, *satisfying*

$$(119)\qquad \left|\begin{array}{l} \text{i) } \Delta u + k^2u = 0 \text{ in } R^n\backslash\Gamma_0\,, \\ \text{ii) } [u]_{\Gamma_0} = \rho\,,\ [\frac{\partial u}{\partial n}]_{\Gamma_0} = \rho'\,, \\ \text{iii) } \textit{the outgoing Sommerfeld condition at infinity,} \end{array}\right.$$

has a unique solution u in $X^1_{\Gamma_o}$, *is given by* $u = P\rho + L\rho'$, *and the traces satisfy:*

$$(120)\qquad \gamma_{\Gamma_{0+}}u = (u, \frac{\partial u}{\partial n})|_{\Gamma_{0+}} \text{ and } \gamma_{\Gamma_{0-}}u = (u, \frac{\partial u}{\partial n})|_{\Gamma_{0-}} \in H^{1/2}(\Gamma_0) \times (H^{1/2}_{o\,o}(\Gamma_o))'.$$

Remark 7. We note that the mapping $S_{\Gamma_0} : (\rho,\rho') \rightarrow \gamma_{\Gamma_{0+}}u + \gamma_{\Gamma_{0-}}u$ is continuous and injective from $H^{1/2}_{o\,o}(\Gamma_o) \times H^{-1/2}(\Gamma_0)$ into $H^{1/2}(\Gamma_0) \times (H^{1/2}_{o\,o}(\Gamma_o))'$, and related to the operator S for Γ by: $S_{\Gamma_0}(\rho,\rho') = S(\tilde{\rho},\tilde{\rho}')|_{\Gamma_0}$. The operators $P_{\Gamma_{0+}}$, $P_{\Gamma_{0-}}$ defined by $P_{\Gamma_{0+}}(\rho,\rho') = \gamma_{\Gamma_{0+}}u$, $P_{\Gamma_{0-}}(\rho,\rho') = \gamma_{\Gamma_{0-}}u$, satisfy $P_{\Gamma_{0\pm}} = \frac{1}{2}(I \pm S_{\Gamma_0})$, but they are not projectors! With the notations $K_{\Gamma_0}\rho = (K\rho)|_{\Gamma_0}, \ldots$, these mappings are continuous in the spaces:

$$(121)\qquad \left|\begin{array}{ll} H^{1/2}_{o\,o}(\Gamma_o) \overset{K_{\Gamma_o}}{\rightarrow} H^{1/2}(\Gamma_o), & H^{1/2}_{o\,o}(\Gamma_o) \overset{R_{\Gamma_o}}{\rightarrow} (H^{1/2}_{o\,o}(\Gamma_o)') \\ H^{-1/2}(\Gamma_o) \overset{L_{\Gamma_o}}{\rightarrow} H^{1/2}(\Gamma_o), & H^{-1/2}(\Gamma_o) \overset{J_{\Gamma_o}}{\rightarrow} (H^{1/2}_{o\,o}(\Gamma_o))' \end{array}\right.$$

and then we have:

$$(122)\qquad S_{\Gamma_0} = \begin{pmatrix} K_{\Gamma_0} & 2L_{\Gamma_0} \\ 2R_{\Gamma_0} & J_{\Gamma_0} \end{pmatrix}.$$

⊗

b) *The Maxwell case*

Since the elements of $H_0^{-1/2}(\mathrm{div},\Gamma_0)$ can be extended by 0 out of Γ_0 into elements of $H^{-1/2}(\mathrm{div},\Gamma)$ (see (85) chap. 2) we have from Proposition 7:

Proposition 15. *Let* u_{Γ_0} be in $H_0^{-1/2}(\mathrm{div},\Gamma_0)$, *then* $L_m u_{\Gamma_0}$ *and* $P_m u_{\Gamma_0}$ *are in the space* $H^1_{loc}(\mathrm{curl},\overline{R^3\backslash\Gamma_0})$.

Proposition 16. *Let* $(J_{\Gamma_0}, M_{\Gamma_0}) \in H_0^{-1/2}(\mathrm{div},\Gamma_0) \times H_0^{-1/2}(\mathrm{div},\Gamma_0)$. *Then the problem: find* (E, H) *with locally finite energy on* $R^3\backslash\Gamma_0$, *satisfying*

$$(123)\quad \left|\begin{array}{l} \text{i) } \mathrm{curl}\, H + i\omega\varepsilon E = 0, \\ \quad -\mathrm{curl}\, E + i\omega\mu H = 0 \ \text{ in } R^3\backslash\Gamma_0, \\ \text{ii) } [n\wedge H]_{\Gamma_0} = -J_{\Gamma_0},\ [n\wedge E]_{\Gamma_0} = M_{\Gamma_0}, \\ \text{iii) } \textit{the Silver-Müller condition at infinity,} \end{array}\right.$$

has a unique solution given by (96) in $H^1_{loc}(\mathrm{curl},\overline{R^3\backslash\Gamma_0}) \times H^1_{loc}(\mathrm{curl},\overline{R^3\backslash\Gamma_0})$.

We denote by $\gamma_{\Gamma_{0+}}, \gamma_{\Gamma_{0-}}$ the trace mappings: $\gamma_{\Gamma_{0\pm}}(E,H) = (n\wedge E,\ n\wedge H)|_{\Gamma_{0\pm}}$ on each side of Γ_0. We have:

Proposition 17. *The mapping* $S_{\Gamma_0} : (J_{\Gamma_0}, M_{\Gamma_0}) \to \gamma_{\Gamma_{0+}}(E,H) + \gamma_{\Gamma_{0-}}(E,H)$ *from* $H_0^{-1/2}(\mathrm{div},\Gamma_0) \times H_0^{-1/2}(\mathrm{div},\Gamma_0)$ into $X(\mathrm{div},\Gamma_0) \times X(\mathrm{div},\Gamma_0)$ *with* $X(\mathrm{div},\Gamma_0)$ *defined by* (86) *chap.* 2, *is related to the mapping* S by:

$$(124)\quad S_{\Gamma_0}(J_{\Gamma_0}, M_{\Gamma_0}) = S(\tilde{J}_{\Gamma_0}, \tilde{M}_{\Gamma_0})\,|_{\Gamma_0}.$$

With the operators T_{Γ_0}, R_{Γ_0} from $H_0^{-1/2}(\mathrm{div},\Gamma_0)$ into $X(\mathrm{div},\Gamma_0)$ such that:

$$(125)\quad T_{\Gamma_0} u_{\Gamma_0} = T(\tilde{u}_{\Gamma_0})|_{\Gamma_0},\quad R_{\Gamma_0} u_{\Gamma_0} = R(\tilde{u}_{\Gamma_0})|_{\Gamma_0},$$

we also have:

$$(126)\quad S_{\Gamma_0} = \begin{pmatrix} -T_{\Gamma_0} & ZR_{\Gamma_0} \\ -\frac{1}{Z}R_{\Gamma_0} & -T_{\Gamma_0} \end{pmatrix},$$

and the operators $P_{\Gamma_{0\pm}}$ defined by $P_{\Gamma_{0\pm}}(J_{\Gamma_0}, M_{\Gamma_0}) = \gamma_{\Gamma_{0\pm}}(E,H)$ are also given by:

$$(127)\quad P_{\Gamma_{0\pm}} = \frac{1}{2}(I \pm S_{\Gamma_0}),\ \text{ with } \mathrm{Im}\, P_{\Gamma_{0\pm}} \subseteq X(\mathrm{div},\Gamma_0) \times X(\mathrm{div},\Gamma_0).$$

1.3.5. Incoming waves

In scattering problems we often disregard incoming waves. However they intervene at different stages, involving holomorphic properties with respect to the wavenumber k: for the spectral resolution of the Laplacian, like in the Limiting Absorption Principle (see sections 3.2, 3.4, and Wilcox [1]), or for resonance states (see chap. 6.2), or for inverse problems (to find the source of a wave from its behavior at infinity). Incoming waves also appear as conjugated to outgoing waves.

For the same evolution in time (here given by $e^{-i\omega t}$), the incoming solution of the Helmholtz equation (4) (resp. Maxwell equations (23)) is given by:

(128) $u^{in} = \Phi^{in} * f$ (resp. $E^{in} = \Phi^{in} * j$, $H^{in} = \Phi^{in} * f$, with (27)).

The relations with the conjugated functions are given for real k, ε, μ by:

(129) $\overline{u^{out}(f)} = u^{in}(\bar{f})$, $\overline{E^{out}(M,J)} = -E^{in}(-\bar{M}, \bar{J})$, $\overline{H^{out}(M,J)} = H^{in}(-\bar{M}, \bar{J})$,

and the behavior at infinity of incoming waves is given by

(130) $u^{in}(r\alpha) = \Phi^{in}(r)\,\hat{f}(-k\alpha)$ (thus $E^{in}(r\alpha) = \Phi^{in}(r)\,\hat{j}(-k\alpha)$, $H^{in}(r\alpha) = \Phi^{in}(r)\,\hat{m}(-k\alpha)$).

Since $\Phi^{in} = \bar{\Phi}$, the integral operators for the Helmholtz equation K, J, L, R and $S = S^{out}$ (see (70), (71)) are changed into their conjugated, thus $S^{in} = \overline{S^{out}}$.

In the Maxwell case, the integral operators $T = T^{out}$, $R = R^{out} = ikR_0$, and $S = S^{out}$ (see (100)', (102)) are changed into $T^{in} = \bar{T}$, $R^{in} = -\bar{R} = ik\,\bar{R}_0$, and:

$$(131)\quad S^{in} = \begin{pmatrix} -T^{in} & ZR^{in} \\ -\frac{1}{Z}R^{in} & -T^{in} \end{pmatrix} = \begin{pmatrix} -\bar{T} & ik\,Z\bar{R}_0 \\ -\frac{ik}{Z}\bar{R}_0 & -\bar{T} \end{pmatrix} = \theta'\,\bar{S}\,\theta', \text{ if } \theta' = \begin{pmatrix} -1 & 0 \\ 0 & 1 \end{pmatrix}.$$

Note that in the decomposition (66)' of Y (or (94)' of Y_{div}, the interior space G_i does not depend on the outgoing or incoming condition, contrary to G_e and the Calderon projectors. Thus we have the decompositions

(132) Y (rep. Y_{div}) $= G_i \oplus G_e^{out} = G_i \oplus G_e^{in}$, with $G_e^{out} = G_e$,

associated with $I = P_i^{in} + P_e^{in} = P_i^{out} + P_e^{out}$, with $P_{i,e}^{out} = P_{i,e}$. Thus

(133) $G_i = \ker(I - P_i^{in}) = \ker(I - P_i^{out})$.

This implies the following relations:

(134) $P_e^{in}.P_i^{out} = 0$, $P_e^{out}.P_i^{in} = 0$.

Then :

(135) $P_e^{in}.(P_i^{out} - P_i^{in}) = P_e^{out}.(P_i^{out} - P_i^{in}) = (P_i^{out} - P_i^{in}).P_e^{out} = (P_i^{out} - P_i^{in}).P_e^{in} = 0$,

thus

(135)' $P_e^{in}.(S^{out} - S^{in}) = P_e^{out}.(S^{out} - S^{in}) = 0, \ (S^{out} - S^{in}).P_i^{out} = (S^{out} - S^{in}).P_i^{in} = 0.$
and thus:

(136) $(S^{out} - S^{in})^2 = 0, \ (2I - (S^{out} + S^{in}))(S^{out} - S^{in}) = 0.$

Therefore:

(137) $S^{in}.S^{out} + S^{out}.S^{in} = 2I, \qquad S^{in}.S^{out} - S^{out}.S^{in} = 2(S^{out} - S^{in}).$

Of course, this gives relations between K, J, L, R (resp. T, R) and their conjugated operators. Relations (135)' imply:

(138) $\operatorname{Im}(S^{out} - S^{in}) \subset G_i \subset \ker(S^{out} - S^{in}).$

Remark 8. We see that the operator $S^{out} - S^{in}$ have a leading role in all these questions. This is due to the following properties.
Let Φ_{en} be the entire solution of Helmholtz equation defined by:

(139) $\Phi_{en} = \frac{1}{2i}(\Phi^{out} - \Phi^{in}) = \frac{1}{2i}(\Phi - \bar{\Phi}) = \frac{\sin kr}{4\pi r}, \ r = |x|.$

Then for all sources f (or (M,J)) with compact support, the function $U_{en} = \Phi_{en} * f$ is an entire solution of the Helmholtz equation (resp. $E_{en} = \Phi_{en} * j$, $H_{en} = \Phi_{en} * m$ for Maxwell equations). When f is given as in (61), U_{en} tends to 0 at infinity as $O(1/r)$, and is decomposed from the surface Γ into the sum of an incoming wave and an outgoing wave. Furthermore its traces on Γ are given by

$$\begin{pmatrix} U_{en}|_\Gamma \\ \frac{\partial U_{en}}{\partial n}|_\Gamma \end{pmatrix} = \frac{1}{2i}(P_i^{out} - P_i^{in})\begin{pmatrix} \rho \\ \rho' \end{pmatrix} = \frac{1}{2i}(P_e^{in} - P_e^{out})\begin{pmatrix} \rho \\ \rho' \end{pmatrix} = \frac{1}{4i}(S_e^{out} - S_e^{in})\begin{pmatrix} \rho \\ \rho' \end{pmatrix}$$

with a similar result in the Maxwell case. Now we can also obtain this function U_{en} from the behavior at infinity of the solution u of (61) given by (15). Let:

(140) $Uf(x) = \int_{S^2} \hat{f}(k\alpha)\, e^{ik\alpha.x}\, d\alpha = \int_{S^2} \langle f, e^{-ik\alpha.y}\rangle e^{ik\alpha.x}\, d\alpha = \langle f, \int_{S^2} e^{ik\alpha.(x-y)}\, d\alpha\rangle.$

Now with $x = r\beta$,

(141) $\int_{S^2} e^{ik\alpha.x}\, d\alpha = \int_{S^2} e^{ikr\alpha.\beta}\, d\alpha = 2\pi \int_{-1}^{1} e^{ikru}\, du = 4\pi \frac{\sin kr}{kr} = \frac{(4\pi)^2}{k}\Phi_{en}(r).$

Thus $Uf = (4\pi)^2 k^{-1} U_{en}$, with a similar result in the Maxwell case. ⊗

Let G^{en} be the space of traces:

(142) $G^{en} = \{(U_{en}|_\Gamma, \frac{\partial U_{en}}{\partial n}|_\Gamma), \ U_{en} = \Phi_{en} * f, \ f = \rho'\delta_\Gamma + \operatorname{div}(\rho n \delta_\Gamma), \ (\rho, \rho') \in Y \text{ (see (59))}\}.$

Then G^{en} is regular (C^∞ if Γ is C^∞, analytic if Γ is analytic) and contained in G_i.

Proposition 18. *The operator* $(S^{out} - S^{in})$ *is a regularizing (thus compact) operator with the essential following properties (in the Helmholtz and the Maxwell cases), with* $(S^{out} - S^{in})^2 = 0$

$\ker(S^{out} - S^{in}) = G_i$, $\operatorname{Im}(S^{out} - S^{in}) = G^{en} = P_i^{in}(G_e^{out}) = P_i^{out}(G_e^{in})$ *is dense in* G_i.
Moreover $(S^{out} - S^{in})$ *is a one-to-one mapping from* G_e^{in} *and* G_e^{out} *onto* G^{en}.

PROOF (for Helmholtz). We first prove that $G_e^{in} \cap G_e^{out} = \{0\}$. Let $(v_0, v_1) \in G_e^{in} \cap G_e^{out}$. Let u be the solution of the exterior Helmholtz problem with $(u|_\Gamma, \frac{\partial u}{\partial n}|_\Gamma) = (v_0, v_1)$. Then the behavior of u at infinity is (up to o(1/r)) given by $\Phi(r)\hat{f}(k\alpha)$ and $\overline{\Phi(r)}\,\hat{f}(-k\alpha)$; thus $u(r\alpha) = o(1/r)$; hence $u = 0$ in Ω' thanks to the Rellich Lemma, thus $(v_0, v_1) = 0$.

i) Let $U = (u_0, u_1) \in \ker(S^{out} - S^{in})$. Thus $(P_e^{in} - P_e^{out}) U = 0$; hence $P_e^{in} U = P_e^{out} U$ is in $G_e^{in} \cap G_e^{out} = \{0\}$. Thus $U \in G_i$.

ii) Density of G^{en} in G_i. This result will be obtained by duality.
a) The Helmholtz case. In $Y = H^{1/2}(\Gamma) \times H^{-1/2}(\Gamma)$ we introduce the duality pairing (so that we can identify Y with its dual space Y') by

$$(143) \qquad \left\langle \begin{pmatrix} u^0 \\ u^1 \end{pmatrix}, \begin{pmatrix} v^0 \\ v^1 \end{pmatrix} \right\rangle_{Y,Y} = \langle u^0, v^1 \rangle - \langle u^1, v^0 \rangle,$$

where $\langle , \rangle$ denotes the duality $H^{1/2}(\Gamma)$, $H^{-1/2}(\Gamma)$.
b) The Maxwell case. Now in $V = H^{-1/2}(\text{div}, \Gamma)$ we introduce the duality pairing (so that we can identify V with its dual space $V' = H^{-1/2}(\text{curl}, \Gamma)$) by:
$$(144) \qquad \langle u, v \rangle = \langle u, v \rangle_{V,V} = \langle u, n \wedge v \rangle_{V,V'} = \int_\Gamma u.n \wedge v \, d\Gamma, \quad \forall u, v \text{ in } V.$$

Then we define the duality pairing in Y_{div} (see (81)) (so that we can identify Y_{div} with its dual space $Y' = Y_{curl}$) by

$$(145) \left\langle \begin{pmatrix} M \\ J \end{pmatrix}, \begin{pmatrix} M^0 \\ J^0 \end{pmatrix} \right\rangle_{Y,Y} = \langle M, J^0 \rangle + \langle J, M^0 \rangle = \langle M, n \wedge J^0 \rangle_{V,V'} + \langle J, n \wedge M^0 \rangle_{V,V'}.$$

Now if (E^1, H^1), (E^2, H^2) are two solutions of Maxwell equations (both incoming or outgoing in Ω'), we have on each side of Γ, using Green formula:
$$(145)' \qquad \int_{\Gamma_\pm} (n \wedge E^1 . H^2 + n \wedge H^1 . E^2) \, d\Gamma = 0.$$

Taking the difference, we obtain with the notations $\gamma_m E = (\gamma_i + \gamma_e)E$ (see (103)), ...
$$\int_\Gamma \{\gamma_m E^1.[H^2] + \gamma_m H^1.[E^2]. + [n \wedge E^1] \gamma_m H^2 + [n \wedge H^1] \gamma_m E^2) \, d\Gamma = 0;$$

thus using (103), the pairing (145), (144) and taking $M^j = [n \wedge E^j]$, $J^j = -[n \wedge H^j]$

$$\left\langle S\begin{pmatrix} M^1 \\ J^1 \end{pmatrix}, \begin{pmatrix} M^2 \\ J^2 \end{pmatrix} \right\rangle + \left\langle \begin{pmatrix} M^1 \\ J^1 \end{pmatrix}, S\begin{pmatrix} M^2 \\ J^2 \end{pmatrix} \right\rangle = 0.$$

Thus in the Maxwell and the Helmholtz cases we have:
$$(146) \qquad {}^t S = -S \text{ and thus: } {}^t P_e = P_i, \; {}^t P_i = P_e, \text{ with } S = S^{in} \text{ or } S^{out}$$

and ${}^t L = L$, ${}^t R = R$, ${}^t K = -J$, ${}^t J = -K$, and also: ${}^t T = -T$, ${}^t R = -R$.
Now (146) implies that the polar set of $G^{en} = \text{Im}(S^{out} - S^{in})$ is $\ker(S^{out} - S^{in})$, i.e. G_i, and the polar set of G_i is again G_i. This proves the density of G^{en} in G_i. ⊗

Remark 9. We can also use an antiduality pairing by

(144)' $\langle u,v\rangle = \langle u, n\wedge\bar{v}\rangle_{V,V'} = \int_\Gamma u.n\wedge\bar{v}\, d\Gamma$,

giving:

(146)' $S^* = (S^{out})^* = -S^{in}$, and thus $T^* = (T^{out})^* = -T^{in}$, $R^* = (R^{out})^* = -R^{in}$.

These formulas are again obtained using Green formulas in Ω and Ω', but with two waves, one incoming and the other outgoing. ⊗

2. SCATTERING PROBLEMS WITH A COMPLEX WAVENUMBER k

Except in the free space case, physical media are dissipative (see chap. 1 section 3.1.6, with the H14) hypothesis). Of course dissipative media are bounded but in a first step we assume that such a medium occupies the whole space. This allows us to define the Calderon projectors and then to tackle physical problems in free space by the "limiting absorption principle".

2.1. Helmholtz equation in R^3 with a complex wavenumber

We again consider the Helmholtz equation (1) with a complex wavenumber k.

Proposition 19. *When* Im k > 0, *the Helmholtz equation* (1), *where* f *is a distribution with compact support, or a tempered distribution in* R^3, *has a unique solution* u *in the space of tempered distributions, with* u *in* $H^2(R^3)$ *when* f *is in* $L^2(R^3)$.

PROOF. Using a Fourier transform method we have:

(147) $(-\xi^2 + k^2)\hat{u} = -\hat{f}$ in R^3.

Since $(-\xi^2+k^2) \neq 0$ for all $\xi \in R^3$, with $(-\xi^2+k^2)^{-1} \to 0$ when $|\xi| \to \infty$, we can divide by $(-\xi^2+k^2)$, thus obtaining:

(148) $\hat{u} = -(-\xi^2+k^2)^{-1}\hat{f} \in \mathcal{S}'(R^3)$

with $\hat{u} \in L^2(R^3)$ when $\hat{f} \in L^2(R^3)$. The inverse Fourier transform is the solution u of (1) in $\mathcal{S}'(R^3)$; for f in $L^2(R^3)$, we write:

(149) $u = R_\lambda f$, with $R_\lambda = (-\Delta - \lambda)^{-1}$, $\lambda = k^2$;

R_λ is the resolvent of the positive selfadjoint operator $(-\Delta)$ since λ is not in the spectrum of $(-\Delta)$. When f is in $\mathcal{S}'(R^3)$, we also have by convolution

(150) $u = \Phi_k * f$, with $\Phi_k = -\mathcal{F}^{-1}(-\xi^2+k^2)^{-1}$, $\Phi_k(x) = \frac{e^{ikr}}{4\pi r}$, $r = |x|$;

Φ_k is the tempered elementary solution of (1). With Im k > 0, it is exponentially decreasing with (high) r; then for f in $\mathcal{E}'(R^3)$, u satisfies, with η = Im k:

(151) $u(x) = O\left(\frac{e^{-\eta r}}{4\pi r}\right), \quad \left(\frac{\partial u}{\partial n} - iku\right)(x) = o\left(\frac{e^{-\eta r}}{4\pi r}\right), \quad \text{when } r \to \infty.$

There are no conditions at infinity to impose on u: Sommerfeld conditions appear when λ tends to a real number with Im λ positive or negative. This is easily seen on the elementary solution:

Lemma 2. *The elementary solution of the Helmholtz equation satisfies*

$$(152) \quad \left| \begin{array}{l} \Phi_k \to \Phi_{k_0} \text{ in } \mathcal{D}'(R^3) \text{ or } \mathcal{S}'(R^3) \textit{ when } \lambda = k^2 \to k_0^2 \textit{ with } \operatorname{Im}\lambda > 0, \\ \Phi_k \to \Phi_{k_0}^{in} = \Phi_{-k_0} \text{ in } \mathcal{D}'(R^3) \text{ or } \mathcal{S}'(R^3) \textit{ when } \lambda = k^2 \to k_0^2 \textit{ with } \operatorname{Im}\lambda < 0. \end{array} \right.$$

Then using the continuity properties of the convolution product (see Schwartz [1] p. 157 and 247) we obtain:

Proposition 20. *Limiting Absorption Principle in* R^3. *When the complex number* k *tends to a real value* k_0, *with* Im k^2 *positive (resp. negative), the solution* u_k *of the Helmholtz equation* (1) (*with* f *a distribution with compact support*) *tends to the solution of the same equation for* k_0 *with the outgoing (resp. incoming) Sommerfeld condition in* $S'(R^3)$; *thus*

(153) $u_k \overset{\text{def}}{=} \Phi_k * f \to u_{k_0} \overset{\text{def}}{=} \Phi_{k_0} * f \text{ in } \mathcal{S}'(R^3) \text{ when } k \to k_0 \text{ with } \operatorname{Im} k > 0.$

When f *is in* $L^2(R^3)$ *with compact support*, (u_k) *converges in* $H^2_{loc}(R^3)$.

Obviously we cannot have $R_\lambda f \to R_{\lambda_0} f$ in $L^2(R^3)$ for all $f \in L^2(R^3)$ when $\lambda_0 = k_0^2$ is in $\sigma(-\Delta)$, the spectrum of operator $(-\Delta)$.

PROOF of Lemma 2. As an exercise, we prove (152) by a Fourier transform method, to specify the Fourier transform of the elementary solutions. For all ϕ in $\mathcal{S}(R^3)$, we have:

(154) $<\hat{\Phi}_k, \phi> = \int_{R^3} \frac{1}{\xi^2 - k^2}\, \phi(\xi)\, d\xi = \int_{R^+ \times S^2} \frac{1}{\rho^2 - k^2}\, \phi(\rho\alpha)\, \rho^2\, d\rho\, d\alpha,$

with $\rho = |\xi|$, Im $k \neq 0$. Let $\theta(\rho)$ be defined on R by:

(155) $\theta(\rho) = \rho^2 M\phi(|\rho|), \quad \text{with } M\phi(\rho) = \int_{S^2} \phi(\rho\alpha)\, d\alpha, \quad \rho > 0.$

Then:

(156) $<\hat{\Phi}_k, \phi> = \frac{1}{2k} \int_R \frac{\theta(\rho)}{\rho - k}\, d\rho.$

The limit for $k \to k_0$ with Im $k > 0$ or Im $k < 0$ is given by the Plemelj formulas with respect to the real axis in C (see Dautray-Lions [1] chap. 11A.2; we could also use the variable $t = \rho^2$, and the half-line R^+ instead of R). When $k_0 \neq 0$, we have

(157) $\lim\limits_{k=k_0+i\varepsilon,\ \varepsilon\to 0,\ \varepsilon>0} <\hat{\Phi}_k,\phi> = <\hat{\Phi}_{k_0},\phi> = \frac{1}{2k_0}\,\mathrm{pv}\int_R \frac{\theta(\rho)}{\rho-k_0}\,d\rho + \frac{1}{2k_0}\, i\pi\theta\,(k_0),$

and

(158) $\lim\limits_{k=k_0-i\varepsilon,\ \varepsilon\to 0,\ \varepsilon>0} <\hat{\Phi}_k,\phi> = <\hat{\Phi}^{E}_{k_0},\phi> = \frac{1}{2k_0}\,\mathrm{pv}\int_R \frac{\theta(\rho)}{\rho-k_0}\,d\rho - \frac{1}{2k_0}\, i\pi\theta\,(k_0).$

Thus

(159) $<\hat{\Phi}_{k_o},\phi> = <\mathrm{pv}\frac{1}{\xi^2-k_o^2},\phi> + i\pi <\delta(\xi^2-k_0^2),\phi>,$

and

(159)' $\hat{\Phi}_{k_0}(\xi) = \mathrm{pv}\frac{1}{\xi^2-k_0^2} + i\pi\,\delta(\xi^2-k_0^2).$

The incoming elementary solution $\Phi^{in}_{k_o}$ is given by:

(160) $\hat{\Phi}^{in}_{k_o}(\xi) = \mathrm{pv}\frac{1}{\xi^2-k_0^2} - i\pi\,\delta(\xi^2-k_0^2).$ ⊗

PROOF of Proposition 20. Lemma 2 implies (153), and thanks to (158)', the Fourier transform of u_{k_0} is:

(161) $\hat{u}_{k_0}(\xi) = \hat{\Phi}_{k_0}(\xi)\,\hat{f}(\xi) = \mathrm{pv}\frac{1}{\xi^2-k_0^2}\,\hat{f}(\xi) + i\pi\,\delta(\xi^2-k_0^2)\,\hat{f}(\xi).$

Note that this implies when $f\in L^2(R^3)$ with bounded support:

(162)
$$\left|\begin{aligned} &\mathrm{Re}\,(u_{k_0},f) = (2\pi)^{-3}\,\mathrm{Re}\,(\hat{u}_{k_0},\hat{f}) = (2\pi)^{-3}\,\mathrm{pv}\int_{R^3}\frac{1}{\xi^2-k_0^2}\,|\hat{f}(\xi)|^2\,d\xi,\\ &\mathrm{Im}\,(u_{k_0},f) = (2\pi)^{-3}\,\mathrm{Im}\,(\hat{u}_{k_0},\hat{f}) = (2\pi)^{-3}\,\frac{\pi}{2k_0}\int_{S^2}|\hat{f}(k_0\alpha)|^2\,d\alpha > 0.\end{aligned}\right.$$

Now we prove the convergence of (u_k) in $H^2_{loc}(R^3)$. Let B_a be a ball which contains the support of f. We have:

(163) $(u_k - u_{k_0})(x) = \int_{B_a}\frac{e^{ikr}-e^{ik_0r}}{4\pi r}\,f(x')\,dx', \quad r=|x-x'|,$

For $k=k_0+i\eta$, with $\eta>0$, we have: $\left|\frac{e^{ikr}-e^{ik_0r}}{r}\right|^2 \le \left|\frac{e^{-\eta r}-1}{r}\right| \le \eta$, thus:

(164) $|(u_k-u_{k_0})(x)|^2 \le C\int_{B_a}\left|\frac{e^{ikr}-e^{ik_0r}}{4\pi r}\right|^2 |f(x')|^2\,dx' \le C\,\eta^2\int_{B_a}|f(x')|^2\,dx',$

hence $u_k \to u_{k_0}$ in $L^\infty(R^3)$, and in $L^2_{loc}(R^3)$. Thus: $\Delta(u_k-u_{k_0})$ converges to 0 in $L^\infty(R^3)$, hence in $L^2_{loc}(R^3)$, since $\Delta(u_k-u_{k_0}) = -k^2(u_k-u_{k_0}) + (k_0^2-k^2)\,u_{k_0}$. Furthermore:

(165) $\mathrm{grad}\,(u_k-u_{k_0}) = \int_{B_a}\alpha\frac{\phi(r)}{4\pi}\,f(x')\,dx',$

with: $\alpha = \frac{x - x'}{r}$, and $\phi(r) = \frac{e^{ikr}}{r}(ik - \frac{1}{r}) - \frac{e^{ik_0 r}}{r}(ik_0 - \frac{1}{r})$. But we have

(166) $$|\phi(r)| \leq [\frac{1 - e^{-\eta r} - \eta r e^{-\eta r}}{r^2} - ik_0 \frac{1 - e^{-\eta r}}{r}] = \eta\, O(1),$$

thus grad $(u_k - u_{k_0}) \rightarrow 0$ in $L^\infty(R^3)$, hence in $L^2_{loc}(R^3)$.

⊗

2.2. Maxwell equations with complex coefficients

Proposition 21. *Let* ω *(or* ε*, or* μ*) be complex, with* Im $k^2 \neq 0$. *Then the Maxwell equations* (22) *with* J *and* M *given in* $E'(R^3)^3$ *or in* $S'(R^3)^3$ *have a unique solution* (E,H) *with* E, H *in* $S'(R^3)^3$. *For* M, J *in* $L^2(R^3)^3$ *(resp. in* H(div,R^3)) *we have* E, H *in* H(curl,R^3) *(resp. in* $H^1(R^3)^3$). *When* ω *tends to a real* ω_0 *with* Im $\omega > 0$ *(resp.* < 0), *and* M, J *are in* $L^2(R^3)^3$ *with compact support, the solution* $(E,H) = (E_\omega, H_\omega)$ *of* (22) *converges in* $S'(R^3)^3 \times S'(R^3)^3$ *to the outgoing (resp. incoming) solution* (E_0, H_0) *of* (22) *for* ω_0 *and* E_0, H_0 *are in* $H_{loc}(\mathrm{rot}, R^3)$.

PROOF. There are two ways to prove this proposition either by a Fourier transform method, or by convolution like in section 1.2.1 reducing to the Helmholtz equation (see (28)') with j and m in $E'(R^3)^3$ (or in $H^{-2}(R^3)^3$).

⊗

2.3. The Calderon projectors for a complex wavenumber k

We consider the "standard" problems (61) in the Helmholtz case, and (86)' in the Maxwell case with charges concentrated on a regular surface Γ as in 1.3.1.

Proposition 22. *Let* ω *(or* ε*, or* μ*) be complex with* Im $k^2 \neq 0$. *Then Proposition 3 in the Helmholtz case and Theorem 3 in the Maxwell case are also valid without condition at infinity if we replace the space* X *by the space*:

(167)
i) *Helmholtz case:* $\tilde{X} = \{u \in D'(R^3),\ u|_\Omega \in H^1(\Delta,\Omega),\ u|_{\Omega'} \in H^1(\Delta,\Omega')\}$

ii) *Maxwell case:* $\tilde{X} = \tilde{X}_0 \times \tilde{X}_0$, with:
$\tilde{X}_0 = \{u \in D'(R^3)^3,\ u|_\Omega \in H(\mathrm{curl},\Omega),\ u|_{\Omega'} \in H(\mathrm{curl},\Omega'),$
$\mathrm{div}\, u|_\Omega \in L^2(\Omega),\ \mathrm{div}\, u|_{\Omega'} \in L^2(\Omega')\}$.

Then this allows us to define the Calderon projectors P_i and P_e like in (65) and in (93) and thus the operators S (see (21) and (102)) with the same properties. Furthermore we have for instance in the Helmholtz case as a consequence of the limiting absorption principle:

Corollary 3. *When the complex number k tends to a real value* k_0, *with* $\operatorname{Im} k^2$ *positive (resp negative), then*

$$(168)\quad S_k\begin{pmatrix}\rho\\ \rho'\end{pmatrix} \rightarrow S_{k_o}\begin{pmatrix}\rho\\ \rho'\end{pmatrix}\ (\text{resp. } S^{in}_{k_o}\begin{pmatrix}\rho\\ \rho'\end{pmatrix})\ \text{in } H^{1/2}(\Gamma)\times H^{-1/2}(\Gamma),$$

and thus the corresponding Calderon projectors and the integral operators K_k, L_k, R_k *and* J_k *strongly converge in their natural functional spaces.*

PROOF. Let $u = u_k$ be the solution of (61) with k a complex number with $\operatorname{Im} k > 0$. We use a lifting U of the jump conditions (63), with U in $H^1(\Delta,\Omega)$ and in $H^1(\Delta,\Omega')$ with compact support. Then $v_k = u - U$ satisfies:

$$(169)\quad \Delta v_k + k^2 v_k = F, \qquad \text{with } F \in L^2(R^3)\cap \mathcal{E}'(R^3).$$

Then $v_k = R_{k^2}F$ converges to a limit denoted by $R^+_{k_0^2}F$ thanks to Proposition 20.

⊗

3. VECTOR HELMHOLTZ EQUATION, KNAUFF-KRESS CONDITIONS

Using the (vector) Laplace operator Δ (as in chap. 2 section 11.2) instead of the curl curl operator in electromagnetism leads to some differences in the scattering theory by integral methods. We again assume that k is a real positive wavenumber.

3.1. Knauff-Kress conditions at infinity

Let f be a given vector distribution with compact support in R^3; then let u be the unique (vector) solution of the Helmholtz problem (13) with Sommerfeld conditions. Its behavior at infinity is given by (15); then we have:

$$(170)\qquad \operatorname{curl} u = \Phi * \operatorname{curl} f, \qquad \operatorname{div} u = \Phi * \operatorname{div} f,$$

and thus their behaviors at infinity are given by:

$$(171)\qquad (\operatorname{curl} u)(r\alpha) \approx \Phi(r).\, ik\alpha\wedge \hat{f}(k\alpha), \quad (\operatorname{div} u)(r\alpha) \approx \Phi(r)\, ik\alpha.\hat{f}(k\alpha);$$

then we obtain:

$$(172)\quad (\alpha \operatorname{div} u - \alpha\wedge\operatorname{curl} u - iku)(r\alpha) \approx \Phi(r)\, ik\,[-\hat{f} + \alpha(\alpha.\hat{f}) - \alpha\wedge(\alpha\wedge\hat{f})] = 0.$$

Thus u = E (or H) must satisfy the following, uniformly in α:

$$(173)\qquad (\alpha \operatorname{div} u - \alpha\wedge\operatorname{curl} u - iku)(r\alpha) = o(\tfrac{1}{r}) \quad \text{when } r \rightarrow \infty.$$

These conditions are called *Knauff-Kress conditions at infinity*.

3.2. Vector Helmholtz problems with jumps conditions

Let u be a solution of the (homogeneous) vector Helmholtz equations in Ω (a bounded open set in R^3) and its complement Ω'. We suppose that u satisfies

(174) $$u|_{\Omega} \in H^1(\Omega)^3, \quad u|_{\Omega'} \in H^1_{loc}(\bar{\Omega}')^3,$$

with the Knauff-Kress conditions at infinity.

Recalling the jumps formulas (53), (54), and then applying the curl and div operators to these formulas we obtain, with the same notations:

(175) $$\begin{cases} \text{curl curl } u = (\text{curl curl } u) - [n \wedge \text{curl } u]_\Gamma \delta_\Gamma - \text{curl }([n \wedge u]_\Gamma \delta_\Gamma), \\ \text{grad div } u = (\text{grad div } u) - [\text{div } u]_\Gamma \, n\delta_\Gamma - \text{grad }([n.u]_\Gamma \delta_\Gamma). \end{cases}$$

With the usual relation: curl curl u = $-\Delta u$ + grad div u, we obtain that u satisfies in $D'(R^3)^3$:

(176) $$\begin{cases} \Delta u + k^2 u = g\,, \\ g = [n \wedge \text{curl } u]_\Gamma \, \delta_\Gamma - [\text{div } u]_\Gamma \, n\delta_\Gamma + \text{curl }([n \wedge u]_\Gamma \, \delta_\Gamma) - \text{grad }([n.u]_\Gamma \, \delta_\Gamma). \end{cases}$$

Then u is of the following form:

(177) $$u = -\,\Phi * g.$$

This formula is analogous to the Kirchhoff formula (62) or the Stratton-Chu formula (88). We give some mathematical frameworks for these jumps:

i) *In the "regular" case* (based on the H^2 regularity):

(178) $$[n \wedge u]_\Gamma \in H_t^{3/2}(\Gamma),\ [n.u]_\Gamma \in H^{3/2}(\Gamma),\ \ [n \wedge \text{curl } u]_\Gamma \in H_t^{1/2}(\Gamma),\ [\text{div } u]_\Gamma \in H^{1/2}(\Gamma).$$

ii) *In the "variational" case* (based on the H^1 regularity):

(179) $$[n \wedge u]_\Gamma \in H_t^{1/2}(\Gamma),\ [n.u]_\Gamma \in H^{1/2}(\Gamma),\ \ [n \wedge \text{curl } u]_\Gamma \in H_t^{-1/2}(\Gamma),\ [\text{div } u]_\Gamma \in H^{-1/2}(\Gamma).$$

Note that the space $H_t^{1/2}(\Gamma)$ and its dual space $H_t^{-1/2}(\Gamma)$ are so that:

$$H_t^{1/2}(\Gamma) = H^{-1/2}(\text{div},\Gamma) \cap H^{-1/2}(\text{curl},\Gamma), \quad H_t^{-1/2}(\Gamma) = H^{-1/2}(\text{div},\Gamma) + H^{-1/2}(\text{curl},\Gamma).$$

Of course we can develop the integral method in these frameworks and define Calderon projectors. We can also consider complex wavenumbers k. Then we have the analogue of Proposition 19 in a domain Ω.

4. BOUNDARY PROBLEMS WITH REAL WAVENUMBER k

In chap. 2, section 12, we first studied some boundary problems which are "coercive" or "elliptic", that is, they may be solved thanks to a variational method and to Lax-Milgram lemma. This is possible for pure imaginary wavenumbers k and more generally for complex wavenumbers, but the case of real wavenumbers is more difficult. We can study it from various methods; here we develop the method based on the limiting absorption principle.

4.1. Limiting absorption principle

We first consider "exterior" problems, i.e., in the complement Ω' of a regular bounded set Ω in R^3. We assume that Ω' is a *connected* set. Either in the (scalar) Helmholtz case with the Neumann (or also Dirichlet) condition or in the (vector) Helmholtz case, these problems can be written as:

$$(180) \qquad Au - \lambda u = f,$$

where A is a positive selfadjoint operator with a continuous spectrum only and λ is in the spectrum of A. We specify in the scalar case, then in the vector case, with the definition (3) chap. 2:

$$(180)' \qquad \left| \begin{array}{l} \text{i) } A = -\Delta, \text{ with } D(\Delta) = \{u \in H^1(\Delta,\Omega'),\ u|_\Gamma = 0 \text{ or } \frac{\partial u}{\partial n}|_\Gamma = 0\}, \\ \text{ii) } A = A^- \text{ defined in (147) chap. 2.} \end{array} \right.$$

When f is a square integrable function with bounded support we have the following result, for which we refer to Wilcox [1], Bendali [2], Hörmander [2].

Lemma 3. *Let* R_0, R *be real positive numbers* $R_0 < R$ *with* Ω *contained in the ball* B_{R0}. *Let* $0 < a < b$, $\sigma > 0$, $J = [a,b] \times (0,\sigma]$. *Then there exists* $M = M(J, R_0, R)$, $M > 0$, *so that:*

$$(181) \quad \| R(z)f \|_{H^1(\Delta,\Omega'_R)} \le M \| f \|_{L^2(\Omega')}, \quad \forall f \in L^2(\Omega'),\ \text{supp}\, f \subseteq \bar{\Omega}'_{R_0},\ \forall z \in J,$$

with $R(z) = (A - zI)^{-1}$ *the resolvent of* A, and with: $\Omega'_R = \Omega' \cap B_R$.

Moreover the mapping: $z \in J \to R(z)f \in H^1_{loc}(\Delta, \bar{\Omega}')$ *is uniformly continuous.*

Theorem 4. *Limiting absorption principle (scalar case). The following problem: find* u in $H^1_{loc}(\Delta,\bar{\Omega}')$, *satisfying, with* k *real and given f with compact support:*

$$(182) \qquad \left| \begin{array}{l} \text{i) } \Delta u + k^2 u = -f \ \text{ in } \Omega' = R^3 \backslash \bar{\Omega},\ \ f \in L^2(\Omega') \cap \mathcal{E}'(R^3) \\ \text{ii) } u|_\Gamma = 0 \ \text{ (resp. } \frac{\partial u}{\partial n}|_\Gamma = 0\text{)}, \\ \text{iii) \textit{the outgoing Sommerfeld condition,}} \end{array} \right.$$

has a unique solution u^+*, given thanks to the resolvent* $R(z)$ *of* $A = -\Delta$ *by:*

(183) $$u^+ = R_k^+ f = \lim_{z = k^2 + i\sigma,\ \sigma > 0,\ \sigma \to 0} R(z) f \text{ in } H^1_{loc}(\Delta, \bar{\Omega}').$$

Obviously the theorem is true if we substitute incoming for outgoing and $-\sigma$ for σ in (183). In the vector case we have:

Theorem 4'. *Limiting absorption principle, vector case. The following problem:*

find u in $H^1_{loc}(\Delta, \bar{\Omega}')^3$, *solution of the Helmholtz problem with* k *real and given* f

(182)' $$\left| \begin{array}{l} \text{i) } \Delta u + k^2 u = -f \text{ in } \Omega', \quad f \in L^2(\Omega')^3 \cap E'(R^3)^3, \\ \text{ii) } n \wedge u|_\Gamma = 0, \quad (\operatorname{div} u)|_\Gamma = 0, \\ \text{iii) } u \textit{ satisfies the Knauff-Kress conditions } (173), \end{array} \right.$$

has a unique solution u^+ *which is given thanks to the resolvent* $R(z)$ *of* $A = A^-$ *by*

(183)' $$u^+ = \lim_{z = k^2 + i\sigma,\ \sigma > 0,\ \sigma \to 0} R(z) f \quad \text{in } H^1_{loc}(\Delta, \bar{\Omega}')^3.$$

PROOF. i) We first prove uniqueness in the scalar case. Using the Green formula on the domain $\Omega'_R = \Omega' \cap B_R$, we have (with $f = 0$)

(184) $$\operatorname{Re}\left(-ik \int_{\Omega'_R} (|\operatorname{grad} u|^2 - k^2 |u|^2)\, dx\right) = \operatorname{Re}\left(-ik \int_\Gamma \frac{\partial u}{\partial n} \bar{u}\, d\Gamma\right) - \operatorname{Re}\left(-ik \int_{S_R} \frac{\partial u}{\partial n} \bar{u}\, dS_R\right) = 0.$$

Then using the Sommerfeld condition we have

(185) $$\operatorname{Re}\left(-ik \int_{S_R} \frac{\partial u}{\partial n} \bar{u}\, dS_R\right) = k^2 \int_{S_R} |u|^2\, dS_R + o(1) = 0.$$

We conclude using the Rellich lemma.
ii) In the vector Helmholtz case we can develop a similar proof, using the Green formula with:

(186) $$\operatorname{Re}\left(-ik \int_{\Omega'_R} (|\operatorname{curl} u|^2 + |\operatorname{div} u|^2 - k^2 |u|^2)\, dx\right) = 0,$$

and then using the Knauff-Kress conditions and the Rellich Lemma.
iii) The existence of the solution follows easily from lemma 3: $u_z = R(z)f$ tends to a limit u^+ when $z \to k^2$; u_z satisfies all the properties of the theorem but the conditions at infinity. These conditions are deduced from the Kirchhoff formula (obtained from (62) and (177)) which gives the solution u_z thanks to its traces on a sphere S_R (successively in the scalar then in the vector case):

$$(187)\quad \left|\begin{array}{l} \text{i) } u(x) = \int_{S_R} \left(u(y)\dfrac{\partial \Phi}{\partial n_y}(x-y) - \dfrac{\partial u}{\partial n_y}\Phi(x-y)\right) d\Gamma_y, \\ \text{ii) } u(x) = -\left\{\int_{S_R} (n \wedge \operatorname{curl} u + n \operatorname{div} u)_y \, \Phi(x-y)\, dS_R(y) \right. \\ \qquad \left. + \operatorname{curl} \int_{S_R} n \wedge u_y \, \Phi(x-y)\, dS_R(y) + \operatorname{grad} \int_{S_R} n.u_y \, \Phi(x-y)\, dS_R(y)\right\}, \end{array}\right.$$

with $u = u_z$ and $\Phi = \Phi_z$ (in order to simplify). Passage to the limit in these formulas when $z \to k^2$, gives the corresponding (Kirchhoff) formulas for k^2 that is u^+ satisfies the required condition at infinity.

⊗

4.2. Exterior boundary problems

Then we can solve some "usual" exterior boundary problems with inhomogeneous boundary conditions:

Theorem 5. *Exterior scalar Helmholtz problem. The following problem: find* u in $H^1_{loc}(\Delta, \bar{\Omega}')$ *satisfying the Helmholtz problem with* k *real and given* u_0, or u_1

$$(188)\quad \left|\begin{array}{l} \text{i) } \Delta u + k^2 u = 0 \text{ in } \Omega', \\ \text{ii) } u|_\Gamma = u_o \in H^{1/2}(\Gamma) \ (\text{resp. } \dfrac{\partial u}{\partial n}|_\Gamma = u_1 \in H^{-1/2}(\Gamma)), \\ \text{iii) } \textit{the outgoing Sommerfeld condition,} \end{array}\right.$$

has a unique solution.

PROOF. Uniqueness is proved in Theorem 4. In order to prove existence of the solution, we use a lifting to the boundary conditions obtained by the (variational) solution U in $H^1(\Omega')$ of the problem (188) when $k = i$ (for instance).

Let ζ be in $D(R^3)$ with $\zeta(x) = 1$ in a neighborhood of Γ. Let f be defined by: $f = [2 \operatorname{grad} \zeta . \operatorname{grad} U + U(\Delta\zeta + (1 + k^2)\zeta)]$, then f is in $L^2(\Omega')$ with compact support, Thus $v = u - \zeta U$ is in $H^1_{loc}(\Delta, \bar{\Omega}')$, and satisfy:

$$(189)\quad \left|\begin{array}{l} \text{i) } \Delta v + k^2 v = -f, \\ \text{ii) } v|_\Gamma = 0 \quad (\text{resp. } \dfrac{\partial v}{\partial n}|_\Gamma = 0), \\ \text{iii) the outgoing Sommerfeld condition.} \end{array}\right.$$

Then the limiting absorption principle gives the result.

⊗

Theorem 5'. *Exterior vector Helmholtz problem. The following problem: find* u in $H^1_{loc}(\Delta, \bar{\Omega}')^3$ *satisfying the Helmholtz equation with* k *real and with given* (m,g)

$$(190)\quad \begin{cases} \text{i) } \Delta u + k^2 u = 0 \quad \text{in } \Omega', \\ \text{ii) } n \wedge u|_\Gamma = m,\ m \in H_t^{1/2}(\Gamma),\ (\operatorname{div} u)|_\Gamma = g,\ g \in H^{-1/2}(\Gamma), \\ \text{iii) } \textit{the Knauff-Kress conditions } (173), \end{cases}$$

has a unique solution.

PROOF. This is similar to the scalar case: uniqueness is proved in Theorem 4'. Existence is proved thanks to a lifting of the boundary conditions, which is the (variational) solution of (174) chap. 2 and then applying Theorem 4'.

⊗

Then we can solve exterior Maxwell problem in the free space:

Theorem 6. *Exterior Maxwell problem. Let* ε, μ *be positive real numbers. The problem: find* (E,H) *satisfying, with a given* m

$$(191)\quad \begin{cases} \text{i) } \operatorname{curl} H + i\omega\varepsilon E = 0, \quad \text{ii) } -\operatorname{curl} E + i\omega\mu H = 0 \text{ in } \Omega', \\ \text{iii) } n \wedge E|_\Gamma = m,\quad m \in H^{-1/2}(\operatorname{div},\Gamma), \\ \text{iv) } \textit{the Silver-Müller conditions } (43) \textit{ at infinity} \\ \text{v) } (E,H) \in H_{loc}(\operatorname{curl},\bar{\Omega}') \times H_{loc}(\operatorname{curl},\bar{\Omega}'), \end{cases}$$

has a unique solution.

PROOF. i) We first prove uniqueness directly for zero boundary conditions. Using the Green formula on Ω'_R, like in (48), with the Maxwell equations, we have:

$$(192)\quad \begin{aligned} \operatorname{Re}\Big(\int_{\Omega'_R} \operatorname{div}(E \wedge \bar{H})\, dx\Big) &= \operatorname{Re}\Big(\int_{S_R} n.\, E \wedge \bar{H}\, dS_R - \int_\Gamma n.\, E \wedge \bar{H}\, d\Gamma\Big) \\ &= \operatorname{Re}\Big(\int_{\Omega'_R} [i\omega\mu\, |H|^2 - i\omega\bar{\varepsilon}\, |E|^2]\, dx\Big) = 0. \end{aligned}$$

Since we have $\int_\Gamma n.\, E \wedge \bar{H}\, d\Gamma = 0$, then $\int_{S_R} n.\, E \wedge \bar{H}\, dS_R = 0$. Thus using the Silver-Müller conditions (43) we obtain:

$$(193)\quad \int_{S_R} |\pi_\Gamma H|^2\, dS_R = \int_{S_R} |\pi_\Gamma E|^2\, dS_R = 0.$$

Since we also have (50), then: $\int_{S_R} |H|^2\, dS_R = \int_{S_R} |E|^2\, dS_R = 0$. We can apply Rellich lemma and we have $E = H = 0$.

ii) The existence of the solution is proved thanks to a lifting of the boundary condition. Using the solution (u,v) of problem (151)' chap. 2 with $\kappa = 1$, and a regular function ζ with compact support equal to 1 in a neighborhood of Γ, we define:

(194) $\quad \tilde{E} = E - \zeta u, \ \tilde{H} = H - \zeta v.$

Let (J,M) be:

$$J = (-i\omega\varepsilon + 1)\zeta u - (\mathrm{grad}\,\zeta)\wedge v, \qquad M = (-i\omega\mu + 1)\zeta v + (\mathrm{grad}\,\zeta)\wedge u.$$

Then $(\tilde{E}, \tilde{H})$ must satisfy:

(195)
$$\left|\begin{array}{l} \text{i) } -\mathrm{curl}\,\tilde{E} + i\omega\mu\,\tilde{H} = M, \quad \text{ii) } \mathrm{curl}\,\tilde{H} + i\omega\varepsilon\,\tilde{E} = J \ \text{ in } \Omega', \\ \text{iii) } n\wedge\tilde{E}|_{\Gamma} = 0, \\ \text{iv) the Silver-Müller conditions (43) at infinity,} \\ \text{v) } \tilde{E} \text{ and } \tilde{H} \in H_{loc}(\mathrm{curl}, \bar{\Omega}'). \end{array}\right.$$

We note that M and J are in H(curl,Ω') with bounded supports. Furthermore, if Γ is smooth (C^∞):

(196) $\quad \mathrm{div}\,J$ and $\mathrm{div}\,M \in H^s(\Omega'), \quad \forall s \in R;$

$\tilde{E}$ must satisfy the vector Helmholtz problem, with f given in $L^2(\Omega')^3 \cap E'(R^3)^3$:

(197)
$$\left|\begin{array}{l} \text{i) } \Delta\tilde{E} + k^2\tilde{E} = -f \ \text{ in } \Omega', \\ \text{ii) } n\wedge\tilde{E}|_{\Gamma} = 0, \quad \mathrm{div}\,\tilde{E}|_{\Gamma} = 0, \\ \text{iii) the Knauff-Kress conditions (173)} \\ \text{iv) } \tilde{E} \in H_{loc}(\mathrm{curl}, \bar{\Omega}'). \end{array}\right.$$

Since $\mathrm{div}\,\tilde{E} = \frac{1}{i\omega\varepsilon}\mathrm{div}\,J \in L^2(\Omega')$ with the boundary condition ii), then $\tilde{E} \in H^1_{loc}(\bar{\Omega}')^3$. Thus we can apply the limiting absorption principle (theorem 4'), which proves the existence of the solution of the Maxwell problem. ⊗

Remark 10. *An application to scattering*. Solving the boundary value problem (191) amounts to solve the following scattering problem: given an incident electromagnetic wave (E_I, H_I) on a (regular) bounded obstacle Ω which we assume to be a perfect conductor, find the reflected wave (E_r, H_r).

Since the total electromagnetic field (E_t, H_t) (the sum of (E_I, H_I) and (E_r, H_r)) must satisfy, on the boundary Γ of Ω: $n\wedge E_t|_{\Gamma} = 0$, then the reflected wave (E_r, H_r) must be a solution of problem (191), with:

$$n\wedge E_r|_{\Gamma} = m = -n\wedge E_I|_{\Gamma} \text{ given.}$$

We give below (see 4.8) another method (a surface integral method) to compute this electromagnetic field. ⊗

4.3. Some consequences; the exterior Calderon operator

We note that in the Maxwell problem (191), the magnetic field H also satisfies

(198) $\quad n\wedge H|_{\Gamma} \in H^{-1/2}(\mathrm{div}, \Gamma).$

Definition 4. *The exterior (outgoing) Calderon operator. The mapping:*

(199) C^e (or $C^{e\omega}_{\varepsilon,\mu}$): $m = n \wedge E|_\Gamma \rightarrow n \wedge H|_\Gamma$

with (E,H) *the solution of* (191), *is called the exterior (outgoing) Calderon operator.*

We also denote by C^e_k the mapping $m = n \wedge E|_\Gamma \rightarrow \frac{1}{ik}(n \wedge \text{curl}\, E|_\Gamma)$, thus $C^e_k = Z\, C^e$ does not depend on Z the impedance of the (exterior) medium, see (34), but only on the wavenumber k.
There are many properties of these Calderon operators. First we have:

Proposition 23. *The exterior Calderon operator is an isomorphism in the space* $H^{-1/2}(\text{div},\Gamma)$, *with:*

(200) $(C^e)^2 = -\frac{1}{Z^2} I, \ (C^e_k)^2 = -I, \quad \text{and}\ (C^{e\omega}_{\varepsilon,\mu})^{-1} = -\ C^{e\omega}_{\mu,\varepsilon}.$

Furthermore the exterior Calderon operator is an isomorphism in $H^s(\text{div},\Gamma)$ *for all real* s *if* Γ *has a* C^∞ *regularity.*

PROOF. First we note that if we exchange ε and μ in the problem (191), then the solution (E,H) of (191) is changed into (H, − E); this easily gives (200). Then we prove that C^e is an isomorphism in $H^{1/2}(\text{div},\Gamma)$, and the regularity result:

If $m \in H^{1/2}(\text{div},\Gamma)$, then the solution (E,H) of (191) is in $H^1_{loc}(\bar{\Omega}')^3 \times H^1_{loc}(\bar{\Omega}')^3$.

This is a consequence of Theorem 6 chap. 2, first applied to ζE with $\zeta \in D(R^3)$, $\zeta = 1$ in a neighborhood of Γ (since $n \wedge E|_\Gamma$ is in $H^{1/2}_t(\Gamma)$), then to H using (28), (30) chap. 2

$$i\omega\mu\, n.H|_\Gamma = n.\text{curl}\, E|_\Gamma = \text{curl}_\Gamma\, \pi_\Gamma E = -\,\text{div}_\Gamma (n \wedge E|_\Gamma) \in H^{1/2}(\Gamma).$$

Then we have the regularity result for all half-integer s thanks to Proposition 7 chap. 2 and then for all real s by interpolation and duality.

⊗

Proposition 24. *Positivity properties. The exterior Calderon operator satisfies:*

(201) $\text{Re} \int_\Gamma C^e m \,.\, (n \wedge \bar{m})\, d\Gamma > 0, \quad \forall\, m \in H^{-1/2}(\text{div},\Gamma), \quad m \neq 0.$

PROOF. We apply the Green formula like in (192). Then we have:

(202) $\text{Re} \int_\Gamma n.\, E \wedge \bar{H}\, d\Gamma = \text{Re} \int_{S_R} n.\, E \wedge \bar{H}\, dS_R > 0,$

from the Silver-Müller conditions (and the Rellich lemma). This gives (201) (thanks to Propositions 3 and 4 chap. 2). We can give a more precise result: thanks to the formulas (30), (33), we have

(202)' $\mathrm{Re}\int_\Gamma n.\,E\wedge\bar{H}\,d\Gamma = \frac{1}{(4\pi)^2 Z}\int_{S^2} |\hat{j}(k\alpha)|^2\,d\alpha$,

with

$\hat{j}(k\alpha) = ik\,[\alpha\wedge\hat{J}(k\alpha) + \frac{1}{Z}\Pi_\alpha\,\hat{M}(k\alpha)],\quad J = n\wedge H|_\Gamma = C^e(n\wedge E|_\Gamma),\quad M = -\,n\wedge E|_\Gamma\,.$

⊗

We can define, instead of (199), other Calderon operators such as: $m \longrightarrow \pi_\Gamma H$, which is an isomorphism from $H^{-1/2}(\mathrm{div},\Gamma)$ into $H^{-1/2}(\mathrm{curl},\Gamma)$, and also an exterior incoming Calderon operator $C^{e,in}$ using incoming instead of outgoing conditions in Definition 4. Then using for instance the antiduality pairing (144)' we have the "reciprocity formula":

(203) $\langle n\wedge E^1,\, C^{e,in}(n\wedge E^2)\rangle = -\,\langle C^e(n\wedge E^1), n\wedge E^2\rangle,\quad n\wedge E^{1,2}\in H^{-1/2}(\mathrm{div},\Gamma)$,

thus $C^e = C^{e,out} = -\,(C^{e,in})^*$, with $C^{e,in} = -\,C^e_{-k} = -\,\overline{C^e_k}$. With the duality pairing (144), we obtain: ${}^tC^e = -\,C^e$.

We can also define an exterior Calderon operator $C^e{}_k$ for complex k, with similar properties, and the limiting absorption principle gives:

(204) $$\lim_{k^2 = k_0{}^2 + i\sigma,\ \sigma>0,\ \sigma\to 0} C^e_k m = C^e_{k_0} m,\quad \forall m\in H^{-1/2}(\mathrm{div},\Gamma).$$

The *exterior Calderon operator* above defined in (199) has numerous properties and is a basic tool in many applications to the scattering in electromagnetism (see below).

4.4. Interior boundary problems

We first consider the inhomogeneous vector Helmholtz problem in a (*regular*) *bounded domain* Ω. Since the selfadjoint operator A^- defined in (147) chap. 2 has a compact resolvent (because the natural injection of the Sobolev space $H^1(\Omega)$ into $L^2(\Omega)$ is compact), we know that the following interior Helmholtz or Maxwell problems depend on the Fredholm alternative:

Theorem 7. *Interior vector Helmholtz problem. The following problem: given a function* f *in* $L^2(\Omega)^3$ *and a complex number* k, *find* u *in* $H^1(\Omega)^3$ *satisfying*

(205) $$\left|\begin{array}{l} \text{i) } \Delta u + k^2 u = -\,f \text{ in } \Omega, \\ \text{ii) } n\wedge u|_\Gamma = 0,\quad (\mathrm{div}\,u)|_\Gamma = 0, \end{array}\right.$$

depends on the Fredholm alternative: either k^2 *is not an eigenvalue of the selfadjoint operator* A^-, *then there exists a unique solution; or* k^2 *is an eigenvalue of the selfadjoint operator* A^-, *then there exists a solution* u *if and only if* f *is orthogonal to the corresponding eigenvector space, and then* u *is determined up to this eigenvector space.*

Theorem 7'. *Interior inhomogeneous vector Helmholtz problem. The following problem with given boundary conditions* (m,g): *find* u *in* $H^1(\Omega)^3$ *satisfying*:

$$(206)\qquad \left| \begin{array}{l} \text{i) } \Delta u + k^2 u = 0 \text{ in } \Omega, \\ \text{ii) } n \wedge u|_\Gamma = m, \quad (\text{div}\, u)|_\Gamma = g, \text{ with } m \in H_t^{1/2}(\Gamma),\ g \in H^{-1/2}(\Gamma), \end{array} \right.$$

depends on the Fredholm alternative: either k^2 *is not an eigenvalue of the selfadjoint operator* A^-, *then there exists a unique solution; or* k^2 *is an eigenvalue of the selfadjoint operator* A^-, *then there exists a solution u if and only if* (m, g) *satisfies*:

$$(207)\qquad \int_\Gamma (m.\text{curl}\, \bar{v} + g\, n.\bar{v})\, d\Gamma = 0,$$

for all eigenvectors v *relative to* k^2, *and then* u *is determined up to this eigenvector space.*

The proof is classical using the Green formula:

$$(208)\qquad -\int_\Gamma (\Delta u + k^2 u)\, \bar{v} - u\, (\Delta \bar{v} + k^2 \bar{v})\, dx = \int_\Gamma [(n \wedge \text{curl}\, u).\bar{v} - n \wedge u.\ \text{curl}\, \bar{v} \\ - \text{div}\, u\, n.\bar{v} + n.u\, \text{div}\, \bar{v}]\, d\Gamma.$$

Theorem 8. *Interior Maxwell problem. The problem: find the electromagnetic field* (E, H) *with finite energy, i.e.,* E, H *in* H(curl,Ω), *satisfying, with given* m

$$(209)\qquad \left| \begin{array}{l} \text{i) curl } H + i\omega\varepsilon E = 0, \qquad \text{ii) } -\text{curl } E + i\omega\mu H = 0 \text{ in } \Omega, \\ \text{iii) } n \wedge E|_\Gamma = m, \quad m \in H^{-1/2}(\text{div}, \Gamma), \end{array} \right.$$

depends on the Fredholm alternative: either $k^2 = \omega^2\varepsilon\mu$ *is not an eigenvalue of the selfadjoint operator* A^- *or of* A *(see* (134) chap. 2), *then there exists a unique solution; or* k^2 *is an eigenvalue of the selfadjoint operator* A^- (or A), *then there exists a solution* (E, H) *if and only if* m *satisfies*:

$$(210)\qquad \int_\Gamma (m.\text{curl}\, \bar{v})\, d\Gamma = 0,$$

for all eigenvectors v *relative to* k^2, *and then* E *is determined up to this eigenvector space.*

PROOF. Thanks to a lifting of the boundary condition, using the solution (u,v) of (151)' chap. 2, we can reduce this problem to an inhomogeneous Maxwell problem which is directly relevant of the Fredholm alternative, like for the exterior Maxwell problem (see (191)). We can also reduce (209) to the vector Helmholtz problem (205). ⊗

If ε or μ is not a real number (i.e. if the medium is dissipative), the problem (209) has a unique solution for all m as in (209)iii). We can also solve the interior Maxwell problem with some heterogeneous media with ε and μ matrix valued and dependent on x, with the hypotheses H17) chap.1 only. We can also solve problem (209) using a variational method on the unknown vector field H.

4.5. Some consequences; the interior Calderon operator

First we note that in the Maxwell problem (209), the magnetic field H also satisfies (198), then we can define in the "*regular*" case, (i.e., when $k^2 = \omega^2\varepsilon\mu$ is not an eigenvalue of the selfadjoint operator A^- or of A):

Definition 5. *The interior Calderon operator. The mapping*:

(211) $$C^i \text{ (or } C^{i\,\omega}_{\varepsilon,\mu}\text{)}: \ m = n \wedge E|_\Gamma \rightarrow n \wedge H|_\Gamma$$

with (E,H) *the solution of* (209), *is called the interior Calderon operator*.

We also denote by C^i_k the mapping $j = n \wedge E|_\Gamma \rightarrow \frac{1}{ik}(n \wedge \text{ curl } E|_\Gamma)$, thus $C^i_k = Z\, C^i$ does not depend on Z the impedance of the (interior) medium, see (34), but only on the wavenumber k, and satisfies $(C^i_k)^2 = -I$.

Note that it is possible to define interior Calderon operators when k^2 is an eigenvalue of the selfadjoint operator A, we only have to eliminate the corresponding eigenspace but with these technical difficulties, it is less useful.

There are also many properties of these Calderon operators. First, Proposition 23 is true for the interior Calderon operators in the regular case. Then, we also have positivity properties:

Proposition 25. *Positivity properties. The interior Calderon operator satisfies*

(212) $$-\text{Re} \int_\Gamma C^i m \,.\, (n \wedge \bar{m})\, d\Gamma \geq 0\,, \qquad \forall\, m \in H^{-1/2}(\text{div},\Gamma).$$

PROOF. This follows from the Green formula:

(212)' $$\text{Re} \int_\Gamma n.\, E \wedge \bar{H}\, d\Gamma = \int_\Omega [\text{Re}\,(-i\omega\bar{\varepsilon})E.\bar{E} + \text{Re}\,(i\omega\mu)H.\bar{H}\,]\, dx \leq 0.$$

If we have a conservative medium (the free space), then (212) is always zero. ⊗

From (201), we have the positivity property of the operator R (see (100)) for $k > 0$

$$-\text{Re}\int_\Gamma RM.n \wedge \bar{M} d\Gamma = 2k\int_{\Gamma\times\Gamma} \frac{\sin kr}{4\pi r}[M(x).\bar{M}(y) - \frac{1}{k^2}\,\text{div}_\Gamma M(x)\,\text{div}_\Gamma \bar{M}(y)]d\Gamma_x d\Gamma_y > 0.$$

For a dissipative medium ($\varepsilon'' = \text{Im}\,\varepsilon > 0$, $\mu'' > 0$), then with (209) we have

$$(213) \quad -\text{Re}\int_\Gamma C^i m \,.\, (n \wedge \bar{m})\, d\Gamma \geq C_o \|E\|^2_{H(\text{curl},\Omega)}, \quad \forall m \in H^{-1/2}(\text{div},\Gamma),$$

with C_o a constant, and thus with the quotient norm:

$$(214) \quad -\text{Re}\int_\Gamma C^i m \,.\, (n \wedge \bar{m})\, d\Gamma \geq C_o \|m\|^2_{H^{-1/2}(\text{div},\Gamma)}, \quad \forall m \in H^{-1/2}(\text{div},\Gamma).$$

Therefore *the sesquilinear form*

$$(215) \qquad a(m,m') = -\int_\Gamma C^i m \,.\, (n \wedge \bar{m}')\, d\Gamma$$

is coercive on $H^{-1/2}(\text{div},\Gamma)$ *for a dissipative medium.*

Remark 11. *Calderon operators for the Helmholtz equation.*
i) First in the *scalar case*, we can define the *exterior Calderon operator* C^e by:

$$(216) \qquad C^e(u_o) = \frac{\partial u}{\partial n}\Big|_\Gamma$$

u being the solution of the Helmholtz problem (188); n is the *exterior* normal to the *bounded* domain Ω; C^e is an isomorphism from $H^{1/2}(\Gamma)$ onto $H^{-1/2}(\Gamma)$, and from $H^s(\Gamma)$ onto $H^{s-1}(\Gamma)$ for all s if Γ is C^∞; C^e has the positivity property (from Green formula, since C^e corresponds to *outgoing* waves), for all $k > 0$,

$$(217) \qquad \text{Im}\,(C^e u_o, u_o) = \text{Im}\int_\Gamma \frac{\partial u}{\partial n}\bar{u}\, d\Gamma > 0, \quad \forall u_o \in H^{1/2}(\Gamma), \quad u_o \neq 0.$$

Then we can define the *interior Calderon operator* C^i like in (216), u being the solution of the *interior* Helmholtz problem corresponding to (188), generally with complex k. When k^2 is not an eigenvalue of the interior Helmholtz problem, neither for the Dirichlet condition, nor for the Neumann condition, C^i is an isomorphism from $H^{1/2}(\Gamma)$ onto $H^{-1/2}(\Gamma)$, with the same regularity property. Furthermore it has the positivity property, from the Green formula

$$(218) \quad \text{Im}\,(C^i u_o, u_o) = -2k'k''\int_\Omega |u|^2\, dx, \quad \forall u_o \in H^{1/2}(\Gamma), \text{ with } k = k' + i\,k''.$$

This implies the positivity properties for the operators L and R of (71) for $k > 0$

$$\text{Im}\,(L\rho',\rho') > 0, \;\; \forall \rho' \in H^{-1/2}(\Gamma),\; \rho' \neq 0, \quad \text{Im}\,(R\rho,\rho) < 0, \;\; \forall \rho \in H^{1/2}(\Gamma),\; \rho \neq 0.$$

We can verify easily using (71), (72), that the Calderon operators C^i and C^e are related to the integral operators L, J, K, R by:

$$(219) \quad C^e = -\tfrac{1}{2}L^{-1}(I + K) = -2\,(I + J)^{-1}R, \quad C^i = -\tfrac{1}{2}L^{-1}(-I + K) = -2\,(-I + J)^{-1}R.$$

ii) Then in the *vector case*, we can define *Calderon operators*. The *exterior Calderon operator* C^e is given, with u the solution of the Helmholtz problem (190)) by:

(220) $$C^e(m,g) = (n \wedge \text{curl}\, u|_\Gamma, n.u|_\Gamma);$$

C^e is an isomorphism from $H_t^{1/2}(\Gamma) \times H^{-1/2}(\Gamma)$ onto $H_t^{-1/2}(\Gamma) \times H^{1/2}(\Gamma)$, and also from $H_t^{3/2}(\Gamma) \times H^{1/2}(\Gamma)$ onto $H_t^{1/2}(\Gamma) \times H^{3/2}(\Gamma)$.

4.6. Integral equations and boundary problems

We can reduce the exterior or interior Maxwell problem (191) or (209) to an integral equation on the boundary in many ways. Here we give some of them. The general method is to choose an extension to the whole space, of the unknown (E,H) solution of the Maxwell equation. Then we only have to determine the jumps of E and H across the boundary Γ, or also the surface electric and magnetic currents J_Γ and M_Γ such that:

(221) $$J_\Gamma = -[n \wedge H]_\Gamma, \quad M_\Gamma = [n \wedge E]_\Gamma.$$

i) *Solution with an electric layer only.* The chosen extension (E,H) is such that there is no jump of the tangential electric field E across the boundary Γ: thus we have to find the surface electric current J_Γ, so that (using (101))

(222) $$ZR\, J_\Gamma = m, \quad \text{with } M_\Gamma = [n \wedge E]_\Gamma = 0.$$

Then (E,H) is obtained from J_Γ by (the restriction to Ω' or to Ω of)

(223) $$E = Z\mathcal{L}_m J_\Gamma, \quad H = -\mathcal{P}_m J_\Gamma.$$

We note that this is valid in the exterior case as well as in the interior case.

ii) *Solution with a magnetic layer only.* The chosen extension (E,H) is such that there is no jump of the tangential magnetic field H across the boundary Γ: thus we have to find the surface magnetic current M_Γ, so that

a) for the *exterior boundary problem* (using (89), (93)):

(224) $$\Pi_1 P_e \begin{pmatrix} M_\Gamma \\ 0 \end{pmatrix} = -m, \quad \text{with } J_\Gamma = -[n \wedge H]_\Gamma = 0,$$

Π_1 being the projection on the first component, so that with (104) and (102):

(224)' $$\frac{1}{2}(I + T)\, M_\Gamma = -m.$$

b) for the *interior boundary problem:*

(225) $$\frac{1}{2}(I - T)\, M_\Gamma = m.$$

Then in the two cases, (E,H) is given by (the restriction to Ω or Ω' of)

(226) $$E = Z\,\mathcal{P}_m\,M_\Gamma, \quad H = \frac{1}{Z}\mathcal{L}_m\,M_\Gamma.$$

iii) ***Solution by extension by*** 0

a) for the ***exterior boundary problem***, we have to find

(227) $$J_\Gamma = n \wedge H|_{\Gamma_e}, \qquad \text{with } M_\Gamma = -\,m.$$

Then (M_Γ, J_Γ) must satisfy:

(228) $$P_i\begin{pmatrix} M_\Gamma \\ J_\Gamma \end{pmatrix} = \begin{pmatrix} I - T & Z\,R \\ -\frac{1}{Z}R & I - T \end{pmatrix}\begin{pmatrix} M_\Gamma \\ J_\Gamma \end{pmatrix} = 0.$$

b) for the ***interior boundary problem***, we have to find

(227)' $$J_\Gamma = -\,n \wedge H|_{\Gamma_i}, \qquad \text{with } M_\Gamma = m.$$

Then (M_Γ, J_Γ) must satisfy

(229) $$P_e\begin{pmatrix} M_\Gamma \\ J_\Gamma \end{pmatrix} = \begin{pmatrix} I + T & -\,ZR \\ \frac{1}{Z}R & I + T \end{pmatrix}\begin{pmatrix} M_\Gamma \\ J_\Gamma \end{pmatrix} = 0,$$

and we obtain the electromagnetic field in Ω or Ω' by formulas (96)'.

Remark 12. There are other possibilities; for instance (see remark 10), when the exterior boundary problem is associated with a scattering problem of an incident wave in free space (which is naturally defined on the domain of the obstacle, like a plane wave) by a perfect conductor, then we can extend the reflected electromagnetic field by $(-E_I, -H_I)$ in the obstacle. Then we are reduced to find the jump:

$$J_\Gamma = -\,[n \wedge H]_\Gamma = n \wedge H_t|_\Gamma, \qquad \text{whereas: } M_\Gamma = [n \wedge E]_\Gamma = -\,n \wedge E_t|_\Gamma = 0,$$

and then J_Γ must satisfy:

(230) $$P_i\begin{pmatrix} 0 \\ J_\Gamma \end{pmatrix} = -\begin{pmatrix} n \wedge E_I|_\Gamma \\ -\,n \wedge H_I|_\Gamma \end{pmatrix},$$

thus

(230)' $$Z\,RJ_\Gamma = -\,2\,n \wedge E_I|_\Gamma, \qquad (I - T)\,J_\Gamma = 2\,n \wedge H_I|_\Gamma.$$

⊗

Remark 13. In fact for a given m, it is not obvious that the solution of the exterior problem is obtained only by an electric layer or a magnetic layer, and then it is possible that equations (222) and (224) have no solution, whereas method iii) (extension by 0) is always allowed. Therefore the system (228) is always solvable (but it is a system!). The difference is due to the eigenvalues of the selfadjoint operator A (see Theorem 8); these values are called irregular frequencies (for the exterior scattering problem). Thus to apply these methods, we have to avoid these irregular frequencies; for instance in a scattering problem, we use a linear combination of (230)' (see below), with a complex constant. ⊗

4.7. Some consequences for the integral operators

i) From Theorem 6 (existence and uniqueness of the solution of the exterior boundary problem) we deduce

(231) $\quad \mathrm{Im}\,(I - T) = \mathrm{Im}\,R, \qquad \ker R \cap \ker (I - T) = \{0\}.$

ii) *The "regular" case*. From Theorem 8 (existence and uniqueness of the solution of the interior boundary problem *in the "regular" case*) we deduce:

(232) $\quad \mathrm{Im}\,(I + T) = \mathrm{Im}\,R, \qquad \ker R \cap \ker (I + T) = \{0\}.$

PROOF. Let G_e be the set of the exterior boundary values for the electromagnetic field: $G_e = \mathrm{Im}\,P_e = \ker P_i$. Then for all u in $X = H^{-1/2}(\mathrm{div},\Gamma)$, there exists v in X so that (u,v) is in G_e, thus:

(233) $\quad P_i\begin{pmatrix} u \\ 0 \end{pmatrix} + P_i\begin{pmatrix} 0 \\ v \end{pmatrix} = 0.$

Therefore the mappings:

$$u \in X \rightarrow P_i\begin{pmatrix} u \\ 0 \end{pmatrix} \quad \text{and} \quad v \in X \rightarrow P_i\begin{pmatrix} 0 \\ v \end{pmatrix}$$

are injective and have the same range. Thanks to (102), (104) this gives (231). The proof of (232) is similar.

⊗

Proposition 26. *In the regular case* (k^2 *is not an eigenvalue of* A), R, (I + T) *and* (I − T) *are isomorphisms in the space* $X = H^{-1/2}(\mathrm{div},\Gamma)$, *and more generally* (*if* Γ *has the* C^∞ *regularity*) *in* $H^s(\mathrm{div},\Gamma)$ *for all real* s.

PROOF. Thanks to relations (106) we have to prove the properties for the R operator only. That R is onto is a consequence of equation (222) and theorems 6 and 8 (or also of (231), (232), since $X = \mathrm{Im}\,(I + T) + \mathrm{Im}\,(I - T)$).

Let j be such that $Rj = 0$; then we have: $m = n \wedge E|_\Gamma = 0$, thus $E = 0$ in Ω and Ω' (in the regular case), thus $H = 0$ and $j = 0$.

⊗

iii) *The "singular" case* (k^2 is an eigenvalue of A)
We define the following spaces:

(234)
$$\begin{aligned} \Xi_k &\overset{\text{def}}{=} \{ n \wedge H|_\Gamma,\ H = i \,\text{curl}\, E,\ E \in H^1(\Omega)^3,\ \Delta E + k^2 E = 0, \\ &\qquad\qquad \text{div}\, E = 0,\ n \wedge E|_\Gamma = 0\}, \\ &= \{ n \wedge H|_\Gamma,\ H \in H^1(\Omega)^3,\ \Delta H + k^2 H = 0,\ \text{div}\, H = 0,\ n \wedge \text{curl}\, H|_\Gamma = 0\}, \end{aligned}$$

and

(234)' $$\tilde{\Xi}_k \overset{\text{def}}{=} \{ n \wedge c,\ c \in \Xi_k\}, \qquad \overset{\sim o}{\Xi}_k \overset{\text{def}}{=} \{u \in X,\ \int_\Gamma u.\bar{v}\, d\Gamma = 0, \quad \forall v \in \tilde{\Xi}_k\}.$$

Then instead of (232), we have

(235)
$$\begin{cases} \ker R \cap \ker (I + T) = \Xi_k, \quad R(\overset{\sim o}{\Xi}_k) = (I + T)\, \overset{\sim o}{\Xi}_k, \quad \text{Im}\, R = \overset{\sim o}{\Xi}_k \\ \text{Im}\, R + \text{Im}\,(I - T) \supseteq \overset{\sim o}{\Xi}_k . \end{cases}$$

As a consequence we have:

(236) $$\ker R = \ker (I + T) = \Xi_k, \qquad \text{Im}\, R = \text{Im}\,(I - T) = \overset{\sim o}{\Xi}_k$$

(and also: $\ker (I - T) = C^e\, \Xi_k$, $\text{Im}\,(I + T) = C^e\, \overset{\sim o}{\Xi}_k$).

Remark 14. From the properties of the S operator, we also have:

(237) $$\ker R \cap \ker T = \{0\}, \qquad \text{Im}\, R + \text{Im}\, T = X.$$

⊗

Remark 15. Let G_e and G_i be the sets $G_e = \text{Im}\, P_e = \ker P_i$, $G_i = \text{Im}\, P_i = \ker P_e$, corresponding to the exterior and interior boundary values for the electromagnetic field, then ***their projections on their components are***:

(238) $\Pi_1 G_e = \Pi_2 G_e = X$, with:
i) in the "regular" case: $\Pi_1 G_i = \Pi_2 G_i = X$,
ii) in the "singular" case: $\Pi_1 G_i = \Pi_2 G_i = \overset{\sim o}{\Xi}_k$.

These results are easily proved from the "reciprocity" formula:

(239) $$\int_\Gamma n \wedge E.\ \bar{H}'\, d\Gamma = \int_\Gamma n \wedge H.\ \bar{E}'\, d\Gamma$$

for all (E,H), (E',H') satisfying the Maxwell equations in Ω; this is proved using the Green formula, with:

(240) $\quad \mathrm{div}\,(E \wedge \bar{H}') = \mathrm{div}\,(H \wedge \bar{E}')$.

Then taking E' an eigenvector of A, gives (238)ii).

⊗

In the "regular" case, we can give the Calderon operators as function of the operators R, T, thanks to the formulas (228):

$$(241)\quad \left| \begin{array}{l} C_k^e = R^{-1}(I-T) = -(I-T)^{-1}R = (I+T)R^{-1} = -R(I+T)^{-1}, \\ C_k^i = -R^{-1}(I+T) = (I+T)^{-1}R = -(I-T)R^{-1} = R(I-T)^{-1}. \end{array} \right.$$

⊗

4.8. On the numerical solution of some scattering problems

4.8.1. The problem of irregular frequencies

In order to solve the scattering problem of the remarks 10 and 12 thanks to an integral method, we have to solve equations (230)'. If k^2 is (near to) an eigenvalue of the operator A, then we cannot solve only one of the two (vector) equations (230) since R and $(I-T)$ are not invertible. This is the problem of *"irregular" frequencies*. We develop a method implemented by Bonnemason-Stupfel [1]. Let α be a (complex) number with $\mathrm{Re}\,\alpha \neq 0$. Then we replace (230) by the linear combination

$$(242)\quad -(I-T)J_\Gamma + \alpha Z\, n \wedge RJ_\Gamma = 2f_I \overset{\mathrm{def}}{=} 2\,[-n \wedge H_I|_\Gamma + \alpha\, \Pi_\Gamma E_I\,].$$

First we know, thanks to Theorem 6 and Remark 10, that (242) has a solution. Then we prove that this problem always has at most one solution.

i) First we consider the interior problem: find (E,H) with E in $H^1(\Omega)^3$ so that

$$(243)\quad \left| \begin{array}{ll} \text{i) } \mathrm{curl}\, H + i\omega\varepsilon\, E = 0, & \text{ii) } -\mathrm{curl}\, E + i\omega\mu\, H = 0 \ \text{ in } \Omega, \\ \text{iii) } n \wedge H|_\Gamma - \alpha\, \Pi_\Gamma E = f_I, & \end{array} \right.$$

with f_I given in $H_t^{1/2}(\Gamma)$ (for instance), with ε and μ real positive numbers.

Let a(E,E') be the sesquilinear form on $H^1(\Omega)^3$

$$(244)\quad a(E,E') = \int_\Omega -\,[\frac{1}{i\omega\mu}\,\mathrm{curl}\, E\,.\mathrm{curl}\, \bar{E}' + i\omega\varepsilon\, E\,.\,\bar{E}'\,]\,dx - \alpha \int_\Gamma \Pi_\Gamma E\,.\,\bar{E}'\,d\Gamma.$$

Then we can write problem (243) in the variational form: find E in $H^1(\Omega)^3$ with

$$(245)\quad a(E,E') = \int_\Gamma f_I\,.\,\bar{E}'\,d\Gamma, \quad \forall E' \in H^1(\Omega)^3.$$

Then taking the real part of (245) with E' = E, we obtain:

$$(246)\quad \mathrm{Re}\, a(E,E) = \mathrm{Re}\,(-\alpha)\int_\Gamma |\Pi_\Gamma E|^2\, d\Gamma = \mathrm{Re}\,(\int_\Gamma f_I\,.\,\bar{E}\, d\Gamma).$$

This implies uniqueness: when $f_I = 0$, (246) implies $\Pi_\Gamma E = 0$. Thus from (243)iii) $n \wedge H|_\Gamma = 0$, and then, from section 13 chap. 2, $E = H = 0$ in Ω.

ii) Let Q_o be defined by:

(247) $$Q_o\begin{pmatrix} M_\Gamma \\ J_\Gamma \end{pmatrix} = -J_\Gamma + \alpha\, n \wedge M_\Gamma,$$

and thus

(248) $$Q_o\begin{pmatrix} n \wedge E|_\Gamma \\ -n \wedge H|_\Gamma \end{pmatrix} = n \wedge H|_\Gamma - \alpha\, \Pi_\Gamma E.$$

From to i), we know that $Q_o\begin{pmatrix} n \wedge E|_\Gamma \\ -n \wedge H|_\Gamma \end{pmatrix} = 0$ implies $E = H = 0$.

iii) Then applying Q_o to equation (230) gives:

(249) $$Q_o P_i\begin{pmatrix} 0 \\ J_\Gamma \end{pmatrix} = -Q_o\begin{pmatrix} n \wedge E_I|_\Gamma \\ -n \wedge H_I|_\Gamma \end{pmatrix} = f_I,$$

and thus we obtain equation (242). Therefore uniqueness of the solution J_Γ of (242) is deduced from that of (243) in the following way:

(250) $$Q_o P_i\begin{pmatrix} 0 \\ J_\Gamma \end{pmatrix} = 0 \text{ implies } P_i\begin{pmatrix} 0 \\ J_\Gamma \end{pmatrix} = 0, \text{ thus } \begin{pmatrix} 0 \\ J_\Gamma \end{pmatrix} \in G_e,$$

and then $J_\Gamma = 0$ from Theorem 6.
We can show that problem (243) (and thus (242)) depends on the Fredholm alternative. This gives another proof of Theorem 6.

⊗

4.8.2. A saddle point method

Here we briefly give a variational (and integral) method for solving a scattering problem, which is due to Bendali [1]. We consider the scattering problem of Remark 10, which is equivalent to (230). We only take the first equation of (230) (we assume that k^2 is not near an eigenvalue). Let

(251) $$c \overset{\text{def}}{=} \frac{i}{2kZ}\Pi_\Gamma E_I, \qquad p \overset{\text{def}}{=} J_\Gamma.$$

Then we have to find p with given c in $H^{-1/2}(\text{curl},\Gamma)$ so that

(252) $$-n \wedge Rp = 2ik\, c.$$

Thanks to the definition (100) with (95) (see also (144)), we have to solve

(253) $$\langle Lp, q\rangle - \frac{1}{k^2}\langle L\,\text{div}_\Gamma p, \text{div}_\Gamma q\rangle = \langle c, q\rangle, \quad \forall q \in H^{-1/2}(\text{div},\Gamma).$$

Then we define:

(254) $$\lambda = -\frac{1}{k^2}\text{div}_\Gamma\, p,$$

and

(255) $H_*^{-1/2}(\Gamma) = \{\phi \in H^{-1/2}(\Gamma), \int_{\Gamma_i} \phi \, d\Gamma = 0$ on all connected component Γ_i of $\Gamma\}$.

Thus $\lambda \in H_*^{-1/2}(\Gamma)$. Then we define, with $\Phi_0(x) = \frac{1}{4\pi r}$, $r = |x|$:

$$L_0\lambda(x) = \int_\Gamma \Phi_0(x-y)\,\lambda(y)\,d\Gamma_y; \tag{256}$$

L_0 is a continuous operator from $H^{-1/2}(\Gamma)$ into $H^{1/2}(\Gamma)$. Then with these notations, problem (253) is: find $p \in H^{-1/2}(\text{div},\Gamma)$ and $\lambda \in H_*^{-1/2}(\Gamma)$ so that:

$$\text{(257)} \quad \left| \begin{array}{l} \text{i) } \langle Lp, q\rangle + \langle L\lambda, \text{div}_\Gamma q\rangle = \langle c, q\rangle, \\ \text{ii) } \langle L_o \text{div}_\Gamma\, p, \phi\rangle + k^2 \langle L_o\lambda, \phi\rangle = 0, \end{array} \right.$$

for all $q \in H^{-1/2}(\text{div},\Gamma)$, and all $\phi \in H_*^{-1/2}(\Gamma)$. Using the sesquilinear forms:

$$\text{(258)} \quad \left| \begin{array}{ll} a(p, q) = \langle Lp, q\rangle, & a_o(p,q) = \langle L_o p, q\rangle, \\ b(\lambda, q) = \langle L\lambda, \text{div}_\Gamma\, q\rangle, & b_o(\lambda, q) = \langle L_o\lambda, \text{div}_\Gamma\, q\rangle, \\ c(\lambda, \phi) = k^2 \langle L_o\lambda, \phi\rangle, & \end{array} \right.$$

with $p, q \in H^{-1/2}(\text{div},\Gamma)$, $\lambda, \phi \in H_*^{-1/2}(\Gamma)$, we can write (257) in the form: find (p,λ) so that for all $q \in H^{-1/2}(\text{div},\Gamma)$, and $\phi \in H_*^{-1/2}(\Gamma)$,

$$\text{(259)} \quad \left| \begin{array}{l} \text{i) } a(p, q) + b(\lambda, q) = \langle c, q\rangle \\ \text{ii) } \overline{b_0(\phi,p)} + c(\lambda, \phi) = 0. \end{array} \right.$$

This system is solved by perturbation of the system: find $(\tilde{p},\tilde{\lambda})$ so that, for all q in $H^{-1/2}(\text{div},\Gamma)$, and ϕ in $H_*^{-1/2}(\Gamma)$,

$$\text{(260)} \quad \left| \begin{array}{l} \text{i) } a_0(\tilde{p}, q) + b_o(\tilde{\lambda}, q) = \langle c, q\rangle \\ \text{ii) } \overline{b_0(\phi,\tilde{p})} = 0 \text{ (or } = \langle \chi, \phi\rangle), \end{array} \right.$$

with given c in $H^{-1/2}(\text{curl},\Gamma)$ and χ in the (anti) dual space of $H_*^{-1/2}(\Gamma)$ that is:

$$(H_*^{-1/2}(\Gamma))' = H_*^{1/2}(\Gamma) = \{\mu \in H^{1/2}(\Gamma), \int_{\Gamma_i} L_0^{-1}\mu \, d\Gamma, \ \forall \Gamma_i \text{ component of } \Gamma\}.$$

Then we prove (see Bendali [1]) that the system (260) satisfies the "inf-sup condition" or "Brezzi condition" (see for instance Girault-Raviart [1] p. 39) and thus (260) has a unique solution for all given (c,χ) in $H^{-1/2}(\mathrm{curl},\Gamma) \times H_*^{1/2}(\Gamma)$. Then we obtain the solution of the system (259) by perturbation of the system by a (regularizing) compact operator, and thus we can apply the Fredholm alternative.

⊗

5. SCATTERING PROBLEMS BY A DIELECTRIC OBSTACLE

Let (E_I,H_I) be an incident electromagnetic wave with an angular frequency ω, on a (regular) bounded obstacle Ω. Let the medium inside Ω be a dielectric with permittivity and permeability ε_1 and μ_1; then let k_1 be the wavenumber inside Ω; outside Ω is the free space (or some "conservative" medium) with permittivity and permeability $\varepsilon = \varepsilon_0$ and $\mu = \mu_0$ and wavenumber k. The problem is to find the electromagnetic wave (E_r,H_r) reflected by this obstacle. Let (E,H) be the electromagnetic field inside Ω; (E_r,H_r) must satisfy the Maxwell equations in free space, with the Silver-Müller conditions at infinity and with locally finite energy; the problem is then reduced to find (E,H) satisfying

$$(261)\qquad \left| \begin{array}{l} \text{i)}\ \mathrm{curl}\, H + i\omega\varepsilon_1 E = 0, \\ \text{ii)}\ -\mathrm{curl}\, E + i\omega\mu_1 H = 0 \text{ in } \Omega, \end{array} \right.$$

and the continuity relation on the boundary Γ of Ω

$$(262)\qquad n \wedge E_r + n \wedge E_I = n \wedge E, \qquad n \wedge H_r + n \wedge H_I = n \wedge H \text{ on } \Gamma.$$

The fact that (E_r,H_r) is a reflected wave, gives us a relation between their boundary values on Γ as seen in section 4.2. We can obtain it in two ways.

5.1. A general variational formulation

Here we could assume that inside Ω the medium is inhomogeneous and even anisotropic with usual dissipative conditions, see H14) chap. 1:

$$(263)\qquad \mathrm{Re}\,(-i\omega\varepsilon_1) > 0, \qquad \mathrm{Re}\,(-i\omega\mu_1) > 0.$$

Using the exterior Calderon operator C^e, we have:

$$(264)\qquad n \wedge H_r|_\Gamma = C^e(n \wedge E_r|_\Gamma),$$

and then with (262), we obtain the boundary condition on Γ:

$$(265)\qquad n \wedge H - C^e(n \wedge E) = f_I \overset{\mathrm{def}}{=} n \wedge H_I - C^e(n \wedge E_I).$$

The problem: find (E,H) in Ω satisfying (261) and (265), is solved by a variational method. Let a(E,E') be the sesquilinear form

(266) $$a(E,E') = \int_\Omega -\left[\frac{1}{i\omega\mu}\operatorname{curl} E \,.\, \operatorname{curl}\bar{E}' + i\omega\varepsilon\, E\,.\,\bar{E}'\right]dx - \int_\Gamma C^e(n\wedge E)\,.\,\bar{E}'\, d\Gamma,$$

for E, E' in H(curl,Ω). The problem (261), (265) is equivalent to: find E $\in$ H(curl,Ω) so that

(267) $$a(E,E') = \int_\Gamma f_I \,.\, \bar{E}' d\Gamma, \quad \forall E' \in H(\operatorname{curl},\Omega).$$

The proof is "classical" with the Green formula. Thanks to (263) (if the medium is inhomogeneous we have to assume a slightly stronger hypothesis), and to the positivity property of the exterior Calderon operator C^e (see (201)), a(E,E') is coercive: there is a constant α_o so that

(268) $$\operatorname{Re} a(E,E) \geq \int_\Omega \left(\operatorname{Re}\left(-\frac{1}{i\omega\mu}\right)|\operatorname{curl} E|^2 + \operatorname{Re}(-i\omega\varepsilon)|E|^2\right)dx \geq \alpha_o \|E\|^2_{H(\operatorname{curl},\Omega)}.$$

Since a(E,E') and the form: $E' \rightarrow \int_\Gamma f_I.\bar{E}'d\Gamma$ are continuous on H(curl,Ω), we can apply the Lax-Milgram lemma: thus the scattering problem has a unique solution.

Remark 16. We can solve other scattering problems, for instance when there is a perfect conductor with domain Ω_1 in the obstacle. Then the electric field must satisfy the boundary condition on Γ_1, the boundary of Ω_1: $n \wedge E|_{\Gamma_1} = 0$.

We apply the variational method with $V = \{E \in H(\operatorname{curl},\Omega\backslash\bar{\Omega}_1),\ n\wedge E|_{\Gamma_1} = 0\}$.

⊗

5.2. Solution using an integral method

Let (M,J) be defined by:

(269) $$M = n\wedge E|_\Gamma, \qquad J = -n\wedge H|_\Gamma,$$

Let P_i^k be the *interior Calderon projector associated to the exterior wavenumber k.* Then we have (by definition):

(270) $$P_i^k\begin{pmatrix} n\wedge E_r|_\Gamma \\ -n\wedge H_r|_\Gamma \end{pmatrix} = 0.$$

Then using (262) and (269) we have:

(271) $$P_i^k\begin{pmatrix} M \\ J \end{pmatrix} = \begin{pmatrix} n\wedge E_I|_\Gamma \\ -n\wedge H_I|_\Gamma \end{pmatrix},$$

thus (with the integral operators T, R associated to k, and the impedance Z of free space):

$$(271)'\quad \left|\begin{array}{l} \text{i) } (I - T)\,M + ZR\,J = 2\,n \wedge E_I\,, \\ \text{ii) } -\frac{1}{Z} R\,M + (I - T)\,J = -\,2n \wedge H_I \ \text{ on } \Gamma. \end{array}\right.$$

Now we assume that the medium is homogeneous with constant ε_1 and μ_1.
Let P_e^1 be the exterior Calderon projector associated to the interior wavenumber k_1. Then we have (by definition):

$$(272)\qquad P_e^1 \begin{pmatrix} M \\ J \end{pmatrix} = 0,$$

that is also with T_1, R_1, and Z_1 relative to the interior medium:

$$(272)'\quad \left|\begin{array}{l} \text{i) } (I + T_1)\,M - Z_1\,R_1\,J = 0, \\ \text{ii) } \frac{1}{Z_1} R_1\,M + (I + T_1)\,J = 0. \end{array}\right.$$

Thus the scattering problem amounts to find (M,J) solution of the integral equations (271) and (272), that is also (271)' and (272)'.

We can take (270) into account in a weak form, (thanks to (100) with (95), see (253) for R): taking the scalar product of (271)'i) with $n \wedge J'$, and (271)'ii) with $n \wedge M'$ and adding, give us for all M', J' in $H^{-1/2}(\mathrm{div},\Gamma)$

$$(273)\quad \left|\begin{array}{l} iZk \int_\Gamma\int_\Gamma [J(x).J'(y) - k^{-2}\,\mathrm{div}_\Gamma J(x).\mathrm{div}_\Gamma J'(y)]\,\Phi(x-y)\,d\Gamma_x d\Gamma_y \\ -\frac{ik}{Z}\int_\Gamma\int_\Gamma [M(x).M'(y) - k^{-2}\,\mathrm{div}_\Gamma M(x).\mathrm{div}_\Gamma M'(y)]\,\Phi(x-y)\,d\Gamma_x d\Gamma_y \\ -\int_\Gamma\int_\Gamma [J(x)\wedge M'(y) + M(x)\wedge J'(y)].\,\mathrm{grad}\,\Phi\,(x-y)\,d\Gamma_x d\Gamma_y \\ -\frac{1}{2}\int_\Gamma [M(x)\wedge J'(x) + J(x)\wedge M'(x)].n\,d\Gamma_x = \int_\Gamma [-\,H_I(x).M'(x) + E_I(x).J'(x)]\,d\Gamma_x. \end{array}\right.$$

We can also write equations (272)' in a weak form. The case of a dielectric layer against a perfect conductor would be solved in the same way.
These problems depend on the Fredholm alternative, and uniqueness follows from the Rellich lemma. For some numerical developments on these methods, we refer to Bendali [1] and Heliot [1].

5.3. Behavior at infinity. Radar cross section. Optical theorem

The behavior at infinity of the reflected electromagnetic field (E_r, H_r) is given by the formulas (30), (31) with:

$$(274)\qquad M = -\,n \wedge E_r|_\Gamma \delta_\Gamma\,,\quad J = n \wedge H_r|_\Gamma\,\delta_\Gamma.$$

Then we define the *differential cross section for the direction* α by

$$(275)\qquad \sigma(\alpha) \overset{\text{def}}{=} \lim_{\rho\to\infty} 4\pi\rho^2 \frac{|E_r(\rho\alpha)|^2}{|E_I|^2}, \qquad \text{with } \rho = |x|.$$

Thus using (30) and (33), we have:

(276) $$\sigma(\alpha) = \frac{1}{4\pi}\frac{|\hat{j}(k\alpha)|^2}{|E_I|^2} = \frac{1}{4\pi}\frac{|\hat{m}(k\alpha)|^2}{|H_I|^2},$$

since the incident wave is a plane wave, that is (E_I, H_I) is given by:

(277) $$E_I(x) = \tilde{E}_I\, e^{ik\alpha_o.x}, \quad H_I(x) = \tilde{H}_I\, e^{ik\alpha_o.x},$$

where $\alpha_o \in S^2$ is the direction of the propagation of the plane wave, and $\tilde{E}_I$, $\tilde{H}_I$ are constant vectors in C^3, Z being the free space impedance:

(278) $$\alpha_o.\tilde{E}_I = 0, \; \alpha_o.\tilde{H}_I = 0, \; \tilde{H}_I = \frac{1}{Z}\alpha_o \wedge \tilde{E}_I, \; \tilde{E}_I = -Z\,\alpha_o \wedge \tilde{H}_I.$$

The *radar cross section, or back scattering* σ_r is defined by $\sigma_r = \sigma(-\alpha_o)$ where α_o is the incident direction of the plane wave. It is a surface and its unit is m^2 (in the S.I.) but it is generally expressed in dbm^2 (db decibels) by $10 \log_{10} \sigma_r$.

Then the *total cross section* or *scattering coefficient* σ_S is defined by:

(279) $$\sigma_S = \int_{S^2} \sigma(\alpha)\, d\alpha = \frac{1}{4\pi}\frac{1}{|E_I|^2}\int_{S^2} |\hat{j}(k\alpha)|^2\, d\alpha.$$

We also define the following powers:
i) the *scattered power* P_s, as the real part of the flux of the complex Poynting vector of the reflected wave on the surface Γ:

(280) $$P_s = \mathrm{Re}\,(\int_\Gamma n.\, E_r \wedge \bar{H}_r\, d\Gamma) = \frac{1}{(4\pi)^2 Z}\int_{S^2} |\hat{j}(k\alpha)|^2\, d\alpha,$$

(see (202)'), and thus $\sigma_S = 4\pi Z\, |E_I|^{-2} P_s$ is proportional to the scattered power;

ii) the *absorbed power* P_a, as the real part of the flux of the complex Poynting vector of the interior wave, on the surface Γ:

(281) $$P_a = -\,\mathrm{Re}\int_\Gamma n.E \wedge \bar{H}\, d\Gamma = \int_\Omega [\mathrm{Re}\,(-\frac{1}{i\omega\mu_1})\, |\mathrm{curl}\, E|^2 + \mathrm{Re}\,(-i\omega\varepsilon_1)\, |E|^2]\, dx \geq 0$$

(with H13, chap. 1); if μ_1 is real, P_a corresponds to the dissipation by Joule effect.
iii) the *total power* $P_{tot} = P_s + P_a$. Using relations (262), we have:

(282) $$P_{tot} = \mathrm{Re}\int_\Gamma [n \wedge H_r.\, \bar{E}_I - n \wedge E_r.\, \bar{H}_I]\, d\Gamma = \mathrm{Re}\int_\Gamma [n \wedge H.\, \bar{E}_I - n \wedge E.\, \bar{H}_I]\, d\Gamma,$$

since the flux of the Poynting vector of the incident wave plane on the obstacle is zero: $\mathrm{Re}\,(\int_\Gamma n.E_I \wedge \bar{H}_I\, d\Gamma) = 0$. Then with (30), (31) and (269), we have:

(283)
$$\left| \begin{aligned} P_{tot} &= -\operatorname{Re} \int_\Gamma [J.\,\tilde{\bar{E}}_I + M.\tilde{\bar{H}}_I]\, e^{-ik\alpha_o.x}\, d\Gamma = -\operatorname{Re}\, [\hat{J}(k\alpha_o).\tilde{\bar{E}}_I + \hat{M}(k\alpha_o).\tilde{\bar{H}}_I] \\ &= \frac{1}{\omega\mu} \operatorname{Im}\, (\hat{j}(k\alpha_o).\tilde{\bar{E}}_I) = \frac{1}{\omega\varepsilon} \operatorname{Im}\, (\hat{m}(k\alpha_o).\tilde{\bar{H}}_I). \end{aligned} \right.$$

This relation which gives the total power as a function of the forward scattering is generally called the *optical theorem*. Since the total power P_{tot} is greater than P_s, we have the inequality

(284)
$$\frac{1}{(4\pi)^2} \int_{S^2} |\hat{j}(k\alpha)|^2\, d\alpha \le \frac{1}{k} \operatorname{Im}\, (\hat{j}(k\alpha_o).\tilde{\bar{E}}_I).$$

Note that in the scalar Helmholtz case, we also define, with usual notations, a scattered power, an absorbed power, and a total power by

$$\left| \begin{aligned} & P_S = \operatorname{Im} \int_\Gamma \frac{\partial u_r}{\partial n} \bar{u}_r\, d\Gamma, \quad P_a = -\operatorname{Im} \int_\Gamma \frac{\partial u}{\partial n} \bar{u}\, d\Gamma = (\operatorname{Im} k_1^2) \int_\Omega |u|^2\, dx, \\ & P_t = P_S + P_a = -\operatorname{Im} \int_\Gamma \Big(\frac{\partial u}{\partial n} \bar{u}_I - u \frac{\partial \bar{u}_I}{\partial n}\Big) d\Gamma = \operatorname{Im} \int_\Omega (k_1^2 - k_0^2)\, u\, \bar{u}_I\, dx, \end{aligned} \right.$$

which we can also write in a form similar to (283) for an incident wave plane.

Remark 17. Is it possible to have a differential cross section vanishing for all directions? First note that the reflected electromagnetic field would be $O(1/r^2)$ at infinity, and thus would be zero from Rellich lemma. Then the electromagnetic field (E, H) in Ω satisfies:

(285)
$$\left| \begin{aligned} & \text{i) } \operatorname{curl} H + i\omega\varepsilon_1 E = 0, \\ & \text{ii) } -\operatorname{curl} E + i\omega\mu_1 H = 0, \\ & \text{iii) } n \wedge E|_\Gamma = n \wedge E_I|_\Gamma, \quad n \wedge H|_\Gamma = n \wedge H_I|_\Gamma, \end{aligned} \right.$$

which is a Cauchy problem (see chap. 2). Let: $E_1 = E - E_I$, $H_1 = H - H_I$ in Ω. Then (E_1, H_1) satisfy:

(286)
$$\left| \begin{aligned} & \text{i) } \operatorname{curl} H_1 + i\omega\varepsilon_1 E_1 = i\omega(\varepsilon - \varepsilon_1) E_I, \\ & \text{ii) } -\operatorname{curl} E_1 + i\omega\mu_1 H_1 = i\omega(\mu - \mu_1) H_I, \ \text{ in } \Omega, \\ & \text{iii) } n \wedge E_1|_\Gamma = 0, \ n \wedge H_1|_\Gamma = 0 \end{aligned} \right.$$

Thus we can extend (E_1, H_1) by 0 to the whole space, and then apply formulas (30) and (31) with (M, J) given (thanks to the characteristic function of Ω, 1_Ω) by

(287)
$$M \overset{\text{def}}{=} i\omega(\varepsilon - \varepsilon_1) E_I 1_\Omega, \quad J \overset{\text{def}}{=} i\omega(\mu - \mu_1) H_I 1_\Omega.$$

Then $\hat{j}(k\alpha) = 0$, $\hat{m}(k\alpha) = 0$, $\forall \alpha \in S^2$, imply with (277):

(288) $\Pi_\alpha\,[\omega\mu\,(\varepsilon-\varepsilon_1)\,\tilde{E}_I - k\,(\mu-\mu_1)\,\alpha\wedge\tilde{H}_I]\,\hat{1}_\Omega(k\alpha)=0,\quad \forall\,\alpha\in S^2,$

when ε_1, μ_1 are constant; thus $\varepsilon_1 = \varepsilon$, $\mu_1 = \mu$, i.e., Ω is occupied by free space ! If the domain Ω is occupied by a dissipative medium (with (213)), we can verify using the power balance (that is the variational method), that the problem (285) cannot have a solution; (E,H) must satisfy

(289) $\int_\Gamma [\mathrm{Re}\,(i\omega\mu)\,H.\bar{H} + \mathrm{Re}\,(-\,i\omega\bar{\varepsilon})\,E.\bar{E}]\,dx = \mathrm{Re}\,(\int_\Gamma n.\,E_I\wedge\bar{H}_I)d\Gamma = 0,$

that implies E = H = 0 (at least where the medium is dissipative), and then E = H = 0 anywhere, in contradiction to (285)iii)!

⊗

Remark 18. It is possible to have a differential cross section vanishing for certain directions (that is for isolated directions) but it cannot be zero in the neighborhood of a direction (that is, in an open cone). This means that there is no perfect shadow domain. In other words:

(290) $\hat{j}(k\alpha)=0\quad \forall\,\alpha\in S^2$ so that $\alpha.\alpha_1>\eta$ for a $\alpha_1\in S^2$, and $\eta>0$,

also implies $E_r = H_r = 0$!
We can prove this result, as an easy consequence of Müller's lemma 115, 117, pp. 341 and 342 (see Müller [1]). As a matter of fact, (290) implies:

(291) $\hat{j}^{\,2}(k\alpha)=0$ and $n.(\hat{j}(k\alpha)\wedge\bar{\hat{j}}(k\alpha))=0,\quad \forall\,\alpha\in S^2$ such that $\alpha.\alpha_1>\eta$

(that is, the wave is circularly and linearly polarized respectively) and this implies thanks to analyticity properties (see Müller [1]) that (291) is satisfied for all α of the unit sphere. This immediately implies that (290) is also satisfied for all α of the unit sphere which proves the property.

⊗

Remark 19. In radar problems, we are interested in having a zero radar cross section. This is possible for particular geometries (symmetry invariance with respect to an axis), as asserted by the well-known (among radarists at least) Weston Theorem; here we give a (small) generalization.

Proposition 27. "*Weston Theorem*". *Given an incident electromagnetic wave* (E_I, H_I) *with an angular frequency* ω, *on a (regular) bounded obstacle* Ω *which is invariant under a 90° rotation* S *about some axis. We assume that the direction of incident propagation is parallel to this axis. Furthermore we assume an "impedance" condition on the boundary* Γ *of* Ω:

(292) $n\wedge H_t|_\Gamma = C\,(n\wedge E_t|_\Gamma)\quad$ so that $\quad \mathcal{S}^{-1}C\mathcal{S} = -\,Z^{-2}\,C^{-1},$

with (E_t, H_t) ***the total electromagnetic field,*** Z ***the impedance of free space,*** C ***an operator on the boundary and*** S ***the rotation. Then the radar cross section is zero.***

We note that this contains the case where the medium inside Ω is a dielectric with permittivity and permeability ε_1 and μ_1 with $\varepsilon_1/\varepsilon_0 = \mu_1/\mu_0$, since the operator C is then the interior Calderòn operator; thus the impedance of the inside medium must be equal to that of free space. This also contains the case of the following usual impedance condition (which requires $\eta = 1$):

(293) $\quad \pi_\Gamma E = \eta\, Z\, n \wedge H \quad \text{on } \Gamma, \quad \eta \in C,$

PROOF. First we note that the differential cross section is given by the following formula for all directions α, where (E,H) denotes the reflected field

(294) $$\sigma(\alpha) = \frac{Z}{|E_I|^2} \lim_{r\to\infty} 4\pi r^2\, \alpha \wedge E(r\alpha).\bar{H}(r\alpha) = \frac{-Z}{|E_I|^2} \lim_{r\to\infty} 4\pi r^2\, E(r\alpha).\alpha \wedge \bar{H}(r\alpha).$$

This a simple consequence of (30), (31). Then we note that the problem relative to the reflected field is invariant under the 90° rotation about the incident axis so that $(\mathcal{S}E, \mathcal{S}H)$ defined by $(\mathcal{S}E)(x) = SE(Sx)$, $(\mathcal{S}H)(x) = SH(Sx)$, and $(ZH, -(1/Z)E)$ are solutions of the same problem. From uniqueness of the solution we deduce that $(\mathcal{S}E, \mathcal{S}H) = (ZH, -(1/Z)E)$. Then this relation implies:

(295) $$\left| \begin{aligned} \sigma(-\alpha_0) &= \frac{Z}{|E_I|^2} \lim_{r\to\infty} 4\pi r^2\, E(-r\alpha_0).\mathcal{S}\bar{H}(-r\alpha_0) \\ &= \frac{Z}{|E_I|^2} \lim_{r\to\infty} \left(-\frac{4\pi r^2}{Z} E(-r\alpha_0).\bar{E}(-r\alpha_0)\right) < 0\,! \end{aligned} \right.$$

⊗

6. SCATTERING AND INFLUENCE COEFFICIENTS FOR SEVERAL OBSTACLES

Let Γ be a "regular" bounded nonconnected surface with p components (Γ_λ), λ in I. We assume that Γ is the boundary of a bounded domain Ω, with N components, $\Omega_1, .., \Omega_N$. Let Ω' be the (open) complement of Ω with $N'+1$ components Ω'_0, $\Omega'_1, .., \Omega'_{N'}$ (with Ω'_0 unbounded). Note that Ω'_j is not the complement of Ω_j, nor is Γ_j the boundary of Ω_j. We denote by $\partial\Omega_j$ (resp. $\partial\Omega'_j$) the boundary of Ω_j (resp. Ω'_j). From Alexander's relation (see for instance Dautray-Lions [1] chap. 2.4.5) we have $p = N + N'$. We assume that there are electric and magnetic surface currents on Γ, that is, on each surface Γ_λ, (M_λ, J_λ); we assume that the whole space outside Γ is occupied by free space. We study the electromagnetic field due to these currents in R^3 and its trace on the surfaces Γ_λ, and the influence of currents on Γ_λ onto another surface Γ_μ.

6.1. Decomposition of the trace spaces
The trace space $X = H^{-1/2}(\mathrm{div},\Gamma)$ has the following decomposition:

$$X = H^{-1/2}(\mathrm{div},\Gamma) = \underset{\lambda\in I}{\oplus} X_\lambda = \underset{\lambda\in I}{\oplus} H^{-1/2}(\mathrm{div},\Gamma_\lambda). \tag{296}$$

Let $(M,J) = (M_\lambda, J_\lambda)$, $\lambda \in I$, be magnetic and electric currents on Γ in this space. The electromagnetic field (E,H) induced by (M,J) is given by formulas (96):

$$\left| \begin{array}{l} E = \sum\limits_\lambda E_\lambda, \quad H = \sum\limits_\lambda H_\lambda, \\ \\ \text{with } \; E_\lambda = \mathcal{P}_m M_\lambda + Z\mathcal{L}_m J_\lambda, \quad H_\lambda = \frac{1}{Z}\mathcal{L}_m M_\lambda - \mathcal{P}_m J_\lambda. \end{array} \right. \tag{297}$$

Then taking the traces on the boundary, we have (see (101) to (103)):

$$\begin{pmatrix} (\gamma_i^\mu + \gamma_e^\mu)E \\ -(\gamma_i^\mu + \gamma_e^\mu)H \end{pmatrix} = \sum_\lambda S_{\mu\lambda} \begin{pmatrix} M_\lambda \\ J_\lambda \end{pmatrix}, \text{ with } S_{\mu\lambda} = \begin{pmatrix} -T_{\mu\lambda} & ZR_{\mu\lambda} \\ -\frac{1}{Z}R_{\mu\lambda} & -T_{\mu\lambda} \end{pmatrix}. \tag{298}$$

We see that S is decomposed into $(S_{\mu\lambda})$; and $S_{\mu\lambda}$ is a continuous operator from $X_\lambda \times X_\lambda$ into $X_\mu \times X_\mu$, and it is regularizing for $\lambda \neq \mu$, that is $\operatorname{Im} S_{\mu\lambda}$ is in $H_t^s(\Gamma_\mu) \times H_t^s(\Gamma_\mu)$ for all s if the surface Γ_μ is C^∞. This implies that $T_{\mu\lambda}$, $R_{\mu\lambda}$ are also regularizing.
Relation (105) implies:

$$\sum_\lambda S_{\gamma\lambda} S_{\lambda\mu} = \delta_{\gamma\mu}, \qquad \forall \gamma, \mu \in I, \tag{299}$$

and on each surface Γ_λ:

$$(S_{\lambda\lambda})^2 = I. \tag{300}$$

Remark 20. *On the orientation of the normals to the surfaces* Γ_λ. Let us denote by Ω_λ and Ω'_λ the domain of R^3 inside and outside of Γ_λ. We note that the natural orientation of the normal to the surface Γ_λ (from inside to outside) does not necessarily agree with that of the domain Ω.
Let $S^{(\varepsilon)}$ be the matrix operator S given by (103) with (100) for a surface Γ with orientation ε of the normal n. We have: $S^{(-\varepsilon)} = -S^{(\varepsilon)}$, and thus $\mathcal{S} = \varepsilon S^{(\varepsilon)}$ is orientation independent. Choosing: $\varepsilon' = -\varepsilon$ for i, $\varepsilon' = \varepsilon$ for e, relations (93) and (104) are written:

$$\begin{pmatrix} n_\varepsilon \wedge E|_{\Gamma_{\varepsilon'}} \\ -n_\varepsilon \wedge H|_{\Gamma_{\varepsilon'}} \end{pmatrix} = -\varepsilon'\varepsilon \, \frac{1}{2}(I - \varepsilon' S^{\varepsilon}) \begin{pmatrix} M \\ J \end{pmatrix}, \tag{301}$$

and thus the Calderon projectors are given by

$$P_{\varepsilon'\varepsilon} = \frac{1}{2}(I - \varepsilon' S^{(\varepsilon)}) = \frac{1}{2}(I - \varepsilon'\varepsilon\mathcal{S}), \quad \text{with } \varepsilon'\varepsilon = -1 \text{ for i}, \; \varepsilon'\varepsilon = +1 \text{ for e}. \tag{302}$$

We denote by S^λ the operator $S = \mathcal{S}$ relative to the surface Γ_λ (for the natural orientation), and by $P^\lambda_{e,i}$ the corresponding Calderon projectors. ⊗

Writting $S_\lambda \stackrel{def}{=} S_{\lambda\lambda}$, we have $S_\lambda = \varepsilon\, S^\lambda$ according to the orientation of the normal to Γ_λ with Ω. Thus the operators

(303) $$P_i^\lambda = \frac{1}{2}(I + S^\lambda) = \frac{1}{2}(I + \varepsilon S_\lambda),\quad P_e^\lambda = \frac{1}{2}(I - S^\lambda) = \frac{1}{2}(I - \varepsilon S_\lambda)$$

are the (exterior and interior for Ω_λ) Calderon projectors relatively to the surface Γ_λ. Then the (total) Calderon projectors relatively to the domain Ω are decomposed into operators $(P_{e,i})_{\lambda,\mu}$ but only their diagonal terms are projectors. We denote by $G_{e,i}$ (resp. $G^\lambda{}_{e,i}$) the graphs of the operators $P_{e,i}$ (resp. $P^\lambda{}_{e,i}$), by Π_λ the restriction to the surface Γ_λ. Then we have:

(304) $$G_i^\lambda \oplus G_e^\lambda = X_\lambda x X_\lambda,\ G_i^\lambda \subseteq \Pi_\lambda G_i,\ G_e^\lambda \subseteq \Pi_\lambda G_e,\quad \text{or}\quad G_i^\lambda \subseteq \Pi_\lambda G_e,\ G_e^\lambda \subseteq \Pi_\lambda G_i,$$

according to whether the natural orientation of the normal to the surface Γ_λ is the same as that for Ω, or not. Note that generally we don't have equality in (304) ! We easily see that the graphs $G_{e,i}$ have the decompositions:

(305) $$G_i = \bigoplus_{j=1 \text{ to } N} G_i(\Omega_j),\ G_e = G_e(\Omega_0) \oplus \bigoplus_{j=1 \text{ to } N'} G_i(\Omega'_j),\ \text{with } \Omega_0 = R^3 \backslash \bar{\Omega}'0,$$

(with obvious notations), but $G_i(\Omega_j)$ is different from the sum $\bigoplus_{\Gamma_\lambda \subseteq \partial\Omega_j} G_i^\lambda$.

6.2. Screen effect. Extinction theorem

In the general case, if (M,J) is in G_e (resp. G_i) then the electromagnetic field (E,H) due to (M,J) is zero in the domain Ω (resp.Ω'). This is generally known as the *extinction theorem*.

Thus for each λ, if (M_λ, J_λ) is in $G^\lambda{}_e$ (resp. $G^\lambda{}_i$), the electromagnetic field (E_λ, H_λ) satisfies:

(306) $$E_\lambda(x) = 0,\quad H_\lambda(x) = 0 \quad \text{for } x \in \Omega_\lambda \text{ (resp. } \Omega'_\lambda).$$

Thus currents in $G^\lambda{}_i$ (resp. $G^\lambda{}_e$) do not influence the exterior (resp.interior) of Ω_λ. This implies:

(307) $$S_{\mu\lambda} P_i^\lambda = 0 \text{ if } \Gamma_\mu \subseteq \Omega'_\lambda,\qquad S_{\mu\lambda} P_e^\lambda = 0 \text{ if } \Gamma_\mu \subseteq \Omega_\lambda.$$

More generally, if (M_λ, J_λ) is in $G_i(\Omega_j)$ for all λ so that $\Gamma_\lambda \subseteq \partial\Omega_j$, then the field:

$(E,H) = \sum_\lambda (E_\lambda, H_\lambda)$ is zero in $R^3 \backslash \Omega_j$. Let Ω'_{jo} be its unbounded component.

Thus *currents on its boundary prevent interior currents* (on the boundary of Ω_j) *to act on points of* Ω'_{jo}.

6.3. Coefficients of mutual influence of antennas

If k^2 is not an eigenvalue of the operator A in any bounded domain (Ω_j or Ω'_i), then we can define Calderon operators $C^{e,i}$ as above. We can write, using standard notation, for the exterior and interior domains respectively:

$$(308) \qquad C^{e,i}((n \wedge E|_{\Gamma_\lambda})_{\lambda \in I}) = (n \wedge H|_{\Gamma_{\lambda e,i}})_{\lambda \in I});$$

thus C^e and C^i are matrices: $C^{e,i} = (C^{e,i}_{\lambda\mu})_{\lambda,\mu \in I x I}$. We have the decompositions:

$$(309) \quad C^i = \underset{j=1,\dots,N}{\oplus} C^i(\Omega_j), \quad C^e = C^e(\Omega_0) \oplus \underset{l=1,\dots,N'}{\oplus} C^i(\Omega'_l), \quad \text{but } C^i(\Omega_j) \neq \underset{\lambda}{\oplus} C^i(\Omega_\lambda).$$

Notice that the exterior Calderon operator for Ω_0 (the complementary set of the unbounded component of Ω') always exists. This operator takes into account the mutual influence of the different bounded obstacles, i.e., the components of Ω_0.

7. MUTUAL INFLUENCE OF SHEETS

7.1. Introduction

Let Γ be a (Lipschitz) connected surface which is the boundary of a bounded open set Ω in R^3. Let Γ_λ, $\lambda \in I$, be a partition of Γ. This partition can be made in order to apply a numerical method, or to decompose Γ into more regular pieces (Γ being a polyhedral surface), or into surfaces enlightened by a wave plane or in the shadow, in order to consider high frequencies problems.

Then to a current J in $H^{-1/2}(\text{div},\Gamma)$, corresponds the family (J_λ) $\lambda \in I$, with $J_\lambda \in X(\text{div},\Gamma_\lambda)$ (see (87) chap. 2) with matching conditions in the case of two pieces see (89), in the more general case these matching conditions are more tedious and technical to write (obviously $H^{-1/2}(\text{div},\Gamma)$ is not the sum of the spaces $X(\text{div},\Gamma_\lambda)$). If we assume that each current on Γ_λ is regular, that is in $H^s(\text{div},\Gamma_\lambda)$ for some positive number s, then these conditions are easy to write:

Proposition 28. *Let* (Γ_λ), $\lambda \in I$, *be a partition of* Γ, *with* Γ_λ *regular. Let* (J_λ) $\lambda \in I$, *be regular currents on* Γ_λ (J_λ *in* $H^s(\text{div},\Gamma_\lambda)$ *with positive* s). *Then* (J_λ) $\lambda \in I$ *defines a current* J *in* $H^{-1/2}(\text{div},\Gamma)$ *if the following matching conditions are satisfied*

$$(310) \qquad \nu_\lambda . J_\lambda - \nu_\mu . J_\mu = 0 \text{ on } \Lambda_{\lambda\mu} = \bar{\Gamma}_\lambda \cap \bar{\Gamma}_\mu, \ \forall \lambda,\mu \in I,$$

with ν_λ *the exterior unit normal in* Γ_λ *to its boundary.*

PROOF. This results from the usual jump formula of the divergence on surfaces (the analogue to (54))

$$(311) \qquad \text{div}_\Gamma J = \sum_\lambda (\text{div}_\Gamma J_\lambda) + \sum_\lambda \varepsilon_\lambda \nu_\lambda . J_\lambda,$$

with ($\text{div}_\Gamma J_\lambda$) in the classical sense, $\varepsilon_\lambda = +1$ or -1 according to the orientation of $\partial\Gamma_\lambda$. Then we have J in $L^2_t(\Gamma)$ and div_Γ J in $L^2(\Gamma)$, thus J is in $H^0(\text{div},\Gamma)$, and therefore in $H^{-1/2}(\text{div},\Gamma)$. ⊗

Obviously we have an analogous proposition by changing div_Γ into curl_Γ, with the matching conditions:

(312) $\nu_\lambda \wedge J_\lambda - \nu_\mu \wedge J_\mu = 0 \;\text{ on } \Lambda_{\lambda\mu} = \bar{\Gamma}_\lambda \cap \bar{\Gamma}_\mu, \quad \forall \lambda,\mu \in I.$

7.2. Electromagnetic field due to currents on a sheet

Here we study the electromagnetic field produced by a regular current J on a sheet Γ_0 which does not satisfy $\nu.J = 0$ on the boundary Λ of Γ_0, and thus is not in the space $H_0^{-1/2}(\text{div},\Gamma_0)$ (see (85) and (92) chap. 2), so that we know that (E,H) is not of locally finite energy in the neighborhood of the boundary Λ of Γ_0.
The electromagnetic field (E,H) produced by J is given by (96) (with M = 0). Its expression will be given thanks to the jump formula (which is a result of differential geometry, see (94), (273) and (278) in the Appendix and also for instance Schwartz [1] (IX,3;24))

(313) $\text{curl}\, J = (\text{curl}\, J)_{cl} - \nu \wedge J_\Lambda\, \delta_\Lambda$, ν the exterior unit normal to Λ in Γ_0,

where $(\;)_{cl}$ is the "classical" part (*on* Γ_0), whereas the last term is concentrated on Λ. We can also prove (313) using the formula (see (310) in the Appendix)

(314) $<\text{curl}\, J, \zeta> \overset{\text{def}}{=} <J, \text{curl}\,\zeta> = \int_{\Gamma_0} J.\,[\frac{\partial}{\partial n}(n \wedge \zeta) - \overrightarrow{\text{curl}}_\Gamma\, \zeta^n + 2R_m\, n \wedge \zeta\,]\, d\Gamma,$

with $\zeta \in D(R^3)^3$. Thus we can separate curl J into its transverse and tangential part. Applying the Stokes formula, we obtain the boundary term, and the expression of $(\;)_{cl}$ is given by:

(315) $<(\text{curl}\, J)_{cl}\,, \zeta> = \int_{\Gamma_0} (\text{curl}_{\Gamma_0} J\,.\zeta^n + J.\frac{\partial}{\partial n}(n \wedge \zeta) + 2R_m J.\, n \wedge \zeta)\, d\Gamma_0\,,$

for all $\zeta \in D(R^3)^3$. For the gradient we also have:

(316) $\text{grad}\,\rho = (\text{grad}\,\rho)_{cl} - \nu\,\rho_\Lambda\,\delta_\Lambda\,,$
with

(317) $<(\text{grad}\,\rho)_{cl}, \zeta> = \int_{\Gamma_0} (\text{grad}_{\Gamma_0}\rho.\,\zeta - \rho\frac{\partial\zeta^n}{\partial n} - 2R_m\rho\zeta^n)\, d\Gamma_0, \;\; \forall \zeta \in D(R^3)^3.$

The expressions of $(\text{grad}\,\rho)_{cl}$ and $(\text{curl}\, J)_{cl}$ contain the transverse distributions $G_1\rho$ and $G_2 J$:

(318) $<G_1\rho,\zeta> = -\int_{\Gamma_0} \rho\frac{\partial\zeta^n}{\partial n}\, d\Gamma_0, \quad <G_2 J,\zeta> = \int_{\Gamma_0} J\frac{\partial\zeta^n}{\partial n}\, d\Gamma_0, \quad \forall \zeta \in D(R^3)^3,$

with the properties, for all regular ρ and J:

$$\phi\,(\Phi * G_1\rho) \in H^1(R^3\backslash\Gamma_0),\quad \phi\,(\Phi * G_2 J) \in H^1(R^3\backslash\Gamma_0)^3,\quad \forall \phi \in \mathcal{D}(R^3).$$

Proposition 29. *The electromagnetic field due to a "regular" electric current J on a sheet Γ_0 which does not satisfy $\nu.J = 0$ on the boundary Λ of Γ_0 is decomposed into its regular part (E_{reg}, H_{reg}) and its singular part (E_{sing}, H_{sing}) according to:*

$$(319)\quad \left| \begin{array}{l} (E,H) = (E_{reg}, H_{reg}) + (E_{sing}, H_{sing}), \\ E_{reg} = ikZ\,(\mathcal{L}J + \frac{i\omega}{k^2}\mathcal{L}\,(\text{grad}\,\rho)_{cl}),\ H_{reg} = \mathcal{L}\,(\text{curl}\,J)_{cl},\ \text{with } i\omega\rho = (\text{div}_\Gamma J)_{cl}, \\ E_{sing} = -\,i\,\frac{Z}{k}\,\mathcal{L}\,[i\omega\nu\rho_\Lambda\delta_\Lambda + \text{grad}\,(\nu.J_\Lambda\delta_\Lambda)],\quad H_{sing} = -\,\mathcal{L}\,(\nu\wedge J_\Lambda\delta_\Lambda). \end{array} \right.$$

with

$$(320)\quad \left| \begin{array}{l} \phi\,E_{reg} \text{ and } \phi\,H_{reg} \in H^1(R^3\backslash\Gamma_0)^3,\quad \forall\phi \in \mathcal{D}(R^3), \\ E_{sing} \text{ and } H_{sing} \in C^\infty(R^3\backslash\Lambda)^3, \\ \text{with } E_{sing}(x) = O(\frac{1}{r}),\ H_{sing} = O(\log r),\ r = d(x,\Lambda). \end{array} \right.$$

and thus

$$(321)\qquad \phi\,E \text{ and } \phi\,H \in H^1(R^3\backslash\Gamma_0)^3,\quad \forall\phi \in \mathcal{D}(R^3\backslash\Lambda).$$

PROOF. The electric field E is given by (96):

$$(322)\quad \left| \begin{array}{l} E = ikZ\,(\mathcal{L}J + \frac{1}{k^2}\,\text{grad}\,\mathcal{L}\,\text{div}_\Gamma J\,) \\ \quad = ikZ\,(\mathcal{L}J + \frac{1}{k^2}\,\text{grad}\,\mathcal{L}\,(\text{div}_\Gamma J)_{cl}) - i\frac{Z}{k}\,\text{grad}\,\mathcal{L}(\nu.J_\Lambda\,\delta_\Lambda). \end{array} \right.$$

We also have:

$$(323)\quad \text{grad}\,\mathcal{L}\,(\text{div}_\Gamma J)_{cl} = i\omega\,\text{grad}\,\mathcal{L}\rho = i\omega\,\Phi * \text{grad}\,\rho = i\omega\,\Phi * [(\text{grad}\,\rho)_{cl} - \nu\,\rho_\Lambda\delta_\Lambda],$$

which gives (319) for the electric field E. For H, this results from:

$$(324)\qquad H = \text{curl}\,\mathcal{L}J = \text{curl}\,\Phi * J = \Phi * \text{curl}\,J,$$

and then we apply (313). Properties (320) on (E_{reg}, H_{reg}) follow from the usual regularity properties of the single layer potentials (see section 1.3.1). For (E_{sing}, H_{sing}), this is a consequence of the properties of the Newtonian potentials relative to a charge density concentrated on a closed line (see below). We have a more precise result

$$(325)\quad E_{sing}(x) = -\,\frac{iZ}{2\pi k}\,(\nu.J_\Lambda)\,\frac{\alpha}{r} + O\,(\log r),\quad H_{sing}(x) = -\,\frac{1}{2\pi}\,\nu\wedge J_\Lambda\,\log r + O\,(1),$$

with $\alpha = \frac{1}{r}\,(x - \pi_\Lambda x)$, $r = d(x,\Lambda)$, $\pi_\Lambda x$ = the projection of x on Λ.

⊗

7.3. Matrix elements; influence coefficients

On the *sheet* Γ_0, we can decompose RJ and TJ into their regular and singular parts

$$(326)\quad \begin{cases} RJ = (RJ)_{reg} + (RJ)_{sing}, & TJ = (TJ)_{reg} + (TJ)_{sing}, \\ (RJ)_{reg} = 2ik\, n \wedge [\, LJ + \frac{i\omega}{k^2} L(\mathrm{grad}_\Gamma \rho)_{cl}], & (TJ)_{reg} = -\,2n \wedge L(\mathrm{curl}\, J)_{cl}, \\ (RJ)_{sing} = -\frac{2i}{k} n \wedge L[\, i\omega\nu\rho_\Lambda\delta_\Lambda + \mathrm{grad}\,(\nu.J_\Lambda\delta_\Lambda)], & (TJ)_{sing} = 2n \wedge L(\nu \wedge J_\Lambda\delta_\Lambda). \end{cases}$$

When we have to calculate the matrix elements of the Calderon projectors, or of the S operator, we have to calculate matrix elements of R and T (see (102), (104) and (273)) between the sheets Γ_λ and Γ_μ (giving the *mutual influence* between Γ_λ and Γ_μ)

$$(327)\quad \begin{cases} <RJ, n \wedge \tilde{J}> = \sum_{\lambda\mu} <RJ_\lambda, n \wedge \tilde{J}_\mu>, \quad <TJ, n \wedge \tilde{J}> = \sum_{\lambda\mu} <TJ_\lambda, n \wedge \tilde{J}_\mu>, \\ \qquad \forall\, J, \tilde{J} \text{ piecewise "regular" currents on } \Gamma. \end{cases}$$

We have four types of elements according to:

i) $\bar{\Gamma}_\lambda \cap \bar{\Gamma}_\mu = \varnothing$, ii) $\Gamma_\lambda \cap \Gamma_\mu$ has one point, iii) $\bar{\Gamma}_\lambda$ and $\bar{\Gamma}_\mu$ has a common line, iv) $\Gamma_\lambda = \Gamma_\mu$. In cases ii) to iv) there are singular integrals to calculate for the matrix elements of T. We can use Stokes formula:

$$(328)\quad \begin{aligned} <TJ_\lambda, n \wedge M_\mu> &= \int_{\Gamma_\lambda x \Gamma_\mu} \mathrm{grad}_{\Gamma,y} \Phi(x-y)\,(J_\lambda(y) \wedge M_\mu(x))\, d\Gamma_y\, d\Gamma_x \\ &= \int_{\Gamma_\lambda x \Gamma_\mu} \Phi(x-y)\, \mathrm{div}_{\Gamma,y} (J_\lambda(y) \wedge M_\mu(x))\, d\Gamma_y\, d\Gamma_x \\ &\quad + \int_{\partial\Gamma_\lambda x \Gamma_\mu} \Phi(x-y)\, \nu_y (J_\lambda(y) \wedge M_\mu(x))\, dL_y\, d\Gamma_x \end{aligned}$$

and then with $\Lambda_\lambda = \partial\Gamma_\lambda$, the last term can be calculated by:

$$(329)\quad \int_{\Lambda_\lambda} dL_y\,(n_y \wedge J_\lambda(y)).\int_{\Gamma_\mu} \Phi(x-y)\, M_\mu(x)\, d\Gamma_x.$$

We refer for instance to Bendali [1], Heliot [1] for the explicit computation of such matrix elements with P_1 functions.

8. FIELDS DUE TO CURRENTS ON A LINE

Let $J = J_\Lambda\, \delta_\Lambda$ be an electric current concentrated on a (closed, i.e. without boundary) bounded regular curve Λ, with J_Λ regular, tangent to this curve. Then the electromagnetic field (E,H) due to this current in free space, which satisfies the Silver-Müller conditions at infinity is given by (44) with M = 0, that is:

$$(330)\quad E = i\omega\mu\,(A + k^{-2}\,\mathrm{grad\, div}\, A)\,,\quad H = \mathrm{curl}\, A\,,\quad \text{with } A = \Phi * J = \mathcal{L}J = \mathcal{L}(j\, e_\Lambda \delta_\Lambda),$$

with $J_\Lambda = j\, e_\Lambda$, e_Λ unit vector tangent to Λ; A is called the *Hertz potential*. We have:

$$\text{(331)} \qquad \operatorname{div} J = \frac{\partial j}{\partial s}\,\delta_\Lambda ,$$

with s the usual curvilinear coordinate on Λ; we can write (330) in the form

$$\text{(332)} \qquad E = i\omega\mu\,(LJ + k^{-2}\operatorname{grad} L(\frac{\partial j}{\partial \Lambda}\delta_\Lambda)), \qquad H = \operatorname{curl}(LJ).$$

We know that (E,H) cannot be of finite energy in the neighborhood of Λ, so we have to study the behavior of (E,H) in the neighborhood of Λ. This follows from the Lemma:

Lemma 4. *Let $\rho_0 d\Lambda$ be a distribution concentrated on the closed line Λ with regular density. The behavior of the Newtonian potential u of $-\rho_0 d\Lambda$, in the neighborhood of Λ is given by (with $x = (\rho\alpha, s)$ in local coordinates):*

$$\text{(333)} \qquad u(x) = \frac{1}{4\pi}\int_\Lambda \frac{1}{|x - y(s)|}\rho_0(y(s))\,ds = -\rho_0(\pi_\Lambda x)\frac{1}{2\pi}\operatorname{Log}\rho + O(1),$$

$\pi_\Lambda x$ being the projection of x onto Λ, $\rho = d(x,\Lambda) = |x - \pi_\Lambda x|$,

$$\text{(334)} \qquad \left| \begin{array}{l} \alpha.\operatorname{grad} u(x) = -\rho_0(\pi_\Lambda x)\dfrac{\alpha}{2\pi\rho} + O(\log\rho), \ \ \alpha \textit{ unit vector orthogonal to } \Lambda, \\[2ex] \alpha.\operatorname{grad} u(x) = -\rho_0'(\pi_\Lambda x)\dfrac{1}{2\pi}\log\rho, \ \ \alpha = e_\Lambda(s) \textit{ along the tangent to } \Lambda. \end{array} \right.$$

PROOF. First with the Lipschitz hypothesis on ρ_0 we have:

$$\text{(335)} \quad 4\pi\, u(x) - \rho_0(\pi_\Lambda x)\int_\Lambda \frac{1}{|x - y(s)|}\,ds = \int_\Lambda \frac{1}{|x - y(s)|}(\rho_0(y(s)) - \rho_0(\pi_\Lambda x))\,ds \le C.$$

Therefore the behavior of u(x) in the neighborhood of Λ is given by that of v(x):

$$\text{(336)} \qquad v(x) = \rho_0(\pi_\Lambda x)\frac{1}{4\pi}w(x) \qquad \text{with } \ w(x) = \int_\Lambda \frac{1}{|x - y(s)|}\,ds\,;$$

with $a < 0 < b$ and $y(0) = 0$, $y'(0) = (0,0,1) = e_\Lambda(0) = e_3$ (choosing $s = 0$ for $\pi_\Lambda x$), we have: $w(x) \approx \int_a^b (\rho^2 + y_3^2)^{-1/2}\,dy_3 \approx -2\log\rho$.

The gradient of u satisfies:

$$\text{(337)} \qquad \frac{\partial u}{\partial x_i}(x) = \frac{1}{4\pi}\int_\Lambda |x - y(s)|^{-2}\beta_i\,\rho_0(y(s))\,ds,$$

with $\beta_i = \dfrac{y_i - x_i}{r}$, $r = |x - y|$, and thus (with $y = y(s)$):

$$\text{(338)} \quad 4\pi\frac{\partial u}{\partial x_i}(x) - \rho_0(\pi_\Lambda x)\int_\Lambda |x - y|^{-2}\beta_i\,ds = \int_\Lambda |x - y|^{-2}\beta_i\,[\rho_0(y) - \rho_0(\pi_\Lambda x)]ds.$$

If ρ_0 is $C^{1,1}$ on Λ, we have

thus:
$$\rho_o(y(s)) - \rho_0(y(0)) = \rho_0(y) - \rho_0(\pi_\Lambda x) = \rho_0'(\pi_\Lambda x)\, s + O(s^2),$$

$$(339)\quad 4\pi \frac{\partial u}{\partial x_i}(x) - \rho_0(\pi_\Lambda x) \int_\Lambda \beta_i\, |x-y|^{-2}\, ds - \rho_0'(\pi_\Lambda x) \int_\Lambda \beta_i\, |x-y|^{-2}\, s\, ds = O(1).$$

The gradient of w(x), given by $\frac{\partial w}{\partial x_i} = \int_\Lambda |x-y|^{-2} \beta_i\, ds$, $i = 1$ to 3, satisfies:

$$(340)\quad \begin{cases} \dfrac{\partial w}{\partial x_i} \approx -2\dfrac{\alpha_i}{\rho} \text{ for } i = 1, 2,\ x_i \text{ orthogonal to } \Lambda,\ \alpha_i = (x - \pi_\Lambda x)_i / \rho, \\ \dfrac{\partial w}{\partial x_3} \approx 0 \text{ along the tangent to } \Lambda\ (\beta_3 \text{ is to first order an odd function of s}). \end{cases}$$

Then we have for $i = 3$:

$$(341)\quad \int_\Lambda \beta_i\, |x-y|^{-2}\, s\, ds \approx 2\int_0^a \frac{y_3}{\rho^2 + y_3^2}\, dy_3 = \int_0^{a^2} \frac{1}{\rho^2 + t}\, dt \approx -2\log\rho,$$

and for $i = 1, 2$:

$$(341)'\quad \int_\Lambda \beta_i\, |x-y|^{-2}\, s\, ds \approx 2\int_0^a \frac{\rho\alpha_i}{(\rho^2 + y_3^2)^{3/2}}\, dy_3 = -2\alpha_i \frac{d}{d\rho} \int_0^a \frac{1}{(\rho^2 + y_3^2)^{1/2}}\, dy_3 \approx \frac{2\alpha_i}{\rho}$$

from which we obtain (334).

⊗

Proposition 30. *The electromagnetic field* (E,H) *due to a line electric current* J *on a closed bounded curve, by* (332) *has the following behavior in the neighborhood of the line, with notations of Lemma 4 and with local coordinates* $x = (\rho\alpha, s)$*:*

$$(342)\quad \begin{cases} E_{tr}(x) \approx -i\omega\mu\, k^{-2} \dfrac{1}{\rho} \dfrac{\partial j}{\partial s} \dfrac{\alpha}{2\pi}, & H_{tr}(x) \approx \dfrac{1}{2\pi\rho}(\alpha \wedge j\, e_s) \\ E_s(x) = i\omega\mu \dfrac{1}{2\pi} \log\rho\, [-j - k^{-2} \cdot \dfrac{\partial^2 j}{\partial s^2}], & H_s(x) = O(\log\rho), \end{cases}$$

with $J = je_s\,\delta_\Lambda$, $e_s = e_\Lambda(s)$ *a unit vector along the line* Λ*, and* (E_s, H_s) *the components of* (E,H) *along* e_s, (E_{tr}, H_{tr}) *the transverse components.*

PROOF. Formula (332) for H gives:

$$(343)\quad H(x) = \int_\Lambda \frac{\partial \Phi}{\partial r}(\beta \wedge j\, e_r)\, ds, \quad \text{with } \frac{\partial \Phi}{\partial r} = \frac{1}{4\pi} \frac{e^{ikr}(1 - ikr)}{r^2}, \quad r = |x - y|.$$

Then we obtain (342) for H, and also for E due to Lemma 4.

⊗

The electromagnetic field (E,H) due to a line current J has the following behavior at infinity:

$$(344)\quad \begin{cases} E(r\alpha) \approx i\omega\mu\,[\hat{J}(k\alpha) - \alpha\,(\alpha.\hat{J}(k\alpha))]\,\Phi(r), & \alpha \in S^2, \\ H(r\alpha) \approx ik\alpha \wedge \hat{J}(k\alpha)\,\Phi(r) & \text{for } r = |x| \to \infty, \end{cases}$$

with $\hat{J}(k\alpha) = \langle e^{-ik\alpha.x}, J\rangle = \int_\Lambda e^{-ik\alpha.x(s)}\, j\,(s)\, e_s\, ds.$

These formulas are an easy consequence of (30), and of:

$$(345)\quad \begin{cases} (\widehat{\mathrm{div}}\, J)(k\alpha) = \langle e^{-ik\alpha.x}, \mathrm{div}\, J\rangle = ik\alpha \langle e^{-ik\alpha.x}, J\rangle = ik\alpha\,\hat{J}(k\alpha), \\ (\widehat{\mathrm{curl}}\, J)(k\alpha) = \langle e^{-ik\alpha.x}, \mathrm{curl}\, J\rangle = ik\alpha \wedge \hat{J}(k\alpha). \end{cases}$$

Remark 21. So far we have described a *radiating antenna*. The modelling by a line of a thin *receiving antenna* with a very conducting medium and the modelling of the induced current by a Dirac distribution concentrated on a line (according to the Faraday's effect) is outside the scope of this book.

⊗

9. SCATTERING PROBLEMS BY A CHIRAL OBSTACLE

We consider the scattering problem of section 5, but with a linear chiral obstacle (see chap. 1.3.3) instead of a dielectric medium. Then we have to find an electromagnetic field described by (D,B), (E,H) in the obstacle Ω satisfying:

$$(346)\quad \begin{cases} \text{i) } \mathrm{curl}\, H + i\omega D = 0, \\ \text{ii) } -\mathrm{curl}\, E + i\omega B = 0 \ \text{ in } \Omega, \end{cases}$$

with the constitutive relations (see (96) chap. 1, where ε, μ, $\tilde{\varepsilon}$, $\tilde{\mu}$ may be matrices):

$$(347)\quad \begin{cases} \text{i) } D = \varepsilon\, E + \tilde{\varepsilon}\, H, \\ \text{ii) } B = \tilde{\mu}\, E + \mu\, H, \end{cases}$$

with hypothesis **H18)** chap. 1 of a dissipative medium, and with the boundary conditions (262), giving (265) on E. We emphasize that the boundary conditions are unchanged with the constitutive relations (at least if the medium is at rest), because they are due to the curl operator!

We can easily prove uniqueness (at most) of the solution of this problem as follows.

Multiply (346)i) by $\bar{E}$ and ii) by $\bar{H}$, integrate and add, we obtain:

$$(348)\quad \left|\begin{aligned} &\mathrm{Re} - i\omega\,[(D,E)+(B,H)] = \mathrm{Re}\int_\Omega (\mathrm{curl}\,H.\,\bar{E} - \mathrm{curl}\,E.\,\bar{H})\,dx \\ &\qquad = -\,\mathrm{Re}\int_\Omega \mathrm{div}\,(E\wedge\bar{H})\,dx = -\,\mathrm{Re}\int_\Gamma n.\,E\wedge\bar{H}\,d\Gamma. \end{aligned}\right.$$

To prove uniqueness of the solution, we can take $E_I = H_I = 0$ and thus

$$\mathrm{Re}\int_\Gamma n.E\wedge\bar{H}\,d\Gamma = \mathrm{Re}\int_\Gamma n\wedge E.\bar{H}\,d\Gamma = \mathrm{Re}\int_\Gamma n\wedge E_\Gamma.\bar{H}_\Gamma\,d\Gamma \geq 0$$

(see (202)). But from the dissipativity hypothesis **H18)** chap. 1 the l.h.s. of (348) is strictly positive except for $E = H = 0$, which proves uniqueness!

Then proving that this problem depends on the Fredholm alternative (like in section (4.8.2)) implies existence and uniqueness of the solution.

Here we develop a variational method with (slightly) different hypotheses. We write Maxwell equations (346) with (347) with E and H only:

$$(349)\quad \left|\begin{aligned} &\text{i) } \mathrm{curl}\,H + i\omega\tilde{\varepsilon}\,H + i\omega\varepsilon\,E = 0 \\ &\text{ii) } -\,\mathrm{curl}\,E + i\omega\tilde{\mu}\,E + i\omega\mu\,H = 0 \text{ in } \Omega. \end{aligned}\right.$$

We can eliminate H applying $-(\mathrm{curl} + i\omega\tilde{\varepsilon})\,(i\omega\mu)^{-1}$ to (349)ii) adding to i)

$$(350)\qquad -\,(\mathrm{curl} + i\omega\tilde{\varepsilon})(i\omega\mu)^{-1}(-\,\mathrm{curl} + i\omega\tilde{\mu})E + i\omega\varepsilon\,E = 0.$$

It is better to use a variational framework. We define the sesquilinear form

$$(351)\quad \left|\begin{aligned} &a(E,F) = -\int_\Omega [(i\omega\mu)^{-1}(\mathrm{curl}\,E - i\omega\tilde{\mu}\,E)\,.\,(\mathrm{curl}\,\bar{F} + i\omega\,{}^t\tilde{\varepsilon}\,\bar{F}) + i\omega\varepsilon\,E\,.\,\bar{F}\,]\,dx \\ &\qquad\qquad - \int_\Gamma C^e(n\wedge E).\,\bar{F}\,d\Gamma, \end{aligned}\right.$$

with ${}^t\tilde{\varepsilon}$ the transposed matrix of $\tilde{\varepsilon}$, and E, F in H(curl, Ω). The scattering problem is

$$(352)\qquad a(E,F) = \int_\Gamma f_I\,.\,\bar{F}\,d\Gamma, \qquad \forall\, F \in H(\mathrm{curl},\Omega).$$

We assume that

$$(353)\qquad \tilde{\varepsilon} = \tilde{\mu}^*, \quad \text{and} \quad \mathrm{Re}\,(-\,i\omega\,\kappa\xi\,.\,\bar{\xi}) \geq \alpha\,|\xi|^2, \quad \forall\,\xi\in C^3, \text{ for } \kappa = \varepsilon, \mu,$$

with $\alpha > 0$ (see H17 chap.1). Thanks to (201) (and to Remark 8 chap.1) we have

$$(354)\qquad \mathrm{Re}\,a(E,E) \geq \alpha\,[\|\mathrm{curl}\,E - i\omega\tilde{\mu}\,E\|^2 + \|E\|^2], \qquad \forall\, E \in H(\mathrm{curl},\Omega).$$

But we can obtain easily: there are positive numbers α_o, β, so that

$$(355)\quad \|\mathrm{curl}\,E - i\omega\tilde{\mu}\,E\|^2 \leq \alpha_o\,[\|\mathrm{curl}\,E\|^2 + \|E\|^2] \leq \beta\,[\|\mathrm{curl}\,E - i\omega\tilde{\mu}\,E\|^2 + \|E\|^2]$$

for all $E \in H(\mathrm{curl},\Omega)$, and thus the norms:

$$\|E\|_1 \stackrel{\text{def}}{=} [\|\text{curl } E - i\omega\tilde{\mu} E\|^2 + \|E\|^2]^{1/2} \text{ and } \|E\|_{H(\text{curl},\Omega)} \stackrel{\text{def}}{=} [\|\text{curl } E\|^2 + \|E\|^2]^{1/2}$$

are equivalent. Therefore the sesquilinear form a(E, F) is coercive on H(curl, Ω). The Lax-Milgram lemma implies existence and uniqueness of the solution of the problem (352) like in section 5.1.

Remark 22. The constitutive relations are often written in different forms, for instance:

$$(356) \qquad D = \varepsilon_c E + \gamma_{em} B, \quad H = \frac{1}{\mu_c} B - \gamma_{me} E,$$

with $\gamma_{em} = -\gamma_{me}$, which is equivalent to (347) with:

$$(357) \qquad \mu = \mu_c, \ \tilde{\mu} = \mu_c \gamma_{me}, \ \varepsilon = \varepsilon_c + \mu_c \gamma_{me}, \ \tilde{\varepsilon} = \gamma_{em} \mu_c = -\tilde{\mu}.$$

⊗

10. CONCLUSION ON THE CALDERON OPERATORS FOR SCATTERING PROBLEMS

Let Ω be a bounded domain occupied by a homogeneous linear isotropic dielectric with permittivity ε and permeability μ. Then *all the electromagnetic properties of this obstacle* (with respect to an incident wave at frequency ω) *are contained in the interior Calderon operator* C^i (see (211)), if ω is a "regular" frequency, that is $k^2 = \omega^2 \varepsilon\mu$ is not an eigenvalue of the operator A defined by:

$$Au = -\Delta u, \ D(A) = \{u \in L^2(\Omega)^3, \ \text{curl } u \in L^2(\Omega)^3, \ \text{div } u = 0, \ n \wedge u|_\Gamma = 0\}.$$

Using a more physical minded terminology we can call it a surface admittance operator. The essential property of C^i is that its graph corresponds (up to the vector product with the normal) to the set of tangential boundary values of the electromagnetic field in Ω. We recall also the positivity property (212).
Less usual for physicists is the similar notion for the exterior domain, which is generally occupied by free space. Then *all the electromagnetic properties of the exterior domain* (at frequency ω) *are contained in the exterior Calderon operator* C^e: this is a surface admittance operator for the free space domain; when ω = 0, this corresponds to the "usual" notion of a capacity operator. Like for C^i the essential property of C^e is that its graph corresponds (up to the vector product with the normal) to the set of tangential boundary values of the electromagnetic field in the exterior domain Ω'(taking into account the Silver-Müller condition and the local finite energy up to the boundary Γ of Ω'). We recall also the positivity property (201).
The main differences between C^i, C^e and the Calderon projectors P_i, P_e are:
i) P_i and P_e keep informations on both sides of Γ, and exist in all frequencies,
ii) P_i and P_e are integral operators (whereas C^i and C^e are *pseudodifferential operators of zeroth order*; for this notion, see for instance Hörmander [1], Gilkey [1], Taylor [1]),

iii) the range spaces $G^e = \text{Im } P^e$ and $G^i = \text{Im } P^i$ are (up to a minus sign) respectively the graph of C^e and C^i for regular frequencies.
Let Γ be a regular surface which is the boundary of a bounded domain Ω.
The space Y of magnetic and electric currents (M,J) on Γ which produce an electromagnetic field (E,H) (at angular frequency ω) in free space, with locally finite energy on both sides of Γ and satisfying the outgoing wave condition at infinity, is decomposed into spaces corresponding to tangential interior and exterior boundary values of the electromagnetic field (E,H) (up to the vector product with the normal) $G^i_{\ o}$ and $G^e_{\ o}$ according to:

(358) $$Y = H^{-1/2}(\text{div},\Gamma) \times H^{-1/2}(\text{div},\Gamma) = G^e_o \oplus G^i_o .$$

Now let the interior domain Ω be occupied by a medium with characteristics (ε_1,μ_1) at ω, and let $G^i_{\ 1}$ be the space corresponding to boundary values of the electromagnetic field in Ω.
Let (E_I,H_I) be an incident wave at frequency ω. The scattering problem is to find the reflected wave (E_r,H_r) and the transmitted wave (E_t,H_t) inside Ω so that:

(359) $$n\wedge E_I = n\wedge E_t - n\wedge E_r,\quad n\wedge H_I = n\wedge H_t - n\wedge H_r ,$$

that is to decompose space Y into the direct sum:

(360) $$Y = G^e_o \oplus G^i_1 .$$

Using the interior Calderon projector P^o_i for the exterior wavenumber k_o, we see from (271) section 3.5 that this problem is reduced to:

the restriction of P^o_i to G^i_1 *is an isomorphism from* G^i_1 onto G^i_o, i.e.,
there exists $\alpha > 0$ such that $\|P^o_i(M,J)\| \geq \alpha\, \|(M,J)\|, \quad \forall (M,J) \in G^1_i$.

We can use the Calderon operators $C^{i,e}$ to solve numerically scattering problems by using special basis. We develop this in the Helmholtz case, for $C^e = C^e_o$.
Let $\Phi^{x_o}(x) = \Phi(x - x_o)$ be the outgoing elementary solution of the Helmholtz equation centered at x_o. Let F be a subset of R^3. We obtain the following, setting:

(361) $$M_F = \{(\Phi^{x_o}|_\Gamma, \frac{\partial \Phi^{x_o}}{\partial n}|_\Gamma), x_o \in F\}.$$

Proposition 31. *We assume that* F *is a dense subset of the boundary* Γ_o *of an open subset* Ω_o *of* Ω, *strictly contained in* Ω, *and such that* $-k^2$ *is not an eigenvalue of the Dirichlet Laplacian in* Ω_o. *Then* M_F *is a total family in* $G_e = G(C^e)$, i.e. *the vector space* $[M_F]$ *generated by finite linear combinations of elements of* M_F *is dense in* G_e.

Proposition 31 implies that given (u^o,u^1) in G_e, (thus $u^1 = C^e u^o$) and $\eta > 0$, there exist (x_j) in Γ_o and complex numbers (α_j) such that

(362) $$\|u^o - \sum_j \alpha_j \Phi^{x_j}|_\Gamma\| \leq \eta,\quad \|u^1 - \sum_j \alpha_j \frac{\partial \Phi^{x_j}}{\partial n}|_\Gamma\| \leq \eta.$$

PROOF of Proposition 31 (adapted from Petit-Cadilhac [2]). We show that the polar set $[M_F]^o$ of $[M_F]$ for the pairing (143) is equal to the polar set of G_e, which is G_e. Let (v^o,v^1) be in $[M_F]^o$; (v^o,v^1) satisfies:

(363) $$<\Phi^{x_o}|_\Gamma, v^1> - <\frac{\partial\Phi^{x_o}}{\partial n}|_\Gamma, v^o> = 0, \quad \forall x_o \in F,$$

thus

(363)' $$\int_\Gamma (v^1(y)\Phi(y-x_o) - v^o(y)\frac{\partial\Phi}{\partial n}(y-x_o))\, d\Gamma_y = 0, \quad \forall x_o \in F.$$

Setting $f = (v^1\,\delta_\Gamma + \text{div}(v^o\, n\,\delta_\Gamma))$, and $u = \Phi * f$, (363)' is: $u(x_o) = 0, \ \forall x_o \in F$.

Thus u satisfies the Helmholtz equation in Ω, hence in Ω_o, with $u|_{\Gamma_o}$. Since $-k^2$ is not an eigenvalue of the Dirichlet Laplacian in Ω_o by hypothesis, this implies that $u = 0$ in Ω_o, thus in Ω. Therefore $(v^o,v^1) \in G_e$, i.e. the Proposition.

⊗

For the interior Calderon operator we have the analogue of Proposition 31:
Let F *be a dense set in the boundary* Γ_1 *of an open set* Ω_1 *with* Ω *strictly contained in* Ω_1. *Then* M_F *is a total family in* $G_i = G(C^i)$.

Instead of elementary solutions we can use plane waves with wavenumber k for G_i; we prove like in (363)' (but with Rellich lemma)

(364) *the set* $M = \{(e^{ik\alpha.x}|_\Gamma, ik\alpha.n\, e^{ik\alpha.x}|_\Gamma), \alpha \in S^2\}$ *is a total family in* G_i.

Numerical methods based on expansions with elementary solutions such as (361) are called Fictitious Sources Methods (see Petit-Cadilhac [2]).

⊗

Calderon operators are useful for modelling many physical situations (see section 3.9 for example). We give another example.

Excitation of a proper mode of a cavity by an incident wave. We assume that the cavity Ω is occupied by the free space (or a conservative medium) and is bounded by a perfectly conducting medium (occupying a bounded domain Ω_1) with a hole Γ_o so that an exterior incident wave (E_I, H_I) at frequency ω can go into the cavity through the hole. Then the electromagnetic field (E,H) in Ω is a solution with finite energy of the Maxwell equations

(365) $$\text{i) curl } H + i\omega\varepsilon E = 0, \quad \text{ii) } - \text{curl } E + i\omega\mu H = 0 \ \text{ in } \Omega.$$

Let Γ_1 be the boundary of the conducting medium for Ω. Let C^e be the Calderon operator exterior to the domain $\tilde\Omega$ occupied by the cavity and the conducting medium, with boundary $\tilde\Gamma$. Then let C^e_o be defined by

$$C^e_o(n \wedge E|_{\Gamma_o}) = C^e(n \wedge E|_{\tilde\Gamma})|_{\Gamma_o}.$$

Recall that $n \wedge E$ is zero on $\tilde{\Gamma}\backslash\Gamma_0$. With the notations (85), (86) of chap. 2 we have

$$n \wedge E|_{\Gamma_0} \in H_0^{-1/2}(\mathrm{div},\Gamma_0), \text{ and } C_0^e(n \wedge E|_{\Gamma_0}) \in X(\mathrm{div},\Gamma_0).$$

Then (E, H) must satisfy the boundary conditions

$$(366)\qquad \left| \begin{array}{l} \text{i) } n \wedge E|_{\Gamma_1} = 0, \\ \text{ii) } n \wedge H|_{\Gamma_0} - C_0^e(n \wedge E|_{\Gamma_0}) = f_I|_{\Gamma_0}, \\ \text{with } f_I = n \wedge H_I - C^e(n \wedge E_I|_{\tilde{\Gamma}}). \end{array} \right.$$

We have proved (see section 5) that the scattering problem for a conducting obstacle has a unique solution and thus problem (365), (366) also has a unique solution. Now we assume that $k^2 = \omega^2\varepsilon\mu$ is a simple eigenvalue for the operator A (see (147) chap.2). Let (E_0, H_0) be a corresponding eigenmode with unit energy

$$\int_\Omega (\varepsilon\, |E_0|^2 + \mu\, |H_0|^2)\, dx = 1.$$

Then (E, H) has an orthogonal decomposition:

$$(E, H) = c_0\,(E_0, H_0) + (\tilde{E}, \tilde{H}), \text{ with } (\tilde{E}, \tilde{H}) \text{ orthogonal to } (E_0, H_0).$$

Then in many physical situations we are interested either to generate at best the eigenmode (E_0, H_0), that is to maximize the modulus of c_q over all possible holes and incident waves with given energy or to minimize it in order to protect the interior from all incident beams.
We note that the problem: find (E, H) satisfying Maxwell equations

$$\text{i) } \mathrm{curl}\, H + i\omega\varepsilon\, E = J, \qquad \text{ii) } -\mathrm{curl}\, E + i\omega\mu\, H = 0 \quad \text{in } \Omega,$$

with the boundary condition on $\Gamma = \partial\Omega$: $n \wedge E|_\Gamma = 0$ or $n \wedge E|_\Gamma = n \wedge E_0|_\Gamma$ and with given J and E_0, depends on the Fredholm alternative and does not generate eigenmodes!

⊗

Remark 23. Interior Calderon operator and its Fourier transform.
Let (E, H) be an electromagnetic field in a domain Ω occupied by a medium with (complex) characteristics (ε, μ). Let $M = n \wedge E|_\Gamma$, $J = -\, n \wedge H|_\Gamma$.
Generalizing (31) to complex k, we have

$$(367)\qquad -ik\alpha \wedge \hat{M}(k\alpha) + i\omega\mu\, \Pi_\alpha \hat{J}(k\alpha) = 0$$

that is also, with the (complex) impedance $Z = \omega\mu/k$,

$$(367)'\qquad \Pi_\alpha \hat{J}(k\alpha) = Z^{-1}\, \alpha \wedge \hat{M}(k\alpha), \quad \text{with } J = -\, C^i M.$$

This can be obtained from the behavior at infinity of the extension of (E, H) by 0 out of Ω. In other words, the mapping: $(M, -J) \rightarrow \hat{j}(k\alpha)$ (and $\hat{m}(k\alpha)$), given by (31) vanish on the space G^i (for the wavenumber k only).
Of course we have similar properties in the scalar case for Helmholtz equation. ⊗

Remark 24. The Calderon operator for Helmholtz or for the Laplacian is also called the Dirichlet-Neumann operator or the Poincaré-Stekloff operator. The Calderon projector for a domain Ω is defined for instance in Chazarain-Piriou [1]. ⊗

Finally note that the Calderon operators and projectors have all the symmetries of the problem. The Calderon projectors well adapted to computation when we have a few number of homogeneous and linear media.

11. MULTIPOLE EXPANSIONS. RAYLEIGH SERIES

Here we develop essentially a spectral method, based on expansions which are very familiar to physicists. But there are fine questions of convergence, well studied by Müller [1], which we follow. We begin by the simplest case, that is the scalar case with Helmholtz equation in two dimensions.

11.1. Multipole expansions for Helmholtz in R^2

Let u be a solution of the (homogeneous) Helmholtz equation for the wavenumber $k > 0$ in a domain Ω in R^2. We assume that 0 is in Ω, and thus there is a disc B_a of radius a contained in Ω. Since u is regular on circles C_r of radius $r \leq a$, its restriction to C_r may be developed in Fourier series, and we have, in polar coordinates:

$$(368) \qquad u(r,\theta) = \sum_{n \in Z} c_n(r)\, e^{in\theta},$$

where $c_n(r)$ satisfies the differential equation:

$$(369) \qquad \frac{d^2 c_n}{dr^2} + \frac{1}{r}\frac{dc_n}{dr} + \left(k^2 - \frac{n^2}{r^2}\right) c_n = 0 \qquad \text{for } r < a,$$

and thus c_n is proportional to J_n the Bessel function of order n:

$$(370) \qquad c_n(r) = c_n\, J_n(kr), \quad \text{with } c_n \in C.$$

Furthermore we have:

$$(371) \qquad \sum |c_n(r)|^2 = \sum |c_n\, J_n(kr)|^2 < +\infty \quad \text{for } r \leq a.$$

When u is a solution of Helmholtz equation in the domain Ω' (the complement of a bounded domain Ω), and satisfies a Sommerfeld condition at infinity, $c_n(r)$ is proportional to the Hankel function of order n:

(370)' $c_n(r) = c_n\, H_n^{(\kappa)}(kr)$, with $c_n \in \mathbb{C}$,

$\kappa = 1$ for the outgoing condition, $\kappa = 2$ for the incoming condition. We also have (371) (replacing the Bessel functions by the Hankel functions) for $r > a$.
Let u_n be defined by:

(372) $u_n(r,\theta) = J_n(kr)\, e^{in\theta}$ (resp. $H_n^{(1)}(kr)\, e^{in\theta}$).

Thus u_n satisfies Helmholtz equation in $\mathbb{R}^2$ (resp. $\mathbb{R}^2\backslash\{0\}$, and $(\Delta + k^2)u_n = \rho_n$ is concentrated at the origin); $u = \Sigma\, c_n u_n$ is called the *multipole expansion* of u (or also a *Rayleigh series*).
If u does not satisfy the homogeneous Helmholtz equation in the whole space $\mathbb{R}^2$ (or $\mathbb{R}^2\backslash\{0\}$), then we see that the expansion $u = \Sigma\, c_n u_n$ cannot be convergent everywhere. Thus we have the question of finding the domain of convergence of this series. The answer is similar to that for entire series $\Sigma\, r^n e^{in\theta}$: there is a radius of convergence ρ so that the series is convergent inside (resp. outside) the disc B_ρ for the expansion with Bessel functions (resp. Hankel functions), and is divergent outside (resp. inside) this disc. We recall that the Rayleigh hypothesis is that these series are convergent everywhere inside (resp. outside) Ω, for any domain Ω; we know that this Rayleigh hypothesis is wrong.
We study this question of radius of convergence for three dimensions only.

11.2. Multipole expansions for Helmholtz in $\mathbb{R}^3$

First we recall some essential properties of spherical harmonics. For the study of such properties, we refer to Dautray-Lions [1] chap. 2.7.3 and 9.B 1.1, to Müller [1], and to Stein-Weiss [1]. These properties originate from the representation of the rotation group in $\mathbb{R}^3$.

We use a spherical coordinate system; let $\alpha = (\alpha_1, \alpha_2, \alpha_3) \in S^2$, with

$$\alpha_1 = \sin\theta\cos\phi,\quad \alpha_2 = \sin\theta\sin\phi,\quad \alpha_3 = \cos\theta,\qquad 0<\theta<\pi,\ 0<\phi<2\pi.$$

The set of functions $(K_{n,j})$, $n \in \mathbb{N}$, $-n \le j \le n$, defined with the Legendre functions P_n^j by:

(373) $K_{n,j}(\alpha) = (2\pi)^{-1/2}\, P_n^j(\cos\theta)\, e^{ij\phi}$ (also often denoted by $Y_{l,m}(\theta,\phi)$, $l = n$, $m = j$)

which are the traces on the unit sphere S^2 of the homogeneous harmonic polynomials in $\mathbb{R}^3$, *is an orthonormal basis of* $L^2(S^2)$. Furthermore *they are the eigenfunctions of the Laplace-Beltrami operator* on S^2 according to

$$\Delta_{S^2} K_{n,j}(\alpha) = -\, n(n+1)\, K_{n,j}(\alpha),$$

and thus the eigenspace V_n corresponding to the eigenvalue $n(n+1)$ has $2n+1$ dimensions. The elements of V_n are called *spherical harmonics of order* n and often denoted by K_n. Using spherical harmonics allows us to find the solutions of the Helmholtz equation with separate variables (r,α) thanks to the usual expression of the Laplacian in (polar) spherical coordinates:

$$\Delta u\,(r\alpha) = \frac{\partial^2 u}{\partial r^2} + \frac{2}{r}\frac{\partial u}{\partial r} + \frac{1}{r^2}\Delta_\alpha u\,, \quad \text{with } \Delta_\alpha = \Delta_{S^2},\ r = |x|,\ \alpha = \frac{x}{r}, \tag{374}$$

we see that $c_n(r)K_n(\alpha)$ is a solution of the Helmholtz equation if $y = c_n(r)$ satisfies:

$$\frac{d^2y}{dr^2} + \frac{2}{r}\frac{dy}{dr} + (k^2 - \frac{n(n+1)}{r^2})\,y = 0. \tag{375}$$

The solutions of this equation are well known: they are the "Bessel" function $j_n(kr)$ (for the regular solution up to 0) and the Hankel functions $h_n(kr)$, related to the previous $J_n(kr)$ and $H_n(kr)$ by:

$$j_n(r) = (\frac{\pi}{2r})^{1/2} J_{n+\frac{1}{2}}(r), \quad h_n^{(\kappa)}(r) = (\frac{\pi}{2r})^{1/2} H^{(\kappa)}_{n+\frac{1}{2}}(r), \quad \text{with } \kappa = 1, 2. \tag{376}$$

Then let u_n be the function:

$$u_n(r\alpha) = h_n^{(\kappa)}(kr)K_n(\alpha) \quad (\text{resp. } j_n(kr)K_n(\alpha)). \tag{377}$$

Expansions of the form $\Sigma\, u_n$ are called *Rayleigh series (or also multipole expansions)*. We recall the main property of convergence of these series (see Müller [1], Theorems 11, 12) which will be a consequence of the following Lemmas.

Theorem 9. *Let* (K_n) *be a sequence of spherical Harmonics. Let*

$$c_n^2 = \int_{S^2} |K_n(\alpha)|^2 d\alpha\,, \quad c_n \geq 0. \tag{378}$$

We assume that there exists r_0 *so that the series*:

$$\sum_n |h_n^{(\kappa)}(kr_0)|^2\, c_n^2 < +\infty \quad (\text{resp. } \sum_n |j_n(kr_0)|^2\, c_n^2 < +\infty). \tag{379}$$

Then the series:

$$u(r\alpha) = \sum_n h_n^{(\kappa)}(kr)\, K_n(\alpha) \quad (\text{resp. } u(r\alpha) = \sum_n j_n(kr)\, K_n(\alpha)) \tag{380}$$

converges absolutely and uniformly on every compact set K in $R^3\backslash\overline{B}_{r_0}$ *(resp. in* B_{r_0}*).*

This is equivalent to: if the series (380) is convergent in $L^2(S_{r_0})$, then it is also convergent on every compact set K as above. Furthermore it is also convergent in all "reasonable" topologies, for all Sobolev spaces for instance.
This implies that the series u *given by* (380) *satisfies the Helmholtz equation with wavenumber* k.
In the "exterior" case, that is for the expansion with Hankel functions, we will verify that such a expansion satisfies the Sommerfeld condition (outgoing or incoming for $\kappa = 1$ or 2).

Lemma 5. *On any finite interval* [a,b] *with* $0 < a < b < \infty$ (*resp.* [0,D]) *we have* (*denoting by* Γ *the Euler function*):

$$(381)\quad h_n^{(\kappa)}(r) \approx (-1)^\kappa \frac{i}{2\sqrt{\pi}}\Gamma(n+\tfrac{1}{2})\left(\frac{2}{r}\right)^{n+1}, \quad (\text{resp. } j_n(r) \approx \frac{1}{2}\frac{\Gamma(1/2)}{\Gamma(n+(3/2))}\left(\frac{r}{2}\right)^n),$$

uniformly for $n \to \infty$, $a \le r \le b$ (*resp.* $0 \le r \le D$) *and then*:

$$(381)'\quad \frac{h_n^{(\kappa)}(r)}{h_n^{(\kappa)}(r_0)} \approx \left(\frac{r_0}{r}\right)^{n+1}, \quad \frac{j_n(r)}{j_n(r_0)} \approx \left(\frac{r}{r_0}\right)^n, \text{ when } n \to \infty.$$

Corollary 4. *With hypotheses* (379) *the coefficients* c_n *of the expansions* (380) *satisfy respectively*

$$(382)\quad c_n = o\left(\frac{2\sqrt{\pi}}{\Gamma(n+(1/2))}\left(\frac{r_0}{2}\right)^{n+1}\right) \quad (\text{resp. } c_n = o\left(2\frac{\Gamma(n+(3/2))}{\Gamma(1/2)}\left(\frac{r_0}{2}\right)^{-n}\right)).$$

Lemma 6. *The spherical harmonics satisfy the following inequality*:

$$(383)\quad |K_n(\alpha)|^2 \le \frac{2n+1}{4\pi}\int_{S^2} |K_n(\beta)|^2 d\beta, \quad \forall n \in N.$$

PROOF of theorem 9. From the above lemmas 5, 6, there are an index N_0 and a constant C such that for $n > N_0$ we have:

$$|h_n^{(\kappa)}(r)K_n(\alpha)| \le C\sqrt{2n+1}\left(\frac{r_0}{r}\right)^{n+1} \quad (\text{resp. } |j_n(r)K_n(\alpha)| \le C\sqrt{2n+1}\left(\frac{r}{r_0}\right)^n),$$

which implies the convergence of the series in the conditions of the theorem.

⊗

Remark 25. It is possible to replace the topology of the L^2 norm on the sphere of radius r_0 by some other topology with the same convergence results; for instance we can assume that the expansion (380) converges in the sense of distributions on the sphere. In R^2 this is equivalent to: there exists a number p such that

$$c_n |J_n(kr_0)| (1+n^2)^{-p} \to 0 \text{ when } |n| \to \infty.$$

⊗

Then we define the *radius of convergence* of the series (with (378)) by:

$$(384)\quad \left|\begin{array}{l} R_{ce} \overset{\text{def}}{=} \inf\{r, \ \sum_n |h_n^{(\kappa)}(kr)|^2 c_n^2 < +\infty\}, \\ R_{ci} \overset{\text{def}}{=} \sup\{r, \ \sum_n |j_n(kr)|^2 c_n^2 < +\infty\}. \end{array}\right.$$

Then if Ω is a bounded obstacle which contains the origin and if u satisfies the Helmholtz equation outside Ω (resp. inside Ω) with the Sommerfeld condition at infinity, then u has an expansion given by (380) in the exterior (resp. interior) of all balls B_ρ which contain (resp. is contained in) Ω, and thus:

$$R_e \le R_M = \sup_{x \in \Omega} |x|, \quad (\text{resp. } R_i \ge R_m = \inf_{x \in \Omega'} |x|).$$

When u is due to a distribution f with compact support (see (1)), we have a similar result with (supp f) replacing Ω in the expression of R_M.

EXAMPLE 1. *Elementary solution relative to a point* x_0 *not at the origin.*

Let $\Phi_{x_0} = \Phi(x - x_0)$ be the elementary outgoing solution of:

$$(385) \qquad \Delta u + k^2 u = -\delta_{x_0}.$$

Then we can prove that $\Phi(x - x_0)$ has the expansions

$$(386) \qquad \Phi(x - x_o) = \begin{cases} \dfrac{ik}{4\pi} \displaystyle\sum_n (2n+1)\, j_n(kr_0)\, h_n^{(1)}(kr)\, P_n(\alpha\alpha_0) & \text{if } r > r_o, \\ \dfrac{ik}{4\pi} \displaystyle\sum_n (2n+1)\, j_n(kr)\, h_n^{(1)}(kr_0)\, P_n(\alpha\alpha_0) & \text{if } r < r_o, \end{cases}$$

with $r = |x|$, $r_0 = |x_0|$, $\alpha = x/r$, $\alpha_0 = x_0/r_0$, P_n the Legendre polynomial of degree n.

We can find these formulas, using that $\Phi(x-x_0)$ has a Rayleigh expansion with Bessel functions for $r < r_0$ and with Hankel functions for $r > r_0$, and then using that the jumps of $\Phi(x-x_0)$ and its normal derivative across the sphere of radius r_0 are (in the sense of distributions on the sphere):

$$(387) \quad [\Phi(x - x_0)]_{S_{r_0}} = \lim_{r \uparrow r_0} \Phi_{x_0}(x) - \lim_{r \downarrow r_0} \Phi_{x_0}(x) = 0, \quad \left[\frac{\partial \Phi(x - x_0)}{\partial r}\right]_{S_{r_0}} = \frac{1}{r_0^2}\, \delta(\alpha - \alpha_0)$$

with

$$\delta_{x_0}(x) = \delta_{r_0}(r) \sum_n \frac{(2n+1)}{4\pi} P_n(\alpha\alpha_0),$$

which is due to the relations:

$$(388) \qquad \delta_{x_0}(x) = \delta_{r_0}(r)\, \delta(\alpha - \alpha_0), \qquad \delta(\alpha - \alpha_0) = \sum_n \frac{(2n+1)}{4\pi} P_n(\alpha\alpha_0).$$

This follows from the relation, for all regular functions f on the unit sphere:

$$(389) \qquad f(\alpha_0) = \sum_n \int_{S^2} \frac{2n+1}{4\pi} P_n(\alpha\alpha_0)\, f(\alpha)\, d\alpha$$

(see for instance Müller [1] p. 53). Then we can see *that the first series in* (386) *is divergent for* $r < r_0$, *and the second series for* $r > r_0$.

Indeed the generic term S_n of the first series is equivalent (thanks to lemma 5), for high n, to:

$$S_n \approx C\left(\frac{r_0}{r}\right)^n P_n(t), \quad t = \alpha.\alpha_0, \quad C \text{ a constant.}$$

Then the convergence is a consequence of the usual formula:

(390) $$(1 - 2zt + z^2)^{-1/2} = \sum_n P_n(t)\, z^n,$$

which defines a convergent series when $|z| < 1$, since the zeros of the polynomial $(1 - 2zt + z^2)$ are $z = e^{i\theta}$ and $e^{-i\theta}$ for $t = \cos\theta$.
Thus the radius of convergence ρ of the entire series (390) is equal to 1 that is:

$$\rho = \limsup_{n\to\infty} |P_n(t)|^{1/n} = 1, \quad \forall t \in [-1,1].$$

⊗

EXAMPLE 2. *Entire solutions of the Helmholtz equation. Plane waves.*
Expansions of the form (380) (with Bessel functions) with an infinite radius of convergence define entire solutions of the Helmholtz equation. An example is given by plane waves, where the expansions are uniformly convergent on compact sets

(391) $$e^{ik\alpha_0.x} = e^{ikr\alpha_0.\alpha} = \sum_n i^n (2n+1) P_n(\alpha_0.\alpha) j_n(kr), \quad \text{with } \alpha_0 \in S^2.$$

Replacing in (391) the Bessel functions by Hankel functions, and using:

$$j_n(kr) = \frac{1}{2}(h_n^{(1)}(kr) + h_n^{(2)}(kr)),$$

we might think that we obtain a decomposition of the plane wave into the sum of an outgoing and an incoming waves, but this is false since the expansions

$$\sum_n i^n (2n+1) P_n(\alpha_0.\alpha) h_n^{(\kappa)}(kr), \ \kappa = 1,2, \ \text{with} \int_{S^2} (P_n(\alpha.\alpha_0))^2\, d\alpha = \frac{4\pi}{2n+1}$$

are nowhere convergent!

⊗

Then before studying the behavior at infinity of Rayleigh expansions, we will answer the question: what is the source of a wave u_n given by (377)?

11.3. The source of a wave u_n

First note that $u_n(r\alpha) = h_n^{(1)}(kr) K_n(\alpha)$ satisfies $(\Delta + k^2)\, u_n = 0$ in $D'(R^3\backslash 0)$.

Furthermore the Hankel function of order n being equivalent in the neighborhood of 0 (up to a constant in r) to r^{-n}, it can be extended (using finite parts, see Schwartz [1]) to a distribution on R^3; thus

(392) $$-\rho_n \overset{\text{def}}{=} (\Delta + k^2) u_n \quad (\text{in } D'(R^3))$$

is a distribution with support $\{0\}$. Thus ρ_n is of the form $\rho_n = Q_n(D)\,\delta$, with Q_n a polynomial. Since u_n satisfies the outgoing condition at infinity, it is given by the convolution product

(393) $$u_n = \Phi * Q_n(D)\,\delta = Q_n(D)\,\Phi.$$

Thus we have to find Q_n so that:

(393)' $$Q_n(D)\Phi = Q_n(D)\frac{e^{ikr}}{4\pi r} = h_n^{(1)}(kr)\,K_n(\alpha).$$

We can see that this implies that Q_n is a homogeneous polynomial of order n, so that its restriction to the unit sphere is proportional to K_n:

(394) $$Q_n(i\xi) = Q_n(i|\xi|\alpha) = \nu_n\,|\xi|^n\,K_n(\alpha).$$

We can also find this result by Fourier transformation (see for instance Stein-Weiss [1] p. 158, thanks to the commutation of the Fourier transformation with rotations !). Then ν_n is easily obtained from the behavior at infinity of u_n and of the Hankel function (see Müller [1] p. 74, Petiau [1]); we have:

$$u_n(r\alpha) = h_n^{(1)}(kr)K_n(\alpha) \approx \frac{e^{ikr}}{ikr}\,e^{-i\pi n/2}\,K_n(\alpha) = \Phi(r)\,\hat{\rho}_n(k\alpha), \quad r \to \infty,$$

with $\alpha \in S^2$, therefore:

(395) $$\hat{\rho}_n(k\alpha) = Q_n(ik\alpha) = \frac{4\pi}{k}(-i)^{n+1}K_n(\alpha),$$

then $\nu_n = \frac{4\pi}{ik}\left(\frac{-i}{k}\right)^n$, and thus:

(395)' $$Q_n(i|\xi|\alpha) = \frac{4\pi}{ik}\left(\frac{-i|\xi|}{k}\right)^n K_n(\alpha).$$

We can also find another expression for ρ_n using distributions in polar coordinates up to 0; for this we must specify the space of distributions.

Space of distributions in polar coordinates in $R^n\backslash\{0\}$ and R^n.

The mapping $h: x \to (r,\alpha)$ is a diffeomorphism from $R^n\backslash\{0\}$ onto $R^+ \times S^{n-1}$ (with $R^+ = (0,\infty)$) which allows us to identify distributions on $R^n\backslash\{0\}$ with distributions on $R^+ \times S^{n-1}$. But here we have precisely distributions on R^n and not only on $R^n\backslash\{0\}$. Let u be a regular function on R^n. Then we denote:

(396) $$U(r,\alpha) = u(r\alpha) = u(x), \text{ i.e., } U = u \circ h^{-1}.$$

Lemma 7. *Let* u *be a regular* ($C^\infty(R^n)$) *function. Then* U *satisfies:*

(397) $$\frac{\partial^p U}{\partial r^p}(0,\alpha) = \sum_{|\lambda| = p} \frac{p!}{\lambda!} \alpha^\lambda D^\lambda u(0), \quad \forall \lambda = (\lambda_1, .., \lambda_n) \in N^n,$$

with $|\lambda| = \lambda_1 + \ldots + \lambda_n$, $\lambda! = \lambda_1!\ldots\lambda_n!$. *Thus* $\frac{\partial^p U}{\partial r^p}(0,\alpha)$ *is a spherical harmonic of order* p *and* $U(0,\alpha) = u(0)$, *is independent of* α.

PROOF. We have, with standard notation:

(398) $$\frac{\partial U}{\partial r}(r,\alpha) = \sum_j D_j u(x).\frac{\partial x_j}{\partial r} = \sum_j \alpha_j D_j u(x),$$

with $x_j = r\alpha_j$, $D_j = \frac{\partial}{\partial x_j}$, hence $i)_1$. Then we easily verify that:

(399) $$(\sum_j \alpha_j D_j)^p u = \sum_{j_1,\ldots,j_p} \alpha_{j_1}\ldots\alpha_{j_p} D_{j_1}\ldots D_{j_p} u,$$

hence (397). ⊗

Then we see that, if u is a $C^\infty(R^n)$ *function*, U is also a $C^\infty([0,\infty) \times S^{n-1})$ function. We define its extension to $R \times S^{n-1}$ (also denoted by U) by

(400) $$U(-r,\alpha) = U(r,-\alpha), \quad \forall r > 0, \ \forall \alpha \in S^{n-1}.$$

Lemma 8. *Let* u *be a regular* ($C^\infty(R^n)$) *function. Then its extension* U *to* $R \times S^{n-1}$ *defined by* (400) (*and* (396)) *is such that*: $U \in C^\infty(R \times S^{n-1})$; furthermore $u \in D(R^n)$ (resp. $S(R^n)$) implies $U \in D(R \times S^{n-1})$ (resp. $S(R \times S^{n-1})$).

PROOF. We only have to verify that $\lim_{r \uparrow 0} \partial_r^p U(r,\alpha) = \lim_{r \downarrow 0} \partial_r^p U(r,\alpha)$, which follows from (397) and (400).

⊗

Lemma 9. *Let* u *be in* $D(R^3)$ *or* $S(R^3)$. *Then* $U(r,\alpha)$ *has an expansion in spherical harmonics*

(401) $$U(r,\alpha) = \sum_{n,j} U_{n,j}(r) K_{n,j}(\alpha), \quad n \in N, \ -n \le j \le n,$$

such that each $U_{n,j}$, *satisfies:*

(401)'i) $$U_{n,j}(0) = 0, \ \ldots, \ \frac{\partial^p U_{n,j}}{\partial r^p}(0) = 0, \quad p = 0 \text{ to } n-1,$$

and also

(401)'ii) $$\frac{\partial^p U_{n,j}}{\partial r^p}(0) = 0 \quad \text{if } (-1)^n \neq (-1)^p.$$

PROOF. (401)'i) is a simple consequence of Lemma 7, since we have:

$$\frac{\partial^p U_{n,j}}{\partial r^p}(0) = \int_{S^2} \frac{\partial^p U}{\partial r^p}(0)\, K_{n,j}(\alpha)\, d\alpha = \sum_{|\lambda|=p} \frac{p!}{\lambda!} D^p u(0) \int_{S^2} \alpha^\lambda K_{n,j}(\alpha)\, d\alpha = 0.$$

We also have $K_{n,j}(-\alpha) = (-1)^n K_{n,j}(\alpha)$, with (400) which implies (401)'ii). Then (401)' with $U_{n,j} \in C^\infty(R^+)$ implies that $U_{n,j}(r) = r^n \zeta_{n,j}(r^2)$, with $\zeta_{n,j} \in C^\infty(R)$.

⊗

Notation 1. *Let* $D_0(R \times S^{n-1})$ *and* $S_0(R \times S^{n-1})$ *be the subspaces of functions* U *in* $D(R \times S^{n-1})$ *and* $S(R \times S^{n-1})$ *with* (400) *and*:

(401)" $\dfrac{\partial^p U}{\partial r^p}(0,\alpha)$ *is a spherical harmonic of order* p.

We can identify the spaces of regular functions $D(R^n)$ and $S(R^n)$ with the spaces $D_0(R \times S^{n-1})$ and $S_0(R \times S^{n-1})$, and then their dual spaces for the duality:

$<U,\phi> = \int_{R^+ \times S^{n-1}} U(r,\alpha)\, \phi(r,\alpha)\, r^{n-1}\, dr d\alpha$, resp. $<U,\phi>_0 = \int_{R^+ \times S^{n-1}} U(r,\alpha)\, \phi(r,\alpha)\, dr d\alpha$

with subspaces of distributions on $R \times S^{n-1}$ with support in $[0,\infty)$.

Notation 2. *We denote by* $D_0'(R \times S^{n-1})$ *and* $S_0'(R \times S^{n-1})$ *(respectively* $D_0{}'_0(R \times S^{n-1})$ *and* $S_0{}'_0(R \times S^{n-1})$*) these subspaces.*

A distribution U on $R \times S^{n-1}$ induces a distribution u on R^n by $<u,\phi> = <U,\Phi>$, with $\Phi(r,\alpha) = \phi(r\alpha)$, but the mapping p: $U \to u \in D_0(R \times S^{n-1})$ or $S_0(R \times S^{n-1})$, has a kernel. In the space $D'_+(R \times S^{n-1})$ of distributions with support in $[0,\infty)$ (with pairing <,>), this kernel is the space of distributions U so that:

$U(r,\alpha) = \sum_\nu \delta^{(\nu)}(r) T_\nu(\alpha)$, with $<T_\nu(\alpha), K_p(\alpha)> = 0$ for every spherical harmonic K_p of order p, $\nu \neq p + n - 1$. Note that if T is a distribution on R^n which is concentrated on {0}, it is a finite sum of Dirac derivatives, thus for n = 3:

$$T = \sum_p \delta^{(p)}(r) K_p(\alpha) = \sum_p \sum_{|\lambda|=p} \frac{p!}{\lambda!} c_{\lambda,p} \delta^{(\lambda)}(x), \text{ with } c_{\lambda,p} = \int_{S^2} \alpha^\lambda K_p(\alpha)\, d\alpha,$$

in $D'_{00}(R \times S^2)$ or $D'(R^3)$, with K_p a spherical harmonic of order p.

Thus $\delta(r) = \delta^{(0)}(r) = 4\pi\, \delta(x)$, $\delta^{(1)}(r)\, \alpha_j = \frac{4\pi}{3} \frac{\partial \delta}{\partial x_j}(x)$.

Lemma 10. *Let* $u_n(r\alpha) = h_n^{(1)}(kr) K_n(\alpha)$ in R^3. *Then* u_n *satisfies:*

(402) $(\Delta + k^2)\, u_n = -\rho_n$,

with $\rho_n = -\nu_n\delta^{(n)}(r)K_n(\alpha)$ in $D'_{oo}(R \times S^2)$, $\rho_n = -\tilde{\nu}_n\delta^{(n+2)}(r)K_n(\alpha)$ in $D'_0(R \times S^2)$,
where

$$\nu_n = \frac{i}{k}\frac{(2n+1)!}{(n!)^2}(-2k)^{-n}, \qquad \tilde{\nu}_n = \frac{i}{k}\frac{(2n+1)!}{n!(n+2)!}(-2k)^{-n}.$$

PROOF. Let $\phi \in D(R^3)$. Then we can define using Lemma 9:

(403) $$I_n(\phi) \overset{\text{def}}{=} \langle(\Delta + k^2)\, u_n, \phi\rangle = \langle u_n, (\Delta + k^2)\,\phi\rangle = \lim_{a\to 0} I_n^a(\phi),$$

with:

(404) $$I_n^a(\phi) = \int_a^\infty h_n^{(1)}(kr)\,[\frac{1}{r^2}\frac{\partial}{\partial r}r^2\frac{\partial}{\partial r}\phi_n + (k^2 - \frac{n(n+1)}{r^2})\,\phi_n]\, r^2 dr,$$

where $\phi_n(r) = \int_{S^2}\phi(r\alpha)\, K_n(\alpha)\, d\alpha$. Then integrate twice by part. We obtain:

(405) $$\begin{aligned} I_n^a(\phi) &= -\,[r^2h_n^{(1)}(kr)\frac{\partial}{\partial r}\phi_n(r) - r^2(\frac{\partial}{\partial r}h_n^{(1)}(kr))\,\phi_n(r)]_{r=a} \\ &= a^{-(n-1)}\,[\frac{\partial}{\partial r}(r^{n+1}h_n^{(1)}(kr))\,\phi_n(r) - h_n^{(1)}(kr)\frac{\partial}{\partial r}(r^{n+1}\phi_n(r))]_{r=a}. \end{aligned}$$

Using Lemma 9 and the recurrence formula for Hankel function:

(406) $$\frac{\partial}{\partial r}(r^{n+1}h_n^{(1)}(r)) = r^{n+1}h_{n-1}^{(1)}(r),$$

we get that the first term in (405) is equivalent to a^2 when $a \to 0$. For the second term, using equivalence (see Abramovitz-Stegun [1] p. 437)

(407) $$h_n^{(1)}(ka) \approx -\frac{\Gamma(2n+1)}{(ka)^{n+1}2^n\Gamma(n+1)} \approx -i\frac{(2n)!}{(ka)^{n+1}2^n\, n!},$$

and also Lemma 9, we have

(408) $$I_n(\phi) = \frac{i}{k}\frac{(2n+1)!}{(n!)^2}\frac{1}{(2k)^n}\frac{\partial^n\phi_n}{\partial r^n}(0),$$

which gives us Lemma 10. At the first order we verify that $h_0^{(1)}(kr) = \frac{4\pi}{ik}\Phi(r)$.

⊗

We note that for $k = 0$, we have:

(409) $$\Delta(r^{-(n+1)}K_n(\alpha)) = -\rho_n,$$

with

$$\rho_n = -\frac{(2n+1)}{n!}\delta^{(n)}(r)\, K_n(\alpha) \text{ in } D'_{oo}(RxS^2), \quad \rho_n = -\frac{(2n+1)}{(n+2)!}\delta^{(n+2)}(r)\, K_n(\alpha) \text{ in } D'_o(RxS^2).$$

⊗

11.4. Multipole expansions and analytical functionals

From Lemma 10 we understand the expression "multipole expansion" for the Hankel functions expansion (376) of the solution u of the Helmholtz equation (1): each term u_n of the expansion is due to a distribution ρ_n of order n concentrated at the origin. The sum $\rho^{an} = \Sigma\rho_n$ cannot be a distribution, but we can define it as an *analytic functional*:

Definition 6. *Analytic functional. An analytic functional is a dual form of the space* $A(C^n)$ *of entire analytic functions on* C^n *(or holomorphic functions on* C^n*) equipped with the topology of uniform convergence on compact sets. We denote by* $A'(C^n)$ *the space of analytic functionals.*

That ρ^{an} is *an analytic functional* results from the Paley-Wiener Theorem for analytical functionals, see below and for instance Hörmander [2], [3], Treves [1] Ex. 22.7. We recall some notions

Definition 7. *An analytic functional* μ *in* C^n *is said to be carried by a compact set* K *(or* K *is a carrier for* μ*) if for every neighborhood* ω *of* K *there is a constant* C_ω *such that*

$$(410) \qquad |\mu(f)| \le C_\omega \sup_\omega |f|, \quad \forall f \in A(C^n).$$

Then we define *the Fourier-Laplace (or Fourier-Borel) transform of an analytic functional* μ in C^n by

$$(411) \qquad \hat{\mu}(\zeta) = FL\mu(\zeta) = \langle \mu_z, e^{-i\langle z,\zeta\rangle} \rangle, \qquad \text{with } \zeta \in C^n, \ \langle z,\zeta\rangle = \sum_j z_j\zeta_j.$$

For all compact set K in C^n, we define the convex function on C^n

$$(412) \qquad H_K(\zeta) = \sup_{z\in K} \mathrm{Re} \langle z,\zeta \rangle.$$

When K is the ball $K = B_a$ in R^n, then $H_K(\zeta) = a\,|\zeta'|$, with $\zeta' = (\zeta'_1,\ldots,\zeta'_n) = \mathrm{Re}\,\zeta$.

Definition 8. *Let* $a = (a_1,\ldots,a_n)$ *with* $a_i \ge 0$. *We denote by* Exp (a) *the space of the entire analytic functions of exponential type* a, i.e., *the space of the entire functions on* C^n *such that there is a constant* A(f) *so that*:

$$|f(z)| \exp(-a_1|z_1| - \ldots - a_n|z_n|) \le A(f).$$

We denote by Exp *the vector space of all entire functions on* C^n *which are of exponential type, i.e., the union of all the spaces* Exp (a), *with* $a_i \ge 0$.

The "Paley-Wiener" theorem for analytical functional is:

Theorem 10. *If μ is an analytic functional in C^n carried by the compact set K then its Fourier-Laplace transform is an entire analytic function of exponential type, and more precisely: for every $\delta > 0$, there is a constant C_δ such that*

$$(413)\qquad |\hat{\mu}(\zeta)| \le C_\delta \exp\left(H_K(-i\zeta) + \delta|\zeta|\right), \qquad \forall \zeta \in C^n.$$

Conversely if K is a convex compact set and f an entire function satisfying (413) *for every $\delta > 0$, there exists an analytical functional μ carried by K whose Fourier-Laplace transform is* f.

Then we consider the expansion

$$(414)\qquad \hat{\rho}^{an}(y) = \sum_n Q_n(iy) = \sum_n Q_n(i|y|\alpha) = \sum_n \frac{4\pi}{ik}\left(\frac{-i|y|}{k}\right)^n K_n(\alpha),$$

where we assume that (c_n), satisfies (379), with $c_n = \int_{S^2} |K_n(\alpha)|^2 d\alpha$.

Theorem 11. *The expansion* (414) *defines a harmonic function on* R^3 *satisfying*

$$(415)\qquad |\hat{\rho}^{an}(y)| = o(r\, e^{r.r_0/2}) \quad \text{for } r = |y| \to \infty,\ y \in R^3,$$

which can be extended to an entire analytic function f *on* C^3 *of exponential type. Thus* f *is the Fourier-Laplace transform of an analytical functional ρ^{an} so that*
i) *ρ^{an} is the source of a wave* u *defined by* (380) *or also by*

$$(416)\qquad u(x) = \langle \rho^{an}(z), \Phi(x-z)\rangle \quad \text{with } x \in R^3,\ |x| > R_{ce},$$

ii) *u is an outgoing wave, and its behavior at infinity is given by*

$$(417)\qquad u(r\alpha) \approx \Phi(r)\, \hat{\rho}^{an}(k\alpha), \quad \text{for } r \to \infty.$$

Furthermore the radius of convergence of the series (380) *is given by:*

$$(418)\ R_{ce} = R^M \overset{\text{def}}{=} \limsup_{r\to\infty}\left(\frac{1}{kr}\log\int_{S^2} |\hat{\rho}^{an}(r\alpha)|^2 d\alpha\right) = \frac{2}{k}\limsup_{n\to\infty}\left(\Gamma(n+\tfrac{1}{2})\, c_n\right)^{1/n}.$$

Remark 26. Note that (416) is a way to write the Helmholtz equation $(\Delta + k^2)u = \rho^{an}$, by the convolution product of Φ with ρ^{an}. But as a function of z, $\Phi(x-z)$ is analytical only if $z \neq x$. Thus (416) has to be feasible, for instance if ρ^{an} can be identified to an analytical functional on an open set Ω in C^3 which does not contain x. Anyway we can define (416) by taking the limit of the series $\Sigma \langle \rho_n, \Phi(x-.)\rangle$ and thus we obtain the series (380).

⊗

Remark 27. Let v be a harmonic function of *exponential type* (a), that is, for every δ there is a constant C_δ such that:

(419) $$|v(y)| \le C_\delta \exp(r(a+\delta)), \quad r = |y|.$$

Then (415) implies: $\hat{\rho}^{an}(y)$ is of exponential type a, with $a = r_0/2$.

⊗

For the proof of Theorem 11, we use the lemma:

Lemma 11. *Let* f *be a harmonic function on* R^n *of exponential type. Then* f *is the restriction to* R^n *of a unique analytic function on* C^n *of exponential type.*

PROOF for n = 3. We use the inequality (due to the Poisson formula), see Chazarain-Piriou [1] p. 23, for every harmonic function f:

(420) $$|\partial^\alpha f(0)| \le (3n)^n r^{-n} \sup_{x \in S_r} |f(x)|, \quad n = |\alpha|,$$

with S_r the sphere of radius r. Thus for $z \in C^3$, $|z| < r_0$

(421) $$\begin{aligned} &|\sum_\alpha \frac{1}{\alpha!} \partial^\alpha f(0) z^\alpha| \le \sum_n \Big(\sum_{|\alpha|=n} \frac{1}{\alpha!} (3n)^n r_0^\alpha \Big) r^{-n} \sup_{x \in S_r} |f(x)|, \\ &\le \sum_n (3n)^n \frac{1}{n!} (3r_0)^n r^{-n} \sup_{x \in S_r} |f(x)| \le \Big[\sum_n \frac{n^n}{n!} \Big(\frac{9r_0}{r}\Big)^n\Big] \sup_{x \in S_r} |f(x)|. \end{aligned}$$

Then using the Stirling formula, we see that for large values of n:

(422) $$a_n \overset{\text{def}}{=} \frac{n^n}{n!} \Big(\frac{9r_0}{r}\Big)^n \approx \frac{1}{\sqrt{\pi n}} (9er_0/r)^n.$$

Thus for $r = Cr_0$ with $C > 9e$ and with relation (419), we obtain:

(423) $$|f(z)| \le C_\delta \exp[r_0(a+\delta)/C], \quad \text{with } |z| < r_0.$$

⊗

Notice that we have, with (419) and (412), (413):

(424) $$a \le \inf_{\alpha \in S^2} H_K(-i\alpha) = \inf_{\alpha \in S^2} H_{\mathrm{Im}\,K}(\alpha).$$

Thus when $\mathrm{Im}\,K = \{\mathrm{Im}\,z,\ z \in K\}$ is a convex set, the ball B_a satisfies $B_a \subseteq \mathrm{Im}\,K$.

⊗

PROOF of Theorem 11. The proof of the first part, that is, that the expansion (414) with (379) or (380) defines a harmonic function satisfying (415) is fairly easy (see Müller [1] p. 89). Thanks to Lemma 11 this implies that f is of exponential type on C^3, with (424) and $a = r_0/2$. The Paley-Wiener theorem 10 implies that ρ^{an} is an analytical functional.

Then we can write (416) (see Remark 25). The behavior at infinity of u can be proved in a way similar to (15), Proposition 1. Thus u is an outgoing wave. There are many other ways to prove it (see for instance Müller [1] p. 90 using a Laplace transformation).
Then we know (see Müller [1] Chap. 3.5, Thms. 16 and 17 then Thm. 18) that for $r > R_{ce}$ the expansion of u defines an entire harmonic function so that $R^M \leq R_{ce}$ and conversely for r greater than R^M (see (418)), the series (380) is convergent, thus $R^M \geq R_{ce}$. Furthermore we can characterize the radius of convergence R_{ce} from the coefficients (c_n) in (414), since we have:

(425) $\quad R_{ce} = \inf\{r, \text{ so that } c_n = o\,[(kr/2)^{n+1} \frac{1}{\Gamma(n+(1/2))}] \text{ for } n \to \infty\},$

thus:

(426) $\quad \frac{1}{2} kR_{ce} = \frac{1}{\rho_c} \overset{\text{def}}{=} \limsup_{n\to\infty} (\Gamma(n+\frac{1}{2})\, c_n)^{1/n};$

ρ_c is the radius of convergence of the entire series $\sum \Gamma(n+1/2)\, c_n r^n$.

⊗

Note that it would be interesting also to use analytic functionals of the variable r in C, with values in $L^2(S^2)$.

⊗

Conclusion. Let X be the space of outgoing waves defined by:

(427) $\quad X = \{u \in \mathcal{D}'(R^3),\ \rho = (\Delta + k^2)u \in \mathcal{E}'(R^3),\ \int_{S^2} |(\frac{\partial u}{\partial n} - iku)(r\alpha)|^2\, d\alpha = o\,(1/r^2)\}.$

With $B'_a = R^3/\bar{B}_a$, we also define $\tilde{X} = \bigcup_{a>0} \tilde{X}_a$, where:

(427)' $\quad \tilde{X}_a = \{u \in \mathcal{D}'(B'_a),\ (\Delta + k^2)\, u = 0 \text{ in } B'_a,\ \int_{S^2} |(\frac{\partial u}{\partial n} - iku)(r\alpha)|^2\, d\alpha = o\,(1/r^2)\}.$

Thus for every u in X, there is an $a > 0$ such that $u|_{B'_a} \in \tilde{X}_a$.

Let Y be the space of traces on the sphere $S_k = kS^2$ of harmonic functions on R^3 of exponential type:

(428) $\quad Y = \{g \in C^\infty(S_k),\ \exists\, f \text{ on } R^3 \text{ with } \Delta f = 0 \text{ and (419)},\ f|_{S_k} = g\}.$

Y is also the space of traces on the sphere $S_k = kS^2$ of the Fourier transforms of distributions with compact support on R^3. Then we have:

Theorem 12. *Trace theorem at infinity. The mapping:* $\gamma_\infty : u \to \lim_{r\to\infty} u(r\alpha)/\Phi(r)$ *is surjective from* X onto Y.

Note that it is not injective! Its kernel is the space $X \cap \mathcal{E}'(R^3)$.

Definition 9. *We denote by* $A'_H(C^3)$ *the space of analytical functionals* ρ^{an} *so that the restriction of* $\hat{\rho}^{an}$ to R^3 *is harmonic of exponential type.*

Theorem 13. *The mapping* $\rho^{an} \rightarrow \hat{\rho}^{an}|_{S_k}$ *is (continuous) one-to-one from* $A'_H(C^3)$ *onto* Y.

We have to compare to the mapping on distributions with compact support:
$\rho \in E'(R^3) \rightarrow \hat{\rho}|_{S_k} \in Y$, which has the kernel $(\Delta + k^2)E'(R^3)$!

Then for every $\rho \in E'(R^3)$, we define $\rho^{an} \in A'_H(C^3)$ with $\hat{\rho}|_{S_k} = \hat{\rho}^{an}|_{S_k}$.
Let $E'_a(R^3)$ be the space of distributions on R^3 with support in the ball $\bar{B}_a$. The mapping $\rho \in E'_a(R^3) \rightarrow u|_{B'_a} \in \tilde{X}_a$ (with $u = \Phi * \rho$), has the kernel $(\Delta + k^2)E'_a(R^3)$, and we can identify the quotient mapping with the mapping $\rho^{an} \in A'_H(C^3) \rightarrow u$, with $u(x) = \langle \rho^{an}, \Phi(x - .)\rangle$.

Note also that the radius of convergence of the multipole expansion of an outgoing wave u is given by the type of the Fourier-Laplace transform of the "source" of u (see Theorem 11). In the case of scattering of an incident wave by an obstacle, a conjecture of Bardos is that the radius of convergence corresponds to the distance of the origin to an envelop of the normals.

11.5. Multipole expansions for the electromagnetic field

We consider Maxwell equations (22), with currents J and M having compact supports; the electromagnetic field (E, H) satisfies Helmholtz equations (28) with (m,j) given by (27). Then we can apply the results we obtained above on Helmholtz equation: let (j^{an}, m^{an}) be the analytical functionals such that

$$(429) \qquad \hat{j}^{an}(k\alpha) = \hat{j}(k\alpha), \quad \hat{m}^{an}(k\alpha) = \hat{m}(k\alpha), \quad \forall \alpha \in S^2,$$

with $(\hat{j}^{an}, \hat{m}^{an})$ entire analytic vector functions on C^3, of exponential type.
Note that, thanks to (33), they also satisfy:

$$(430) \quad \alpha.\hat{j}^{an}(k\alpha) = 0, \quad \alpha.\hat{m}^{an}(k\alpha) = 0, \quad \forall \alpha \in S^2 \ (\text{and } \alpha \wedge \hat{j}^{an}(k\alpha) = Z\,\hat{m}^{an}(k\alpha)).$$

Definition 10. *We denote* $A'^{3}_{H,k}(C^3) \stackrel{def}{=} \{j^{an} \in A'_H(C^3)^3,\ \alpha.\hat{j}^{an}(k\alpha) = 0,\ \alpha \in S^2\}$.

Thus (E, H) is obtained from (j^{an}, m^{an}) by

$$(431) \quad E(x) = \langle j^{an}, \Phi(x - .)\rangle, \quad H(x) = \langle m^{an}, \Phi(x - .)\rangle, \quad \text{with } r = |x| > R.$$

Note that relations (430) imply (from Rellich Theorem) div E = 0, div H = 0 for r > R. Then from the expansion of $\hat{j}^{\,an}$ and $\hat{m}^{\,an}$ into spherical harmonics,

$$(432) \qquad \hat{j}^{\,an}(\rho\alpha) = \frac{4\pi}{ik}\sum_n \left(\frac{-i\rho}{k}\right)^n \vec{K}^0_n(\alpha), \qquad \hat{m}^{\,an}(\rho\alpha) = \frac{4\pi}{ik}\sum_n \left(\frac{-i\rho}{k}\right)^n \vec{K}^1_n(\alpha),$$

we obtain the *multipole expansion* of (E, H):

$$(433) \qquad E(r\alpha) = \sum h^{(1)}_n(kr)\vec{K}^0_n(\alpha), \quad H(r\alpha) = \sum h^{(1)}_n(kr)\vec{K}^1_n(\alpha),$$

with the radii of convergence given by (see (418)):

$$(434)\ R^E_{ce} = \limsup_{r\to\infty}\frac{1}{kr}\log\int_{S^2}|\hat{j}^{\,an}(r\alpha)|^2\,d\alpha\,,\ \ R^H_{ce} = \limsup_{r\to\infty}\frac{1}{kr}\log\int_{S^2}|\hat{m}^{\,an}(r\alpha)|^2\,d\alpha.$$

Obviously, from Maxwell equations, we have $R^E_{ce} = R^H_{ce}$.

Remark 28. First we have to note that the vector product or the scalar product by α of a (vector) spherical harmonic of order n is not a spherical harmonic.

Furthermore $\alpha \wedge \vec{K}_n(\alpha)$ and $\alpha \wedge \vec{K}_m(\alpha)$ (resp. $\alpha.\vec{K}_n(\alpha)$ and $\alpha.\vec{K}_m(\alpha)$) $n \neq m$, are not orthogonal. ⊗

Remark 29. Note also that we have the relations between j and m thanks to (27)

$$(435) \qquad \left|\begin{array}{l} \text{i) } i\omega\mu\, m + \text{curl}\, j = (\Delta + k^2)M, \\ \text{ii) } \text{curl}\, m + i\omega\varepsilon j = -(\Delta + k^2)J\,. \end{array}\right.$$

When M = 0 we have: $i\omega\mu\, m + \text{curl}\, j = 0$, but we do not have: $i\omega\mu\, m^{an} + \text{curl}\, j^{an} = 0$!

The differential operators (here the curl operator) do not operate in $A'_{H,t}(C^3)$! ⊗

EXAMPLE 3. *The (oscillating) dipole*. We define the *dipole* oscillating at the angular frequency ω (and centred at the origin) as the charge density $\rho(x,t)$ and the current $J(x,t)$ given by:

$$(436) \qquad \rho(x,t) = e^{-i\omega t}\rho(x), \ \ \rho(x) = p.\text{grad}\,\delta, \ \ J(x,t) = e^{-i\omega t}J(x), \ \ J(x) = i\omega p.\delta,$$

with $p \in R^3$ the dipole momentum; we have

$$\langle\rho(x)\,,\,x_i\rangle = \langle p.\text{grad}\,\delta\,,\,x_i\rangle = -\langle\delta\,,\,\text{div}\,(px_i)\rangle = -p_i.$$

With the usual notations we have:

(437) $\hat{J}(y) = i\omega p, \quad \hat{j}(y) = i\omega\mu\,[\hat{J} - k^{-2}\,y\,(y.\hat{J})] = -\,\omega^2\,\mu\,[p - k^{-2}\,y\,(y.p)],$

thus $\hat{j}(k\alpha) = -\,\omega^2\mu\,[p - \alpha(\alpha.p)]$, which is also $\hat{j}^{\,an}(k\alpha)$ from the definition. Then we easily find

(438) $\hat{j}^{\,an}(y) = -\,(\omega^2\mu/k^2)\,\{[p\,y^2 - y(y.p)] - \frac{2}{3}\,p\,(y^2 - k^2)\},$

and then: $\hat{m}(k\alpha) = ik\alpha \wedge \hat{J}(k\alpha) = -\,k\omega\,\alpha \wedge p$, giving: $\hat{m}^{\,an}(y) = -\,\omega\,y \wedge p$. We can write (438) also in the form:

(438)' $\hat{j}^{\,an}(\rho\alpha) = -\,(\omega^2\mu/k^2)\,\{\rho^2\,\vec{K}_2(\alpha) + \vec{K}_0(\alpha)\},$

$\vec{K}_2(\alpha)$ and $\vec{K}_0(\alpha)$ spherical Harmonics of order 2 and 0 resp., $\vec{K}_2(\alpha) = \frac{1}{3}\,p - \alpha\,(\alpha.p)$,

$\vec{K}_0(\alpha) = \frac{2}{3}\,p\,k^2$. We have $\alpha.\vec{K}_2(\alpha) = -\,\frac{2}{3}\,\alpha.p = -\,\alpha.\vec{K}_0(\alpha)$; thus $\vec{K}_2(\alpha)$ is not tangent to the sphere S^2, and $\alpha.\vec{K}_2(\alpha)$ is not orthogonal to $\alpha.\vec{K}_0(\alpha)$ in $L^2(S^2)$.

We also verify that:

$$i\omega\mu\,\hat{m}^{\,an}(y) + iy \wedge \hat{j}^{\,an}(y) \neq 0 \quad \text{and} \quad y.\hat{j}^{\,an}(y) \neq 0\,!$$

⊗

Conclusion. In the electromagnetic case we can conclude in a very similar way as in the Helmholtz case.
Let X^{em} be the space of electromagnetic fields (E,H) on R^3, satisfying:
i) the Maxwell equations (22) with (J,M) distributions with compact support.
ii) the Silver-Müller conditions (43) at infinity.

We also define $\tilde{X}^{em} = \bigcup_{a>0} \tilde{X}_a^{em}$, where $\tilde{X}_a^{em}$ is the space of electromagnetic fields (E,H) on B'_a satisfying:
i) the Maxwell equations:

$$\operatorname{curl} H + i\omega\varepsilon\,E = 0\,, \qquad -\operatorname{curl} E + i\omega\mu\,H = 0 \;\text{ in } B'_a$$

ii) the Silver-Müller conditions (43) at infinity.

Thus for every (E,H) in X^{em}, there is an $a > 0$ such that $(E,H)|_{B'_a} \in \tilde{X}_a^{em}$.
Then let Y^{em} be the trace space on S_k

$$Y^{em} = \{g \in C^\infty(S_k)^3,\ \alpha.g(k\alpha) = 0,\ \forall\,\alpha \in S^2,\ \exists\, f: R^3 \to C^3 \text{ harmonic of exponential type, i.e., with (419) and } f|_{S_k} = g\}.$$

(The condition on f is equivalent to $\limsup_{r\to\infty} \frac{1}{kr} \log \int_{S^2} |f(r\alpha)|^2\,d\alpha < \infty$.)

We have the following results:

Theorem 14. *Trace theorem at infinity. The mapping*:

$\gamma_\infty : (E,H) \to \lim_{r\to\infty} E(r\alpha)/\Phi(r)$ *is surjective from* X^{em} (or $\tilde{X}^{em}$) to Y^{em}.

But it is not injective ! Like in the Helmholtz case, the mapping (with (27), (29)) $(J,M) \in E_a'(R^3)^3 \times E_a'(R^3)^3 \to (E,H)|_{B_a'} \in \tilde{X}_a^{em}$, or $(J,M) \to \lim_{r\to\infty} E(r\alpha)/\Phi(r) \in Y^{em}$ is not injective, and we can identify its quotient mapping with the function:

$$j^{an} \in A'^{3}_{H,k}(C^3) \to (E,H) \text{ (with } (E,H) \text{ defined by (431))}.$$

We recall that m^{an} is obtained from j^{an} thanks to (430). Furthermore *the radius of convergence of the multipole expansion of* (E,H) is obtained from j^{an} by (434).

Remark 30. A natural question in scattering problems is to "control" the total P_{tot} (see (283)) by the total cross section σ_S (see (279)). This is important also for variational methods in scattering. The question is equivalent to the following: does there exists a constant $C > 0$ such that

$$(439) \quad |g(\alpha_0)| \le C \left(\int_{S^2} |g(\alpha)|^2 d\alpha\right)^{1/2}, \quad \forall g \in Y\,?$$

(Y being defined by (428) with $k = 1$ to simplify writing). The answer is negative. We can see it by taking the functions: $g_n(\alpha) = P_n(\alpha\beta)$, with a given unit vector β and P_n the Legendre polynomial of degree n. We have (see Müller [1])

$$(440) \quad \int_{S^2} |g_n(\alpha)|^2 d\alpha = 2\pi \int_{-1}^{+1} (P_n(t))^2 dt = \frac{4\pi}{n+1},$$

but $g_n(\beta) = P_n(1) = 1$; for $n \to \infty$, we see that we cannot have $1 \le C\,(4\pi/(n+1))^{1/2}$.

⊗

12. SCATTERING BY A DIELECTRIC BALL

The scattering of an incident wave plane by a dielectric ball B_R is often used as a reference to numerical implementation. Results are well known due to Rayleigh expansions.

12.1. Scattering with Helmholtz equation

12.1.1. Calderon operators and Calderon projectors on a sphere

First we can obtain the Calderon operators C^i, C^e and the Calderon projectors P^i, P^e on the sphere S_R (for a wavenumber k) thanks to the Rayleigh series (380) quite easily. The derivatives of these expressions with respect to r, for r = R, are for the (exterior) outgoing wave, and the interior wave:

$$(441) \quad \frac{\partial u}{\partial r}(R\alpha) = \sum k\, h_n'(kR) K_n(\alpha), \quad \text{resp.} \quad \frac{\partial u}{\partial r}(R\alpha) = \sum k\, j_n'(kR) K_n(\alpha),$$

with $h_n = h_n^{(1)}$; the prime denotes the derivative. Thus if $j_n(kR) \ne 0$, $\forall$ n:

(442) $C^eK_n(\alpha) = \theta_n^e K_n(\alpha),\ C^iK_n(\alpha) = \theta_n^i K_n(\alpha),\ \theta_n^e = k\frac{h_n'}{h_n}(kR),\ \theta_n^i = k\frac{j_n'}{j_n}.$

Thus the Calderon operators are simple multiplication in the space of spherical Harmonics of order n. This is a consequence of the fact that the rotation group commutes with the Calderon operators. Now the exterior Calderon operator has the extra property (with $k > 0$):

$$\mathrm{Re} - (C^e u_0, u_0) \geq 0, \quad \textit{and even} \quad \mathrm{Re} - (C^e u_0, u_0) \geq \frac{1}{R}\|u_0\|^2_{L^2(S_R)}.$$

Using (442), and the decomposition $u_0 = \sum u_{0n}$ into spherical Harmonics, we have

$$\mathrm{Re}\,(C^e u_0, u_0) = \sum \mathrm{Re}\,\theta_n^e\, |u_{0n}|^2,\quad \mathrm{Re}\,\theta_n^e = \frac{1}{|h_n|^2}\frac{\partial}{\partial r}(|h_n(kr)|^2),$$

and $|h_n(kr)|$ is a decreasing function of r, see Abramowitz-Stegun [1] p.439. Then the *Calderon projectors* are obtained in the space of spherical Harmonics of order n also by solving the transmission problem:

$$(443)\quad \left|\begin{array}{l} \text{i) } j_n(kR)\,K_n(\alpha) - h_n(kR)\,\tilde{K}_n(\alpha) = \rho_n(\alpha), \\ \text{ii) } k\,j_n'(kR)\,K_n(\alpha) - k\,h_n'(kR)\,\tilde{K}_n(\alpha) = \rho_n'(\alpha), \end{array}\right.$$

with given $\rho_n(\alpha)$, $\rho_n'(\alpha)$. Thanks to the Wronskian

(444) $W(j_n, h_n)(kR) = [h_n' j_n - h_n j_n'](kR) = i/(kR)^2,$

we obtain,

$$(445)\quad \begin{pmatrix} \tilde{K}_n(\alpha) \\ K_n(\alpha) \end{pmatrix} = -\,ikR^2 \begin{pmatrix} kj_n' & -j_n \\ kh_n' & -h_n \end{pmatrix}\begin{pmatrix} \rho_n(\alpha) \\ \rho_n'(\alpha) \end{pmatrix},$$

with j_n for $j_n(kR)$, h_n for $h_n(kR)$. This gives the Calderon projectors P_n^e, P_n^i by:

$$(446)\quad P_n^e = ikR^2 \begin{pmatrix} kh_nj_n' & -\ h_nj_n \\ k^2h_n'j_n' & -\,kh_n'j_n \end{pmatrix},\quad P_n^i = ikR^2 \begin{pmatrix} -\,k\,h_n'j_n & h_nj_n \\ -k^2h_n'j_n' & kh_nj_n' \end{pmatrix},$$

and:

$$P_n^e + P_n^i = I,\quad P_n^i - P_n^e = S_n,\quad \text{with } S_n = ikR^2 \begin{pmatrix} -\,k\,(h_nj_n)' & 2\,h_nj_n \\ -\ 2k^2h_n'j_n' & k(h_nj_n)' \end{pmatrix}.$$

12.1.2. Scattering of an incident wave by a ball

Let k be the wavenumber of the incident wave u_I, k_1 be the wavenumber of the scattered wave in the ball. We assume that u_I is an entire solution of Helmholtz equation. The usual transmission conditions on the sphere S_R are

(447) $u^i - u^e = u_I,\qquad \frac{\partial u^i}{\partial n} - \frac{\partial u^e}{\partial n} = \frac{\partial u_I}{\partial n}.$

Using the Rayleigh expansion (380) and (442), we have (if $j_n(k_1R) \neq 0$)

$$(448) \qquad u_n^i - u_n^e = u_{I,n}, \qquad \theta_n^{il} u_n^i - \theta_n^e u_n^e = \theta_n^i u_{I,n},$$

with (see 442):

$$(449) \qquad \theta_n^e = k\frac{h_n'}{h_n}(kR), \quad \theta_n^i = k\frac{j_n'}{j_n}(kR), \quad \theta_n^{il} = k_1\frac{j_n'}{j_n}(k_1R).$$

If k_1 is real, θ_n^{il} is also real, thus $\theta_n^e - \theta_n^{il} \neq 0$ since the Wronskian $W(h_n, \bar{h}_n) \neq 0$. Then the solution of (448) is given by

$$(450) \quad u_n^e = R_n u_{In}, \; u_n^i = T_n u_{In}, \; R_n = -\frac{\theta_n^i - \theta_n^{il}}{\theta_n^e - \theta_n^{il}}, \quad T_n = \frac{\theta_n^e - \theta_n^i}{\theta_n^e - \theta_n^{il}}, \; T_n - R_n = 1.$$

Then for $u_I(r\alpha) = \Sigma j_n(kr)K_n^I(\alpha)$, $u_{I,n} = j_n(kR)K_n^I(\alpha)$, the waves u^i and u^e are

$$(451) \quad \begin{cases} u^i(r\alpha) = \sum_n \tilde{T}_n j_n(k_1r)K_n^I(\alpha), & \text{with } \tilde{T}_n = T_n j_n(kR)/j_n(k_1R), \\ u^e(r\alpha) = \sum_n \tilde{R}_n h_n^{(1)}(kr)K_n^I(\alpha), & \text{with } \tilde{R}_n = R_n j_n(kR)/h_n(kR), \text{ for } r \geq R, \end{cases}$$

with $K_n^I(\alpha) = i^n(2n+1)P_n(\alpha.\alpha_0)$ for a plane wave. The series for u^e is convergent in $L^2(S_r)$ for all $r \geq R$. Its radius of convergence R_{ce} is given by (418), with:

$$c_n^2 = \int_{S^2} |\tilde{R}_n|^2 (2n+1)^2 \, |P_n(\alpha.\alpha_0)|^2 \, d\alpha = 4\pi(2n+1)|\tilde{R}_n|^2.$$

Then we can verify that $R_{ce} = 0$: the Rayleigh series converge for all $r > 0$!

12.2. Scattering with Maxwell equations

12.2.1. Debye potential of the electromagnetic field

Let (E,H) be an electromagnetic field in a ball B_a or its complementary, satisfying Maxwell equations

$$(452) \qquad \text{curl } H + i\omega\varepsilon E = 0, \qquad -\text{curl } E + i\omega\mu H = 0 \quad \text{in } B_a \text{ or } B'a = R^3\backslash\bar{B}_a,$$

with constant ε,μ, with (locally) finite energy, and the Silver-Müller conditions at infinity. Then (see Schulenberger [1], Aydin-Hizal [1]), there exist two scalar (outgoing) solutions φ_1, φ_2 of the Helmholtz equation for $k^2 = \omega^2\varepsilon\mu$, such that

$$(453) \text{ i) } E = -\text{curl}(x\varphi_1) + \tfrac{i}{\omega\varepsilon}\text{curl curl}(x\varphi_2), \quad \text{ii) } H = \tfrac{i}{\omega\mu}\text{curl curl}(x\varphi_1) + \text{curl}(x\varphi_2).$$

We get φ_1 and φ_2 up to a function of $r = |x|$ from E or H by solving the equations

$$-i\omega\varepsilon\, x.E = L^2\varphi_2, \qquad -x.\text{curl } E = -i\omega\mu\, x.H = L^2\varphi_1,$$

with $L^2\phi = x.\text{curl curl}(x\phi)$; L^2 is the angular momentum operator (we can identify L^2 with the Laplace-Beltrami operator $-\Delta_{S^2}$); in $L^2(\Omega)$, $\Omega = B_a$ or $R^3\backslash B_a$, it is a

positive selfadjoint operator (see Dautray-Lions [1] chap. 9B.1.1).
The functions ψ_1 and ψ_2 are *called the Debye potentials of the electromagnetic field* (E, H). Furthermore the fields $(E^{(1)}, H^{(1)})$ and $(E^{(2)}, H^{(2)})$ defined by:

$$(454) \quad \begin{cases} E^{(1)} = -\operatorname{curl}(x\psi_1), & H^{(1)} = \frac{i}{\omega\mu} \operatorname{curl}\operatorname{curl}(x\psi_1), \\ E^{(2)} = \frac{i}{\omega\varepsilon} \operatorname{curl}\operatorname{curl}(x\psi_2), & H^{(2)} = \operatorname{curl}(x\psi_2), \end{cases}$$

are *respectively called* TE (*transverse electric*) *fields and* TM (*transverse magnetic*) *fields*. Thanks to relation (28) chap. 2, we have: $n.E^{(1)} = 0$, and $n.H^{(2)} = 0$, when n is the normal to the sphere $\Sigma = S_r$. Then taking the tangential trace of these fields on the sphere Σ, we obtain:

$$(455) \quad \begin{cases} \text{i) } n \wedge E|_\Sigma = \operatorname{grad}_\Sigma(-r\psi_1) + \overrightarrow{\operatorname{curl}}_\Sigma\left(\frac{i}{\omega\varepsilon}(I + D)\psi_2\right), \\ \text{ii) } n \wedge H|_\Sigma = \overrightarrow{\operatorname{curl}}_\Sigma\left(\frac{i}{\omega\mu}(I + D)\psi_1\right) + \operatorname{grad}_\Sigma(r\psi_2), \end{cases}$$

with

$$(456) \quad D\psi = \sum x_k \frac{\partial\psi}{\partial x_k} = r\frac{\partial\psi}{\partial r},$$

or also for the projections on the sphere Σ:

$$(455)' \quad \begin{cases} \text{i) } \pi_\Sigma E = \overrightarrow{\operatorname{curl}}_\Sigma(r\psi_1) + \operatorname{grad}_\Sigma\left(\frac{i}{\omega\varepsilon}(I + D)\psi_2\right), \\ \text{ii) } \pi_\Sigma H = \operatorname{grad}_\Sigma\left(\frac{i}{\omega\mu}(I + D)\psi_1\right) - \overrightarrow{\operatorname{curl}}_\Sigma(r\psi_2). \end{cases}$$

This is due to the relations (28), (30), (30)' chap. 2, with $\operatorname{curl}(x\psi) = -x \wedge \operatorname{grad}\psi$.
The Debye potentials ψ_1 and ψ_2 on Σ are obtained from $n \wedge E|_\Sigma$ by:

$$(457) \quad r\Delta_\Sigma \psi_1 = -\operatorname{div}_\Sigma(n \wedge E|_\Sigma), \quad \Delta_\Sigma(I + D)\psi_2 = -i\omega\varepsilon \operatorname{curl}_\Sigma(n \wedge E|_\Sigma).$$

Formulas (455) or (455)' correspond to the Hodge decomposition of currents on the sphere Σ (see section 6.2, Appendix). Note that the sphere is simply connected, and thus the cohomology space $H^1(\Sigma)$ is reduced to 0.
Let Π_c and Π_g be the orthogonal projectors in the space $L^2_t(\Sigma)$ of square integrable tangent fields on Σ, so that:

$$(458) \quad f = f_c + f_g = \operatorname{grad}_\Sigma \phi_2 + \overrightarrow{\operatorname{curl}}_\Sigma \phi_1 \text{ with } f_g = \Pi_g f = \operatorname{grad}_\Sigma \phi_2, \; f_c = \Pi_c f = \overrightarrow{\operatorname{curl}}_\Sigma \phi_1,$$

and ϕ_1, ϕ_2 are obtained using the inverse G of the (opposite of the) Laplace-Beltrami operator on the sphere in $L^2_0(\Sigma) = \{u \in L^2(\Sigma), \int_\Sigma u \, d\Sigma = 0\}$, by:

$$\phi_1 = -G\operatorname{curl}_\Sigma f, \quad \phi_2 = -G\operatorname{div}_\Sigma f.$$

On the unit sphere, G is the integral operator (see Schulenberger [1]):

$$Gf(\alpha) = \int_\Sigma g(\alpha, \beta) f(\beta)\, d\beta, \quad \text{with } g(\alpha, \beta) = -\frac{1}{4\pi}\log\frac{1 - \alpha.\beta}{2}, \; \alpha, \beta \in \Sigma = S^2.$$

Note that the Debye potentials are related to the radial components E_r, H_r by

(459) $\Delta_{S^2}\,\varphi_1 = i\omega\mu\, rH_r$, $\quad \Delta_{S^2}\,\varphi_2 = i\omega\varepsilon\, rE_r$.

This is easily seen from the radial component of Maxwell equations, with (28), (30) chap. 2. Furthermore $rE_r = x.E$ and $rH_r = x.H$ satisfy the Helmholtz equation and converge to 0 at infinity (see (50)). Thus they can be used as potentials for the electromagnetic field, but it is not very easy to prove that they satisfy the outgoing condition, and the Debye potentials are more regular. For the converse of (459), we only have to eliminate constants functions on the sphere.

12.2.2. The Calderon operators on the sphere

To obtain the Calderon operators on the sphere, we have to inverse the relations between (φ_1, φ_2) and E_Σ. Now D in (455) is also (up to factor r) the exterior or interior Calderon operator on the sphere S_r of the scalar case.

In the exterior case, $B_{er} = (C_r^e + \frac{1}{r})$ is an isomorphism from $H^s(\Sigma)$ onto $H^{s-1}(\Sigma)$. In the interior case, we have to eliminate irregular frequencies, see 4.7, 4.8).

Now *in the exterior case*, we obtain from (455), (457),

(460) $n \wedge H_\Sigma = C^e(n \wedge E_\Sigma) = -\frac{1}{i\omega\mu}\,\overrightarrow{\mathrm{curl}}_\Sigma \Delta_\Sigma^{-1} B_{er}\,\mathrm{curl}_\Sigma E_\Sigma - i\omega\varepsilon\,\mathrm{grad}_\Sigma \Delta_\Sigma^{-1}\, B_{e\,r}^{-1}\,\mathrm{div}_\Sigma\, E_\Sigma$.

We will have a similar formula for waveguides, see (68) chap.4.2. Now we give explicit formulas thanks to Rayleigh series. Using the orthonormal basis $(K_{n,j}(\alpha)$ of spherical Harmonics (see (373)), then:

$$\mathrm{grad}_\Sigma\, K_{n,j},\quad \overrightarrow{\mathrm{curl}}_\Sigma\, K_{n,j},\quad n = 1, 2, \ldots, -n \le j \le n,$$

is an orthogonal family of $L^2_t(\Sigma)$, giving an orthonormal basis by normalization. Now we use differential operators on the unit sphere S^2, instead of Σ, thanks to

(461) $\mathrm{grad}_\Sigma(r\varphi(x)) = \mathrm{grad}_{S^2}\,\Psi(r,\alpha)$, $\quad \overrightarrow{\mathrm{curl}}_\Sigma\,(r\varphi(x)) = \overrightarrow{\mathrm{curl}}_{S^2}\,\Psi(r,\alpha)$,

with $\varphi(x) = \Psi(r,\alpha)$. Thanks to the exterior (resp. interior) Rayleigh series of φ_1, φ_2

(462) $\varphi_j(r\alpha) = \sum_{n>0} h_n^{(1)}(kr)K_n^j(\alpha)$ $\quad$ (resp $\sum_{n>0} j_n(kr)K_n^j(\alpha)$), $j = 1,2$,

in (455)', we obtain the exterior Rayleigh series:

(463)
$$\left|\begin{array}{l} \pi_\Sigma E = \sum_n h_n\,\overrightarrow{\mathrm{curl}}_{S^2}\,K_n^1 + \sum_n \frac{i}{\omega\varepsilon r}\,\tilde{h}_n\,\mathrm{grad}_{S^2}\,K_n^2, \\ \pi_\Sigma H = \sum_n \frac{i}{\omega\mu r}\,\tilde{h}_n\,\mathrm{grad}_{S^2}\,K_n^1 - \sum_n h_n\,\overrightarrow{\mathrm{curl}}_{S^2}\,K_n^2, \end{array}\right.$$

with the notations:

(464) $\quad h_n = h_n^{(1)}(kr), \quad \tilde{h}_n = (I + D)h_n^{(1)}(kr).$

We obtain the *exterior Calderon operator for the TE and TM waves* $C_n^{TE,e}$ and $C_n^{TM,e}$

(465) $\quad C_n^{TE,e} = -\frac{i}{\omega\mu r}\tilde{\theta}_n^e\, \alpha \wedge \Pi_g, \quad C_n^{TM,e} = i\omega\varepsilon r\,(\tilde{\theta}_n^e)^{-1}\, \alpha \wedge \Pi_c,$

with $\alpha \in S^2$ the normal to the sphere, $\tilde{\theta}_n^e = \tilde{h}_n / h_n$, Π_g, Π_c given by (458).
The interior Rayleigh expansion is obtained by substituting j_n for h_n. The *interior Calderon operators for TE and TM waves* are given by:

(466) $\quad C_n^{TE,i} = -\frac{i}{\omega\mu r}\tilde{\theta}_n^i\, \alpha \wedge \Pi_g, \quad C_n^{TM,i} = i\omega\varepsilon r\,(\tilde{\theta}_n^i)^{-1}\, \alpha \wedge \Pi_c,$

with the notations:

(467) $\quad j_n = j_n(kr), \quad \tilde{j}_n = (I + D)j_n(kr), \quad \tilde{\theta}_n^i = \tilde{j}_n / j_n.$

12.2.3. Scattering of an incident wave by a dielectric ball
Now let (E_I, H_I) be an incident wave on a ball B_a, with a dielectric medium of permittivity ε_1 and permeability μ_1. We assume that the incident wave is an entire solution of Maxwell equations in free space (for instance a plane wave). Then the incident wave is given thanks to the Debye potentials φ_{I1} and φ_{I2}, with the Rayleigh expansions:

(468) $\quad \varphi_{Ij}(r\alpha) = \sum j_n(kr)K_n^{I,j}(\alpha), \quad j = 1,2.$

Let $(\varphi_1^i, \varphi_2^i)$ and $(\varphi_1^e, \varphi_2^e)$ be the Debye potentials of the electromagnetic field inside the ball (E^i, H^i), and of the reflected field (E^e, H^e). Let:

(469) $\quad \varphi_j^i(r\alpha) = \sum j_n(k_1 r)K_n^{i,j}(\alpha), \quad \varphi_j^e(r\alpha) = \sum h_n^{(1)}(kr)K_n^{e,j}(\alpha), \quad j = 1,2,$

be their Rayleigh expansions with unknown harmonic functions $K_n^{i,j}$, $K_n^{e,j}$. Then writing the transmission conditions on the sphere Σ, with $r = a$, for TE and TM waves we have:

$$(470)\quad \left|\begin{array}{l} \text{i) } [\varphi_j]_\Sigma = \varphi_j^i - \varphi_j^e = \varphi_j^I,\ j = 1, 2 \\ \text{ii) } [\frac{1}{\omega\mu}(I + D)\varphi_1]_\Sigma = \frac{1}{\omega\mu_1}(I + D)\varphi_1^i - \frac{1}{\omega\mu}(I + D)\varphi_1^e = \frac{1}{\omega\mu}(I + D)\varphi_1^I \\ \text{iii) } [\frac{1}{\omega\varepsilon}(I + D)\varphi_2]_\Sigma = \frac{1}{\omega\varepsilon_1}(I + D)\varphi_2^i - \frac{1}{\omega\varepsilon}(I + D)\varphi_2^e = \frac{1}{\omega\varepsilon}(I + D)\varphi_2^I. \end{array}\right.$$

Now using the Rayleigh expansions of these Debye potentials, and the notations

(471) $\quad j_n^1 = j_n(k_1 r), \quad \tilde{j}_n^1 = (I + D)j_n(k_1 r), \quad \tilde{\theta}_n^{1i} = (\tilde{j}_n^1)/j_n^1,$

we obtain the system of equations:

$$(472)\qquad \begin{cases} j^1_n K^{i,1}_n - h_n K^{e,1}_n = j_n K^{I,1}_n, \\ \frac{1}{\mu_1} \tilde{j}^1_n K^{i,1}_n - \frac{1}{\mu} \tilde{h}_n K^{e,1}_n = \frac{1}{\mu} \tilde{j}_n K^{I,1}_n. \end{cases}$$

We have the same system for the superscript 2, but with ε instead of μ. Their solutions are given by

$$(473)\qquad h_n K^{e,j}_n = R^{(j)}_n j_n K^{I,j}_n, \quad j^1_n K^{i,j}_n = T^{(j)}_n j_n K^{I,j}_n, \qquad j = 1, 2,$$

and thus:

$$(474)\qquad \begin{cases} \tilde{h}_n K^{e,j}_n = \mathcal{R}^{(j)}_n \tilde{j}_n K^{I,j}_n, \quad j = 1, 2, \\ (\mu_1)^{-1} \tilde{j}^1_n K^{i,1}_n = \mathcal{T}^{(1)}_n (\mu^{-1} \tilde{j}_n K^{I,1}_n), \quad (\varepsilon_1)^{-1} \tilde{j}^1_n K^{i,2}_n = \mathcal{T}^{(2)}_n (\varepsilon^{-1} \tilde{j}_n K^{I,2}_n), \end{cases}$$

with

$$(475)\qquad R^{(1)}_n = - \frac{\mu_1^{-1} \tilde{\theta}^{1i}_n - \mu^{-1} \tilde{\theta}^{i}_n}{\mu_1^{-1} \tilde{\theta}^{1i}_n - \mu^{-1} \tilde{\theta}^{e}_n}, \qquad \mathcal{R}^{(1)}_n = \frac{\tilde{\theta}^e_n}{\tilde{\theta}^i_n} R^{(1)}_n,$$

and also:

$$(475)' \qquad T^{(1)}_n = \frac{\mu^{-1} (\tilde{\theta}^i_n - \tilde{\theta}^e_n)}{\mu_1^{-1} \tilde{\theta}^{1i}_n - \mu^{-1} \tilde{\theta}^e_n}, \qquad \mathcal{T}^{(1)}_n = \frac{\mu}{\mu_1} \frac{\tilde{\theta}^{1i}_n}{\tilde{\theta}^i_n} T^{(1)}_n,$$

with similar formulas for $R^{(2)}_n$ and $T^{(2)}_n$, but with ε, ε_1 in place of μ, μ_1. Note that:

$$(476)\qquad T^{(j)}_n - R^{(j)}_n = 1, \qquad \mathcal{T}^{(j)}_n - \mathcal{R}^{(j)}_n = 1, \quad j = 1, 2.$$

We finally obtain the reflected electromagnetic field (E^e, H^e), thanks to (463), by its components:

$$(477)\qquad \begin{cases} \pi_\Sigma E^e{}_{TE,n} = R^{(1)}_n (\pi_\Sigma E_I)_{TE,n}, \quad \pi_\Sigma H^e{}_{TE,n} = \mathcal{R}^{(1)}_n (\pi_\Sigma H_I)_{TE,n}, \\ \pi_\Sigma E^e{}_{TM,n} = \mathcal{R}^{(2)}_n (\pi_\Sigma E_I)_{TM,n}, \quad \pi_\Sigma H^e{}_{TM,n} = R^{(2)}_n (\pi_\Sigma H_I)_{TM,n}, \end{cases}$$

and the field (E^i, H^i) inside the ball by:

$$(478)\qquad \begin{cases} \pi_\Sigma E^i{}_{TE,n} = T^{(1)}_n (\pi_\Sigma E_I)_{TE,n}, \quad \pi_\Sigma H^i{}_{TE,n} = \mathcal{T}^{(1)}_n (\pi_\Sigma H_I)_{TE,n}, \\ \pi_\Sigma E^i{}_{TM,n} = \mathcal{T}^{(2)}_n (\pi_\Sigma E_I)_{TM,n}, \quad \pi_\Sigma H^i{}_{TM,n} = T^{(2)}_n (\pi_\Sigma H_I)_{TM,n}. \end{cases}$$

EXAMPLE. If the incident wave is a plane wave, given by:

(479) $E_I(x) = E_{I0}\, e^{ikr\alpha_0.\alpha}, \quad H_I(x) = H_{I0}\, e^{ikr\alpha_0.\alpha}, \quad x = r\alpha,\ \alpha \in S^2,$

with given $\alpha_0 \in S^2$, E_{I0} and $H_{I0} \in C^3$, with $\alpha_0.E_{I0} = \alpha_0.H_{I0} = 0$, $ZH_{I0} = \alpha_0 \wedge E_{I0}$, and Z the impedance of free space. Taking $\alpha_0 = (0,0,1)$, $E_{I0} = (1,0,0)$, we obtain the Debye potentials of this plane wave by the Rayleigh series (468), with the harmonic functions (373) (see Jones [1] p. 446):

(480) $K_n^{I,1}(\alpha) = -\zeta_n[K_{n,1} - K_{n,-1}], \quad K_n^{I,2}(\alpha) = -\frac{i}{Z}\zeta_n[K_{n,1} + K_{n,-1}],$

with $\zeta_n = (2\pi)^{1/2}\frac{(2n+1)\, i^{n+1}}{2n\,(n+1)}$. This gives us the scattered field thanks to (477).

⊗

3.13. ADDENDUM. COMPACTNESS PROPERTIES IN SCATTERING PROBLEMS. The scattering problems of scalar waves on bounded (regular) obstacles depend on the Fredholm alternative; this is a consequence of the compactness of the operators K, J and L (see (70)) in the framework $H^{-1/2}(\Gamma)$ (or $H^{1/2}(\Gamma)$, or $C^0(\Gamma)$), the range of these operators being in $H^{1/2}(\Gamma)$. Since we can often prove fairly easily uniqueness at most of the solution in scattering problems, then this also implies its existence! In electromagnetism, scattering problems also depend on the Fredholm alternative, and the sum: $C^e + C^i = -R^{-1}T$ (see (241)) is compact in $H^{-1/2}(\text{div},\Gamma)$, as a consequence of the proposition:

Proposition 32. *If Γ is of class $C^{1,1}$, the operator T defined by (100)' is compact in $H^{-1/2}(\text{div},\Gamma)$, with Im T in $H^{1/2}(\text{div},\Gamma)$ if Γ is of class $C^{2,1}$.*

PROOF. i) For all u_Γ in $H^{-1/2}(\text{div},\Gamma)$, we verify that Tu_Γ is given by:

(481) $Tu_\Gamma = \text{grad}_\Gamma\, \phi + v_\Gamma, \quad \phi \in H^{1/2}(\Gamma), \quad v_\Gamma \in L^2(\Gamma)^3$, with $v_\Gamma \in H_t^{1/2}(\Gamma)$ if Γ is $C^{2,1}$.

By developing the double vector product in (100)' and using (70), we have:

(482) $Tu_\Gamma(x) = -Ju_\Gamma(x) - n_x\,(n_x.Ju_\Gamma(x)) - 2\sum_j n_{x,j}\, \text{grad}_{\Gamma,x}\, Lu_{\Gamma,j}(x),$

which is of the form (481) with $\phi = -2n_x.Lu_\Gamma$, thanks to the properties of K and J.
ii) Now we verify that $\text{div}_\Gamma\, Tu_\Gamma \in H^{1/2}(\Gamma)$. Let (E, H) be defined by (223) with $J_\Gamma = u_\Gamma$. Using the notation $\{v\}_\Gamma = v|_{\Gamma_i} + v|_{\Gamma_e}$, we have like for (84) (see also (28), (30) chap.2)

(483) $\text{div}_\Gamma\, Tu_\Gamma = -\text{div}_\Gamma(\{n \wedge H\}_\Gamma) = -i\omega\varepsilon\,\{n.E\}_\Gamma.$
Then using (95)' and (70), we have:

(484) $\{n.E\}_\Gamma = ikZ\,\{n.\mathcal{L}u_\Gamma\}_\Gamma + i\frac{Z}{k}J\,(\text{div}_\Gamma\, u_\Gamma) = 2\, ikZ\, n.Lu_\Gamma + i\frac{Z}{k}J\,(\text{div}_\Gamma\, u_\Gamma).$

Since Lu_Γ and $J\,(\text{div}_\Gamma\, u_\Gamma)$ are in $H^{1/2}(\Gamma)$, this implies the proposition. ⊗

CHAPTER 4

WAVEGUIDE PROBLEMS

1. WAVEGUIDES WITH HELMHOLTZ EQUATIONS

Here we study the propagation of scalar stationary waves in a "waveguide" which is a semi-infinite cylinder. The method that we develop, will allow us to deal with cases of more general geometry and later, with the electromagnetic case.

Let $\Omega_+ = \Omega \times R^+$ be an infinite open set which is a semi-infinite cylinder in R^3 with a regular bounded cross-section Ω in R^2, of boundary Γ. Let $\Gamma_+ = \Gamma \times R^+$ be the lateral boundary of Ω_+. We first consider the Dirichlet (resp. Neumann) problem: find u such that

$$(1)\quad \begin{cases} \text{i) } \Delta u + k^2 u = 0 \text{ in } \Omega_+, \\ \text{ii) } u|_{\Gamma_+} = 0 \ \left(\text{resp. } \frac{\partial u}{\partial n}\Big|_{\Gamma_+} = 0\right), \\ \text{iii) } u(x,0) = u^o(x)\,,\ x \in \Omega \subset R^2, \end{cases}$$

with a real wavenumber k and a given u^o (for instance in $L^2(\Omega)$). We denote by Δ_T the Laplacian in the transverse variables $x_T = (x_1,x_2)$ (in R^2), by x_3 the variable along the axis of the cylinder with origin at the end of the cylinder. Then we can write (1)i) in the form

$$(1)\text{i})' \quad \frac{\partial^2 u}{\partial x_3^2} + \Delta_T u + k^2 u = 0 \ \text{ in } \Omega^+.$$

Let Δ_T^D (resp. Δ_T^N) be the Laplacian with Dirichlet (resp. Neumann) condition. Denoting by A either $-\Delta_T^D$ or $-\Delta_T^N$ (both selfadjoint operators), we can rewrite (1) in the form:

$$(2)\qquad \left|\begin{array}{l} \text{i) } \dfrac{\partial^2 u}{\partial x_3^2} - A\,u + k^2 u = 0, \\ \text{ii) } u(.,0) = u^o. \end{array}\right.$$

Let (λ_n^2, ϕ_n) be a spectral decomposition of A (that is $((\lambda_n^D)^2, \phi_n^D)$ for $A = -\Delta_T^D$, and $((\lambda_n^N)^2, \phi_n^N))$ for $A = -\Delta_T^N$, with ϕ_n an orthonormal basis in $L^2(\Omega)$.
We decompose u^o and $u(.,x_3)$ on this basis:

$$(3)\qquad u^o = \sum_n u_n^o \phi_n, \qquad u(.,x_3) = \sum_n u_n(x_3)\,\phi_n,$$

and then we have to find the unknown coefficients $u_n(x_3)$ so that:

$$(4)\qquad \left|\begin{array}{l} \text{i) } \dfrac{\partial^2 u_n}{\partial x_3^2} + (-\lambda_n^2 + k^2)\,u = 0, \\ \text{ii) } u_n(0) = u_n^o. \end{array}\right.$$

Denoting by α_n and β_n two arbitrary constants, the general solution of (4)i) is:

a) if $\lambda_n^2 > k^2$, $u_n(x_3) = \alpha_n \exp(x_3\sqrt{\lambda_n^2 - k^2}) + \beta_n \exp(-x_3\sqrt{\lambda_n^2 - k^2})$.

If we require that the solution be tempered with respect to x_3, thus $\alpha_n = 0$, then:

$$(5)\qquad u_n(x_3) = \exp(-x_3\sqrt{\lambda_n^2 - k^2})\,u_n^o;$$

b) if $\lambda_n^2 = k^2$, $u_n(x_3) = \alpha_n x_3 + \beta_n$.
If we require that the solution be x_3-bounded, then $\alpha_n = 0$, $u_n(x_3) = u_n^o$;

c) if $\lambda_n^2 < k^2$, $u_n(x_3) = \alpha_n \exp(-ix_3\sqrt{\lambda_n^2 - k^2}) + \beta_n \exp(ix_3\sqrt{\lambda_n^2 - k^2})$.

This is called the propagating part of the wave u. Then (4)ii) implies $\alpha_n + \beta_n = u_n^o$.
Thus α_n, β_n and u_n are not determined in a unique way: we say that problem (1) is ill-posed, i.e., the given u^o does not determine a solution u in a unique way.
We transform problem (1) into a well-posed problem by adding a physical condition concerning the sense of wave propagation: if we suppose that the stationary problem (1) comes from an evolution problem with a solution of the type $u(x,t) = e^{-i\omega t}u(x)$, we say that u *is a wave propagating towards the positive (resp. negative) axis* x_3 *if the propagating part of* u *is a superposition of waves of the type*:

$$(6)\qquad u_n(x_3) = e^{i\mu_n x_3} v_n \quad (\text{resp. } e^{-i\mu_n x_3} v_n), \quad \mu_n \geq 0.$$

For a time evolution $u(x,t) = e^{i\omega t}u(x)$, we make the converse choice. We define:

(7) $\theta_n \overset{\text{def}}{=} \theta_n^+ = (\lambda_n^2 - k^2)^{1/2}$ if $\lambda_n^2 > k^2$, $\theta_n \overset{\text{def}}{=} -i\,\theta_n^- = -i(k^2 - \lambda_n^2)^{1/2}$ if $\lambda_n^2 \leq k^2$.

Thus with the physical hypothesis of positive axis x_3 wave propagation, we have to take $\alpha_n = 0$ in case c), and then we obtain a (unique) solution of (1) in the form

(8) $$u(.,x_3) = \sum_n e^{-\theta_n x_3} u_n^o \phi_n.$$

Theorem 1. *Problem* (1) *with* u^o *given in* $L^2(\Omega)$ *and with the condition*:

(9) u *is an* x_3*-bounded wave, propagating towards the positive axis* x_3,

has a unique solution u in $C^0([0,+\infty[,L^2(\Omega))$, *and it is obtained thanks to a holomorphic contraction semigroup of class* C^0 *in* $L^2(\Omega)$, $(G(x_3))$, $x_3 > 0$, *by* (8).

For these notions of semigroup, we refer to Dautray-Lions [1], Pazy [1]. We recall that a contraction semigroup in a Banach space X satisfies:

$$\|G(t)u^o\| \leq \|u^o\|, \quad \forall u^o \in X, \ t = x_3.$$

We note $G(x_3) = G^D(x_3)$ for the Dirichlet boundary condition, $G(x_3) = G^N(x_3)$ for the Neumann boundary condition. Its infinitesimal generator C (C_D or C_N) is defined (on the basis of the cylinder) in $L^2(\Omega)$ by:

(10) $$\left| \begin{array}{l} \dfrac{\partial u}{\partial x_3}(.,0) = Cu^o \ \text{(with } C = C_D \text{ or } C_N\text{)}, \\ D(C) = \{u^o \in L^2(\Omega),\ u(.,x_3) = G(x_3)u^o \in C^1([0,+\infty), L^2(\Omega))\}. \end{array} \right.$$

Definition 1. *The infinitesimal generator* C *of the semigroup* $(G(x_3))$ *defined by* (10) *is called the Calderon (or also the capacity) operator of the guide at the wavenumber* k.

We can characterize this operator thanks to the spectral decomposition of the Laplacian by:

(11) $$C\phi_n = -\theta_n\phi_n, \quad \forall n, \ \text{and} \ D(C_D) = H_0^1(\Omega),\ D(C_N) = H^1(\Omega).$$

Thus $C^2 = A - k^2$. Hence C *is a square root of the selfadjoint operator* $(A - k^2I)$. *Furthermore it has the following properties* (for all v in D(C)):

(12) $$\text{Re}\,(Cv, v) = \sum_{n,\ \lambda_n^2 \geq k^2} -\theta_n^+ |v_n|^2 \leq 0, \quad \text{Im}\,(Cv, v) = \sum_{n,\ \lambda_n^2 < k^2} \theta_n^- |v_n|^2 \geq 0.$$

Thus $L^2(\Omega)$ has an orthogonal decomposition into: $L^2(\Omega) = H_k^+ \oplus H_k^-$, with:
$H_k^- \overset{\text{def}}{=} \{\sum c_n \phi_n \text{ with } \lambda_n^2 < k^2\}$, of finite dimension (the space of "propagating modes"), and C has the corresponding decomposition:

(13) $\quad C = -C^+ + i\,C^-$,

with C^+ and C^- positive selfadjoint operators on H_k^+ and H_k^-.

Then $C = C_D$ is a continuous operator from $H_0^1(\Omega)$ into $L^2(\Omega)$ and by duality from $L^2(\Omega)$ into $H^{-1}(\Omega)$ then by interpolation from $H_{o\,o}^{1/2}(\Omega)$ into $(H_{o\,o}^{1/2}(\Omega))'$.

More generally, Theorem 1 *is valid if we replace space* $L^2(\Omega)$ by $D(A^s)$ *for any real* s. This will allow us to deal with many examples.

Example 1. *Junctions and Cascades*

A typical junction consists of a bounded domain Ω *in* R^3 with p guides arriving at it with different directions (see for instance Jones[1]). Let (Γ_j), j = 1 to p, be bases for these guides. We can choose Ω and these bases so that Γ_j for j = 1 to p be part of the boundary Γ of the domain Ω. Let Γ_o be the part $\Gamma/\cup\Gamma_j$ of the boundary (the wall of the junction). We assume for instance:

i) that the guides walls and Γ_o are *hard*, i.e., with Dirichlet boundary conditions. We could also assume them *soft*, that is, with Neumann boundary conditions; the terms hard or soft originate from problems in acoustics where pressure is the unknown;

ii) that a given incident wave $u_I = (u_I)_j$ is coming from the guides to the junction;

iii) that the wavenumbers are k in the junction and k_j in each guide (with the same frequency ω).

Then the reflected or transmitted wave in the jth guide (u_{rj}) must satisfy (with the capacity operator C_j of the jth guide):

$$\frac{\partial u_{rj}}{\partial n} = C_j\, u_{rj}. \tag{14}$$

Then using the continuity relations of the wave across the boundary Γ_j:

$$u = u_r + u_I, \qquad \frac{\partial u}{\partial n} = \frac{\partial u_r}{\partial n} + \frac{\partial u_I}{\partial n} \text{ on every } \Gamma_j, \tag{15}$$

we obtain from (14), (15) that the wave u must satisfy:

$$\frac{\partial u}{\partial n} - C_j\, u = f_{Ij} \quad \text{on } \Gamma_j \text{ with } f_{Ij} = \frac{\partial u_{Ij}}{\partial n} - C_j\, u_{Ij}. \tag{16}$$

Thus we have to find the wave u in the junction, satisfying:

$$\left| \begin{array}{l} \text{i) } \Delta u + k^2 u = 0 \text{ in } \Omega, \\ \text{ii) } u|_{\Gamma_o} = 0 \text{ and (16) on } \Gamma_j, \ j = 1 \text{ to } p. \end{array} \right. \tag{17}$$

This problem can be written in a variational form; we define the space:

(18) $H^1_{\Gamma_0}(\Omega) = \{u \in H^1(\Omega),\ \ u|_{\Gamma_0} = 0\}$,

and the sesquilinear form a(u, v) on this space by:

(19) $a(u,v) = \int_\Omega (\text{grad } u \,.\, \text{grad } \bar{v} - k^2 u\bar{v})\, dx - \sum_j \int_{\Gamma_j} C_j u\, \bar{v}\, d\Gamma$,

which is continuous on $H^1_{\Gamma_0}(\Omega)$, and $H^1_{\Gamma_0}(\Omega)$ coercive with respect to $L^2(\Omega)$, thanks to the above properties of the capacity operators.
The variational form of problem (17) is: find u in $H^1_{\Gamma_0}(\Omega)$ satisfying

(20) $a(u,v) = \sum_j \int_{\Gamma_j} f_{Ij} \bar{v}\, d\Gamma, \quad \forall v \in H^1_{\Gamma_0}(\Omega)$.

For a regular domain Ω, the natural injection $H^1_{\Gamma_0}(\Omega) \to L^2(\Omega)$ is compact.
Thus problem (17) depends on the Fredholm alternative: it has a unique solution except for wavenumbers k such that (17) has a nontrivial solution, that is, for the eigenmodes of the junction.
Using the decomposition corresponding to (12) for the jth guide, we denote by:
$C_j = -C_j^+ + i\, C_j^-$ the decomposition of C_j, by V_j^- the space of propagating modes for the jth guide (above denoted by H_k^-), and by P^j the corresponding projection of u^j (the trace of u on Γ_j) onto the space V_j^-.
Then u is an eigenmode of the junction if (20) is satisfied with $f_I = 0$. Taking v = u in (20), and the real and imaginary parts, we easily obtain that u must satisfy:

(21) $$\left|\begin{array}{l} \text{i) } \Delta u + k^2 u = 0 \text{ in } \Omega, \\ \text{ii) } u|_{\Gamma_0} = 0,\ \ \dfrac{\partial u}{\partial n} - C_j^+ u = 0 \text{ on } \Gamma_j,\ j = 1 \text{ to } p, \text{ and } P_j u_j = 0, \text{ (or } u_j \in V_j^+). \end{array}\right.$$

Remark 1. *The case of small junctions (or of weak wavenumber).* If the domain of the junction is small, we can apply the Poincaré inequality:

(22) $\|u\| \leq C\, \|\text{grad } u\|,\ \ \forall u \in H^1_{\Gamma_0}(\Omega)$,

with $C = \lambda_0^{-1}$, λ_0 being the smallest eigenvalue λ of:

(23) $$\left|\begin{array}{l} \text{i) } \Delta u + \lambda u = 0 \text{ in } \Omega, \\ \text{ii) } u|_{\Gamma_0} = 0,\ \ \dfrac{\partial u}{\partial n}\Big|_{\Gamma\setminus\Gamma_0} = 0. \end{array}\right.$$

Then taking the real part of (20) with v = u and f = 0, we obtain with (12):

(24) $\int_\Omega (|\text{grad } u|^2 - k^2 |u|)\, dx \leq 0$;

and using (22) we have:

(25) $\quad (1 - k^2C^2)\int_\Omega |\text{grad}\, u|^2\, dx \le 0\,, \quad \forall u \in H^1_{\Gamma_o};$

thus for $(1 - k^2C^2) > 0$ (that is for $\lambda_o > k$), we have $u = 0$.
This implies that the wave problem (17) in a small junction always has a unique solution.

⊗

Remark 2. *Dissipative junctions.* If the wavenumber k is complex with $\text{Im}\, k^2 > 0$, we also have uniqueness of the solution of problem (20). This is obtained taking the imaginary part of (20) with $v = u$ and $f = 0$ and using (12):

(26) $$\text{Im}\,(-k^2)\,\|u\|^2 = \sum_j \text{Im}\,(C_j u, u) \ge 0\,,$$

thus giving $u = 0$, and we have the same conclusion as in Remark 1.

⊗

Remark 3. *Cascades* (see for instance Jones [1] p. 257). The case when the cross-section of the guide changes (with discontinuities or not) over a finite distance, is a special case of junction (with two different guides arriving to it).

⊗

Remark 4. We can generalize to "inhomogeneous" waveguides that is when there are different media with different wavenumbers in the cross section of the guide (with the *usual transmission conditions* at the interfaces of the fibers).

⊗

Remark 5. We can also generalize to "inhomogeneous" junctions with a wavenumber k dependent on x. The problem can be solved numerically by using finite elements inside the junction and spectral decompositions at the boundary for the capacity operators (for instance).

⊗

2. WAVEGUIDES IN ELECTROMAGNETISM

We consider a semi-infinite cylinder $\Omega_+ = \Omega \times R^+$ in R^3 with x_3 axis, occupied by a conservative medium with permittivity and permeability ε and μ (with positive real ε, μ), bounded by a perfect conductor. We assume that the cross-section Ω is an open regular bounded and connected set (it can be simply connected or not). We first consider the following problem: find the electromagnetic field (E,H) in the waveguide Ω_+, at angular frequency ω, satisfying (with the same notations as in section 1)

(27) $$\begin{cases} \text{i) curl}\, H + i\omega\varepsilon\, E = 0\,, & \text{ii)} - \text{curl}\, E + i\omega\mu\, H = 0 \ \text{ in } \Omega_+\,, \\ \text{iii) } n \wedge E|_{\Gamma_+} = 0\,, & \text{iv) } E_T(0) = E_T^0 \ \text{ on } \Omega, \end{cases}$$

where E_T^0 is a given transverse electric field (for instance in $H_o(\text{curl}_T, \Omega)$).

We denote by (E_T, H_T) the transverse components of the electromagnetic field (E, H) and we use the index T for transverse derivations: let ϕ be a function on Ω, $v_T = (v_1, v_2)$ be a vector function on Ω, we denote:

$$(28) \quad \begin{cases} \text{grad}_T\,\phi = \left(\frac{\partial\phi}{\partial x_1}, \frac{\partial\phi}{\partial x_2}\right) = (\partial_1\phi, \partial_2\phi), \quad \overrightarrow{\text{curl}}_T\,\phi = (\partial_2\phi, -\partial_1\phi) \\ \text{curl}_T\, v_T = \partial_1 v_2 - \partial_2 v_1. \end{cases}$$

Let e_3 be the unit vector along x_3, and let S be the operator:

$$(29) \qquad S\,v_T = S(v_1, v_2) = (-v_2, v_1) = e_3 \wedge v_T.$$

With these notations, the Maxwell equations are:

$$(30) \quad \begin{cases} \text{i) } -\partial_3 H_T + \text{grad}_T\, H_3 + i\omega\varepsilon\, SE_T = 0, \\ \text{ii) } \partial_3 E_T - \text{grad}_T\, E_3 + i\omega\mu\, SH_T = 0, \end{cases}$$

with:

$$(31) \quad \begin{cases} \text{i) } \text{curl}_T\, H_T + i\omega\varepsilon E_3 = 0, \\ \text{ii) } -\text{curl}_T\, E_T + i\omega\mu H_3 = 0, \end{cases}$$

and since $n \wedge E = (n_2, -n_1, n_1E_2 - n_2E_1) = -E_3\,Sn + e_3 n_T \wedge E_T$, the boundary conditions are:

$$(32) \quad \begin{cases} \text{i) } n_T \wedge E_T|_{\Gamma_+} = 0 \text{ and } E_3|_{\Gamma_+} = 0, \\ \text{ii) } E_T(0) = E_T^0. \end{cases}$$

Like in the Helmholtz case we can see that this problem is ill-posed and it becomes a well-posed problem adding to it the physical hypothesis of propagation of the wave (E, H):

(33) (E, H) *is an* x_3*-bounded wave, propagating towards the positive axis* x_3.

We *first formally* replace E_3 and H_3 in (30) by their values in (31); thus:

$$(34) \qquad \text{i) } -\partial_3 H_T + A_1\, E_T = 0, \quad \text{ii) } \partial_3 E_T + A_2\, H_T = 0,$$

with:

$$(35) \quad \begin{cases} \text{i) } A_1 E_T = \frac{1}{i\omega}\,\text{grad}_T \frac{1}{\mu}\,\text{curl}_T\, E_T + i\omega\varepsilon\, SE_T \\ \text{ii) } A_2 H_T = \frac{1}{i\omega}\,\text{grad}_T \frac{1}{\varepsilon}\,\text{curl}_T\, H_T + i\omega\mu\, SH_T\,. \end{cases}$$

We can write this system of equations (34), on E_T or H_T only:

$$(36) \qquad \partial_3^2\, H_T + A_1 A_2\, H_T = 0, \qquad \partial_3^2\, E_T + A_2 A_1\, E_T = 0,$$

with:

$$(37)\quad \left| \begin{array}{l} \text{i) } A_1A_2\, H_T = (\text{grad}_T \frac{1}{\mu} \text{curl}_T\, \mu S + \varepsilon S\, \text{grad}_T \frac{1}{\varepsilon} \text{curl}_T + \omega^2\, \varepsilon\mu I)\, H_T\,, \\ \text{ii) } A_2A_1\, E_T = (\text{grad}_T \frac{1}{\varepsilon} \text{curl}_T\, \varepsilon S + \mu S\, \text{grad}_T \frac{1}{\mu} \text{curl}_T + \omega^2\, \varepsilon\mu I)\, E_T\,, \end{array} \right.$$

or also using relations:

$$(38)\qquad S\, \text{grad}_T = -\, \overrightarrow{\text{curl}}_T\,, \qquad \text{curl}_T\, S = \text{div}_T,$$

$$(39)\quad \left| \begin{array}{l} \text{i) } A_1A_2\, H_T = (\text{grad}_T \frac{1}{\mu} \text{div}_T\, \mu - \varepsilon S\, \overrightarrow{\text{curl}}_T \frac{1}{\varepsilon} \text{curl}_T + \omega^2\, \varepsilon\mu I)\, H_T\,, \\ \text{ii) } A_2A_1\, E_T = (\text{grad}_T \frac{1}{\varepsilon} \text{div}_T\, \varepsilon - \mu S\, \overrightarrow{\text{curl}}_T \frac{1}{\mu} \text{curl}_T + \omega^2\, \varepsilon\mu I)\, E_T\,. \end{array} \right.$$

Now we make the following important hypothesis: *The domain of the guide is occupied by a homogeneous and isotropic medium; thus ε and μ are constants in* Ω. With $k^2 = \omega^2\varepsilon\mu$, (39) is:

$$(40)\quad \left| \begin{array}{l} \text{i) } A_1A_2H_T = (\text{grad}_T\, \text{div}_T - \overrightarrow{\text{curl}}_T\, \text{curl}_T + k^2 I)\, H_T\,, \\ \text{ii) } A_2A_1E_T = (\text{grad}_T\, \text{div}_T - \overrightarrow{\text{curl}}_T\, \text{curl}_T + k^2 I)\, E_T. \end{array} \right.$$

From now on we will not write subscripts T to differential operators grad, etc. The operators A^1A^2 and A^2A^1 are identical to the transverse Laplacian; we have verified only that E_T and H_T satisfy the Helmholtz equation! But these operators differ by their boundary conditions. We define them by the usual variational method. We first define the following spaces:

$$(41)\quad \left| \begin{array}{l} V = \{ v \in L^2(\Omega)^2,\ \text{div}\, v \in L^2(\Omega),\ \text{curl}\, v \in L^2(\Omega)\}, \\ V^- = \{v \in V,\ n \wedge v|_\Gamma = 0\}, \qquad V^+ = \{v \in V,\ n.v|_\Gamma = 0\}. \end{array} \right.$$

From the Appendix (see (121), (122)) we know that spaces V^+ and V^- are contained in $H^1(\Omega)$. Then we define the sesquilinear form:

$$(42)\qquad a(u,v) = \int_\Omega (\text{div}\, u\, \text{div}\, \bar{v} + \text{curl}\, u\, .\, \text{curl}\, \bar{v})\, dx.$$

Thus we can define the operators Δ^+ and Δ^- (see Appendix (156)") by:

$$(43)\qquad D(\Delta^+)\ (\text{resp. } D(\Delta^-)) = \{u \in V^+\ (\text{resp. } V^-),\ v \rightarrow a(u,v) \text{ is continuous on } V^+\ (\text{resp. } V^-) \text{ equipped with the } L^2 \text{ topology}\}$$

and thus:

$$(44)\quad \left| \begin{array}{l} D(\Delta^-) = \{u \in H^1(\Omega)^2,\ \Delta u \in L^2(\Omega)^2,\ n \wedge u|_\Gamma = 0,\ \text{div}\, u|_\Gamma = 0\}, \\ D(\Delta^+) = \{u \in H^1(\Omega)^2,\ \Delta u \in L^2(\Omega)^2,\ n.u|_\Gamma = 0,\ \text{curl}\, u|_\Gamma = 0\}; \end{array} \right.$$

Δ^+ and Δ^- are positive selfadjoint operators, with compact resolvent, thus with discrete spectrum.

Then we verify (at least formally) that if (E, H) is a solution of the problem (27), $(E_T, H_T)(x_3)$ is in $D(\Delta^+) \times D(\Delta^-)$; condition (32) implies $\frac{\partial E_3}{\partial x_3}\Big|_{\Gamma_+} = 0$, and from equation div E = 0, we have div $E_T|_{\Gamma_+} = 0$. Besides, (27)ii) implies $n_T . H_T|_{\Gamma_+} = 0$, and (31)i), (32)i) imply curl $H_T|_{\Gamma_+} = 0$.

Determination of the transverse electric field E_T. We come back to problem (27) which, after (36), (32), and (44), amounts to determine the transverse electric field E_T that satisfies:

$$(45) \qquad \left| \begin{array}{l} \text{i) } \partial_3^2 E_T = (-\Delta^- - k^2 I)\, E_T, \\ \text{ii) } E_T(0) = E_T^0. \end{array} \right.$$

But this is an ill-posed problem as we see it by spectral decomposition of the operator $(\Delta^- + k^2 I)$. It is transformed into a well-posed problem by adding the physical hypothesis (33) for E_T. Using the decomposition of the selfadjoint operator $-(\Delta^- + k^2 I)$ into its positive and negative parts corresponding to the space decomposition of $L^2(\Omega)^2$

$$(46) \qquad L^2(\Omega)^2 = H_+ \oplus H_-, \quad -(\Delta^- + k^2 I) = (-\Delta^- - k^2 I)_+ - (-\Delta^- - k^2 I)_-,$$

we can define (from the symbolic calculus) two "square roots" of $-(\Delta^- + k^2 I)$ denoted by A^+ and A^-, so that:

$$(47) \left| \begin{array}{l} \text{on } H_+ : A^+ = A^- = -(-\Delta^- - k^2 I)_+^{1/2} \text{ negative selfadjoint,} \\ \text{on } H_- : A^+ = -A^- = -i(-\Delta^- - k^2 I)_-^{1/2} \text{ with } (-\Delta^- - k^2 I)_-^{1/2} \text{ bounded, positive.} \end{array} \right.$$

Thus A^+ and A^- are adjoint normal operators with domain V^-, with spectrum

$$\sigma(A^+) \subset R^- \cup -i[0, k] \quad \text{resp. } \sigma(A^-) \subset R^- \cup i[0, k].$$

(The Hodge decomposition allows us a more explicit spectral decomposition using the spectral decomposition of the Dirichlet and the Neumann Laplacian.) Taking (33) into account, we then replace problem (45) by:

$$(48) \qquad \left| \begin{array}{l} \text{i) } \partial_3 E_T = A^- E_T \text{ in } \Omega_+, \\ \text{ii) } E_T(0) = E_T^0. \end{array} \right.$$

The choice A^+ would give the opposite direction of propagation. *From now on we simply denote by* A *the operator* A^-. With the above properties, the operator A is the infinitesimal generator of a contraction (and holomorphic) semigroup of class C^0 in $L^2(\Omega)^2$, denoted by $(G_A(x_3))$, $x_3 > 0$; the solution of (48) is given by

(49) $$E_T(x_3) = G_A(x_3)\, E_T^0, \quad x_3 > 0.$$

For the transverse magnetic field H_T, first assuming that it is given for $x_3 = 0$, and denoting by $(G_B(x_3))$, $x_3 > 0$, the semigroup generated by the square root B^- (like in (47)) of the operator $-(\Delta^+ + k^2 I)$, we have:

(49)' $$H_T(x_3) = G_B(x_3)\, H_T(0), \quad x_3 > 0.$$

Then we easily determine all other unknown functions from the initial values $E_3(0)$, $H_3(0)$ and $H_T(0)$ which we have to obtain from the initial value of E_T.

Determination of H_3. First taking the scalar product of the relation (30)i) with the unit normal to Γ_+ on the boundary Γ_+ gives

(50) $$-\partial_3\, n_T . H_T + \frac{\partial H_3}{\partial n_T} - i\omega\varepsilon\, n_T \wedge E_T = 0 \quad \text{on } \Gamma_+ ,$$

which implies (using (27)iii) and its consequence $n.H|_{\Gamma_+} = 0$):

(51) $$\frac{\partial H_3}{\partial n_T} = 0 \quad \text{on } \Gamma_+ .$$

Furthermore the initial value of H_3 is obtained from (31)ii):

(52) $$H_3(0) = H_3^0 = \frac{1}{i\omega\mu} \operatorname{curl} E_T(0) = \frac{1}{i\omega\mu} \operatorname{curl} E_T^0 .$$

Thus we have to determine H_3 satisfying:

(53)
- i) $\Delta H_3 + k^2 H_3 = 0$ in Ω_+ (with Δ in R^3 !),
- ii) the boundary condition (51) on Γ_+,
- iii) the initial condition (52),
- iv) the propagation condition (33) for H_3.

We have solved this Helmholtz problem in section 1; its solution is given using (a generalization of) Theorem 1 (note that for E_T^0 in $L^2(\Omega)^2$, H_3^0 is not in $L^2(\Omega)$)

(54) $$H_3(x_3) = G^N(x_3)\, H_3^0 .$$

Determination of E_3. We first have from equation $\operatorname{div} E = 0$ in R^3, for $x_3 = 0$:

(55) $$\frac{\partial E_3}{\partial x_3}(0) = -\operatorname{div} E_T(0) = -\operatorname{div} E_T^0 .$$

First assuming that $E_3(0) = E_3^0$ is given (in $L^2(\Omega)$), E_3 must satisfy:

(56)
- i) $\Delta E_3 + k^2 E_3 = 0$ in Ω_+, (with Δ in R^3),
- ii) $E_3|_{\Gamma_+} = 0$ (see (32)),
- iii) $E_3(0) = E_3^0$ on Ω,
- iv) the propagation condition (33) for E_3.

Then, from Theorem 1, we know that this problem has a unique solution, which is given by:

(57) $$E_3(x_3) = G^D(x_3)\, E_3^0.$$

We determine the "initial" condition E_3^0 from (55) by:

(58) $$\frac{\partial E_3}{\partial x_3}(0) = C_D E_3^0 = -\operatorname{div} E_T^0.$$

If k^2 *is not an eigenvalue of* $(-\Delta_T^D)$, C_D is invertible and (58) has a unique solution in $L^2(\Omega)$ when E_T^0 is in $L^2(\Omega)^2$:

(59) $$E_3^0 = -C_D^{-1} \operatorname{div} E_T^0.$$

Determination of the transverse magnetic field H_T. We have to determine the initial condition $H_T(0) = H^0{}_T$. This is obtained from equation (30)ii) for $x_3 = 0$. Using (48) we have

(60) $$AE_T^0 - \operatorname{grad} E_3^0 + i\omega\mu\, SH_T^0 = 0,$$

and thus, with (59) and using the impedance Z with $kZ = \omega\mu$

(61) $$SH_T^0 = \frac{-1}{ikZ}\,[AE_T^0 + \operatorname{grad} C_D^{-1} \operatorname{div} E_T^0].$$

⊗

First properties. Calderon operator C of the guide.
It will be useful to transform the basic relation (61) using the following Hodge decomposition (see chap.2 (114)") of u in $L^2(\Omega)^2$ in the form:

(62) $$u = \operatorname{grad} \Delta_D^{-1} \operatorname{div} u - \overrightarrow{\operatorname{curl}}\, A_N^{-1} L_u + \tilde{u},$$

with $\tilde{u} \in H_2(\Omega)$, where $H_2(\Omega) = \{v \in L^2(\Omega)^2,\ \operatorname{curl} v = \operatorname{div} v = 0,\ n \wedge v|_\Gamma = 0\}$;
A_N is the isomorphism from $H^1(\Omega)/R$ onto $(H^1(\Omega))'$ associated with the Neumann Laplacian and $L_u \in (H^1(\Omega))'$ is defined for u in $L^2(\Omega)^2$ by:

(63) $$L_u(v) = (Su, \operatorname{grad} v), \quad \forall\, v \in H^1(\Omega),$$

thus $w = A_N^{-1} L_u$ is the solution (up to constants) of:

(63)' $$(\operatorname{grad} w, \operatorname{grad} v) = (Su, \operatorname{grad} v), \quad \forall\, v \in H^1(\Omega),$$

or also: $(\overrightarrow{\operatorname{curl}}\, w, \overrightarrow{\operatorname{curl}}\, v) = (u, \overrightarrow{\operatorname{curl}}\, v), \ \forall\, v \in H^1(\Omega).$
For $u \in H(\operatorname{curl},\Omega)^2$, we have $L_u = \operatorname{curl} u - n \wedge u|_\Gamma \delta_\Gamma$ and for $u \in H_0(\operatorname{curl}, \Omega)$, we have

$$A_N^{-1} L_u = \Delta_N^{-1} \operatorname{curl} u.$$

From the spectral decomposition of the operators Δ_D, Δ_N, the space $L^2(\Omega)^2$ splits into H^+ and H^- (see (46)), with:

(64) $H_- = \text{grad}\,(H_k^-)^D \oplus \overrightarrow{\text{curl}}\,(H_k^-)^N \oplus \mathcal{H}_2(\Omega), \quad H_+ = \text{grad}\,(V_k^+)^D \oplus \overrightarrow{\text{curl}}\,(V_k^+)^N$,

where $(H_k^-)^D$ and $(H_k^-)^N$ are finite dimensional spaces corresponding to the negative part of the operators $(-\Delta_D - k^2 I)$ and $(-\Delta_N - k^2 I)$ (see the decomposition in the Helmholtz case), and $(V_k^+)^D$ and $(V_k^+)^N$ are spaces corresponding to the negative part of these operators (in $H_0^1(\Omega)$ or $H^1(\Omega)$).
This decomposition gives us the relation (for u in V^-, with (62)):

(65) $Au = \text{grad}\,\Delta_D^{-1} C_D\, \text{div}\, u - \overrightarrow{\text{curl}}\, C_N (\Delta_N^{-1}\, \text{curl}\, u) + ik\, \tilde{u}$,
with:
(66) $\text{Re}\,(Au,u) \leq 0$ and $\text{Im}\,(Au,u) \geq 0, \quad \forall u \in V^-$.

Yet we return to relation (61); we have: $-\Delta^D - k^2 I = C_D^2$, and thus:

(67) $C_D^{-1} + k^2 \Delta_D^{-1} C_D^{-1} = -\Delta_D^{-1} C_D$.

Therefore (61) becomes (for E_T^0 in V^- then by continuity, for E_T^0 in $H_0(\text{curl},\Omega)$)

(68) $SH_T^0 = \frac{i}{kZ}[-k^2\, \text{grad}\,\Delta_D^{-1} C_D^{-1} \text{div}\, E_T^0 - \overrightarrow{\text{curl}}\, C_N \Delta_N^{-1}\, \text{curl}\, E_T^0 + ik\, \tilde{E}_T^0]$,

with $\tilde{E}_T^0$ the projection of E_T^0 on $\mathcal{H}_2(\Omega)$. A first consequence of (68) is:

(69) $$\begin{aligned}(SH_T^0, E_T^0) = (n \wedge H_T^0, E_T^0) = \frac{i}{kZ}[k^2 &\langle \Delta_D^{-1} C_D^{-1} \text{div}\, E_T^0, \text{div}\, E_T^0\rangle \\ &- \langle C_N \Delta_N^{-1}\, \text{curl}\, E_T^0, \text{curl}\, E_T^0\rangle + ik\langle \tilde{E}_T^0, \tilde{E}_T^0\rangle].\end{aligned}$$

This implies (for all E_T^0 in V^- and then in $H_0(\text{curl},\Omega)$):

(69)' $\text{Re}\,(SH_T^0, E_T^0) = \text{Re}\,(n \wedge H_T^0, E_T^0) \leq 0$.

Then using the term "mild" by reference to a solution of Maxwell equation in a "weak" sense (in the sense of semigroup theory, see Pazy [1]), we can prove:

Theorem 2. *Except when* k^2 *is an eigenvalue of the Laplacian with Dirichlet condition, problem* (27) *with* (33) *and the initial condition*
(70) $E_T^0 \in H_0(\text{curl}_T,\Omega)$
has a unique mild solution (E,H) *which satisfies:*

(71) $$\begin{aligned}&E_T \in \mathcal{C}^0([0,+\infty), H_0(\text{curl}_T,\Omega)), \quad H_T \in \mathcal{C}^0([0,+\infty), H(\text{curl}_T,\Omega)), \\ &E_3 \text{ and } H_3 \in \mathcal{C}^0([0,+\infty), L^2(\Omega)).\end{aligned}$$

This solution is given by (49), (49)', (68), (52), (54), (57), (59).

Remark 6. When k^2 is an eigenvalue of the "Dirichlet Laplacian", we verify easily that the foregoing problem has no unique solution, since the following electromagnetic field (E, H) is a solution for zero "initial" condition:

$E_T^0 = 0$, $H_3 = 0$, but $E_3(x_3) = E_3(0)$ so that $(\Delta + k^2)E_3(0) = 0$ with $E_3(0)$ in $H_0^1(\Omega)$ (an eigenvector of the Dirichlet Laplacian) and $H_T(x_3) = H_T(0) = \frac{1}{i\omega\mu}\,\overrightarrow{\text{curl}}_T\, E_3(0)$.

⊗

Definition 2. *The mapping* C: $SE^0_T \to SH^0_T$ *(or* $\mathcal{C}$: $E^0_T \to H^0_T$*) which is defined by* (68) *is called the Calderon (or also the admittance) operator of the guide at frequency* ω.

Proposition 1. *The Calderon operator* $\mathcal{C}$ *(resp. C) is a continuous mapping from space* $H_0(\text{div},\Omega)$ *onto* $H(\text{div},\Omega)$ *(resp.* $H_0(\text{curl},\Omega)$ *onto* $H(\text{curl},\Omega)$*) with the following properties*:

(72) $\quad \text{Re}\,(CSE_T^0, E_T^0) \le 0\,, \quad \forall\, E_T^0 \in H_0(\text{curl},\Omega)$.

Remark 7. We can give the dependence of the electromagnetic field with respect to x_3 (therefore the semigroups (49) and (49)') thanks to the semigroups $G^D(x_3)$ and $G^N(x_3)$ of section 1; for instance:

$$\left|\begin{aligned} E_T(x_3) &= \text{grad}\, G^D(x_3)\, \Delta_D^{-1}\, \text{div}\, E_T^0 - \overrightarrow{\text{curl}}\, G^N(x_3)\, \Delta_N^{-1}\, \text{curl}\, E_T^0 + e^{ikx_3}\, \tilde{E}_T^0(x_3) \\ H_T(x_3) &= \text{grad}\, G^N(x_3)\, \Delta_N^{-1}\, \text{div}\, H_T^0 - \overrightarrow{\text{curl}}\, G^D(x_3)\, \Delta_D^{-1}\, \text{curl}\, H_T^0 + e^{ikx_3}\, \tilde{H}_T^0(x_3), \end{aligned}\right.$$

for E_T^0 in V^-, with $\tilde{E}_T^0$ in $H_2(\Omega)$ and $\tilde{H}_T^0$ in $H_1(\Omega)$. We note that the relation:

(73) $\quad H_T(x_3) = \mathcal{C}\, E_T(x_3)$ (or $SH_T(x_3) = C\, SE_T(x_3)$),

which is true for $x_3 = 0$, is valid for all $x_3 \ge 0$: the Calderon operator $\mathcal{C}$ commutes with the semigroups of (49), (49)', simply denoted by $(G_A(x_3))$ and $(G_B(x_3))$

(73)' $\quad G_B(x_3)\, \mathcal{C} = \mathcal{C}\, G_A(x_3)\,, \quad \forall\, x_3 \ge 0$.

⊗

Remark 8. Note that when E_T^0 is in $H_0(\text{curl},\Omega)$ and not only in D(A), we have

$$\frac{\partial H_3}{\partial x_3}(x_3) = C_N H_3(x_3) \quad \text{in } (H^1(\Omega))' \text{ (from semigroup theory),}$$

whereas $\text{div}_T\, H_T$ is in $H^{-1}(\Omega)$, and this does not correspond to the usual sense of $\text{div}\, H = 0$ in $\mathcal{D}'(R^3)$. Applying the curl and div operators to (68), we obtain:

(74) $\quad$ i) $\text{curl}\, H_T^0 = \frac{ik}{Z}\, C_D^{-1}\, \text{div}\, E_T^0$, $\quad$ ii) $\text{div}\, H_T^0 = \frac{i}{Zk}\, \Delta\, C_N\, \Delta_N^{-1}\, \text{curl}\, E_T^0$,

that is also $P\partial_3 H_3^0 + \operatorname{div} H_T^0 = 0$, with $P = \Delta A_N^{-1}$ which is the (orthogonal) projection from $(H^1(\Omega))'$ onto $H^{-1}(\Omega)$.
Conversely, thanks to the Hodge decomposition ("conjugated" to (62)):

(62)' $$v = -\operatorname{grad} A_N^{-1} L_{Sv} - \overrightarrow{\operatorname{curl}}\, \Delta_D^{-1} \operatorname{curl} v + \tilde{v},$$

with $\tilde{v} \in \mathcal{H}_1(\Omega)$, where $\mathcal{H}_1(\Omega) = \{v \in L^2(\Omega)^2,\ \operatorname{curl} v = \operatorname{div} v = 0,\ n.v|_\Gamma = 0\}$, applied to $v = H_T^0$, and then using (74), we easily obtain (68).

We can also obtain the inverse relation to E_T^0 with respect to H_T^0 either as above for SH_T^0, or using the Hodge decomposition (62) with the inverse to (74) (this gives us the inverse to the Calderon operator C):

(75) $$\begin{aligned} SE_T^0 &= -\overrightarrow{\operatorname{curl}}\, \Delta_D^{-1} \operatorname{div} E_T^0 - \operatorname{grad} \Delta_N^{-1} \operatorname{curl} E_T^0 + S\tilde{E}_T^0 \\ &= \frac{iZ}{k}[\overrightarrow{\operatorname{curl}}\, \Delta_D^{-1} C_D \operatorname{curl} H_T^0 - k^2 \operatorname{grad} C_N^{-1} A_N^{-1} L_{SH_T^0} - ik\tilde{H}_T^0] \end{aligned}$$

when $H_T^0 \in H(\operatorname{curl},\Omega)$, with

(76) $$\tilde{H}_T^0 = Z^{-1} S \tilde{E}_T^0;$$

$\tilde{H}_T^0$ is the orthogonal projection of H_T^0 onto $\mathcal{H}_1(\Omega)$ whereas $\tilde{E}_T^0$ is that of E_T^0 onto $\mathcal{H}_1(\Omega)$. The inverse relation to (74)ii) is given by

(77) $$\operatorname{curl} E_T^0 = ikZ\, \Delta C_N^{-1} A_N^{-1} L_{SH_T^0}\ .$$

We recall that $w = -A_N^{-1} L_{SH_T^0}$ satisfies $\Delta w = \operatorname{div} H_T^0$ since:

$$(\operatorname{grad} w, \operatorname{grad} v) = (H_T^0, \operatorname{grad} v), \qquad \forall v \in H^1(\Omega).$$

Using (75), we can see that conversely, given $H_T^0 \in H(\operatorname{curl},\Omega)$, we have E_T^0 in $H(\operatorname{curl},\Omega)$ with the boundary condition $n_T \wedge E_T^0 = 0$ on Γ; this is due to the relations $n_T \wedge \overrightarrow{\operatorname{curl}}\, \phi = -n.\operatorname{grad} \phi$ and $n_T \wedge \operatorname{grad} \phi = \tau.\operatorname{grad} \phi$, $\forall \phi$ with $\tau = (-n_2, n_1)$ tangent to Γ. This proves Proposition 1 for the Calderon operator.

⊗

Remark 9. *Transverse-Electric (TE), Transverse-Magnetic (TM) and Transverse-ElectroMagnetic (TEM) waves.*
The Hodge decomposition (62) for the initial transverse electric field gives the "usual" decomposition of the electromagnetic field (E, H) into:

i) the "Transverse-ElectroMagnetic field" (TEM waves) $(\tilde{E}, \tilde{H})$ with $\tilde{E}_3 = \tilde{H}_3 = 0$ and $\tilde{E}_T(x_3) = e^{ikx_3} \tilde{E}_T^0$, $\tilde{H}_T(x_3) = e^{ikx_3} \tilde{H}_T^0$, with $\tilde{E}_T^0 \in \mathcal{H}_2(\Omega)$, $\tilde{H}_T^0 \in \mathcal{H}_1(\Omega)$,

and

$$\tilde{H}\,{}^{0}_{T} = Z^{-1} S \tilde{E}\,{}^{0}_{T};$$

ii) the "Transverse-Electic field" (TE waves) (E_{TE}, H_{TE}) with $E_3 = 0$, therefore:

$$\text{curl}\, H^0_T = 0, \quad \text{div}\, E^0_T = 0, \quad \text{then} \quad H_3(x_3) = \frac{-i}{Zk} G_N(x_3)\, \text{curl}\, E^0_T, \quad \text{and:}$$

$$E_T(x_3) = -\,\overrightarrow{\text{curl}}\, G^N(x_3)\, \Delta_N^{-1}\, \text{curl}\, E^0_T, \quad H_T(x_3) = \frac{i}{Zk}\, \text{grad}\, G^N(x_3)\, C_N\, \Delta_N^{-1}\, \text{curl}\, E^0_T.$$

Thus the *Calderon operator for TE-waves* is given by

$$SH^0_T = \frac{-i}{kZ}\, \overrightarrow{\text{curl}}\, C_N\, \Delta_N^{-1}\, \text{curl}\, E^0_T;$$

iii) the "Transverse-Magnetic field" (TM waves) (E_{TM}, H_{TM}) with $H_3 = 0$, therefore:

$$\text{curl}\, E^0_T = 0, \quad \text{div}\, H^0_T = 0, \quad \text{then} \quad E_3(x_3) = G_D(x_3)\, E^0_3 = -\, G_D(x_3)\, C_D^{-1}\, \text{div}\, E^0_T, \quad \text{and:}$$

$$E_T(x_3) = \text{grad}\, G^D(x_3)\, \Delta_D^{-1}\, \text{div}\, E^0_T, \quad H_T(x_3) = \overrightarrow{\text{curl}}\, G_D(x_3)\, \Delta_D^{-1}\, \text{curl}\, H^0_T$$

with (74). Thus *the Calderon operator for TM-waves* is given by:

$$SH^0_T = \frac{-ik}{Z}\, \text{grad}\, \Delta_D^{-1}\, C_D^{-1}\, \text{div}\, E^0_T.$$

This decomposition of the electromagnetic field is due to the splitting of the domains V^+ and V^- by the Hodge decomposition.

⊗

Yet we give a more useful framework. For usual applications, we have to find an electromagnetic field with locally finite energy in the closure of Ω_+. This implies (see chap. 2.7.2) that in each cross-section of the cylinder we have, with notations (85), (86) chap. 2:

$$(78) \qquad E_T(x_3)|_\Omega \in H_o^{-1/2}(\text{curl}, \Omega), \quad H_T(x_3)|_\Omega \in X(\text{curl}, \Omega), \quad \forall x_3 \geq 0.$$

Thus we have to solve problem (27) with a given initial boundary condition ${E^0}_T$ with (78). These spaces are not very easy to handle, so we extensively use the interpolation theory and the following Hodge decomposition (compare to the Hodge decomposition above):

Lemma 1. *We have equivalence between* i) *and* ii):

i) $u \in H_o^{-1/2}(\text{div}, \Omega)$ (see (85) chap. 2)

ii) $u = \text{grad}\, \phi + \overrightarrow{\text{curl}}\, \psi + \tilde{u}$, with $\phi \in D(A_N^{3/4})$, $\psi \in D(A_D^{1/4})$, and $\tilde{u} \in H_1(\Omega)$

and also, using the S operator, we have equivalence between:

i) $u \in H_o^{-1/2}(\text{curl}, \Omega)$,

ii) $u = \text{grad}\, \phi + \overrightarrow{\text{curl}}\, \psi + \tilde{u}'$, with $\phi \in D(A_D^{1/4})$, $\psi \in D(A_N^{3/4})$, and $\tilde{u}' \in H_2(\Omega)$.

Here A_D and A_N denote the generalizations of the Dirichlet Laplacian and the Neumann Laplacian to the interpolated spaces and the dual spaces.

For regular Ω, we will use the domains $D(A_D) = H^2(\Omega) \cap H_0^1(\Omega)$, $D(A_D^{1/2}) = H_0^1(\Omega)$ and by interpolation: $D(A_D^{1/4}) = H_{00}^{1/2}(\Omega)$, $D(A_D^{3/4}) = H_0^{3/2}(\Omega) = H^{3/2}(\Omega) \cap H_0^1(\Omega)$. We also have $D(A_N) = \{v \in H^2(\Omega), \frac{\partial v}{\partial n}|_\Gamma = 0\}$, $D(A_N^{1/2}) = H^1(\Omega)$ and by interpolation:

$$D(A_N^{1/4}) = H^{1/2}(\Omega), \quad D(A_N^{3/4}) = \{u \in H^{3/2}(\Omega),\ n.\mathrm{grad}\, u \in H_{00}^{1/2}(V_\Gamma)\},$$

V_Γ being a neighborhood of Γ in $\bar{\Omega}$.

(SKETCH OF) PROOF of Lemma 1. i) implies ii). First ϕ is determined (up to a constant) by solving:

$$(79) \qquad A_N\phi = \mathrm{div}\, u \in H^{-1/2}(\Omega) = D(A_N^{-1/4}),$$

which gives ϕ in $D(A_N^{3/4})$. Then since the grad and $\overrightarrow{\mathrm{curl}}$ operators are continuous from $H_0^{3/2}(\Omega)$ into $H^{1/2}(\Omega)^2$, by duality the div and curl operators are continuous from $H^{-1/2}(\Omega)^2$ into $H^{-3/2}(\Omega)$. Thus φ is obtained by solving equation

$$(80) \qquad A_D\varphi = \mathrm{curl}\, u \ \in H^{-3/2}(\Omega),$$

which has a unique solution in $H_{00}^{1/2}(\Omega)$.

ii) implies i) Thanks to the characterization of the interpolated domains, $\phi \in D(A_N^{3/4})$ implies $v = \mathrm{grad}\, \phi \in H_0^{-1/2}(\mathrm{div}, \Omega)$.

⊗

Lemma 1 implies for (locally) finite energy electromagnetic field in R^3:

Theorem 3. *Except when* k^2 *is an eigenvalue of the Laplacian with Dirichlet condition, the problem* (27) *with* (33) *and the initial condition:*

$$(81) \qquad E_T^0 \in H_0^{-1/2}(\mathrm{curl}_T, \Omega)$$

has a unique (mild) solution (E, H) *which satisfies* (*with def.* (86) chap.2)*:*

$$(82) \quad \left| \begin{array}{l} E_T \in C^0([0,+\infty), H_0^{-1/2}(\mathrm{curl}_T, \Omega)), \quad H_T \in C^0([0,+\infty), X(\mathrm{curl}_T, \Omega)), \\ \quad E_3 \in C^0([0,+\infty), (H_{0\,0}^{1/2}(\Omega))') \ \text{ and } \ H_3 \in C^0([0,+\infty), H^{-1/2}(\Omega)). \end{array} \right.$$

Furthermore (E, H) *has a locally finite energy up to the boundary in* $\bar{\Omega}_+$*:*

$$(83) \quad E \text{ and } H \in H_{loc}(\mathrm{curl}, \bar{\Omega}_+), \text{ i.e., } \zeta E, \zeta H \in H(\mathrm{curl}, \Omega_+), \ \ \forall \zeta \in D(R^3).$$

Then we generalize Definition 2 and Proposition 1 on the Calderon operator to the present framework.

Proposition 2. *The Calderon operator* C (*resp.* C) *of the guide at frequency* ω *is a continuous mapping from* $H_0^{-1/2}(\text{div},\Omega)$ *onto* $X(\text{div},\Omega)$ (resp. $H_0^{-1/2}(\text{curl},\Omega)$ *onto* $X(\text{curl},\Omega)$), *with:*

(72)' $\quad \text{Re}\,(CSE_T^0, E_T^0) \leq 0, \quad \forall E_T^0 \in H_0^{-1/2}(\text{curl},\Omega).$

Note that the electromagnetic field in Theorem 3 and the Calderon operator are obtained by the same formulas as above.
Theorem 2 is a regularity result with respect to Theorem 3 for the electromagnetic field. Then we have another regularity result in the H^1 framework:

Theorem 4 *(Regularity). When* k^2 *is not an eigenvalue of the Laplacian with Dirichlet condition, the solution* (E,H) *of problem* (27) *with* (33) *and the "initial" condition:*

(84) $\quad E_T^0 \in H_0^{1/2}(\text{curl}_T,\Omega)$

is such that:

(85) $\quad H_T^0 \in H^{1/2}(\text{curl}_T,\Omega) \cap H_0^{-1/2}(\text{div}_T,\Omega)),\ E_3^0 \in H_{00}^{1/2}(\Omega)$ and $H_3^0 \in H^{1/2}(\Omega)$,

and thus (E,H) *satisfies continuity properties from* x_3 *(in* R^+*) into these spaces. Furthermore* (E,H) *has* H^1 *regularity up to the boundary, that is*:

(86) $\quad E$ *and* $H \in H^1_{loc}(\bar{\Omega}_+)^3$, i.e. $\zeta E, \zeta H \in H^1(\Omega_+)^3, \quad \forall \zeta \in D(R^3)$.

Proposition 3. *The Calderon operator C is a continuous mapping from the space*

$H_0^{1/2}(\text{curl}_T,\Omega)$ *onto* $H^{1/2}(\text{curl}_T,\Omega) \cap H_0^{-1/2}(\text{div}_T,\Omega)$.

Example 2. *Junctions and Cascades in electromagnetism*
We consider the situation of Example 1 with a junction and p guides incoming to it. We assume that:
i) the (side) boundaries of the guides and of the junction Γ_0 are perfectly conducting;
ii) a given regular incident electromagnetic wave (E_I, H_I) with angular frequency ω is coming from the guides to the junction;
iii) the permittivity and the permeability of the jth guide and of the junction are respectively (ε_j, μ_j) and (ε, μ).
Thus the reflected wave (E_{rj}, H_{rj}) in the jth guide must satisfy (from Propositions 2 and 3, using the Calderon operator C_j of the jth guide) on top Γ_j of the guide

(87) $\quad n \wedge H_r|_{\Gamma_j} = C_j\,(n \wedge E_r|_{\Gamma_j}).$

Then using the continuity relations of the wave across the boundary Γ_j:

$$(88)\qquad \left|\begin{array}{l}[n\wedge E]_{\Gamma_j}=0, \qquad [n\wedge H]_{\Gamma_j}=0, \text{ that is:}\\ n\wedge E=n\wedge E_r+n\wedge E_I, \quad n\wedge H=n\wedge H_r+n\wedge H_I, \text{ on } \Gamma_j,\ j=1 \text{ to } p,\end{array}\right.$$

we obtain that the electromagnetic field (E,H) has to satisfy on Γ_j:

$$(89)\qquad n\wedge H|_{\Gamma_j}-C_j\,(n\wedge E|_{\Gamma_j})=f_{Ij}, \quad \text{with } f_{Ij}=n\wedge H_I|_{\Gamma_j}-C_j\,(n\wedge E_I|_{\Gamma_j}).$$

Thus we have to find the electromagnetic field (E,H) with finite energy in the junction satisfying:
i) Maxwell equations in Ω:

$$(90)\qquad \left|\begin{array}{l}\text{i) } \operatorname{curl} H+i\omega\varepsilon\, E=0\,,\\ \text{ii) } -\operatorname{curl} E+i\omega\mu\, H=0 \text{ in } \Omega,\end{array}\right.$$

ii) the boundary condition on Γ_o, then on Γ_j

$$(91)\qquad n\wedge E|_{\Gamma_o}=0\,, \text{ with (89) on } \Gamma_j.$$

We write this problem in a variational form. We define:

$$(92)\qquad H_{\Gamma_o}(\operatorname{curl},\Omega)\overset{\text{def}}{=}\{u\in H(\operatorname{curl},\Omega)\,,\ n\wedge u|_{\Gamma_o}=0\},$$

and the sesquilinear form on $H_{\Gamma_o}(\operatorname{curl},\Omega)$:

$$(93)\quad a(E,\tilde{E})=\int_\Omega\Big(-\frac{1}{i\omega\mu}\operatorname{curl} E\,.\,\operatorname{curl}\bar{\tilde{E}}-i\omega\varepsilon\, E\,.\,\bar{\tilde{E}}\Big)\,dx-\sum_j\int_{\Gamma_j}C_j(n\wedge E|_{\Gamma_j})\,.\,\bar{\tilde{E}}|_{\Gamma_j}\,d\Gamma_j.$$

Then problem (90), (91) is equivalent to find E in $H_{\Gamma_o}(\operatorname{curl},\Omega)$ satisfying:

$$(94)\qquad a(E,\tilde{E})=\sum_j\int_{\Gamma_j}F_{Ij}\,.\,\bar{\tilde{E}}\,dx, \qquad \forall\,\tilde{E}\in H_{\Gamma_o}(\operatorname{curl},\Omega).$$

i) First *if the medium of the junction is dissipative*, that is if

$$(95)\qquad \operatorname{Im}\varepsilon>0, \text{ and } \operatorname{Im}\mu>0,$$

problem (94) has a unique solution; this is easily proved as follows.
Taking the real part of (94) with $E=\tilde{E}$, thanks to (72)', there is a constant C_o so that:

$$(96)\qquad \operatorname{Re} a(E,E)\ \geq\ C_o\|E\|^2_{H_{\Gamma_o}(\operatorname{curl},\Omega)}\,, \quad \forall\,E\in H_{\Gamma_o}(\operatorname{curl},\Omega).$$

Thus the sesquilinear form $a(E,\tilde{E})$ is coercive on $H_{\Gamma_o}(\operatorname{curl},\Omega)$.
Thanks to the Lax-Milgram lemma, problem (94) has a unique solution.
ii) *If the medium of the junction is conservative* (that is when ε and μ are real positive) we will see that problem (94) depends on the Fredholm alternative.

Note that the natural injection $H_{\Gamma_o}(\text{curl},\Omega) \to L^2(\Omega)^3$ is not compact. But for given regular incident electromagnetic field, f_{Ij} is (for instance) in $H^{1/2}(\text{curl},\Gamma_j) \cap H_o^{-1/2}(\text{div},\Gamma_j)$; E must satisfy a variational problem in $H_{\Gamma_o}(\text{curl},\Omega)$ with $\text{div}\, E = 0$ and its boundary value must be in $H^{1/2}(\text{curl},\Gamma)$; thus E is in $H^1(\Omega)^3$. Since the natural mapping from $H^1(\Omega)$ into $L^2(\Omega)$ is compact (for regular Ω), this problem (94) depends on the Fredholm alternative: (94) has a unique solution except for the eigenfrequencies of the junction (and also of the guides corresponding to the eigenvalues of the Dirichlet Laplacian).

Remark 10. In these junction problems we are essentially interested in obtaining the electromagnetic field and above all the propagating modes which are outgoing from the junction through the guides. If the medium of the junction is homogeneous, we can write the problem on the boundary of the junction thanks to the Calderon projectors P_i or P_e (see Def. 3 chap. 3) or the (interior) Calderon operator C^i (see Def. 5 and (211) chap. 3); thus we have to find the boundary value of the electromagnetic field satisfying:

$$(97) \qquad P_e(n \wedge E|_{\Gamma}, - n \wedge H|_{\Gamma}) = 0, \quad \text{or} \quad n \wedge H|_{\Gamma} = C^i(n \wedge E|_{\Gamma}),$$

with (89) and (91). Taking the value of the magnetic field on Γ_j, $j = 1$ to p, gives us a system of equations on the boundary value of the electric field. This is very simply written thanks to the Calderon operator (but less suitable for a numerical solution):

$$(98) \qquad \sum_{j'=1 \text{ to } p} (C^i_{jj'} - C_j\delta_{jj'})\, M_{j'} = f_{Ij}\,,\ j = 1 \text{ to } p, \text{ with } M_j = n \wedge E|_{\Gamma_j} \in H_o^{-1/2}(\text{div}, \Gamma_j).$$

If the medium in the junction is inhomogeneous, then we have to connect (for instance) a finite element method in the domain Ω with a spectral method on the boundary of Ω. Such a method has been used and numerically implemented in order to evaluate the permittivity and permeability of a sample in a wave guide. Obviously there is a great variety of examples and of geometries for this study.

⊗

Remark 11. We haven't studied the more difficult case of waveguides which are inhomogeneous in the fiber, that is, the cross-section of the guide up to infinity is inhomogeneous. It seems necessary to use microlocal analysis for such a case.

⊗

Remark 12. In Examples 1 (with Helmholtz) and 2 (with Maxwell), the Calderon operators (C_j) depend on the angular frequency ω, and even are holomorphic functions in a neighborhood of the real axis (except for $\omega = 0$ in the Maxwell case), see for instance Sanchez Palencia[1] p.350). Thus using a lifting, we can write problem (20) (or (94)) in the form:

$$(99) \qquad A^{(\omega)}u = F_I^{(\omega)} \text{ with given } F_I^{(\omega)},$$

where $(A^{(\omega)})$ is a holomorphic (in ω) family of closed operators in a Hilbert space *H*, with compact resolvents. Therefore we can apply theorem 1.10 (p.371) in Kato [1]. Since outside the real axis (for ω) the operator $A^{(\omega)}$ has a kernel reduced to {0}, we deduce that the problem (99) depends also on the Fredholm alternative, with only a finite number of singular values ω on every finite interval J (in the Maxwell case we have to assume that 0 is not in J).

⊗

CHAPTER 5

STATIONARY SCATTERING PROBLEMS ON UNBOUNDED OBSTACLES

1. PLANE GEOMETRY

The study of stationary waves scattered by infinite obstacles follows naturally from scattering by finite obstacles at high frequency. For a plane geometry, we have an ill-posed problem (as in the waveguides) if we don't take into account the sense of wave propagation. Here we extensively use the Fourier transform method, which replaces the discrete spectral decomposition in waveguides. First we study Helmholtz problems, then electromagnetism with Maxwell equations.

1.1. Plane geometry with Helmholtz equation

First we consider a model Helmholtz problem with a Dirichlet condition (which corresponds to scattering by a soft wall in acoustics) as the starting point to understand the basic problem.

1.1.1. Helmholtz problem in a half-space with Dirichlet condition

Let $R^n_+ = \{X = (x, x_n) \in R^{n-1} \times R,\ x_n > 0\} = R^{n-1} \times R_+$ with (generally) $n = 3$, and let $\Gamma = R^{n-1} \times \{0\}$ be its boundary. We consider the following Dirichlet problem: find u satisfying

$$(1) \qquad \begin{cases} \text{i) } \Delta u + k^2 u = 0 & \text{in } R^n_+ \\ \text{ii) } u|_\Gamma = u_0 & \text{on } \Gamma, \end{cases}$$

with a given wavenumber $k > 0$ and given u_0 on Γ, for instance in $L^2(\Gamma)$. We study this problem thanks to a (transverse) Fourier transformation. Let $\mathcal{F}v$ denote the Fourier transform of a tempered distribution v, with:

(2) $$\hat{v}(\xi) = \mathcal{F}v(\xi) = \int_{\mathbb{R}^{n-1}} v(x)\, e^{-ix.\xi} dx, \quad \text{for } v \in \mathcal{S}(\mathbb{R}^{n-1}).$$

Then with the notation $\hat{u}(x_n) = \hat{u}(.,x_n)$, we write (1) as:

(3) $$\left| \begin{array}{l} \text{i) } \dfrac{d^2\hat{u}}{dx_n^2} + (k^2 - \xi^2)\,\hat{u} = 0 \quad \text{for } x_n > 0, \\ \text{ii) } \hat{u}(.,0) = \hat{u}_0. \end{array} \right.$$

Thus we are reduced to solving a differential equation in terms of the variable x_n only. We get, with constant functions (with respect to x_n) α and β to specify

i) for $|\xi| > k$,

(4) $$\hat{u}(\xi,x_n) = \alpha(\xi)\, e^{\theta_+ x_n} + \beta(\xi)\, e^{-\theta_+ x_n}, \quad \text{with } \theta_+ = \sqrt{\xi^2 - k^2},$$

ii) for $|\xi| < k$,

(5) $$\hat{u}(\xi,x_n) = \alpha(\xi)\, e^{i\theta_- x_n} + \beta(\xi)\, e^{-i\theta_- x_n}, \quad \text{with } \theta_- = \sqrt{k^2 - \xi^2}.$$

Wanting a solution which is tempered with respect to x_n (and ξ) at infinity we have to choose

(6) $\alpha(\xi) = 0$ for $|\xi| > k$, thus: $\beta(\xi) = \hat{u}_0(\xi)$ for $|\xi| > k$,

therefore:

(7) $$\hat{u}(\xi,x_n) = e^{-\theta_+ x_n}\,\hat{u}_0(\xi) \text{ for } |\xi| > k.$$

When $|\xi| < k$, (3)ii) with (5) gives:

(8) $\alpha(\xi) + \beta(\xi) = \hat{u}_0(\xi)$ for $|\xi| < k$,

and this does not allow us to uniquely determine a solution of (1): problem (1) is ill-posed. Similar to waveguides, we transform it into a well-posed problem by adding a physical condition of wave propagation. For a time evolution given by $u(X,t) = e^{-i\omega t}u(X)$, we say that u is *a wave propagating towards the positive (resp. negative) axis* x_n *if the propagating part of u (i.e., (5)) is a superposition of waves of the type*

(9) $$\exp(i\theta_- x_n)\, v(\xi) \quad (\text{resp. } \exp(-i\theta_- x_n)\, v(\xi)), \quad \text{with } \theta_- > 0.$$

Then with the physical choice of waves propagating towards the positive axis x_n, we get a unique solution for (3) (and then for (1)) of the form:

(10) $$\hat{u}(\xi,x_n) = e^{-\theta(\xi)x_n}\hat{u}_0(\xi),$$

with:

(11) $\theta(\xi) = \theta_+ = (\xi^2 - k^2)^{1/2}$ when $|\xi| > k$, $\theta(\xi) = -i\theta_- = -i(k^2 - \xi^2)^{1/2}$ when $|\xi| < k$.

Theorem 1. *The problem* (1) *with* u_0 *given in* $L^2(R^{n-1})$ *and with the condition*

(12) u *is a tempered wave propagating towards the positive axis* x_n,

has a unique solution u in $C^0([0,+\infty), L^2(R^{n-1}))$, *and it is obtained thanks to a contraction (and holomorphic) semigroup of class* C^0 *in* $L^2(R^{n-1})$, $(G(x_n))$, $x_n > 0$, *by its Fourier transform* (10), *with* (11)

(13) $$\hat{u}(\xi, x_n) = \hat{G}(x_n)\hat{u}_0(\xi) = e^{-\theta(\xi)x_n}\hat{u}_0(\xi), \qquad \xi \in R^{n-1}.$$

PROOF. The operator $G(x_n)$ is a contraction in $L^2(R^{n-1})$ since

(14) $$\sup_{\xi \in R^{n-1}} |\exp(-\theta(\xi)x_n)| = 1, \quad \forall x_n \geq 0.$$

Furthermore, we have:

(15) $$\left| \begin{aligned} \|u(.,x_n) - u_0\|^2 &= (2\pi)^{-n} \int_{R^{n-1}} |\hat{u}(\xi,x_n) - \hat{u}_0(\xi)|^2 d\xi \\ &= (2\pi)^{-n} \int_{R^{n-1}} |\hat{u}_0(\xi)|^2 |1 - e^{-\theta x_n}|^2 d\xi. \end{aligned} \right.$$

Since $\hat{u}_0 \in L^2(R^{n-1})$, for all $\varepsilon_1 \geq 0$, there exists $a > 0$ such that: $\int_{|\xi|>a} |\hat{u}_0|^2 d\xi < \varepsilon_1$, and there exists $\eta > 0$ such that $0 \leq x_n \leq \eta$ implies

(16) $$|1 - e^{-\theta x_n}|^2 < \varepsilon_1, \quad \forall \xi \text{ with } |\xi| < a,$$

thus for $0 \leq x_n \leq \eta$,

(17) $$\|u(.,x_n) - u_0\|^2 \leq \varepsilon_1(1 + \|u_0\|^2),$$

which implies that the semigroup is continuous.

⊗

Proposition 1. *The infinitesimal generator* A *of the semigroup* $(G(x_n))$, $x_n > 0$, *in* $L^2(R^{n-1})$ *has the domain* $D(A) = H^1(R^{n-1})$ *and is given by its Fourier transform:*

(18) $$\hat{A}\hat{u}_0(\xi) = -\theta(\xi)\hat{u}_0(\xi) \qquad \textit{with } \theta \textit{ defined by } (11).$$

PROOF. The domain D(A) of the generator is given by its Fourier transform

(19) $$\left| \begin{aligned} D(\hat{A}) &= \{\hat{u}_0 \in L^2(R^{n-1}), \hat{A}\hat{u}_0 \in L^2(R^{n-1})\}, \\ &= \{\hat{u}_0 \in L^2(R^{n-1}), \int_{|\xi|>k} (\xi^2 - k^2) |\hat{u}_0|^2 d\xi < \infty\}, \end{aligned} \right.$$

from which we clearly have $D(A) = H^1(R^{n-1})$.

⊗

We also have:

(20) $\hat{A}^2 \hat{u}_0(\xi) = (\xi^2 - k^2)\hat{u}_0(\xi)$,
therefore:

(21) $A^2 = -(\Delta + k^2)$, with $D(A^2) = H^2(R^{n-1})$.

Thus A is a square root of the selfadjoint (but not positive) operator $-(\Delta + k^2)$. Like in the waveguide case, we can decompose the space $L^2(R^{n-1})$ into:

(22) $L^2(R^{n-1}) = H_k^+ \oplus H_k^-$,
so that A splits into:

(23) $A = -A^+ + i A^-$, A^+, A^- positive selfadjoint operators resp. in H_k^+ and H_k^-,
and A satisfies:

(24) $\mathrm{Re}\,(Au,u) \leq 0$, $\mathrm{Im}\,(Au,u) \geq 0$, $\forall u \in H^1(R^{n-1})$;

A is a normal operator: $D(A) = D(A^*)$ and A commutes with A^*, with spectrum

(25) $\sigma(A) = (-\infty, 0] \cup [0, ik]$.

Definition 1. *The infinitesimal generator* A *of the semigroup* $(G(x_n))$, $x_n > 0$, *defined by* (18) *is called the Calderon* (*or the capacity*) *operator of the half-space*.

We recall that A is the mapping $u_0 \rightarrow \frac{\partial u}{\partial x_n}|_\Gamma$, u the solution of (1) with (12).

Remark 1. We have to verify that choosing waves propagating towards the positive axis x_n as above, agrees with choosing outgoing waves defined by the Sommerfeld condition. It suffices to verify that the elementary outgoing wave $\Phi_k{}^a$ centered at point $X_a = (0,0,-a)$, "a" being positive, satisfies (12). Using a translation, we can take $a = 0$. Then using the Fourier transform of the elementary outgoing wave Φ_k given by (159)' chap. 3, and taking the inverse Fourier transform with respect to variable x_n only, we obtain (for $n = 3$)

(26) $F_{x_3}^{-1} F_X \Phi_k(\xi)(x_3) = F_x \Phi_k(\xi, x_3) = \hat{\Phi}_k(\xi, x_3) = \frac{1}{2\theta(\xi)} \exp(-\theta(\xi)|x_3|)$.
Thus:

(27) $\hat{\Phi}_k^a(\xi, x_3) = \hat{\Phi}_k(\xi, x_3 + a) = \frac{1}{2\theta(\xi)} \exp(-\theta(\xi)|x_3 + a|)$,

which is of the form (13) for $x_n > 0$ (but for $a = 0$, $\hat{u}_0 = 1/(2\theta)$ is not in L^2).

⊗

From formulas (13) and (26), we see that the semigroup is also given by:

(28) $$\hat{G}(x_3)\hat{u}_0(\xi) = -2\frac{\partial\hat{\Phi}_k}{\partial x_3}(\xi, x_3)\hat{u}_0(\xi),$$

that is also with the convolution product $*$ (a priori for u_0 with compact support)

(29) $$G(x_3)u_0 = -2\frac{\partial\Phi_k}{\partial x_3}(.,x_3)\underset{x}{*}u_0.$$

The semigroup $(G(t))$, $t = x_3 > 0$, is also given with $\Phi_k(x,t) = \frac{e^{ikR}}{4\pi R}$, $R = (x^2+t^2)^{1/2}$ by

(30) $$G(t)u_0(x) = -2\int_{R^2}\frac{\partial\Phi_k}{\partial t}(x-x',t)u_0(x')\,dx'.$$

Thus the operator A is obtained by the convolution in R^2

$$Au_0 = a\underset{x}{*}u_0, \text{ with } a(x) = -2\frac{\partial^2\Phi}{\partial t^2}(x,0) = -2\,FP\Big(\frac{1}{r}\frac{\partial}{\partial r}\Big(\frac{e^{ikr}}{4\pi r}\Big)\Big),\ r = |x|, \text{ i.e.}$$

$$\langle a, \phi\rangle = \lim_{\varepsilon\to 0}\int_\varepsilon^\infty \frac{e^{ikr}}{r}\bar{\phi}'(r)\,dr, \text{ with } \bar{\phi}'(r) = \frac{1}{2\pi}\int_0^{2\pi}\phi'(r,\theta)\,d\theta,\ \forall\phi\in D(R^2).$$

Note that $\bar{\phi}'(0) = 0$ from (397) chap.3. When $k = 0$, the semigroup $(G(t))$ is given by

(31) $$G(t)u_0(x) = \frac{1}{2\pi}\int_{R^2}\frac{t}{((x-x')^2+t^2)^{3/2}}u_0(x')\,dx',$$

that is, the Cauchy-Poisson semigroup which is holomorphic in all $L^p(R^2)$, $p \neq \infty$ (see Butzer-Berens [1] p. 248). This seems not to be true for $k \neq 0$.

Remark 2. We can find formula (29) using the "image" method: by symmetry we define in the whole space R^3

(32) $U(x,x_3) = u(x,x_3)$ when $x_3 > 0$, $\quad U(x,x_3) = -u(x,-x_3)$ when $x_3 < 0$.

Then

(33) $\Delta U + k^2U = \mathrm{div}(2u_0 n\delta_\Gamma)$ in R^3, with $n = e_3 = (0,0,1)$.

Thus, for u_0 with compact support, U is an outgoing wave given by

(34) $$U = -\Phi_k\underset{X}{*}\mathrm{div}(2u_0n\delta_\Gamma) = -\mathrm{grad}\,\Phi_k\underset{X}{*}(2u_0n\delta_\Gamma) = -2\frac{\partial\Phi_k}{\partial x_3}(.,x_3)\underset{x}{*}u_0,$$

that is (29); thus U is a double layer potential. ⊗

Remark 3. *Some generalizations to other frameworks.* First note that A is not a bijective mapping from $H^1(R^2)$ onto $L^2(R^2)$. This suggests that it is not the best framework to consider (from the point of view of applications also). We first define spaces with a special weight for tangent (or longitudinal) waves, i.e., for ξ on the sphere S_k of radius k:

(35) $$\left|\begin{array}{l} H_k^{1/2}(R^2) = \{u\in S'(R^2),\ \hat{u}\in L^1_{loc}(R^2),\ \hat{u}\in L^2(R^2, |\theta|\,d\xi)\}, \\ H_k^{-1/2}(R^2) = \{u\in S'(R^2),\ \hat{u}\in L^1_{loc}(R^2),\ \hat{u}\in L^2(R^2, |\theta|^{-1}\,d\xi)\}. \end{array}\right.$$

Note that if $\rho_k(\xi) = d(\xi, S_k)$ is the distance from ξ to the sphere S_k, then in the neighborhood of S_k:

(36) $\quad |\theta_k(\xi)|^2 = |(|\xi|^2 - k^2)| = |(|\xi| - k)|\,(|\xi| + k) \approx 2k\,\delta(\xi, S_k) = 2k\,\rho_k(\xi).$

An example is given by the trace on the plane Γ of a wave due to a point source not in the plane: thanks to (26), for $a \neq 0$, $\Phi_k^a|_\Gamma \in H_k^{1/2}(R^2)$.

Since θ and θ^{-1} are in $L^1_{loc}(R^2)$, then $\hat{u} \in L^2(R^2, |\theta|\,d\xi)$ implies by Cauchy-Schwarz inequality

(37) $\quad \int_K |\hat{u}|\,d\xi \leq (\int_K |\hat{u}|^2 |\theta|\,d\xi)^{1/2} (\int_K |\theta|^{-1} d\xi)^{1/2},$

for all compact sets K in R^2, thus $\hat{u} \in L^1_{loc}(R^2)$. When $\hat{u} \in L^2(R^2, |\theta|^{-1}\,d\xi)$, we have the same result. The condition $\hat{u} \in L^1_{loc}(R^2)$ in (35) is only used to eliminate measures concentrated on the sphere S_k of radius k.

Furthermore, using the inequality: $|\theta| < C_k(1 + |\xi|^2)^{1/2}$, $C_k = \max(1,k)$, we have

(38) $\quad \|u\|^2_{H_k^{1/2}(R^2)} \stackrel{\text{def}}{=} \int_{R^2} |\hat{u}|^2 |\theta|\,d\xi \leq C_k \int_{R^2} |\hat{u}|^2 (1 + |\xi|^2)^{1/2} d\xi = C_k \|u\|^2_{H^{1/2}(R^2)}.$

We deduce the following space inclusions

(39) $\quad H^{1/2}(R^2) \to H_k^{1/2}(R^2)$ and $H_k^{-1/2}(R^2) \to H^{-1/2}(R^2),$

by duality, $H_k^{-1/2}(R^2)$ being identified with the dual space of $H_k^{1/2}(R^2)$.

Then we have by Fourier transformation, duality and interpolation:

Proposition 2. *The mapping* $u \to Au$ *is an isometry from* $H_k^{1/2}(R^2)$ onto $H_k^{-1/2}(R^2)$, *and is a continuous mapping from* $H^{1/2}(R^2)$ into $H^{-1/2}(R^2)$.

Moreover for u_0 in $H_k^{1/2}(R^2)$, *the solution u of* (1), (12), *given by* (13) *is in* $H^1_{loc}(\bar{R}^3_+)$, *that is,* u *is of locally finite energy up to the boundary.*

The framework (35) does not allow us to directly tackle all useful examples such as plane waves. So we have to generalize. The framework of tempered distributions (in R^2) is not adequate since $\exp(-\theta x_3)$ is not a multiplier in $\mathcal{S}'(R^2)$. But we can use other functional spaces, for instance Besov space B used by Hörmander [1], T.1 chap. XIV p. 227 and its dual space B^*. For most applications in view, it seems interesting to work with the space E_k defined by:

(40) $\quad E_k = \{u \in \mathcal{S}'(R^3),\ \mathcal{F}_X u$ is a Radon measure concentrated on the sphere $S_k \subset R^3\}.$

From the Paley-Wiener Theorem, the elements of E_k are entire functions in R^3, which satisfy the Helmholtz equation in the whole space. The set of elements of E_k propagating towards the positive axis x_3 is

(41) $E_k^+ = \{u \in E_k,\ \mathcal{F}_x u$ is a Radon measure concentrated on the half sphere $S_k^+\}$, with $S_k^+ = \{(\xi,\xi_3) \in S_k,$ with $\xi_3 \geq 0\}$.

Of course, there are other functional spaces which may be useful, like:

(42) $X = \{u \in L^2(R^2),\ \hat{u} \in C_0(R^2)\}$, with $C_0(R^2) = \{v \in C(R^2),\ v(\xi) \to 0$ when $|\xi| \to \infty\}$,

and its dual space X' also defined by

(43) $X' = \{f \in \mathcal{S}'(R^2),\ \hat{f} \in \mathcal{M}^1(R^2) + L^2(R^2)\}$, with $\mathcal{M}^1(R^2)$ the set of bounded Radon measures.

We can prove that X' is contained in the space:

(44) $C_b(R^2) + L^2(R^2)$, with $C_b(R^2)$ the space of continuous bounded functions on R^2.

Furthermore, we can prove that the spaces X and X' are stable under the semigroup $(G(t))$, $t > 0$, which is a semigroup of class C^0 in these spaces; X' allows us to tackle the scattering of an incident plane waves by an infinite plane wall. But these spaces are not very easy to handle from the point of view of functional analysis. Plane waves are often used by the spectral theory as generalized eigenvectors (of the operator A).

⊗

Remark 4. The Helmholtz problem (1) with Im k different from 0 ($0 < \arg k < \pi$) is well-posed if we only require that u be a x_3-tempered solution. It is also obtained by a semigroup of class C^0, which is holomorphic, and given by its Fourier transform:

(45) $\hat{u}(\xi,x_3) = \hat{G}(x_3)\hat{u}_0(\xi) = \exp(-\theta x_3)\hat{u}_0(\xi),\quad \theta(\xi) = (\xi^2 - k^2)^{1/2}$ *with* $\mathrm{Re}\,\theta(\xi) \geq 0$.

Here $\theta \neq 0$, thus A is an isomorphism from $H^{1/2}(R^2)$ onto $H^{-1/2}(R^2)$.
The limit of (45) when Im k tends to 0 with Re k positive (resp. negative) gives the wave propagating towards the positive (resp. negative) real axis x_3, following the limiting absorption principle (see chap.3). ⊗

1.1.2. Transmission Helmholtz problems in R^3. Calderon projectors
First we consider the following Helmholtz problem in the whole space: find u in R^3 satisfying (for a given real wavenumber k)

(46)
$$\left|\begin{array}{l} \text{i) } \Delta u + k^2 u = 0 \quad \text{for } x_3 \neq 0, \text{ i.e., in } R^3_+ \text{ and } R^3_-, \\ \text{ii) } [u]_\Gamma \overset{\text{def}}{=} u|_{\Gamma_-} - u|_{\Gamma_+} = \rho, \quad [\frac{\partial u}{\partial x_3}]_\Gamma = \rho' \text{ on } \Gamma, \\ \text{iii) the restrictions of u to each half-space } R^3_+ \text{ and } R^3_- \text{ are "outgoing" waves,} \\ \text{i.e., resp. propagating towards } x_3 > 0 \text{ and } x_3 < 0, \end{array}\right.$$

with given ρ and ρ' on $\Gamma = R^2 \times \{0\}$, with hypotheses to be specified later.
We know (for instance from chap. 3) that this problem has at most one solution. Using the above section, we are led to find the boundary values of the solution on Γ only. We can use different methods. Here we use an integral method, similar to chap. 3. Under usual (regularity) conditions on ρ and ρ' (see chap. 3), i) and ii) are equivalent to

(47) $$\Delta u + k^2 u = -(\rho' \delta_\Gamma + \text{div}(\rho n \delta_\Gamma)) \quad \text{in } \mathcal{D}'(R^3).$$

With the "outgoing" condition iii) (at least when ρ and ρ' have compact supports), problem (1) has a unique solution u given by the convolution product:

(48) $$u = \Phi \underset{X}{*} (\rho' \delta_\Gamma + \text{div}(\rho n \delta_\Gamma)),$$

with $\Phi = \Phi_k$ the elementary outgoing solution (see (8) chap.3), or also

(49) $$u(.,x_3) = \Phi(.,x_3) \underset{x}{*} \rho' + \frac{\partial \Phi}{\partial x_3}(.,x_3) \underset{x}{*} \rho.$$

If we differentiate with respect to x_3, we obtain:

(50) $$\frac{\partial u}{\partial x_3}(.,x_3) = \frac{\partial \Phi}{\partial x_3}(.,x_3) \underset{x}{*} \rho' + \frac{\partial^2 \Phi}{\partial x_3^2}(.,x_3) \underset{x}{*} \rho.$$

Then, when $x_3 \to 0_+$ and $x_3 \to 0_-$, using (29), we have:

(51) $$\begin{pmatrix} -\ u(0_+) \\ -\frac{\partial u}{\partial x_3}(0_+) \end{pmatrix} = P_+ \begin{pmatrix} \rho \\ \rho' \end{pmatrix}, \quad \begin{pmatrix} u(0_-) \\ \frac{\partial u}{\partial x_3}(0_-) \end{pmatrix} = P_- \begin{pmatrix} \rho \\ \rho' \end{pmatrix},$$

with P_+ and P_- given by

(52) $$P_+ = \frac{1}{2}(I - S), \quad P_- = \frac{1}{2}(I + S), \quad S = -\begin{pmatrix} 0 & A^{-1} \\ A & 0 \end{pmatrix}.$$

Thus, we have

(53) $$P_+ + P_- = I, \text{ and } S^2 = I.$$

We can verify that

(54) $$B \overset{\text{def}}{=} -\frac{1}{2} A^{-1},$$

is an integral operator given by the convolution product:

$$(55) \qquad B\rho'(x) = \Phi(.,0) \underset{x}{*} \rho' = \int_{R^2} \frac{e^{ik|x-x'|}}{4\pi|x-x'|} \rho'(x')\, dx'.$$

Proposition 3. *The operator* S *is an isomorphism satisfying* (53), *in the space*

$$(56) \qquad Y_k \overset{def}{=} H_k^{1/2}(R^2) \times H_k^{-1/2}(R^2),$$

and thus P_+ *and* P_- *are continuous complementary projectors in this space, giving the boundary values of the problem* (46) *by* (51) *when* (ρ, ρ') *are given in* Y_k.

Definition 2. *The operators* P_+ *and* P_- *are called Calderon projectors (for the Helmholtz problem) of the half-space at wavenumber* k.

If we compare the half-space problem to that of a bounded domain (see chap. 3 Def. 2 and (71)), we see that we have, with notations of chap. 3:

$$K = 0, \quad J = 0, \quad 2L = -A^{-1}, \quad 2R = -A.$$

1.1.3. Transmission problems with two different media

We assume that the whole space is occupied by two different media, one on each side of a plane. Let k_1 and k_2 be (*real* for instance) wavenumbers in each medium. Then we consider the following transmission problem: find u in R^3 satisfying

$$(57) \qquad \left| \begin{array}{l} \text{i) } \Delta u + k^2 u = 0 \quad \text{for } x_3 \neq 0, \text{ with } k = k_1 \text{ in } R^3_+ \text{ and } k = k_2 \text{ in } R^3_-, \\ \text{ii) } [u]_\Gamma \overset{def}{=} u|_{\Gamma_-} - u|_{\Gamma_+} = \rho, \quad [\frac{\partial u}{\partial x_3}]_\Gamma = \rho' \text{ on } \Gamma, \\ \text{iii) the restrictions of u to each half-space } R^3_+ \text{ and } R^3_- \text{ are "outgoing" waves,} \\ \text{i.e., resp. propagating towards } x_3 > 0 \text{ and } x_3 < 0, \end{array} \right.$$

with given ρ and ρ' on $\Gamma = R^2 \times \{0\}$, with hypotheses to be specified later.
This problem is a scattering problem when we have a given incident wave u_I (in the first medium for instance) either produced by a source of compact support in R^3_+, or by a plane wave. Thus we have to find the reflected wave u_1 in R^3_+ and the diffracted (or transmitted) wave u_2 in R^3_-, satisfying:

$$(58) \qquad u_1 + u_I = u_2, \quad \frac{\partial u_1}{\partial x_3} + \frac{\partial u_I}{\partial x_3} = \frac{\partial u_2}{\partial x_3} \quad \text{on } \Gamma.$$

Thus taking $u = u_1$ in R^3_+, $u = u_2$ in R^3_-, we see that u has to satisfy (57) with:

$$(59) \qquad \rho = u_I|_\Gamma, \quad \rho' = \frac{\partial u_I}{\partial x_3}|_\Gamma.$$

Let A_j and θ_j, j = 1, 2 resp. be the Calderon operator A and the function θ defined by (18) and (11) for k_j. Then problem (57) is reduced to finding the boundary values of u, u_1 and u_2 on Γ, and these are obtained thanks to the Calderon operators of each half spaces from (57)ii) (with i) and iii)) by:

$$(60) \qquad \left| \begin{array}{l} \text{i) } \theta_2\hat{u}_2 + \theta_1\hat{u}_1 = \hat{\rho}' \\ \text{ii) } \hat{u}_2 - \hat{u}_1 = \hat{\rho}. \end{array} \right.$$

The solution of (60) is given by:

$$(61) \qquad \hat{u}_2 = \frac{\hat{\rho}' + \theta_1\hat{\rho}}{\theta_1 + \theta_2}, \qquad \hat{u}_1 = \frac{\hat{\rho}' - \theta_2\hat{\rho}}{\theta_1 + \theta_2}.$$

When ρ and ρ' are given by (59) with an incident wave propagating towards the negative axis x_3, we have

$$(62) \qquad \rho' = -A_1(u_I|_\Gamma) = -A_1(\rho).$$

We obtain the boundary values of the transmitted wave and the reflected wave by

$$(63) \qquad \hat{u}_2 = \hat{T}\hat{u}_I, \qquad \hat{u}_1 = \hat{R}\hat{u}_I \quad \text{on } \Gamma,$$

with

$$(64) \qquad \hat{T} \stackrel{\text{def}}{=} \frac{2\theta_1}{\theta_1 + \theta_2}, \qquad \hat{R} \stackrel{\text{def}}{=} \frac{\theta_1 - \theta_2}{\theta_1 + \theta_2};$$

$\hat{R}$ and $\hat{T}$ are called the *reflection and transmission coefficients*. They satisfy

$$(65) \qquad \hat{T} - \hat{R} = 1.$$

Proposition 4. *We assume that the incident wave* u_I *satisfies*

$$(66) \qquad u_I|_\Gamma \in H^{1/2}_{k_1}(R^2), \quad \text{thus } (u_I, \frac{\partial u_I}{\partial n})|_\Gamma \in Y_{k_1}.$$

On the boundary Γ *of the two media the reflected and the transmitted waves satisfy*

$$(67) \quad u_1|_\Gamma \in H^{1/2}_{k_1}(R^2) \text{ and } u_2|_\Gamma \in H^{1/2}_{k_1}(R^2) \cap H^{1/2}_{k_2}(R^2) \text{ (and even } u_2|_\Gamma \in H^{1/2}(R^2))$$

and thus $(u_1, \frac{\partial u_1}{\partial n})|_\Gamma \in Y_{k_1}$, $(u_2, \frac{\partial u_2}{\partial n})|_\Gamma \in Y_{k_2} \cap Y_{k_1}$.

PROOF. We have

$$(68) \qquad \int_{R^2} |\hat{u}_2|^2 |\theta_2| \, d\xi = \int_{R^2} |\hat{u}_I|^2 |\hat{T}|^2 |\theta_2| \, d\xi = \int_{R^2} |\hat{u}_I|^2 |\theta_1| \tau \, d\xi,$$

with $\tau = \dfrac{4|\theta_1\theta_2|}{|\theta_1+\theta_2|^2}$. We easily see that $\tau \le 2$, and thus: $\int_{R^2}|\hat{u}_2|^2|\theta_2|\,d\xi < \infty$.

Since $|\hat{T}| \le 2$, we also have $\int_{R^2}|\hat{u}_2|^2|\theta_1|\,d\xi < \infty$. Furthermore:

$$\int_{R^2}|\hat{u}_2|^2(1+|\xi|^2)^{1/2}\,d\xi = \int_{R^2}|\hat{u}_I|^2|\theta_1|\,\sigma\,d\xi, \text{ with } \sigma = \frac{4|\theta_1|(1+|\xi|^2)^{1/2}}{|\theta_1+\theta_2|^2}.$$

But σ is bounded on R^2, thus we have $u_2|_\Gamma \in H^{1/2}(R^2)$, and Proposition 3 follows.

⊗

Generalizing to spaces such as that defined by (41), or (43) allows us to deal with plane waves, giving the famous *Snell-Descartes laws*. The incident wave u_I is given by

(69) $$u_I(X) = u_{I0}\,e^{ik_1\alpha.X}, \quad \text{with } u_{I0} \in C,\ \alpha = (\alpha_T, \alpha_3) \in S^2,\ \alpha_3 < 0,$$

whose Fourier transform is the measure

(69)' $$\hat{u}_I(\xi,x_3) = (2\pi)^2\,\delta(\xi - k_1\alpha_T)\,e^{ik_1\alpha_3x_3}\,u_{I0}, \quad \text{with } |\alpha_T|^2 + \alpha_3^2 = 1.$$

Thus ρ and ρ' are given by their Fourier transforms:

(70) $$\hat{\rho}(\xi) = \hat{u}_I(\xi,0) = (2\pi)^2\,\delta(\xi - k_1\alpha_T)\,u_{I0},\quad \hat{\rho}'(\xi) = \frac{\partial\hat{u}_I}{\partial x_3}(\xi,0) = (2\pi)^2\,ik_1\alpha_3\,\delta(\xi - k_1\alpha_T)\,u_{I0}.$$

Here formulas (60) are also valid and imply that $\hat{u}_1$ and $\hat{u}_2$ are Dirac measures concentrated on $\xi = k_1\alpha_T$. Then θ_1 and θ_2 are given by

(71) $$\theta_1(k_1\alpha_T) = ik_1\alpha_3,\quad \theta_2(k_1\alpha_T) = \left|\begin{array}{ll} (k_1^2\alpha_T^2 - k_2^2)^{1/2} & \text{if } k_1|\alpha_T| > k_2, \\ -i\,(k_2^2 - k_1^2\alpha_T^2)^{1/2} & \text{if } k_1|\alpha_T| < k_2. \end{array}\right.$$

Thus with the (relative) index $\eta = k_2/k_1$, we have (63) with:

(72) $$\hat{T} = \frac{2|\alpha_3|}{|\alpha_3| + \eta\,\theta_{12}},\qquad \hat{R} = \frac{|\alpha_3| - \eta\,\theta_{12}}{|\alpha_3| + \eta\,\theta_{12}},$$

and

(73) $$\theta_{12} = i\theta_2/k_2 = i((\alpha_T^2/\eta^2) - 1)^{1/2} \text{ if } |\alpha_T| > \eta,\quad \theta_{12} = (1 - (\alpha_T^2/\eta^2))^{1/2} \text{ if } |\alpha_T| < \eta.$$

The inverse Fourier transforms of the reflected and the transmitted waves are, for *the reflected wave*

(74) $$u_1(x,x_3) = \hat{R}\,u_{I0}e^{ik_1\beta.X} \qquad \text{with } \beta = (\alpha_T, -\alpha_3),$$

and *the transmitted wave*

(74)' $$u_2(x,x_3) = \hat{T}\,u_{I0}e^{ik_2\tilde{\beta}.X} \qquad \text{with } \tilde{\beta} = (\tilde{\beta}_T, \tilde{\beta}_3),$$

where $\tilde{\beta}$ may be complex; we have

i) for $k_1 < k_2$, i.e., $\eta > 1$,

$\tilde\beta \in S^2$, with $\tilde\beta_T = \alpha_T/\eta$, $\tilde\beta_3 = -(1-(\alpha_T^2/\eta^2))^{1/2}$, thus $|\tilde\beta_T| < |\alpha_T|$, $|\tilde\beta_3| > |\alpha_3|$,

or with respect to incident and refracted angles of wave propagations ϕ_I and ϕ_2,

$$|\tilde\beta_T| = \sin\phi_2 = |\alpha_T|/\eta = (\sin\phi_I)/\eta, \quad \phi_2 < \phi_I;$$

ii) for $k_1 > k_2$, i.e., $\eta < 1$,

a) $k_1|\alpha_T| < k_2$,

$$\tilde\beta \in S^2,\ \tilde\beta_T = \alpha_T/\eta, \text{ thus } |\tilde\beta_T| < |\alpha_T|,\ |\tilde\beta_3| < |\alpha_3|,\ \phi_2 < \phi_I;$$

b) $k_1|\alpha_T| > k_2$,

$\tilde\beta_T = \alpha_T/\eta$, but $\tilde\beta_3$ is complex (the transmitted wave is exponentially damped)

$$\tilde\beta_3 = -i((\alpha_T^2/\eta^2) - 1)^{1/2}, \text{ and } \tilde\beta^2 = \tilde\beta_T^2 + \tilde\beta_3^2 = 1, \text{ but } \tilde\beta \text{ is not in } S^2.$$

⊗

1.1.4. Some examples of applications of the Calderon operator

There are very many examples, where we have to use the Calderon operator A in order to obtain a well-posed problem, taking into account the sense of wave propagation, even for geometries which are not simple half-planes. Using the Calderon operator A often gives a boundary condition, which allows us to reduce the domain of the wave problem to a slab or to a bounded domain. Thus, we can treat multilayer scattering problems (that is, for slabs with different homogeneous media in the free space), screen scattering problems, where the screen can be thick or thin, infinite (for instance a half plane) or not, with apertures of arbitrary shape....

Here we give an example with a thin soft screen with bounded apertures. Let Γ_0 be a "flat domain" that is an open subset of $R^2 \times \{0\}$, modelling a thin screen; we assume that its complementary set $\Gamma_1 = R^2 \times \{0\} \backslash \Gamma_0$ is a "regular" bounded set (connected or not), modelling apertures in the screen. We assume that the free space is outside the screen and that a given incident stationary wave u_I with wavenumber k is propagating in R^3_+ towards the screen. Let u_1 and u_2 resp. be the reflected wave in R^3_+, the wave in R^3_- transmitted through the apertures.

Let u be defined in $R^3 \backslash (R^2 x\{0\})$ by $u = u_1$ in R^3_+, $u = u_2$ in R^3_-. Then u must satisfy:

$$(75)\quad \begin{cases} \text{i) } \Delta u + k^2 u = 0 \text{ in } R^3\backslash\Gamma, \text{ with } \Gamma = R^2 \times \{0\}, \\ \text{ii) } [u]_{\Gamma_1} \overset{\text{def}}{=} u_2|_{\Gamma_1} - u_1|_{\Gamma_1} = u_I|_{\Gamma_1},\ [\frac{\partial u}{\partial n}]_{\Gamma_1} = \frac{\partial u_2}{\partial n}|_{\Gamma_1} - \frac{\partial u_1}{\partial n}|_{\Gamma_1} = \frac{\partial u_I}{\partial n}|_{\Gamma_1}, \text{ on } \Gamma_1 \\ \text{iii) } u_1|_{\Gamma_0} = -u_I|_{\Gamma_0},\ u_2|_{\Gamma_0} = 0 \text{ (soft screen conditions)} \\ \text{iv) } u_1 \text{ (resp. } u_2\text{) is propagating towards the positive (resp. negative) axis } x_3. \end{cases}$$

Using the Calderon operator A, this problem is reduced to finding u on both sides of Γ satisfying

(76) $u_2 - u_1 = u_I$ on Γ_1, with (75)iii) on Γ_0, and $(Au_2 + Au_1)|_{\Gamma_1} = -\frac{\partial u_I}{\partial n}|_{\Gamma_1}$ on Γ_1.

Let $w = u_2 + u_1 + u_I$ on Γ. Then, from (76), $w = 2u_2$, and w must satisfy

(77) $\quad Aw|_{\Gamma_1} = f_I|_{\Gamma_1}$, with $w = 0$ on Γ_0,

with $f_I|_{\Gamma_1} = -\frac{\partial u_I}{\partial n}|_{\Gamma_1} - Au_I|_{\Gamma_1} = -2\,Au_I|_{\Gamma_1}$ given on Γ_1. We assume that u_I satisfies

(78) $\quad u_I|_{R^2 x\{0\}} \in H_k^{1/2}(R^2)$, thus $f_I \in H_k^{-1/2}(R^2)$.

Now let $V = \{v \in H_k^{1/2}(R^2),\ v|_{\Gamma_0} = 0\}$. Then we can write (77) in the variational form: find w in V such that

(77)' $\quad (Aw, v) = (f_I, v), \quad \forall v \in V.$

Lemma 1. *The sesquilinear form* a(u,v) *defined by*:

(79) $\quad a(u,v) = (Au,v), \quad u, v \in H_k^{1/2}(R^2),$

is "coercive" on $H_k^{1/2}(R^2)$ *and more precisely*

(80) $\quad |a(u,u)| = \|u\|^2_{H_k^{1/2}(R^2)}, \quad \forall u \in H_k^{1/2}(R^2).$

PROOF. From definition (79), we have

(81) $\quad a(u,u) = -\int_{R^2} \theta\, |\hat{u}|^2\, d\xi,$

and thus we obtain (80) from

(81)' $\quad |a(u,u)| = \int_{R^2} |\theta|\, |\hat{u}|^2\, d\xi.$

⊗

Lemma 2. $H_k^{1/2}(R^2)$ (resp. $H_k^{-1/2}(R^2)$) *is contained in* $H_{loc}^{1/2}(R^2)$ (resp. $H_{loc}^{-1/2}(R^2)$).

PROOF. Working on the Fourier transform of these spaces, we decompose the corresponding L^2 spaces into the sum of two spaces, one W_0 with functions equal to zero in a neighborhood of the sphere S_k and the other W_1 with functions equal to zero outside this neighborhood. Now we see that the inverse Fourier transform of functions in W_0 are in the Sobolev space $H^{1/2}(R^2)$ (resp. into $H^{-1/2}(R^2)$), and that the functions in W_1 having compact support are transformed into regular functions.

⊗

Proposition 5. *Let* u_I *be an incident wave satisfying* (78). *Problem* (77)' (*thus* (77)) *has a unique solution* w (*thus* u_2 *the transmitted wave*) *such that*

(82) $$w \in H^{1/2}_{o\,o}(\Gamma_1),$$

giving the unique solution of (75) (*with bounded apertures* Γ_1 *in a screen in* R^3), *that is the transmitted wave and the reflected wave thanks to the semigroups* $(G(x_3))$ *and* $(G(-x_3))$ *by* (31).

PROOF. First, using (39) and the lemma 2, we can identify $H^{1/2}_{o\,o}(\Gamma_1)$ with the closed subspace V of $H^{1/2}_k(R^2)$ by the mapping $u \longrightarrow \tilde{u}$, $\tilde{u}$ being the extension by 0 of u outside Γ_1. Furthermore, we have $f_I \in V' = (H^{1/2}_{o\,o}(\Gamma_1))'$. Then using Lemma 1, we can conclude thanks to the Lax-Milgram lemma (see for instance Lions [4]).

⊗

1.2. Plane geometry with Maxwell equations

1.2.1. A typical problem in a half-space

We first consider the following problem: find the electromagnetic field (E, H) in free space, at angular frequency ω, satisfying

(83) $$\left|\begin{array}{l} \text{i) } \operatorname{curl} H + i\omega\varepsilon E = 0\,, \\ \text{ii) } -\operatorname{curl} E + i\omega\mu H = 0 \quad \text{in } R^3_+ = R^2_x \times R^+_{x_3}\,, \\ \text{iii) } n \wedge E|_\Gamma = n \wedge E^0_T \quad (\text{or } \pi_\Gamma E = E^0_T)\,, \text{ with given } E^0_T\,, \end{array}\right.$$

with π_Γ the orthogonal projection on the boundary Γ of R^3_+.
Like in the Helmholtz case, we can see by Fourier transformation with respect to x (under usual hypotheses on E^0_T) that this problem is an ill-posed problem, and it becomes well-posed with the following hypothesis

(84) (E, H) *is a wave propagating towards the positive axis* x_3.

Since the components of E and H satisfy the Helmholtz equation, they are obtained thanks to the semigroup $(G(x_3))$ of section 1.1 due to (84), from their boundary values on Γ. With usual notations (see also chap. 4) we write

(85) $$E_T(x_3) = G(x_3)E^0_T,\ \ E_3(x_3) = G(x_3)E^0_3,\ \ H_T(x_3) = G(x_3)H^0_T,\ \ H_3(x_3) = G(x_3)H^0_3,$$

thus we only have to find E^0_3, H^0_3, H^0_T from the transverse component E^0_T of E.

We can do it in many complementary ways. We first write Maxwell equations in the form (30),(31) chap.4, decomposing the electromagnetic field into its transverse and longitudinal (along x_3 axis) components. Using (31) and (38) in (30) gives us (34) with (35) chap. 4, then using the inverse of the infinitesimal generator A of the semigroup $(G(x_3))$, we obtain (with $\omega\mu = kZ$, and $\omega\varepsilon = k/Z$) :

(86)
$$\left|\begin{array}{l} \text{i) } H_T = \frac{i}{Zk} A^{-1} (k^2 + \text{grad}_T \, \text{div}_T) S E_T , \\ \text{ii) } E_T = -\frac{iZ}{k} A^{-1} (k^2 + \text{grad}_T \, \text{div}_T) S H_T . \end{array}\right.$$

Thus we get the following relations

(87)
$$\left|\begin{array}{l} \text{i) } H_T^0 = \mathcal{C} E_T^0 , \quad \text{or } n \wedge H_T^0 = C (n \wedge E_T^0) \, (\text{i.e. } S H_T^0 = C (S E_T^0)), \\ \text{ii) } E_T^0 = - Z^2 \mathcal{C} H_T^0, \text{ or } n \wedge E_T^0 = - Z^2 C (n \wedge H_T^0), \, (\text{i.e. } S E_T^0 = - Z^2 C (S E_T^0)), \end{array}\right.$$

and thus:

(88) $$\mathcal{C}^2 = - Z^{-2} I, \quad C^2 = - Z^{-2} I, \text{ and } \mathcal{C} = - SCS, \ C = - S\mathcal{C}S,$$

with $\mathcal{C}$ (and C) given by (86), that is, in Fourier transform, $\hat{H}_T^0 = \hat{\mathcal{C}} \hat{E}_T^0$,

(89) $$\hat{\mathcal{C}} = \frac{i}{Zk\theta} \begin{pmatrix} \xi_1^2 - k^2 & \xi_1 \xi_2 \\ \xi_2 \xi_1 & \xi_2^2 - k^2 \end{pmatrix} S = \frac{i}{Zk\theta} \begin{pmatrix} \xi_1 \xi_2 & k^2 - \xi_1^2 \\ -(k^2 - \xi_2^2) & -\xi_1 \xi_2 \end{pmatrix},$$

and

$$\hat{C} = -\frac{i}{Zk\theta} \begin{pmatrix} \xi_1 \xi_2 & \xi_2^2 - k^2 \\ (k^2 - \xi_1^2) & -\xi_1 \xi_2 \end{pmatrix}.$$

Now we follow another method. Taking into account that E and H are divergence free (and satisfy Helmholtz equation), we can substitute to (30) chap. 4:

(90) $$\text{div}\, E = 0, \qquad \text{div}\, H = 0,$$

that is we only consider the equations:

(91)
$$\left|\begin{array}{ll} \text{i) } \frac{\partial E_3}{\partial x_3} + \text{div}_T E_T = 0 , & \frac{\partial H_3}{\partial x_3} + \text{div}_T H_T = 0 , \\ \text{ii) } \text{curl}_T H_T + i\omega\varepsilon E_3 = 0 , & - \text{curl}_T E_T + i\omega\mu H_3 = 0. \end{array}\right.$$

In the following we drop subscript T to differential operators (when it is not confusing). From these equations, we easily get (at least formally)

(92) $$H_3^0 = -\frac{i}{kZ} \text{curl}\, E_T^0 , \qquad E_3^0 = - A^{-1} \text{div}\, E_T^0 ,$$

(93) $$\text{div}\, H_T^0 = - A H_3^0 , \qquad \text{curl}\, H_T^0 = -\frac{ik}{Z} E_3^0 ,$$

and thus:

(94) $$\text{div}\, H_T^0 = \frac{i}{kZ} A \, \text{curl}\, E_T^0, \quad \text{curl}\, H_T^0 = \frac{ik}{Z} A^{-1} \text{div}\, E_T^0 ,$$

that is:

(94)' $$\begin{pmatrix} \text{div}\, H_T^0 \\ \text{curl}\, H_T^0 \end{pmatrix} = \frac{i}{Z} \begin{pmatrix} 0 & \frac{1}{k} A \\ k A^{-1} & 0 \end{pmatrix} \begin{pmatrix} \text{div}\, E_T^0 \\ \text{curl}\, E_T^0 \end{pmatrix}.$$

Then we get H_T^0 using the Hodge decomposition:

(95) $$H_T^0 = \text{grad}\,\Delta^{-1}\,\text{div}\,H_T^0 - \overrightarrow{\text{curl}}\,\Delta^{-1}\,\text{curl}\,H_T^0.$$

We define the following Riesz operators (this is useful to specify spaces)

(96) $$\left|\begin{array}{l} R_g\phi = \text{grad}\,(-\Delta)^{-1/2}\,\phi, \quad \text{i.e., } \hat{R}_g\,\hat{\phi}(\xi) = \dfrac{i\xi}{|\xi|}\,\hat{\phi}(\xi), \\ R_c\phi = \overrightarrow{\text{curl}}\,(-\Delta)^{-1/2}\,\phi, \quad \text{i.e., } \hat{R}_c\,\hat{\phi}(\xi) = -\dfrac{iS\xi}{|\xi|}\,\hat{\phi}(\xi), \end{array}\right.$$

and their adjoint operators

(96)' $$\left|\begin{array}{l} R_g^* u = -\,\text{div}\,(-\Delta)^{-1/2}\,u, \quad \text{i.e., } \hat{R}_g^*\,\hat{u}(\xi) = -\dfrac{i\xi.\hat{u}(\xi)}{|\xi|}, \\ R_c^* u = \text{curl}\,(-\Delta)^{-1/2}\,u, \quad \text{i.e., } \hat{R}_c^*\,\hat{u}(\xi) = \dfrac{i\xi \wedge \hat{u}(\xi)}{|\xi|}. \end{array}\right.$$

These operators satisfy:

(97) $$R_g = S\,R_c, \quad R_c = -\,SR_g, \quad R_c^* = R_g^* S, \quad R_g^* = -\,R_c^* S,$$

and also:

(98) $$|\hat{R}_g\hat{\phi}(\xi)| = |\hat{\phi}(\xi)|, \qquad |\hat{R}_c\hat{\phi}(\xi)| = |\hat{\phi}(\xi)|,$$

and thus (with L^2 norm for instance):

(99) $$\|R_g\phi\| = \|\phi\|, \qquad \|R_c\phi\| = \|\phi\|,$$

i.e., R_g and R_c are isometric operators in L^2. Then the (orthogonal) projectors P_g and P_c on the images of the grad and the (vector) curl operators associated to the Hodge decomposition are given by:

(100) $$P_g u = R_g R_g^* u = \text{grad}\,\Delta^{-1}\,\text{div}\,u, \quad P_c v = R_c R_c^* v = -\,\overrightarrow{\text{curl}}\,\Delta^{-1}\text{curl}\,v,$$

or with their Fourier transforms:

(100)' $$\hat{P}_g\,\hat{u}(\xi) = \frac{\xi.\hat{u}(\xi)}{|\xi|}\,\frac{\xi}{|\xi|}, \qquad \hat{P}_c\,\hat{v}(\xi) = \frac{\xi \wedge \hat{v}(\xi)}{|\xi|}\,\frac{S\xi}{|\xi|},$$

which are associated with the decomposition: $\hat{u}(\xi) = \hat{P}_g\,\hat{u}(\xi) + \hat{P}_c\,\hat{u}(\xi)$.
Using these operators, we can write (94) in the form

(101) $$R_g^* H_T^0 = -\frac{i}{kZ}\,A R_c^*\,E_T^0, \qquad R_c^* H_T^0 = -\frac{ik}{Z}\,A^{-1} R_g^*\,E_T^0,$$

and also (since $R_c^*\,P_g = 0$, $R_g^*\,P_c = 0$):

$$(102)\quad \begin{cases} P_g H_T^0 = -\frac{i}{kZ} R_g A R_c^* E_T^0 = -\frac{i}{kZ} A P_g S E_T^0, \\ P_c H_T^0 = -\frac{ik}{Z} R_c A^{-1} R_g^* E_T^0 = \frac{ik}{Z} A^{-1} P_c S E_T^0, \end{cases}$$

and thus the Calderon operator C is also given by:

$$(103)\quad \begin{cases} H_T^0 = C E_T^0 = -\frac{i}{Z} [\frac{1}{k} R_g A R_c^* + k R_c A^{-1} R_g^*] E_T^0 \\ = \frac{i}{Z} [-\frac{1}{k} A P_g + k A^{-1} P_c] S E_T^0. \end{cases}$$

Now we define the following functional space:

$$(104)\quad H_k(R^2) \overset{\text{def}}{=} \{u \in \mathcal{S}'(R^2)^2,\ \hat{u} \in L^1_{loc}(R^2)^2,\ R_g^* u \in H_k^{-1/2}(R^2),\ R_c^* u \in H_k^{1/2}(R^2)\}.$$

This is a Hilbert space with the norm:

$$(105)\quad \begin{cases} \|u\|_{H_k} = (\int_{R^2} [\,|\theta|\,|\xi \wedge \hat{u}(\xi)|^2 + \frac{1}{|\theta|}\,|\xi.\hat{u}(\xi)|^2]\,\frac{1}{|\xi|^2}\,d\xi)^{1/2} \\ = (\|R_c^* u\|^2_{H_k^{1/2}(R^2)} + \|R_g^* u\|^2_{H_k^{-1/2}(R^2)})^{1/2}. \end{cases}$$

We also define the space $\tilde{H}_k(R^2) = S H_k(R^2)$:

$$(104)'\quad \tilde{H}_k(R^2) \overset{\text{def}}{=} \{u \in \mathcal{S}'(R^2)^2,\ \hat{u} \in L^1_{loc}(R^2)^2,\ R_g^* u \in H_k^{1/2}(R^2),\ R_c^* u \in H_k^{-1/2}(R^2)\}.$$

This is a Hilbert space with its natural norm, and S is an isomorphism from $H_k(R^2)$ onto $\tilde{H}_k(R^2)$. Now we assume that:

$$(106)\quad E_T^0 \in H_k(R^2).$$

Theorem 2. *With hypothesis* (106), *problem* (83) *with* (84) *has a unique solution* (E, H) *with locally finite energy (up to the boundary, i.e.,* E *and* H *are in* $H_{loc}(\text{curl}, R^2 \times [0,\infty))$ *and the boundary value of the magnetic field also satisfies*

$$(107)\quad H_T^0 \in H_k(R^2).$$

For the other components of the electromagnetic field, we define, like (35):

$$(108)\quad X_k \overset{\text{def}}{=} \{u \in \mathcal{S}'(R^2),\ \hat{u} \in L^1_{loc}(R^2),\ \int_{R^2} |\hat{u}|^2 (|\theta|/|\xi|^2)\, d\xi < \infty\}.$$

Note that if v satisfies $\int_{R^2} |v|^2 (|\theta|/|\xi|^2)\, d\xi < \infty$, thus $v \in L^1_{loc}(R^2)$, since from Cauchy-Schwarz inequality

$$\int_K |v|\, d\xi < (\int_K |v|^2 (|\theta|/|\xi|^2)\, d\xi)^{1/2}.\ (\int_K (|\xi|^2/|\theta|)\, d\xi)^{1/2},$$

for all compact sets K in R^2. Furthermore since $|\theta|/|\xi|^2 \approx 1/|\xi|$ when $|\xi| \to \infty$,

(108)' $X_k \subseteq H^{-1/2}_{\mathrm{loc}}(R^2)$.

Now with hypothesis (106), we have $(-\Delta)^{-1/2}E^0_3$ and $(-\Delta)^{-1/2}H^0_3 \in H^{1/2}_k(R^2)$, from (92), (93), that is:

(109) E^0_3 and $H^0_3 \in X_k$.

We also have:

(109)' $\tilde{H}_k(R^2) \subseteq H^{-1/2}_{\mathrm{loc}}(\mathrm{div}, R^2)$, $H_k(R^2) \subseteq H^{-1/2}_{\mathrm{loc}}(\mathrm{curl}, R^2)$.

proof of (109)'. Let $u \in \tilde{H}_k(R^2)$, then $R_g R_g^* u \in H^{1/2}_k(R^2)^2$, and $R_c R_c^* u \in H^{-1/2}_k(R^2)^2$, thus: $u = R_g R_g^* u + R_c R_c^* u \in H^{-1/2}_{\mathrm{loc}}(R^2)^2$, thanks to lemma 2.
Furthermore $R_g^* u \in H^{1/2}_k(R^2)$ implies $\mathrm{div}\, u \in X_k$, and by (108)', $\mathrm{div}\, u \in H^{-1/2}_{\mathrm{loc}}(R^2)$. This prove the inclusions in (109)', thanks to operator S.

Then, like in Definition 4 chap. 3 (with (199), (200)), we define the "usual" surface operator (which depends on Z and k):

Definition 3. *The operator* C (*or* C) *defined by* (87), *or* (89), *or* (103) *is called the Calderon operator* (*or the surface admittance operator*) *for the half-space* (*and for* Z, k, *and the chosen sense of propagation*).

As a consequence of above properties, we have:

Proposition 6. *The Calderon operator C* (*resp.* C) *is an isomorphism in* $H_k(R^2)$ (*resp.* $SH_k(R^2)$). *Both commute with the semigroup* $(G(x_3))$, *satisfy* (88), *and*

(110) $\mathrm{Re}\,(CSE^0_T, E^0_T) \leq 0, \quad \forall E^0_T \in H_k(R^2)$.

This inequality is a consequence of the formula:

(111) $$\int_{R^2} n \wedge H^0_T . \bar{E}^0_T \, dx = \int_{R^2} \frac{i}{Z} \left[-\frac{\theta}{k} |\xi \wedge \hat{E}^0_T|^2 + \frac{k}{\theta} |\xi . \hat{E}^0_T|^2 \right] \frac{1}{\xi^2} \, d\xi,$$

which is easily deduced from (103), and the definition (11).

Remark 5. Similar to waveguides the electromagnetic field is decomposed into transverse-electric waves and transverse-magnetic waves, thanks to the Hodge decomposition; using the decomposition

(112) $$\left| \begin{array}{l} H_k(R^2) = H_{k,g}(R^2) \oplus H_{k,c}(R^2), \text{ with} \\ H_{k,g}(R^2) = \{u \in H_k(R^2), \ \mathrm{curl}\, u = 0\}, \ H_{k,c}(R^2) = \{u \in H_k(R^2), \ \mathrm{div}\, u = 0\}, \end{array} \right.$$

we have, with hypothesis (106):

$$(113)\quad \left| \begin{array}{l} E_T^0 \in H_{k,c}(R^2) \Leftrightarrow E_3^0 = 0 \Leftrightarrow H_T^0 \in H_{k,g}(R^2) \text{ for TE waves} \\ E_T^0 \in H_{k,g}(R^2) \Leftrightarrow H_3^0 = 0 \Leftrightarrow H_T^0 \in H_{k,c}(R^2) \text{ for TM waves,} \end{array} \right.$$

and thus the Calderon operator C is an isomorphism from $H_{k,c}(R^2)$ (resp. $H_{k,g}(R^2)$) onto $H_{k,g}(R^2)$ (resp. $H_{k,c}(R^2)$) given by

$$(114)\quad \left| \begin{array}{l} \text{i) for TE waves: } H_T^0 = C\, E_T^0 = \frac{-i}{kZ} S\, A\, E_T^0 \\ \text{ii) for TM waves: } H_T^0 = C\, E_T^0 = \frac{ik}{Z} S\, A^{-1}\, E_T^0. \end{array} \right.$$

Note that there are no transverse electromagnetic (TEM) waves since the whole space is simply connected.

⊗

Remark 6. *Calderon projectors for the half-space*
We can easily verify that the Calderon operator for the inferior half space C_i (or C_-, for the same medium, the same angular frequency) is given by

$$(115)\quad C_i = -C \text{ (or } C_i = -C\text{)}.$$

Then we consider the usual transmission problem (or with given electric and magnetic currents, see (86) chap.3): find (E,H) with locally finite energy in each half space (up to the boundary) so that

$$(116)\quad \left| \begin{array}{l} \text{i) curl } H + i\omega\varepsilon E = 0, \\ \text{ii) } -\text{curl } E + i\omega\mu H = 0 \quad \text{in } R_+^3 \text{ and } R_-^3, \\ \text{iii) } [n \wedge E]_\Gamma = M,\ -[n \wedge H]_\Gamma = J,\ \Gamma = R^2 \times \{0\},\ J \text{ and } M \text{ given in } \tilde{H}_k(R^2), \\ \text{iv) (E,H) is a wave propagating towards the positive (resp. negative) axis} \\ \qquad \text{in } R_+^3 \text{ (resp. } R_-^3\text{)}. \end{array} \right.$$

Thanks to the Calderon operators, this problem is reduced to finding the boundary values of (E,H) satisfying (116)iii) with:

$$(117)\quad n \wedge H|_{\Gamma_+} = C(n \wedge E|_{\Gamma_+}), \quad n \wedge H|_{\Gamma_-} = C(n \wedge E|_{\Gamma_-}).$$

Thus we get a linear system of equations which is very easy to solve explicitly, using (88) and (115). We can write the solution in the usual form

$$(118)\quad -\begin{pmatrix} n \wedge E|_{\Gamma_+} \\ -n \wedge H|_{\Gamma_+} \end{pmatrix} = P_+ \begin{pmatrix} M \\ J \end{pmatrix}, \quad \begin{pmatrix} n \wedge E|_{\Gamma_-} \\ -n \wedge H|_{\Gamma_-} \end{pmatrix} = P_- \begin{pmatrix} M \\ J \end{pmatrix},$$

with

$$(119)\quad P_+ = \frac{1}{2}(I - S), \quad P_- = \frac{1}{2}(I + S),$$

where S is given thanks to the Calderon operator $C = Z^{-1}C_k^+$ by:

$$(120) \qquad S = \begin{pmatrix} 0 & -Z^2C \\ C & 0 \end{pmatrix} = \begin{pmatrix} 0 & -ZC_k^+ \\ Z^{-1}C_k^+ & 0 \end{pmatrix}.$$

[Of course we can also use other forms, for instance:

$$(118)' \quad -\begin{pmatrix} n\wedge E|_{\Gamma_+} \\ n\wedge H|_{\Gamma_+} \end{pmatrix} = \tilde{P}_+ \begin{pmatrix} M \\ J \end{pmatrix}, \quad \begin{pmatrix} n\wedge E|_{\Gamma_-} \\ n\wedge H|_{\Gamma_-} \end{pmatrix} = \tilde{P}_- \begin{pmatrix} M \\ J \end{pmatrix}$$

with:

$$(119)' \qquad \tilde{P}_+ = \frac{1}{2}(I - \tilde{S}), \quad \tilde{P}_- = \frac{1}{2}(I + \tilde{S}), \text{ and } \tilde{S} = -S.]$$

The operators P_+ and P_- are the *Calderon projectors in* $H_k(R^2) \times H_k(R^2)$ for the half space (with ε, μ, ω). Of course we also have $S^2 = I$.

⊗

1.2.2. Scattering problems with two different media

As in section 1.1.3, we assume that the whole space is occupied by two different media on each side of the plane Γ ($x_3 = 0$), one with permittivity and permeability (ε_1,μ_1) in the domain R^3_+, the other with permittivity and permeability (ε_2,μ_2) in the domain R^3_-. We assume that there is an incident electromagnetic field (E_I,H_I) in R^3_+, at angular frequency ω, for instance produced by charges and currents with compact support in R^3_+. We have to find the reflected and transmitted (or refracted) electromagnetic fields, resp. (E_r,H_r) in R^3_+ and (E_t,H_t) in R^3_-.

Let (E,H) be the electromagnetic field defined by: $(E,H) = (E_r,H_r)$ in R^3_+ and (E_t,H_t) in R^3_-. Let (ε,μ) be (ε_1,μ_1) in R^3_+, (ε_2,μ_2) in R^3_-. We first assume that (ε_1,μ_1), (ε_2,μ_2) are positive real numbers (the two media are conservative). Then (E,H) has to satisfy (116), with M and J given by:

$$(121) \qquad M = n\wedge E_I|_\Gamma, \qquad J = -\, n\wedge H_I|_\Gamma.$$

Thanks to the Calderon operators in each half-space, C_1 and $C_2 = -C_{2i}$, we are led to find the boundary values of the field, which satisfy (116)iii), or also

$$(122) \qquad H_{T2} - H_{T1} = H_{TI}, \quad E_{T2} - E_{T1} = E_{TI}.$$

Applying the adjoint Riesz operators (see (96)') to (122), then using (101), we obtain two simple scalar systems. We define:

$$(123) \quad u_j^g = R_g^* E_{Tj},\ j = 1, 2, \quad u_I^g = R_g^* E_{TI}, \quad u_j^c = R_c^* E_{Tj},\ j = 1,2, \quad u_I^c = R_c^* E_{TI},$$

and thus the Fourier transformation of (101) gives, keeping (ε,μ) of the media:

$$(124)\quad \left|\begin{array}{l} \hat{R}\underset{g}{*}\hat{H}_{T1} = \frac{i}{\omega\mu_1}\theta_1\,\hat{u}_1^c,\quad \hat{R}\underset{g}{*}\hat{H}_{T2} = -\frac{i}{\omega\mu_2}\theta_2\,\hat{u}_2^c,\quad \hat{R}\underset{g}{*}\hat{H}_{TI} = -\frac{i}{\omega\mu_1}\theta_1\,\hat{u}_I^c \\ \hat{R}\underset{c}{*}\hat{H}_{T1} = i\omega\varepsilon_1\,\theta_1^{-1}\,\hat{u}_1^g,\quad \hat{R}\underset{c}{*}\hat{H}_{T2} = -i\omega\varepsilon_2\,\theta_2^{-1}\,\hat{u}_2^g,\quad \hat{R}\underset{c}{*}\hat{H}_{TI} = -i\omega\varepsilon_1\,\theta_1^{-1}\,\hat{u}_I^g. \end{array}\right.$$

We have to solve:

$$(125)_g\quad \left|\begin{array}{l} \varepsilon_2\,\theta_2^{-1}\,\hat{u}_2^g + \varepsilon_1\,\theta_1^{-1}\,\hat{u}_1^g = \varepsilon_1\,\theta_1^{-1}\,\hat{u}_I^g \\ \hat{u}_2^g - \hat{u}_1^g = \hat{u}_I^g, \end{array}\right.$$

and

$$(125)_c\quad \left|\begin{array}{l} \frac{1}{\mu_2}\theta_2\,\hat{u}_2^c + \frac{1}{\mu_1}\theta_1\,\hat{u}_1^c = \frac{1}{\mu_1}\theta_1\,\hat{u}_I^c \\ \hat{u}_2^c - \hat{u}_1^c = \hat{u}_I^c. \end{array}\right.$$

The solution of the system $(125)_c$ is given by:

$$(126)\quad \left|\begin{array}{ll} \hat{u}_2^c = \hat{T}_c\,\hat{u}_I^c, & \hat{u}_1^c = \hat{R}_c\,\hat{u}_I^c, \\ \hat{T}_c = \dfrac{2\mu_2\,\theta_1}{\mu_2\theta_1 + \mu_1\theta_2}, & \hat{R}_c = \dfrac{\mu_2\,\theta_1 - \mu_1\theta_2}{\mu_2\theta_1 + \mu_1\theta_2}. \end{array}\right.$$

Thus for an incident field such that:

(127) E_{IT} and $H_{IT} \in H_{k_1}(R^2)$,

we have $u_I^g \in H_{k_1}^{1/2}(R^2)$, and from Proposition 4, we get

(128) $u_1^c \in H_{k_1}^{1/2}(R^2)$ and $u_2^c \in H_{k_1}^{1/2}(R^2) \cap H_{k_2}^{1/2}(R^2)$ (and even $u_2^c \in H^{1/2}(R^2)$).

Then we define

(129) $\hat{v}_j = \theta_j^{-1}\hat{u}_j^g,\ j = 1, 2,\qquad \hat{v}_I = \theta_1^{-1}\hat{u}_I^g,$

so that $(125)_g$ becomes:

$$(130)\quad \left|\begin{array}{l} \text{i)}\ \varepsilon_2\,\hat{v}_2 + \varepsilon_1\,\hat{v}_1 = \varepsilon_1\,\hat{v}_I \\ \text{ii)}\ \theta_2\,\hat{v}_2 - \theta_1\,\hat{v}_1 = \theta_1\,\hat{v}_I. \end{array}\right.$$

The solution of the system (130) is given by

$$(131)\quad \left|\begin{array}{ll} \hat{v}_2 = \hat{T}\,\hat{v}_I, & \hat{v}_1 = \hat{R}\,\hat{v}_I, \\ \hat{T} = \dfrac{2\varepsilon_1\theta_1}{\varepsilon_1\theta_2 + \varepsilon_2\theta_1}, & \hat{R} = \dfrac{\varepsilon_1\theta_2 - \varepsilon_2\theta_1}{\varepsilon_1\theta_2 + \varepsilon_2\theta_1}. \end{array}\right.$$

With the same hypothesis (127), we have $v_I \in H^{1/2}_{k_1}(R^2)$ and from Proposition 4:

(132) $v_1 \in H^{1/2}_{k_1}(R^2)$ and $v_2 \in H^{1/2}_{k_1}(R^2) \cap H^{1/2}_{k_2}(R^2)$ (and even $v_2 \in H^{1/2}(R^2)$).

Thus we have:

(133) $u_1^g \in H^{-1/2}_{k_1}(R^2), \quad u_2^g \in H^{-1/2}_{k_2}(R^2),$

and

(134)
$$\left| \begin{array}{ll} \hat{u}_2^g = \hat{T}_g \, \hat{u}_I^g, & \hat{u}_1^g = \hat{R}_g \, \hat{u}_I^g, \\ \hat{T}_g = \dfrac{\theta_2}{\theta_1} \hat{T} = \dfrac{2\varepsilon_1\theta_2}{\varepsilon_1\theta_2 + \varepsilon_2\theta_1}, & \hat{R}_g = \hat{R} = \dfrac{\varepsilon_1\theta_2 - \varepsilon_2\theta_1}{\varepsilon_1\theta_2 + \varepsilon_2\theta_1}. \end{array} \right.$$

(126) and (134) are *Fresnel's formulas* for electric field. Note that

(135) $\hat{T}_g - \hat{R}_g = 1, \quad \hat{T}_c - \hat{R}_c = 1.$

Thus we get

Theorem 3. *With hypothesis* (127), *the scattering problem* (116) *of an incident wave* (E_I, H_I) *by a (conservative) medium in a half-space (with boundary* Γ*) has a unique solution* (E,H) *of locally finite energy up to the boundary (on each side of* Γ*), which is obtained thanks to two semigroups* $(G_1(x_3))$ $x_3 > 0$, *and* $(G_2(-x_3))$ $x_3 < 0$ *(see* (85) *and* (29)*) associated to the wavenumbers* k_1, k_2 *in each medium, and thanks to the transverse components of the field on* Γ*, so that*:

(136) $E_{Tj}|_\Gamma$ and $H_{Tj}|_\Gamma \in H_{k_j}(R^2)$, $j = 1,2$ *on each side of the plane* Γ.

Moreover the transverse components of the solution on Γ *split into:*

(137)
$$\left| \begin{array}{l} E_{Tj} = E^g_{Tj} + E^c_{Tj}, \quad H_{Tj} = H^g_{Tj} + H^c_{Tj} \\ E^g_{Tj} = R_g R_g^* E_{Tj}, \quad E^c_{Tj} = R_c R_c^* E_{Tj}, \quad H^g{}_{Tj} = R_g R_g^* H_{Tj}, \quad H^c{}_{Tj} = R_c R_c^* H_{Tj}, \end{array} \right.$$

with $(R_g^* E_{Tj}, R_g^* H_{Tj})$, $(R_c^* E_{Tj}, R_c^* H_{Tj})$ *given by* (123), (126) and (134).

Remark 7. The decomposition (137) of the transverse field also corresponds to the decomposition of the total field (E,H) into TE-waves and TM-waves.

When $E_T = E_T^c$ on each side of Γ, then div $E_{Tj} = 0$, thus $E_{3j} = 0$ on Γ and in R^3. Thus $H_{3j} = \dfrac{1}{i\omega\mu_j}$ curl E_{Tj} satisfies, with $\hat{u}_j^c$ given by (126) with respect to $\hat{u}_I^c$:

(138) $\hat{H}_{3j}|_\Gamma = \dfrac{1}{i\omega\mu_j} |\xi| \, \hat{u}_j^c, \quad$ thus $H_{3j}|_\Gamma \in X_{k_j}.$

When $E_T = E_T^g$ on each side of Γ, then curl $E_{Tj} = 0$, thus $H_{3j} = 0$ on Γ and in R^3. Thus $E_{3j} = -A_j^{-1}$ div E_{Tj} satisfies, with $\hat{u}_j^g$ given by (134) with respect to $\hat{u}_I^g$:

$$(139) \qquad \hat{E}_{3j}|_\Gamma = |\xi|\, \theta_j^{-1}\, \hat{u}_j^g, \quad \text{thus } E_{3j}|_\Gamma \in X_{k_j}.$$

⊗

We call $\hat{T}_g$, $\hat{T}_c$, and $\hat{R}_g$, $\hat{R}_c$ *the transmission and the reflection coefficients of the* electric field. We also define the *transmission and reflection coefficients for the magnetic field* $\hat{T}'_g$, $\hat{T}'_c$, and $\hat{R}'_g$, $\hat{R}'_c$ which satisfy (135) also: let

$$(123)' \qquad v_j^g = R_g^* H_{Tj},\ j = 1, 2, \quad v_I^g = R_g^* H_{TI}, \quad v_j^c = R_c^* H_{Tj},\ j = 1,2, \quad v_I^c = R_c^* H_{TI}.$$

Then we get *Fresnel's formulas* for magnetic field:

$$(140) \qquad \left| \begin{array}{l} \hat{v}_2^c = \hat{T}'_c \hat{v}_I^c, \quad \hat{v}_1^c = \hat{R}'_c \hat{v}_I^c, \quad \hat{v}_2^g = \hat{T}'_g \hat{v}_I^g, \quad \hat{v}_1^g = \hat{R}'_c \hat{v}_I^g, \quad \text{with:} \\ \hat{T}'_c = \dfrac{\varepsilon_2 \theta_1}{\varepsilon_1 \theta_2} \hat{T}_g, \quad \hat{T}'_g = \dfrac{\mu_1 \theta_2}{\mu_2 \theta_1} \hat{T}_c, \quad \hat{R}'_c = -\hat{R}_g, \quad \hat{R}'_g = -\hat{R}_c. \end{array} \right.$$

Remark 8. We can extend these results and Theorem 3 to the case where the second medium is dissipative (with complex ε, μ), and also where the two media are dissipative. The main point is to prove that the terms: $\varepsilon_1\theta_2 + \varepsilon_2\theta_1$ and $\mu_1\theta_2 + \mu_2\theta_1$ are never null. This is the case for instance when the first medium is conservative, and when Re ε_2, Re μ_2 (with Im ε_2, Im μ_2) are all positive numbers. Furthermore for a dissipative medium, the wavenumber k_2 has an imaginary part $k_2'' \neq 0$; This implies that the wave is exponentially decreasing from the boundary. We define the *skin depth* δ as the distance in which the wave is attenuated by $1/e$, that is using the notations of (45):

$$(141) \qquad |\hat{u}(\xi,\delta)|/|\hat{u}(\xi,0)| = \exp(-(\text{Re}\,\theta_2)\,\delta) < 1/e, \quad \forall \xi \in R^2.$$

Since Re $\theta_2 \geq$ Im $k_2 = k_2''$, we can take $\delta = 1/k_2''$. For a conductor with $\mu_2 = \mu_0$, we have $\varepsilon_2 = \varepsilon_2' + (i\sigma/\omega) = \varepsilon_2' + i\varepsilon_0\varepsilon_r''$, with $\sigma >> \omega\varepsilon_2'$, so that $k_2'' \approx \omega\sigma\mu_2/2$, thus the skin depth is $\delta = (2/\omega\sigma\mu_0)^{1/2} = (2/\omega^2\mu_0\varepsilon_0\varepsilon_r'')^{1/2} = (2/\varepsilon_r'')^{1/2}\lambda/2\pi$, with the wavelength λ.

⊗

Remark 9. In order to directly tackle the case of an incident plane wave, we have to use a more general framework (see (43) for instance). Here also we are reduced to the scalar case of section 1.1.3, by decomposition of the field into TE and TM waves. Thus we obtain the Snell-Descartes laws for electromagnetism.

⊗

Remark 10. Using the Calderon operator C (or *C*) defined for the half space allows us to tackle many other examples like in the scalar case, section 1.1.4. For instance we can tackle the case of an incident electromagnetic field through a (flat) perfectly conducting screen, thick or thin, with apertures.

The case of a "thin perfectly conducting screen", that is a surface Γ (a part of a plane) on which we take $n \wedge E = 0$ is reduced to the scalar case, since the electromagnetic field (E, H) also splits into TE and TM waves.

⊗

1.3. The slab

1.3.1. The scalar case with Helmholtz equation

First we consider the Helmholtz problem in a slab of thickness τ and with a real wavenumber k, for the Dirichlet condition:

$$(141)' \quad \left| \begin{array}{l} \text{i) } \Delta u + k^2 u = 0 \ \text{ in } \Omega_\tau \subseteq \mathbb{R}^3, \text{ with } 0 < x_3 < \tau, \\ \text{ii) } u(.,0) = u_0, \ \ u(.,\tau) = u_\tau, \end{array} \right.$$

with given u_0, u_τ. It is natural to solve (141) using a Fourier transformation with respect to $x = (x_1, x_2)$ (if u_0, u_τ are tempered distributions). With usual notations we get

$$(142) \quad \left| \begin{array}{l} \text{i) } \dfrac{d^2\hat{u}}{dx_3^2} + (k^2 - \xi^2)\,\hat{u} = 0, \qquad 0 < x_3 < \tau, \\ \text{ii) } \hat{u}(.,0) = \hat{u}_0, \quad \hat{u}(.,\tau) = \hat{u}_\tau. \end{array} \right.$$

We note $t = x_3$. Then (at least formally) the solution of (142) is given by

$$(143) \quad \left| \begin{array}{l} \text{i) } \ \hat{u}(\xi,t) = \hat{u}_0(\xi)\dfrac{\operatorname{sh}((\tau - t)\sqrt{\xi^2 - k^2})}{\operatorname{sh}(\tau\sqrt{\xi^2 - k^2})} + \hat{u}_\tau(\xi)\dfrac{\operatorname{sh}(t\sqrt{\xi^2 - k^2})}{\operatorname{sh}(\tau\sqrt{\xi^2 - k^2})} \text{ when } |\xi| > k, \\ \text{ii) } \ \hat{u}(\xi,t) = \hat{u}_0(\xi)\dfrac{\sin((\tau - t)\sqrt{k^2 - \xi^2})}{\sin(\tau\sqrt{k^2 - \xi^2})} + \hat{u}_\tau(\xi)\dfrac{\sin(t\sqrt{k^2 - \xi^2})}{\sin(\tau\sqrt{k^2 - \xi^2})} \text{ when } |\xi| < k. \end{array} \right.$$

or using $\theta(\xi)$ defined by (11)

$$(143)' \quad \hat{u}(.,t) = \hat{u}_0\,\frac{\operatorname{sh}(\tau - t)\theta}{\operatorname{sh}\tau\theta} + \hat{u}_\tau\,\frac{\operatorname{sh} t\theta}{\operatorname{sh}\tau\theta}.$$

Let χ_0 and χ_τ be functions defined by

$$(144) \qquad \chi_0(\xi,t) = \frac{\operatorname{sh}(\tau - t)\theta}{\operatorname{sh}\tau\theta}, \qquad \chi_\tau(\xi,t) = \frac{\operatorname{sh} t\theta}{\operatorname{sh}\tau\theta}.$$

When $|\xi| > k$, $\chi_0(\xi,t) \leq 1$, $\chi_\tau(\xi,t) \leq 1$. But when $|\xi| < k$, $\chi_0(\xi,t)$ and $\chi_\tau(\xi,t) \to \infty$ when $\tau\theta \to \pi n$, n integer.

Let λ be the wavelength, $\lambda = 2\pi/k$, and $n_\tau = \max n$ with n integer $n\pi \leq \tau k$, thus $n_\tau \leq 2\tau/\lambda$. We also define:

$$(145) \qquad k_n = \sqrt{k^2 - (n\pi/\tau)^2} = k\sqrt{1 - (n\lambda/2\tau)^2}, \qquad S^\tau = \bigcup_{0 < n \leq n_\tau} S_{k_n},$$

with S_{k_n} the sphere of radius k_n, the values α_n^2 with $\alpha_n = n\pi/\tau$, $n \in N$, $n > 0$ being the eigenvalues of $-d^2/dt^2$ on $(0,\tau)$, with Dirichlet condition.
Thus we have to eliminate the "irregular values" ξ on these spheres, i.e., $\xi \in S^\tau$.
Recall that the (generalized) eigenvectors (or eigenmodes of the slab) are:

$$v_{n,\xi}(x,t) = e^{i\xi.x} \sin(t\pi n/\tau), \text{ with } \xi^2 + (\pi n/\tau)^2 = k^2, \; n \in N, \; n > 0.$$

When $k^2 - \xi^2$ is equal to α_n^2, the solution of (142) must satisfy the condition

(146) $$\int_0^\tau (\hat{u}''v - \hat{u}v'')\, dt = [\hat{u}'v - \hat{u}v']_0^\tau = \hat{u}(0)v'(0) - \hat{u}(\tau)v'(\tau) = 0,$$

with $v(t) = \sin(\pi nt/\tau)$, and thus:

(147) $$(-1)^n \hat{u}_\tau - \hat{u}_0 = 0.$$

But since ξ is a continuous parameter, we don't have to impose condition (147) in this form. We have to give a sense to (143) only. Thus when $|\xi| < k$ we write:

(148) $$\hat{u}(\xi,t) = \hat{u}_0(\xi) \cos(t\sqrt{k^2-\xi^2}) - \frac{\hat{u}_0(\xi)\cos(\tau\sqrt{k^2-\xi^2}) - \hat{u}_\tau(\xi)}{\sin(\tau\sqrt{k^2-\xi^2})} \sin(t\sqrt{k^2-\xi^2}).$$

Thus we see that the hypotheses u_0, $u_\tau \in L^2(R^2)$, and:

(149) $$\int_{V_n} \frac{|(-1)^n \hat{u}_0(\xi) - \hat{u}_\tau(\xi)|^2}{\sin^2(\tau\sqrt{k^2-\xi^2})} d\xi < \infty, \text{ for all } n, \; 0 < n \leq n_\tau,$$

with V_n a neighbourhood of the sphere S_{k_n}, will imply that $u(.,t) \in L^2(R^2)$, $0 < t < \tau$.
We can also take space $H_k^{1/2}(R^2)$ instead of $L^2(R^2)$ with (149).
We note that:

i) if the slab is thin enough, that is if $\tau < \lambda/2$, problem (141) has a unique solution (for instance in $L^2(R^2 \times (0,\tau))$ when u_0 and u_τ are in $L^2(R^2)$);

ii) if the wavenumber k is complex, with $\operatorname{Im} k > 0$, we have the same conclusion.

Then we can get the values of the normal derivative of u at the boundary of the slab (at least formally) thanks to (143)'. Their Fourier transforms are in matrix form:

(150) $$\begin{pmatrix} \hat{u}'(.,\tau) \\ \hat{u}'(.,0) \end{pmatrix} = \hat{C}_\tau \begin{pmatrix} \hat{u}(.,\tau) \\ \hat{u}(.,0) \end{pmatrix},$$

with

(151) $$\hat{C}_\tau = \frac{\theta}{\operatorname{sh} \tau\theta} \begin{pmatrix} \operatorname{ch} \tau\theta & -1 \\ 1 & -\operatorname{ch} \tau\theta \end{pmatrix}.$$

We easily see that $\hat{C}_\tau$ satisfies:

(152) $(\hat{C}_\tau)^2 = \theta^2 I$,

and thus $\hat{C}_\tau$ has (at least formally) the inverse: $(\hat{C}_\tau)^{-1} = \theta^{-2}\hat{C}_\tau$.
The operator C_τ defined by (150) is the *Calderon operator of the slab* for k. For other developments on the slab, and for guided modes of the slab, see Petit [2].
It is interesting to consider the Cauchy problem (with $x_3 = t$), also (for instance in order to treat problems with many different slabs):

$$(153)\quad \begin{cases} \text{i) } \dfrac{\partial^2 u}{\partial t^2} + \Delta_x u + k^2 u = 0 \ \text{ in } \Omega_\tau \subseteq R^3, \text{ with } 0 < t < \tau, \\ \text{ii) } u(.,0) = u_0, \ \dfrac{\partial u}{\partial t}(.,0) = u_1, \end{cases}$$

with u_0, u_1 given (at least temperate distributions).
Using as above a Fourier transformation, we obtain the solution in the form:

$$(154)\quad \begin{pmatrix} \hat{u}(t) \\ \hat{u}'(t) \end{pmatrix} = \begin{pmatrix} \text{ch}\,\theta t & \theta^{-1}\,\text{sh}\,\theta t \\ \theta\,\text{sh}\,\theta t & \text{ch}\,\theta t \end{pmatrix} \begin{pmatrix} \hat{u}_0 \\ \hat{u}_1 \end{pmatrix} = G_\theta(t) \begin{pmatrix} \hat{u}_0 \\ \hat{u}_1 \end{pmatrix}.$$

For every given θ, $(G_\theta(t))_{t>0}$ is a semigroup, which splits into:

$$(155)\quad G_\theta(t) = e^{\theta t} P_+^\theta + e^{-\theta t} P_-^\theta,$$

with

$$P_+^\theta = \frac{1}{2}(I + S_\theta), \quad P_-^\theta = \frac{1}{2}(I - S_\theta), \quad S_\theta = \begin{pmatrix} 0 & \theta^{-1} \\ \theta & 0 \end{pmatrix},$$

that is, P_+^θ and P_-^θ are the Calderon projectors, since:

$$S_\theta^2 = I, \text{ thus } (P_+^\theta)^2 = P_+^\theta,\ (P_-^\theta)^2 = P_-^\theta,\ P_+^\theta P_-^\theta = P_+^\theta P_-^\theta = 0.$$

By inverse Fourier transform, $(G_\theta(t))$ does not correspond to a semigroup in the usual L^2 functional spaces, in other words: the Cauchy problem (153) is an ill-posed problem. But (153) is helpful because:
i) we know that there is at most uniqueness of the solution of (153),
ii) very often in physics, we work with given functions (or measures) u_0, u_1 whose Fourier transforms have bounded support, and then the inverse Fourier transform of (154) is well-defined. (Note that when the supports of the Fourier transforms of u_0 and u_1 are contained in the ball of radius k, the semigroup $(G_\theta(t))$ is that of the wave evolution.)
Thus if we have a heterogeneous slab composed of n homogeneous slabs with thickness τ_j and wavenumber k_j , the values of the wave at the end of the heterogeneous slab is obtained in function of the values at the beginning by:

$$(156)\quad \begin{pmatrix} \hat{u}(\tau) \\ \hat{u}'(\tau) \end{pmatrix} = G_{\theta_1}(\tau_1)\ldots G_{\theta_n}(\tau_n) \begin{pmatrix} \hat{u}_0 \\ \hat{u}_1 \end{pmatrix}, \text{ with } \tau = \sum \tau_j.$$

Thus we easily obtain the solution of the scattering problem of a plane wave by a multilayer slab.

1.3.2. The slab with Maxwell equations

Now we consider the Maxwell problem in a slab (with real ε,μ):

$$(157)\qquad \left|\begin{array}{l} \text{i) } \operatorname{curl} H + i\omega\varepsilon E = 0, \quad \text{ii) } -\operatorname{curl} E + i\omega\mu H = 0, \text{ in } \Omega_\tau,\ 0 < x_3 < \tau, \\ \text{iii) } E_T(.,0) = E_T^0, \quad E_T(.,\tau) = E_T^\tau \text{ with given } E_T^0, E_T^\tau. \end{array}\right.$$

Since the components of (E,H) satisfy the Helmholtz equation, we obtain the transverse electric field E_T from the above section.

Thus we have to find the other components of (E,H) from E_T. This is very easy from the usual decomposition into TE and TM waves, as follows:

i) for TE waves, due to (30)ii) chap. 4, we have

$$E_3^0 = 0, \quad \partial_3 E_T + i\omega\mu\, SH_T = 0,$$

and thus using the Calderon operator C_τ given by (151):

$$(158)\qquad H_T = -\frac{i}{\omega\mu} C_\tau S E_T;$$

ii) for TM waves, from (30)ii) chap. 4, we have

$$H_3^0 = 0, \quad \partial_3 H_T - i\omega\varepsilon\, SE_T = 0,$$

and thus:

$$(159)\qquad H_T = i\omega\varepsilon\, C_\tau^{-1} S E_T.$$

These two results are given by:

$$(160)\qquad H_T = \frac{i}{Z}\left[-\frac{1}{k} C_\tau P_g + k C_\tau^{-1} P_c\right] S E_T,$$

which is relation (103) with C_τ in place of A, where E_T and H_T denote the couples:

$$E_T = (E_T(.,0), E_T(.,\tau)), \qquad H_T = (H_T(.,0), H_T(.,\tau)).$$

Furthermore the components E_3, H_3 are obtained thanks to (91), by:

$$(161)\qquad H_3 = \frac{1}{i\omega\mu} \operatorname{curl}_T E_T, \quad E_3 = -C_\tau^{-1} \operatorname{div}_T E_T.$$

There are no new difficulties with respect to the scalar case: we only have to eliminate the "irregular values" of ξ corresponding to the generalized eigenvectors of the slab with Dirichlet condition, thanks to a weight, like in (149).

In order to treat multilayer problem, we consider the Cauchy problem:

$$(162)\quad \left| \begin{array}{l} \text{i) } \operatorname{curl} H + i\omega\varepsilon E = 0, \quad \text{ii) } -\operatorname{curl} E + i\omega\mu H = 0, \quad \text{in } \Omega_\tau, \quad 0 < x_3 < \tau \\ \text{iii) } E_T(.,0) = E_T^0, \;\; H_T(.,0) = H_T^0, \text{ with given } E_T^0, H_T^0. \end{array} \right.$$

This is also (at least formally) easily solved due to the decomposition into TE and TM waves, using the projectors P_g and P_c (see (100)) or the Riesz operators R_g and R_c. We first write Maxwell equations in the form (30), (31), then (34) chap. 4, i.e.,

$$(163)\quad \partial_3 H_T = \frac{i}{\omega\mu}(k^2 + \operatorname{grad}\operatorname{div})SE_T, \quad \partial_3 E_T = -\frac{i}{\omega\varepsilon}(k^2 + \operatorname{grad}\operatorname{div})SH_T.$$

Then applying P_g and P_c to (163), we have:

$$(164)\quad \partial_3 P_g H_T = \frac{i}{\omega\mu}(k^2 + \Delta)P_g SE_T, \quad \partial_3 P_g E_T = -\frac{i}{\omega\varepsilon}(k^2 + \Delta)P_g SH_T,$$

$$(165)\quad \partial_3 P_c H_T = i\omega\varepsilon\, P_c SE_T, \quad \partial_3 P_c E_T = -\, i\omega\mu\, P_c SH_T,$$

Then using the commutation relations:

$$(166)\quad SP_c = P_g S, \quad P_c S = SP_g,$$

and using relations (154) with $u = P_g H_T$, then $P_c E_T$, with (164) and (165), gives us:

$$(167)\quad \begin{pmatrix} P_c \hat{E}_T(\tau) \\ P_g \hat{H}_T(\tau) \end{pmatrix} = \begin{pmatrix} \operatorname{ch}(\theta\tau) & -i\omega\mu\theta^{-1}\operatorname{sh}(\theta\tau)S \\ -\frac{i}{\omega\mu}\theta\operatorname{sh}(\theta\tau)S & \operatorname{ch}(\theta\tau) \end{pmatrix} \begin{pmatrix} P_c \hat{E}_T^0 \\ P_g \hat{H}_T^0 \end{pmatrix}$$

$$(168)\quad \begin{pmatrix} P_g \hat{E}_T(\tau) \\ P_c \hat{H}_T(\tau) \end{pmatrix} = \begin{pmatrix} \operatorname{ch}(\theta\tau) & \frac{i}{\omega\varepsilon}\theta\operatorname{sh}(\theta\tau)S \\ i\omega\varepsilon\,\theta^{-1}\operatorname{sh}(\theta\tau)S & \operatorname{ch}(\theta\tau) \end{pmatrix} \begin{pmatrix} P_g \hat{E}_T^0 \\ P_c \hat{H}_T^0 \end{pmatrix}.$$

We write (167),(168) globally in the form:

$$(169)\quad \begin{pmatrix} \hat{E}_T(\tau) \\ \hat{H}_T(\tau) \end{pmatrix} = G_\theta(\tau) \begin{pmatrix} \hat{E}_T(0) \\ \hat{H}_T(0) \end{pmatrix}.$$

Then we have the same conclusion as in the Helmholtz case. The formulas (167), (168) (or (169)) allow us to treat very easily the multilayer case like in (156). There are many applications of these formulas to scattering (Salisbury screen, Jaumann screen).

2. PERIODIC GEOMETRY, 2-D GRATINGS

Periodic structures appear in numerous applications, especially in scattering by gratings (see for instance Petit [1]). As usual we first develop the scalar case, then electromagnetism.

2.1. Periodic geometry with Helmholtz equation
We first define the notion of a quasiperiodic function:

Definition 4. *Let* $K = (K_1, K_2)$ *with* K_1, K_2 *real numbers and* $L = (L_1, L_2)$ *with* L_1 *and* L_2 *positive real numbers, be given. Let* u *be a function (or a distribution) in* R^2 *which satisfies "Bloch condition":*

(170) $u(x_1 + L_1, x_2 + L_2) = \exp i(K_1L_1 + K_2L_2)\, u(x_1, x_2),\quad x = (x_1, x_2) \in R^2,$
i.e.:
(170)' $u(x + L) = \exp(iK.L)\, u(x).$

Then u *is called a quasiperiodic function (or distribution) of type* K-L, *or a* K-L *(quasiperiodic) function (or distribution).*

When $K = (0,0)$ or when $K.L \in 2\pi N$, we obtain the L-periodic functions (distributions). Conversely given a quasiperiodic function u we obtain a periodic function v by:

(171) $v(x) = \exp(-iK.x)u(x).$

This allows us to obtain the main properties of the quasiperiodic functions from those of the periodic functions. Let P denote the "*elementary cell*", that is the rectangle $P = [0, L_1] \times [0, L_2]$ with identified opposite faces; P is identified with the 2-D torus T^2. Let $|P|$ denote its area, $|P| = L_1L_2$. We also define:

Definition 5. *The space of* K-L *"regular"* (C^∞) *quasiperiodic functions in* R^2 *(with the topology of uniform convergence of functions and all their derivatives) is denoted by* $D_{K,L}(R^2)$ *(or* $D_{K,L}(P)$*); then its dual space, i.e., the space of quasiperiodic distributions is denoted by* $D'_{K,L}(R^2)$, *and the space of quasiperiodic functions in* R^2 *which are square integrable on the period is denoted by* $L^2_{K,L}(R^2)$ *or simply* $L^2_{K,L}$. *We also denote by* $H^s_{K,L}(R^2)$ or $H^s_{K,L}$ *the Sobolev space of order s (real) of quasiperiodic functions (distributions), i.e., for s = m an integer:*

(172) $$H^m_{K,L}(R^2) = \{u \in L^2_{K,L}(R^2),\ \frac{\partial^J u}{\partial x^J} \in L^2_{K,L}(R^2),\ \forall J = (J_1, J_2) \in N^2,\ \text{with } |J| \overset{\text{def}}{=} J_1 + J_2 \le m\}.$$

An orthonormal basis (up to a constant) of $L^2_{K,L}$ is given by the functions:

(173) $$\phi_J(x) = \exp(iK.x)\exp 2\pi i\left(\frac{J_1}{L_1}x_1 + \frac{J_2}{L_2}x_2\right),\quad \text{with } J = (J_1, J_2) \in Z^2.$$

We also denote:

(173)' $\phi_J(x) = \exp i\xi_J x$, with $\xi_J = K + \xi_J^0$ and $\xi_J^0 = 2\pi\left(\frac{J_1}{L_1}, \frac{J_2}{L_2}\right)$, $J = (J_1, J_2) \in Z^2$.

By expansion of quasiperiodic distributions on this basis, we have:

(172)' $H^s_{K,L}(R^2) = \{v \in D'_{K,L}(R^2),\ v = \sum_J v_J \phi_J,\ \sum_J |K + \xi_J^0|^{2s} |v_J|^2 < \infty\}$, $s \in R$.

We consider the standard problem in the half-space: find u satisfying

(174)
$$\begin{cases} \text{i) } \Delta u + k^2 u = 0 \text{ in } R^3_+, \\ \text{ii) } u|_\Gamma = u_0 \text{ given on the boundary } \Gamma \text{ of } R^3_+, \end{cases}$$

with k a given wavenumber ($k \in R^+$) and u_0 a K-L quasiperiodic function.
The Helmholtz equation and the domain being invariant by (transverse) translations, we look for K-L quasiperiodic solutions. Thus we develop the unknown function u (and the given function u_0) on the above basis:

(175) $$u(x, x_3) = \sum_J u_J(x_3)\, \phi_J(x), \qquad u_0 = \sum_J u_{0J}\, \phi_J,$$

and problem (174) is reduced to finding $(u_J(x_3))$ satisfying (with $\xi_J = K + \xi_J^0$):

(176)
$$\begin{cases} \text{i) } \dfrac{\partial^2 u_J}{\partial x_3^2} + (k^2 - \xi_J^2)\, u_J = 0, \ x_3 \geq 0, \\ \text{ii) } u_J(0) = u_{0J}. \end{cases}$$

When $|\xi_J| \geq k$, there is only one bounded solution of (176):

(177) $$u_J(x_3) = u_{0J} \exp(-\theta_J x_3), \qquad \text{with } \theta_J = \theta_J^+ = (\xi_J^2 - k^2)^{1/2}.$$

When $|\xi_J| < k$, the general solution of (176) is (with $\alpha_J + \beta_J = u_{0J}$):

(178) $$u_J(x_3) = \alpha_J \exp(-i\theta_J^- x_3) + \beta_J \exp(i\theta_J^- x_3), \quad \text{with } \theta_J^- = (k^2 - \xi_J^2)^{1/2}.$$

Thus we cannot determine in a unique way the solution of (174), which is an ill-posed problem. In order to obtain a well-posed problem, we have to impose a sense to the wave propagation, like for waveguides, such as

(179) u *is a wave propagating towards the positive axis* x_3, i.e., *the propagating part of* u (178) *is a superposition of waves*

$$u_J(x_3)\phi_J(x) = e^{i\mu_J x_3}\phi_J(x) \quad \text{with } \mu_J > 0.$$

Then taking:

(180) $\quad \theta_J = (\xi_J^2 - k^2)^{1/2}$ when $\xi_J^2 \geq k^2$, $\quad \theta_J = -\, i(k^2 - \xi_J^2)^{1/2}$ when $\xi_J^2 \leq k^2$,

we obtain a unique solution of (174) with (178) by the *Rayleigh series*:

(181) $$u(x,x_3) = \sum u_{0J}\, e^{-\theta_J x_3} \phi_J(x) \quad \text{with } J \in Z^2.$$

We have a situation quite similar to that of the waveguide, replacing the boundary condition on the waveguide by quasiperiodic conditions. Thus we have due to the Fourier expansion, or to the semigroup method

Theorem 4. *Problem* (174) *with a given quasiperiodic function* u_0 *in* $L^2_{K,L}(R^2)$ *and with condition* (179) *has a unique solution* u *in* $C^0([0,\infty), L^2_{K,L}(R^2))$, *given by* (181) *with a contraction holomorphic semigroup* $(G(x_3))$, $x_3 > 0$, *in* $L^2_{K,L}(R^2)$.

The infinitesimal generator A of the semigroup in $L^2_{K,L}(R^2)$ is defined by

(182) $$Au_0 = \frac{\partial u}{\partial x_3}(.,0), \quad D(A) = \{u_0 \in L^2_{K,L}(R^2),\ u \in C^1([0,\infty), L^2_{K,L}(R^2))\},$$

and using the spectral decomposition of the quasiperiodic Laplacian:

(182)' $$Au_0 = -\sum_J \theta_J\, u_{0J} \phi_J, \text{ with } u_0 = \sum_J u_{0J}\phi_J, \quad D(A) = H^1_{K,L}(R^2).$$

Definition 6. *The infinitesimal generator* A *of the semigroup* $(G(x_3))$ *is called the Calderon operator for the quasiperiodic functions and the wavenumber* k.

This operator has the same properties as that of the waveguides; A is a normal operator, a square root of the selfadjoint operator $-(\Delta + k^2 I)$ in $L^2_{K,L}(R^2)$. Let:

(183) $$\left| \begin{array}{l} U_k^- = \{J \in Z^2 \text{ so that } |\xi_J| < k\},\ U_k^+ = \{J \in Z^2 \text{ so that } |\xi_J| > k\}, \\ U_k^0 = \{J \in Z^2 \text{ so that } |\xi_J| = k\}, \end{array} \right.$$

and let:

(183)' H_k^-, H_k^+, H_k^0 the Hilbert spaces generated by ϕ_J, $J \in U_k^-$, U_k^+, U_k^0 (resp.).

Thus: $\ker A = H_k^0$, and Im A is its orthogonal space.

Furthermore A satisfies the following inequalities for all v in D(A):

(184) $$\left| \begin{array}{l} \mathrm{Re}\,(Av,v) = -\,|P| \sum \theta_J^+ |v_J|^2 \leq 0, \text{ (the sum being on J in } U_k^+,\ \theta_J^+ = \theta_J), \\ \mathrm{Im}\,(Av,v) = |P| \sum \theta_J^- |v_J|^2 \geq 0, \text{ (the sum being on J in } U_k^-,\ \theta_J^- = |\theta_J|\,). \end{array} \right.$$

According to (183), we have the decomposition of $L^2_{K,L}(R^2)$ and of A into:

(185) $L^2_{K,L}(R^2) = \mathcal{H}^+_k \oplus H^-_k$, with $\mathcal{H}^+_k = H^0_k \oplus H^+_k$, and $A = -A^+ + iA^-$,

with A^+ and A^- positive selfadjoint operators in the orthogonal spaces $\mathcal{H}^+_k$ and H^-_k, where H^-_k is of finite dimension and corresponds to the propagating modes.

Remark 11. The $L^2_{K,L}(R^2)$ framework is not "optimal" for all applications we have in view. More generally Theorem 4 is true in the quasiperiodic Sobolev spaces $H^s_{K,L}(R^2)$, with real s, and in this framework the Calderon operator A is continuous from $H^{s+1}_{K,L}(R^2)$ into $H^s_{K,L}(R^2)$.

The case $s = -1/2$ corresponds to waves u with locally finite energy.

⊗

Example 1. *Scattering of a plane wave by a periodic soft wall.*
Let Γ be a (regular) periodic surface in R^3 (with period $L = (L_1, L_2)$) which is a modelling of a periodic soft wall (or a 2D-grating) with free space on one side. We assume that Γ is contained in the half-space $x_3 < 0$. Let u_I be an incident plane wave in the free space Ω given by

(186) $$u_I(X) = u_I(x, x_3) = u_{I0}\, e^{ik\alpha.X} = u_{I0}\, e^{ik(\alpha_T.x + \alpha_3.x_3)}, \quad \alpha \in S^2,\ \alpha_3 < 0.$$

Therefore u_I satisfies the K-L quasiperiodic condition with $K = k\alpha_T$ (thus $|K| < k$) and $u_I(.,0) = u_{I0}\phi_{(0,0)}$ (with (173)). We have to find the wave u_r reflected by this structure. This is a K-L quasiperiodic function satisfying the following on the plane Γ_0 ($x_3 = 0$):

(187) $$\frac{\partial u_r}{\partial n} - Au_r = 0 \quad \text{on } \Gamma_0.$$

The Calderon operator A reduces the problem in the infinite domain Ω to the bounded domain Ω_0 with disjoint boundaries Γ_0 and Γ and L-periodic with respect to x. Let $u = u_I + u_r$ be the total wave. The transmission conditions on Γ_0 are:

(187)' $$\frac{\partial u}{\partial x_3} = \frac{\partial u_r}{\partial x_3} + \frac{\partial u_I}{\partial x_3}, \qquad u = u_r + u_I \ \text{on } \Gamma_0.$$

Thus u has to satisfy on the boundary Γ_0:

(187)" $$\frac{\partial u}{\partial x_3} - Au = f_I, \qquad \text{with } f_I = \frac{\partial u_I}{\partial x_3} - Au_I.$$

Therefore we have to find the total wave u, which is a K-L quasiperiodic function in Ω_0 with finite energy (in Ω_0) satisfying

(188) $$\left|\begin{array}{l} \text{i) } \Delta u + k^2 u = 0 \text{ in } \Omega_0, \\ \text{ii) } u|_\Gamma = 0, \text{ and u satisfies (187)" on } \Gamma_0. \end{array}\right.$$

We can write this problem in a variational framework. Let V be defined by:

(188)' $V = \{u \in H^1(\Omega_0),\ u|_\Gamma = 0,\ u(.,x_3)$ is a K-L quasiperiodic function in $\Omega_0\ \forall x_3\}$.

Then let a(u,v) be the sesquilinear form on V:

(189) $$a(u,v) = \int_{\Omega_0} (\text{grad}\, u.\text{grad}\, \bar{v} - k^2 u\, \bar{v})\, dx - \int_{\Gamma_0} Au.\bar{v}\, d\Gamma_0.$$

The problem (188) amounts to finding u in V satisfying (for all v in V)

(190) $$a(u,v) = \int_{\Gamma_0} f_I \bar{v}\, d\Gamma_0.$$

Since the inclusion of V in $L^2(\Omega_0)$ is compact, this problem depends on the Fredholm alternative. But we prove below (with fairly general hypotheses) that there is at most one solution, which implies its existence. ⊗

Remark 12. We can treat more general scattering problems, for instance with a "hard" wall (with the Neumann condition) or a wall coated with some medium which may be heterogeneous (but periodic). If there are dissipative inclusions in this medium, that is domains where the wavenumber is not real, then these problems depend on the Fredholm alternative, and we can prove that there is at most one solution, thus we have a unique solution also.

⊗

Example 2. *Scattering by a slab with periodic inclusions, or by a grid*
We consider a system of periodic inclusions in a slab placed in free space, and with an incident plane wave arriving from one side. The inclusions can be hard, or soft, or made of a dissipative material. We can consider the case of a grid with thickness d, or the scattering by a dielectric medium also. We have to solve the standard problem in a (fictitious) slab:

(191)
$$\left|\begin{array}{l} \text{i) } \Delta u + \tilde{k}^2 u = 0 \quad \text{in } \Omega_\tau = R^2 \times (-\tau, 0), \text{ or } P \times (-\tau,0), \\ \text{ii) } \dfrac{\partial u}{\partial t} - Au = f_I \quad \text{on } \Gamma_0,\ \ f_I \text{ given by (187)''},\ \ t = x_3 \\ \text{iii) } \dfrac{\partial u}{\partial t} + A_1 u = 0 \quad \text{on } \Gamma_1 = \Gamma_{-\tau}, \\ \text{iv) } u \text{ is a K-L quasiperiodic function with finite energy in } P \times (-\tau, 0), \end{array}\right.$$

where Γ_0, Γ_1 are the boundaries of Ω_τ; A_1 is the Calderon operator for the domain with $x_3 < -\tau$ where the wavenumber is k_1 (here k_1 is a real number).
In the simple dielectric grating case, we have

(192) $$\bar{\Omega}_\tau = \bar{\Omega}_0 \cup \bar{\Omega}_1, \quad \text{with } \tilde{k} = k \text{ in } \Omega_0,\ \ \tilde{k} = k_1 \text{ in } \Omega_1.$$

Like in Example 1, we can write this problem in a variational framework. We define (with usual notations) the space on the periodic domain Ω_τ:

(193) $V = H^1_{K,L}(\Omega_\tau) = H^1_{K,L}(P\times(-\tau,0))$,

and the sesquilinear form a(u,v) on V:

(194) $a(u,v) = \int_{\Omega_\tau} (\text{grad}\, u.\text{grad}\, \bar{v} - k^2 u.\bar{v})\, dx - \int_{\Gamma_0} Au.\bar{v}\, d\Gamma_0 - \int_{\Gamma_1} A_1 u.\bar{v}\, d\Gamma_1.$

(If we have hard inclusions occupying a domain O in Ω_0, we replace V by the subspace of functions which are null in O). Then problem (191) is written:

(195) $a(u,v) = \int_{\Gamma_0} f_I\, \bar{v}\, d\Gamma_0, \quad \forall v \in V.$

Since the natural inclusion of V in $L^2(\Omega_\tau)$ is compact, this problem depends on the Fredholm alternative also. Under supplementary hypotheses, we can prove that there is at most one solution, which implies its existence.

⊗

Proposition 7. *We assume that one of the following hypotheses is satisfied:*
i) *the grating is thin: the depth* d *(*Γ *is contained in a slab* Ω_d *of depth* d) *is small with respect to the wavelength* λ ($d < \lambda/4$);
ii) *the grating is given by a Lipschitz periodic function* $x_3 = g(x)$, *except on vertical parts*;
iii) the grating is coated by a (heterogeneous) medium with a dissipative part.
Then the scattering problems (188) *and* (191) *have a unique solution u with locally finite energy up to the boundary.*

PROOF. Thanks to the variational method, we can prove that problems (190) and (195) with $f_I = 0$ implies $u = 0$. Here we only give the proof in two cases.

PROOF FOR (188), with hypothesis i). i) Taking first the real part of (190) with $v = u$ gives, using (184),

(196) $\int_{\Omega_0} (|\text{grad}\, u|^2 - k^2|u|^2)\, dx \leq \text{Re}\, a(u,u) = 0.$

Now assuming that Γ is above a plane Γ_{-d} we can extend u by 0 up to Γ_{-d}, and this extension $\tilde{u}$ is in $H^1_{K,L}(\Omega_d)$, $\Omega_d = P\times(-d,0)$, thus:

(197) $\int_{\Omega_0} |\text{grad}\, u|^2\, dx \geq \kappa_1^2 = \inf \int_{\Omega_0} |\text{grad}\, w|^2\, dx,$

the inf being taken on $\{w \in H^1_{K,L}(\Omega_d),\ w = 0 \text{ on } \Gamma_{-d},\ \|w\| = 1\}$, and it is reached at the solution in this space, of

$$(198)\qquad \begin{cases} \text{i) } \Delta w + \kappa_1^2 w = 0 \ \text{ in } \Omega_d, \\ \text{ii) } w|_{\Gamma_{-d}} = 0, \quad \dfrac{\partial w}{\partial n}\Big|_{\Gamma_0} = 0. \end{cases}$$

Thus we have $\kappa_1 \geq \pi/2d$. Since $k = 2\pi/\lambda$, when d and λ satisfy:

(199) $\kappa_0^2 - k^2 \geq (\pi/2d)^2 - (2\pi/\lambda)^2 > 0$, i.e., $d < \lambda/4$,

we have from (196):

(200) $$\int_{\Omega_0} |u|^2 dx \geq \int_{\Omega_0} |\text{grad } u|^2 dx \geq \int_{\Omega_0} \kappa_1^2 |u|^2 dx,$$

which implies u = 0 from (199). We note that this is a version of the Poincaré inequality.

PROOF for (188) with hypothesis ii). This is adapted from Cadilhac's proof (see Petit [1]). First taking the imaginary part of (190) with v = u gives

(201) $$u_0 \overset{\text{def}}{=} u|_{\Gamma_0} \in H_k^+$$

(see (185), u is not propagating). Then applying the Green formula to u and v (another solution of the Helmholtz equation satisfying (201) also), we have

(202) $$\int_{\Omega_0} (\Delta u \bar{v} - u \Delta \bar{v}) dx = -\int_{\Gamma} \frac{\partial u}{\partial n} \bar{v} \, d\Gamma + \int_{\Gamma_0} \left(\frac{\partial u}{\partial t} \bar{v} - u \frac{\partial \bar{v}}{\partial t}\right) = 0.$$

But the integral on Γ_0 is zero thanks to (201). Thus

(203) $$\int_{\Gamma} \frac{\partial u}{\partial n} \bar{v} \, d\Gamma = 0.$$

Then taking $v = \frac{\partial u}{\partial t}$ in (203), and since $\frac{\partial u}{\partial t} = e_3.\text{grad } u = n_3 \frac{\partial u}{\partial n}$, with $n_3 \geq 0$ we obtain $\frac{\partial u}{\partial n} = 0$ at last on a part of Γ, and thus (with u = 0 on Γ) we have u = 0 on Ω_0.

⊗

Now given a regular periodic surface Γ in R^3, we can define Sobolev spaces of quasiperiodic functions $H^s_{K,L}(\Gamma)$ for real s, like in (172) or (172)'.
We take the situations of examples 1 and 2, but with inhomogeneous boundary conditions on obstacles. Let Ω be a regular connected periodic domain with one of the following hypotheses, according to example 1 or 2:

i) $\Omega \subseteq R^2 \times (-\tau, +\infty)$, with boundary $\Gamma \subseteq R^2 \times (-\tau, 0)$ (grating with one side)

ii) $\Omega' = R^3 \backslash \bar{\Omega} \subseteq R^2 \times (-\tau, 0)$; Ω' may be connected (grid) or not (inclusions).

Proposition 8. *We assume that one of the hypotheses of Proposition 7 is satisfied. Then the problem: find a* K-L *quasiperiodic function* u *with locally finite energy (up to the boundary) satisfying*

(204)
$$\left| \begin{array}{l} \text{i) } \Delta u + k^2 u = 0 \text{ in } \Omega, \textit{ with real } k, \\ \text{ii) } u|_{\Gamma} = u_0 \textit{ given in } H^{1/2}_{K,L}(\Gamma) \\ \text{iii) } u \textit{ is a wave propagating to } x_3 > 0 \text{ in } R^3_+, \text{ (also to } x_3 < 0 \text{ in } R^3_-), \end{array} \right.$$

has a unique solution. Furthermore we have:

(205) $\frac{\partial u}{\partial n}|_\Gamma \in H^{-1/2}_{K,L}(\Gamma).$

This is a consequence of Proposition 7 and Theorem 4. Then Proposition 8 implies that the mapping $A = A_\Gamma$ defined by:

(206) $u_0 \rightarrow \frac{\partial u}{\partial n}|_\Gamma$, with u the solution of (204),

is continuous from $H^{1/2}_{K,L}(\Gamma)$ into $H^{-1/2}_{K,L}(\Gamma)$.

The operator A of Definition 6 is a particular case of this mapping when Ω is the half space. If Γ separate R^3 into two connected components Ω_1 and Ω_2 (the grating case), we have to consider two operators $A_{\Gamma+}$ and $A_{\Gamma-}$.

Definition 7. *The operator* A_Γ *defined by* (206) *is called the Calderon operator of the 2-D periodic surface* Γ.

This operator has the following main property:

(207) $\text{Im}\,(A_\Gamma v, v) \geq 0, \quad \forall v \in H^{1/2}_{K,L}(\Gamma).$

The proof is an easy consequence of the property (184) for the half space, thanks to the Green formula. Furthermore (in the case of a grating for instance) if we define the sesquilinear form

(208) $a_\Gamma(u_0, v_0) = -\frac{i}{2}[(A_\Gamma u_0, v_0) - (u_0, A_\Gamma v_0)] \quad \text{on } H^{1/2}_{K,L}(\Gamma),$

which satisfies:

$$a_\Gamma(u_0, u_0) = \text{Im}\,(A_\Gamma u_0, u_0) \geq 0,$$

we obtain due to the Green formula that

(209) $a_\Gamma(u_0, v_0) = |P| \sum i\theta_J u_J . \bar{v}_J, \quad \text{with } J \in U^-_k \text{ only},$

u_J being the J *component of the trace of* u *on the plane* Γ_0 in the basis (173). Taking the conjugate of v_0 in (209) leads to the *reciprocity relations* (see Cadilhac in Petit [1]). But note that if u is K-L quasiperiodic, its conjugate is not K-L quasiperiodic, but it is quasiperiodic for $(-K)$.

Remark 13. In the case of 1-D gratings, that is if we assume that Γ is a periodic line in R^2, as in Petit [1], all this is true under very little change: in particular we have to change double Fourier series with subscripts J in Z^2 to simple Fourier series with subscripts n in Z. This corresponds to problems in R^3 with cylindrical gratings (with axis x_2 for instance) with given data which are independent of x_2. But if the data depend on x_2, without x_2 periodicity, we have to make a Fourier transform with respect to x_2 and we obtain a situation which is in-between the plane case and the periodic case.

We have to change the mathematical framework: we define spaces of K_1-L_1 quasiperiodic functions with respect to x_1, using an orthonormal basis (up to a constant) (ϕ_n) of K_1-L_1 quasiperiodic square integrable functions

$$\phi_n(x_1) = \exp i\xi_1 x_1 , \quad \text{with } \xi_1 = K_1 + 2\pi\,(n/L_1), \quad n = J_1 \in Z,$$

with a Fourier transformation with respect to x_2 and then we have to substitute the following space for the Sobolev space (172)' with $s = 1/2$:

$$\text{(210)}\quad H^{1/2}_{k,K_1,L_1}(R^2) = \{u = u(x_1,x_2) = \sum_{n \in Z} u_n(x_2)\,\phi_n(x_1),\ \sum_n \int_R |\theta_n(\xi)|\ |\hat{u}_n(\xi)|^2 d\xi < \infty\}$$

where $\theta_n = (\xi^2 + \xi_n^2 - k^2)^{1/2}$, $\xi_n = K_1 + 2\pi\,(n/L_1)$, $n \in Z$, and where $\hat{u}_n(\xi)$ is the Fourier transform of $u_n(x_2)$. We have a space in-between $H^{1/2}_k(R^2)$ (see (37)) and $H^{1/2}_{K,L}(R^2)$ (see (172)'), which has the $H^{1/2}_{loc}(R^2)$ regularity.

⊗

2.2. An integral method for gratings

Similar to bounded obstacle we can solve the scattering problems with a periodic obstacle from an integral method. Here we use the space $D'_{K,L}(P \times R)$ of distributions which are K-L quasiperiodic with respect to x. We will assume (with notations (173) and (183)) that K, L, k satisfy:

$$\text{(211)}\quad \left|\begin{array}{l} |\xi_J| \neq k,\ \forall J \in Z^2, \text{ i.e., } U^0_k = \varnothing, \text{ or also:} \\ (\Delta_x + k^2)u = 0 \text{ in P, (with } u \in L^2_{K,L}(R^2)) \text{ implies } u = 0. \end{array}\right.$$

First we have to define an "elementary outgoing solution" of the Helmholtz equation in $D'_{K,L}(P \times R)$. Let $\delta^{(p)}(x)$ be the Dirac (quasi) periodic distribution. We look for $\Phi^{(p)}(x,x_3)$ the "outgoing" quasiperiodic solution of:

$$\text{(212)}\quad \left|\begin{array}{l} \text{i) } (\Delta + k^2)\,\Phi^{(p)}(x,x_3) = -\,\delta^{(p)}(x)\,\delta(x_3), \\ \text{ii) } \Phi^{(p)} \text{ is a wave propagating from } x_3 = 0 \text{ towards } x_3 > 0 \text{ (resp. } x_3 < 0). \end{array}\right.$$

We develop $\Phi^{(p)}$ on the basis (173):

$$\text{(213)}\quad \Phi^{(p)}(x,x_3) = |P|^{-1} \sum \Phi^{(p)}_J(x_3)\phi_J(x), \quad J \in Z^2,\ |P| = L_1 L_2;$$

applying (212)i) to ϕ_j, we see that $\Phi^{(p)}_J(x_3)$ must satisfy, with $t = x_3$:

$$\text{(214)}\quad \frac{\partial^2 \Phi^{(p)}_J}{\partial t^2} - \theta_J^2\,\Phi^{(p)}_J = -\,\delta(t), \quad \text{with } \theta_J^2 = \xi_J^2 - k^2,$$

and thus it has jumps across $t = x_3 = 0$:

(215) $[\Phi_J^{(p)}]_0 = 0, \qquad [\frac{\partial \Phi_J^{(p)}}{\partial t}]_0 = -1,$

with the outgoing condition. Since $\theta_J \neq 0$, $\Phi_J^{(p)}$ is given by:

(216) $\Phi_J^{(p)}(x_3) = \frac{1}{2\theta_J} \exp(-\theta_J |x_3|),$

with θ_J given by (180). Thus $\Phi^{(p)}$ is given by its (double) Fourier series:

(217) $\Phi^{(p)}(x,x_3) = |P|^{-1} \sum \frac{1}{2\theta_J} \exp(-\theta_J |x_3|)\, \phi_J(x).$

We can split $\Phi^{(p)}$ into two parts with J in U_k^+ and J in U_k^-:

(218) $\Phi^{(p)} = \Phi_-^{(p)} + \Phi_+^{(p)},$

the first term corresponding to the propagating modes, and the second to the diffusion (or evanescent modes). We easily see that:

(219) $\Phi^{(p)} \in L^2_{loc}(P \times R)$, i.e., $\Phi^{(p)}|_{P \times [a,b]} \in L^2(P \times [a,b]), \quad \forall\, a, b \in R,$

since $\Phi_-^{(p)}$ is a finite sum of terms in $L^2_{loc}(P \times R)$, thus $\Phi_-^{(p)}$ is in $L^2_{loc}(P \times R)$, and $\Phi_+^{(p)}$ is in $L^2(P \times R)$, since $\int_{P \times R} |\Phi_+^{(p)}|^2 \, dx dx_3 = |P|^{-1} \sum 2\,(2\theta_J)^{-3} < \infty$, the sum being on $J \in U_k^+$.

$\Phi^{(p)}$ is the *elementary outgoing solution* of Helmholtz equation in $D'_{K,L}(P \times R)$.

Remark 14. We can see that the two notions of outgoing waves mentioned in chaps. 3 and 5 agree: if we compare $\Phi^{(p)}$ with the usual elementary outgoing solution Φ (see (8) chap. 3), from (29) and (214) we have:

(220) $\Phi_J^{(p)}(x_3) = \hat{\Phi}(\xi_J, x_3)$, and thus $\Phi^{(p)}(x,x_3) = |P|^{-1} \sum \hat{\Phi}(\xi_J, x_3)\, \phi_J(x)$,

with $\hat{\Phi}$ the Fourier transform of Φ. ⊗

Remark 15. Let χ be a smooth function (C^∞) with $\chi(x,x_3) = 1$ in a ball B_r of R^3, and $\chi(x,x_3) = 0$ in the complementary of the ball B_{2r}, with $4r < \min(L_1, L_2)$.
Then $\Phi\chi$ is a distribution with compact support in R^3, and we can define its corresponding periodic distribution by translations $(\Phi\chi)^{(p)} = \sum \tau_J \Phi\chi$. We have defined a "*parametrix*" which is equal to $\exp(-iK.x)\Phi^{(p)}$ up to a regular function. Thus $\Phi^{(p)}$ has a $1/r$ singularity at the origin in $P \times R$ and it is C^∞ outside. ⊗

Proposition 9. *Let* Γ *be a regular (Lipschitz)* 2-D *periodic surface in* R^3 *which is contained in a (fictitious) slab of finite thickness* d, $R^2 \times (-d,0)$. *With hypothesis* (211), *the problem: find* u *a* K-L *quasiperiodic function with locally finite energy up to* Γ, *satisfying*

$$(221) \quad \left| \begin{array}{l} \text{i) } \Delta u + k^2 u = -f \text{ in } R^3 \quad \text{with } f = \rho' \delta_\Gamma + \operatorname{div}(\rho n \delta_\Gamma) \\ \text{with given } (\rho, \rho') \in H^{1/2}_{K,L}(\Gamma) \times H^{-1/2}_{K,L}(\Gamma) \\ \text{ii) u } \textit{satisfies the "outgoing wave condition", i.e., the propagation condition} \\ (179) \textit{ towards } x_3 > 0 \text{ or } < 0 \textit{ out of the slab,} \end{array} \right.$$

has a unique solution u *which is given by the Kirchhoff formula*

$$(222) \qquad u = \Phi^{(p)} \underset{X}{*} f = \Phi^{(p)} \underset{X}{*} (\rho' \delta_\Gamma + \operatorname{div}(\rho n \delta_\Gamma)),$$

and the jumps of u and its normal derivative across Γ *are given by*

$$(223) \qquad [u]_\Gamma = \rho, \quad [\frac{\partial u}{\partial n}]_\Gamma = \rho'.$$

This proposition is similar to Proposition 3 chap. 3 for a bounded surface Γ (which is the boundary of a bounded domain). We have to note that the convolution product in (231) is that of $D'_{K,L}(P \times R)$, that is with respect to x (for quasiperiodic distributions) and to x_3; this makes sense since ρ and ρ' have compact supports with respect to x_3.
Then we can define single and double layer potentials like in (67) chap. 3 by

$$(224) \qquad L^{(p)}\rho' = \Phi^{(p)} \underset{X}{*} (\rho'\delta_\Gamma), \quad P^{(p)}\rho = \Phi^{(p)} \underset{X}{*} \operatorname{div}(\rho n \delta_\Gamma),$$

with the "usual" properties (see (69) chap. 3). We also define the Calderon projectors $P_i^{(p)}$, $P_e^{(p)}$ (or $P_-^{(p)}$, $P_+^{(p)}$ for gratings), like in (65) chap. 3 in:

$$(225) \qquad Y^{(p)}_{K,L} = H^{1/2}_{K,L}(\Gamma) \times H^{-1/2}_{K,L}(\Gamma).$$

They will satisfy (72) chap. 3 with an operator $S = S^{(p)}$ so that $S^2 = I$, given by the matrix (71) chap. 3 with four integral operators $L^{(p)}$, $K^{(p)}$, $J^{(p)}$ and $R^{(p)}$ which are obtained thanks to $\Phi^{(p)}$ by Fourier expansions. For their particular expressions in 1-D gratings, we refer to Maystre-Vincent and Petit in Petit [1].
This allows us to solve scattering problems with gratings like in chap. 3. Here we do not develop this topic. We only point out that the difficulties are quite similar to those of the bounded obstacles, in particular with respect to convergence of the Rayleigh series: for gratings we know that the solution of the usual scattering problem has a Rayleigh expansion (181), which is convergent in all (reasonable) senses, above 0 (and below $-d$ also if we substitute + for $-$ in (181)). For the questions of convergence of these expansions, see Cadihlac-Petit [1] (quite similar to those of chap. 3). ⊗

Remark 16. *Singular cases in periodic geometry.* We often assume the wavenumber k, K and period L are such that (211) is satisfied, that is the kernel of the operator A is reduced to 0. Here we study what happens if it is not true. First we consider the Neumann problem in a half-space: find u with locally finite energy (up to the boundary) satisfying

$$(226)\quad \left|\begin{array}{l} \text{i) } \Delta u + k^2 u = 0 \text{ in } R^3_+, \\ \text{ii) } \dfrac{\partial u}{\partial n}(.,0) = u_1, \\ \text{iii) the propagation condition (178),} \end{array}\right.$$

with u_1 given in ker A. It is an ill-posed problem, since the solution of (226) is a priori given by $u(.,x_3) = x_3 u_1 + C$, C constant, and is not bounded !
We have to eliminate this case, and thus to take:

$$(227)\quad u_1 \in (\ker A)^{\perp} = \operatorname{Im} A, \quad \text{i.e., } \int_P u_1 \bar{\phi}_J \, dx = 0, \quad \forall J \in U^0_k.$$

This condition written for the half-space allows us to obtain well-posed problems in singular cases. First in problems with given jumps (ρ,ρ') of u and its normal derivative (see (221)) on $\Gamma = P \times \{0\}$, ρ' must also satisfy (227), and then the solution u of (221) is determined up to an element of ker A. As a consequence we have to change the definition of the Green function (212), since the Dirac (quasi)periodic distribution is not orthogonal to ker A.
Instead of (212)i) we have:

$$(228)\quad (\Delta + k^2)\Phi^{(p)}(x,x_3) = F, \quad \text{with } F = -\delta^{(p)}(x)\,\delta(x_3) + |P|^{-1} \sum_{J \in U^0_k} \phi_J(0)\,\phi_J(x)\,\delta(x_3).$$

Then $\Phi^{(p)}_J$ must satisfy (224) for J out of U^0_k, and:

$$(229)\quad [\Phi^{(p)}_J]_0 = 0, \quad [\frac{\partial \Phi^{(p)}_J}{\partial t}]_0 = 0, \text{ for J in } U^0_k,\ t = x_3.$$

We can take $\Phi^{(p)}_J = 0$ for J in U^0_k, so that $\Phi^{(p)}$, $\dfrac{\partial \Phi^{(p)}}{\partial t}$ are orthogonal to ker A.

Now if we go back to the jump problem (221) we have:

i) in the simple case where $\Gamma = P \times \{0\}$, we have a unique solution in the space(s) orthogonal to ker A, when the given data satisfy:

$$(230)\quad (\rho,\rho') \in H^{1/2}_{K,L} \times H^{-1/2}_{K,L} \quad \text{with } \int \rho \bar{\phi}_J \, dx = 0,\ \int \rho' \bar{\phi}_J \, dx = 0,\ \forall J \in U^0_k.$$

ii) in the general case of a grating Γ, we can prove (using the Green formula successively in domains above and below the grating with u and ϕ_J then with $x_3\phi_J$), that the orthogonality of ρ and ρ' to ker A must be replaced by the following conditions:

(231) $\quad \int_\Gamma (\rho'\bar{\Phi}_J - \rho \frac{\partial \bar{\Phi}_J}{\partial n})\, d\Gamma = 0, \qquad \int_\Gamma [(\rho'\bar{\Phi}_J - \rho \frac{\partial \bar{\Phi}_J}{\partial n}) x_3 - \rho n_3 \bar{\Phi}_J]\, d\Gamma = 0,$

and then the boundary values of u and its normal derivative on each side of the grating will also satisfy these conditions (231).

Thus the Calderon projectors $P_+^{(p)}$ and $P_-^{(p)}$ act in the subspace $V_{K,L}$ defined by:

$$V_{K,L} = \{ (\rho,\rho') \in H^{1/2}_{K,L}(\Gamma) \times H^{-1/2}_{K,L}(\Gamma), \text{ satisfying (231)} \}.$$

⊗

2.3. Periodic geometry with Maxwell equations

First we consider like in section 1.2.1 (or in chap. 4 section 2) the following problem: find the electromagnetic field (E, H) in free space at angular frequency ω satisfying

(232) $\quad \begin{cases} \text{i) } \operatorname{curl} H + i\omega\varepsilon E = 0 \\ \text{ii) } -\operatorname{curl} E + i\omega\mu H = 0 \quad \text{in the half-space } R^3_+, \\ \text{iii) } E_T(.,0) = E^0_T \quad (\text{or } n \wedge E = n \wedge E^0_T) \end{cases}$

with $E^0{}_T$ a given K-L quasiperiodic transverse electric field.
Similar to the Helmholtz case, using Fourier series, we see that (232) is an ill-posed problem (in K-L quasiperiodic spaces), and it becomes (generally) well-posed thanks to the usual physical assumption:

(233) $\quad$ (E, H) *is a wave propagating towards the positive* x_3 axis.

Since each component of E and H satisfies the Helmholtz equation, they will be determined by the semigroup $(G(x_3))$ of section 2.1 from their boundary values on the boundary Γ ($x_3 = 0$) at least formally by:

(234) $\quad E_T(x_3) = G(x_3)E^0_T,\ E_3(x_3) = G(x_3)E^0_3,\ H_T(x_3) = G(x_3)H^0_T,\ H_3(x_3) = G(x_3)H^0_3,$

and thus we only have to find E^0_3, H^0_3 and H^0_T from E^0_T.
At first we specify the framework according to the usual given data.

2.3.1. Mathematical framework

A *"regular" assumption* is:

(235) $\quad E^0_T \in H^1_{KL}(P)^2.$

Since $(G(x_3))$ is also a continuous semigroup in this space, this implies: $E_T(x_3)$ is in $H^1_{KL}(P)^2$, for all $x_3 > 0$. Then (91) (for instance) and (92) imply:

(236) $\quad H^0_3 \in L^2_{KL}(P),\ H_3(x_3) \in L^2_{KL}(P),\ E^0_3 \in H^1_{KL}(P), \text{ and } E_3(x_3) \in H^1_{KL}(P),$

thus (thanks to (30) chap.4): $H_T(x_3)$ and $H^0_T \in L^2_{KL}(P)^2$.
We see that hypothesis (235) implies that the regularity of the electric field $E(x_3)$ is H^1, whereas that of the magnetic field $H(x_3)$ is L^2.

A *less regular assumption* is:

(237) $E^0_T \in H^{-1/2}_{KL}(\text{curl}, P)$.

This space corresponds to the "usual" regularity (see $H_k(R^2)$ (104)). Using (double) Fourier series, it is defined by:

(238) $$H^{-1/2}_{KL}(\text{curl}, P) = \{u = \sum u_J\phi_J\, ,\ \sum \frac{1}{|\xi_J|}(|u_J|^2 + |\xi_J \wedge u_J|^2) < \infty\},$$

whereas: $H^1_{KL}(P)^2 = \{u = \sum u_J\phi_J,\ \sum (|\xi_J . u_J|^2 + |\xi_J \wedge u_J|^2) < \infty\}$.

Then (237) implies for the other components:

(239) $$H^0_T \in H^{-1/2}_{KL}(\text{curl}, P),\quad E^0_3 \in H^{-1/2}_{KL}(P),\quad H^0_3 \in H^{-1/2}_{KL}(P),$$

and $E_T(x_3)$, $E_3(x_3)$, $H_T(x_3)$, $H_3(x_3)$, will be in these spaces for all $x_3 > 0$.

With assumption (237), *the electromagnetic field* (E,H) *solution of* (232) *with* (233) *is of locally finite energy up to the boundary.*
More generally the hypothesis (for any real s):

(240) $E^0_T \in H^s_{KL}(\text{curl}, P)$

implies $H^0_T \in H^s_{KL}(\text{curl}, P)$; then E and H have the same regularity.

2.3.2. The Calderon operator

We first assume that the operator A (see (182)) is *invertible, i.e.,* (211) *is satisfied*. Then we have the usual relations between the transverse components of E and H given by (86) with $A = A(p)$. Using Fourier series decomposition, we have

(241) $$H_{TJ} = \frac{-i}{Zk\theta_J}[\xi_J(\xi_J \wedge E_{TJ}) + k^2 SE_{TJ}].$$

Definition 8. *The mapping* $E^0{}_T \to H^0{}_T$ (*resp.* $SE^0{}_T \to SH^0{}_T$) *defined by* (86) (*with invertible* A *given by* (182)'), *or in Fourier series by* (241), *is called the Calderon operator* (*or admittance operator*) *C or* $\mathcal{C}^{(p)}$ (*resp.* C *or* $C^{(p)}$) *for the half-space in* K-L *quasiperiodic spaces* (*at angular frequency* ω).

Proposition 10. *The Calderon operator C* (*resp.* C) *is a continuous operator in*: $H^{-1/2}_{K,L}(\text{curl}, P)$ (resp. $H^{-1/2}_{K,L}(\text{div}, P)$) *and in* $H^s_{K,L}(\text{curl}, P)$ (resp. $H^s_{K,L}(\text{div}, P)$), $s \in R$.
Moreover it satisfies:

(242) $\mathcal{C}^2 = -Z^{-2} I,\quad C^2 = -Z^{-2} I,\quad \text{and}\quad \mathcal{C} = -SCS,\quad C = -S\mathcal{C}S,$

and also:

(243) $\mathrm{Re}\,(CSE_T^0, E_T^0) = \mathrm{Re}\,(\mathcal{SC}E_T^0, E_T^0) \leq 0,$
with:

(244) $$(CSE_T^0, E_T^0) = \sum SH_{TJ}.\bar{E}_{TJ} = \sum_J \frac{i}{Z|\xi_J|^2}\left[-\frac{\theta_J}{k}|\xi_J \wedge E_{TJ}|^2 + \frac{k}{\theta_J}|\xi_J.E_{TJ}|^2\right]$$

giving only a finite sum ($J \in U_k^-$) ***in the real part*** (243).

Furthermore from (241), we see that the Calderon operator $\mathcal{C}$ (or C) is naturally decomposed by Fourier series into $\mathcal{C} = \Sigma\, \mathcal{C}_J$ (resp. $C = \Sigma\, C_J$) with (see (89)):

(245) $$\mathcal{C}_J = \frac{i}{Zk\theta_J}\begin{pmatrix} \xi_{J1}^2 - k^2 & \xi_{J1}\xi_{J2} \\ \xi_{J2}\xi_{J1} & \xi_{J2}^2 - k^2 \end{pmatrix} S, \quad C_J = -\frac{i}{Zk\theta_J}\begin{pmatrix} \xi_{J1}\xi_{J2} & \xi_{J2}^2 - k^2 \\ k^2 - \xi_{J1}^2 & -\xi_{J1}\xi_{J2} \end{pmatrix}$$

and E_3^0, H_3^0 are given by (92), or with the Fourier series:

(246) $$E_{3J}^0 = i\,\theta_J^{-1}\,\xi_J.E_{TJ}^0, \quad H_{3J}^0 = \frac{1}{kZ}\xi_J \wedge E_{TJ}^0,$$

and thus we can write formulas (94), (94)' between the div and curl:

(247) $$\begin{pmatrix} \xi_J \cdot H_{TJ}^0 \\ \xi_J \wedge H_{TJ}^0 \end{pmatrix} = -\frac{i}{Z}\begin{pmatrix} 0 & k^{-1}\theta_J \\ k\theta_J^{-1} & 0 \end{pmatrix}\begin{pmatrix} \xi_J \cdot E_{TJ}^0 \\ \xi_J \wedge E_{TJ}^0 \end{pmatrix}.$$

Thus we obtain another expression of the Calderon operator (see (103)), corresponding to the Hodge decomposition of the transversal fields, or also of the total electromagnetic field into its TE (transverse electric, with $E_3 = 0$) and TM (transverse magnetic, with $H_3 = 0$) components, which is given with the Fourier series decomposition by:

(248) $$H_{TJ} = (\xi_J.H_{TJ})\frac{\xi_J}{|\xi_J|^2} + (\xi_J \wedge H_{TJ})\frac{S\xi_J}{|\xi_J|^2} = \frac{-i}{Z|\xi_J|^2}\left[\frac{\theta_J}{k}(\xi_J \wedge E_{TJ})\,\xi_J + \frac{k}{\theta_J}(\xi_J.E_{TJ})\,S\xi_J\right]$$

thanks to the relation $S\xi_J.u = \xi_J \wedge u$, for all vectors u. This gives (244) directly.
The Hodge decomposition of the space (237) into:

(249) $$\left|\begin{array}{l} H_{KL}^{-1/2}(\mathrm{curl}, P) = H_{KL,g}(P) \oplus H_{KL,c}(P) \text{ (like in (112),} \\ H_{KL,g}(P) \text{ (resp. } H_{KL,c}(P)) = \{u \in H_{KL}^{-1/2}(\mathrm{curl}, P),\ \mathrm{curl}\, u = 0,\ (\text{resp. } \mathrm{div}\, u = 0)\} \end{array}\right.$$

gives the TE and TM decomposition of (E,H) according to (113) and the decomposition of the Calderon operator $\mathcal{C}$ according to (114), given (from Fourier decomposition) by

(250) $$H_{TJ}^0 = \frac{i}{kZ}\theta_J\, SE_{TJ}^0 \text{ for TE waves}, \quad H_{TJ}^0 = -\frac{ik}{Z}\theta_J^{-1}\, SE_{TJ}^0 \text{ for TM waves}.$$

Now we assume that the operator A *is not invertible*.
With notations (183), (185), we assume:
i) $E^0_{TJ} = \text{grad}\, \phi_J$, with $\phi_J \in H^0_k$ (thus $A\phi_J = 0$, and $H^0_3 = 0$).
This implies $E_{TJ}(x_3) = E^0_{TJ}$, and $\partial_3 E^0_3 = -\,\text{div}\, E_{TJ} = -\,\Delta\phi_J = k^2\,\phi_J$, whose solution is the (unphysical) field $E_3(x_3) = k^2\phi_J\, x_3 + \text{constant}$.
ii) $E^0_{TJ} = \overrightarrow{\text{curl}}\, \phi_J$, with $\phi_J \in H^0_k$, thus $\text{div}\, E^0_{TJ} = 0$, giving: E_3 does not depend on x_3 and E^0_3 is not determined by E^0_{TJ}. (E^0_3 is of the form: $E^0_3 = \sum C_J\, \phi_J$ with $\phi_J \in H^0_k$ and C_J constant). Furthermore we have $E_{TJ}(x_3) = E^0_{TJ}$, and using (30) chap. 4, we have $H_T = \frac{i}{\omega\mu}\, S\, \text{grad}\, E_3 = -\frac{i}{\omega\mu}\, \overrightarrow{\text{curl}}\, E_3$ which is not determined by E^0_{TJ}.
Thus in the two cases, we have an ill-posed problem. To eliminate these cases we assume that:

(251) $\qquad E^0_T$ is orthogonal to $\ker A \times \ker A$.

With this assumption, we can develop the theory as in the regular case; for instance we can define the Calderon operator C in the space (and it is an isomorphism in this space):

(252) $\qquad H^{-1/2}_{KL}(\text{curl}, P) \cap (\ker A)^{\perp} \times (\ker A)^{\perp}$.

Remark 17. In these spaces of KL-quasiperiodic functions, we easily verify (for instance thanks to a Fourier expansion) that:

(253) $\qquad u \in L^2_{KL}(P)$, $\text{curl}\, u = 0$, $\text{div}\, u = 0$ implies $u = 0$,

except in the trivial case where $K.L/2\pi$ is an integer (then any constant vector u is a solution). This means that the cohomology space $H_{KL}(P)$ is reduced to $\{0\}$ (thus there are no TEM waves), except for periodic functions where $\dim H_{KL}(P) = 2$.

⊗

2.3.3. Some scattering problems
Thanks to sections 2.3.1, 2.3.2, we can treat the scattering of an incident plane wave (E_I, H_I) in free space by a P-periodic structure, like in Examples 1, 2.
From the geometrical point of view, we can tackle two types of scattering problems with a P-periodic structure:
i) *one-sided problems* with a perfectly conducting grating, with a connected boundary Γ so that the domain of the scattering problem is on one side of this grating. The grating can be coated with some dielectric medium or not.
ii) *two-sided problems* in one of the following cases: dielectric grating (in a domain which contains a half space), periodic dielectric medium contained in a slab, periodic dielectric (or perfectly conducting) inclusions or grid in a slab of finite thickness.

From the above theory, we reduce these scattering problems in an infinite domain to problems in a finite domain with quasiperiodic conditions and with one or two Calderon operators at the boundary.

Now *from the mathematical point of view*, for the solution of these problems we have the following cases:

i) We are reduced to a problem in a bounded domain Ω occupied by a lossy medium, that is, with a permittivity ε and a permeability μ, with $\varepsilon'' = \text{Im}\,\varepsilon > 0$, $\mu'' = \text{Im}\,\mu > 0$ (more generally the medium can be anisotropic, inhomogeneous). In this case we can apply a variational method in the usual space of electromagnetic fields with finite energy in Ω, with a coercive sesquilinear form on a subspace V of the natural Hilbert space $H_{KL}(\text{curl},\Omega)$ of K-L quasiperiodic electrical fields with finite energy. Thus from the Lax-Milgram lemma, there is a unique solution in Ω, and then for the scattering problem.

ii) We are reduced to a problem in a bounded domain Ω occupied in part only by a lossy medium, so that we don't have a coercive sesquilinear form any more. But we can prove that the problem depends on the Fredholm alternative; this is generally due to the fact that the natural injection of V in $L^2_{KL}(\Omega)^3$ is compact, but it is not true for $H_{KL}(\text{curl},\Omega)$ and thus we have to substitute the Sobolev space of quasiperiodic fields with free divergence in $H^1(\Omega)^3$ (see chap.2 section 11.2), to $H_{KL}(\text{curl},\Omega)$. Then we have one of the following situations:
either we can prove that the problem has at most one solution: for instance this is the case where there is really a part with a lossy medium in Ω and a part with free space (or a conservative medium), and very likely in each of the cases of the Proposition 7). This implies (thanks to the Fredholm alternative) that there exists a unique solution in Ω and then in R^3 to the scattering problem;
or we cannot have uniqueness; the periodic structure has eigenmodes.

From a *numerical point of view*, we often have to couple a spectral method (in order to take account of the Calderon operator) with a finite element method for instance to describe an inhomogeneous periodic medium, like in waveguides. But there are numerous other methods for gratings (see Petit [1]). We can also replace the boundary condition with the Calderon operator by "absorbing conditions", or we can use an integral method also (see Nedelec-Starling [1]) briefly described below.

Example 3. "*One sided problem*". We assume that we have a lossy dielectric (with permittivity ε_1 and permeability μ_1) in a P-periodic domain Ω contained in $R^2 \times (-\tau, 0)$. This domain Ω is bounded on one side Γ by a perfectly conducting medium, on the other side Γ_0 by free space. We will assume (in order to simplify) that Γ_0 is the plane $R^2 \times \{0\}$. We have to find the electromagnetic field (E, H) in Ω due to the incident field (E_I, H_I) with angular frequency ω:

$$(254)\quad \left| \begin{array}{l} u_I(x,x_3) = u_{I0} e^{ik(\alpha_T.x + \alpha_3 x_3)},\ u_{I0} \in \mathbb{C}^3,\ \alpha = (\alpha_T, \alpha_3) \in S^2,\ \alpha_3 < 0\,, \\ u_I = E_I \text{ or } H_I,\ E_{I0} = -\frac{k}{\omega\varepsilon}\alpha \wedge H_{I0},\ H_{I0} = \frac{k}{\omega\mu}\alpha \wedge E_{I0},\ \alpha.E_{I0} = 0,\ \alpha.H_{I0} = 0\,. \end{array} \right.$$

Now we write the boundary conditions on Γ_0, like in the Helmholtz case, see (187) Example 1: the reflected field (E_r, H_r) satisfies thanks to C the Calderon operator of the half-space:

$$(255)\qquad n \wedge H_r = C\,(n \wedge E_r) \text{ on } \Gamma_0.$$

Then the electromagnetic field (E,H) in Ω satisfies the continuity relations on Γ_0:

$$(256)\qquad n \wedge H = n \wedge H_r + n \wedge H_I,\quad n \wedge E = n \wedge E_r + n \wedge E_I.$$

Thus (E,H) must satisfy the boundary conditions on Γ_0:

$$(257)\qquad n \wedge H - C(n \wedge E) = f_I,\quad \text{with } f_I = n \wedge H_I - C(n \wedge E_I) \text{ on } \Gamma_0.$$

Thus we have to find (E,H) in Ω satisfying:

$$(258)\quad \left| \begin{array}{l} \text{i) curl } H + i\omega\varepsilon_1 E = 0,\quad \text{ii) } -\text{curl } E + i\omega\mu_1 H = 0 \ \text{ in } \Omega\,, \\ \text{iii) } n \wedge E|_\Gamma = 0, \text{ and (257) on } \Gamma_0. \\ \text{iv) } E \text{ and } H \in H_{KL}(\text{curl}, \Omega), \end{array} \right.$$

i.e., (E,H) is a K-L quasiperiodic field with finite energy; *we save notation Ω for the elementary periodic domain*. We can write this problem in a variational framework, with the sesquilinear form:

$$(259)\quad a(E,\tilde{E}) = \int_\Omega \left(-\frac{1}{i\omega\mu_1}\text{curl } E.\ \text{curl } \bar{\tilde{E}} - i\omega\varepsilon_1 E.\ \bar{\tilde{E}}\right) dx - \int_{\Gamma_0} C\,(n \wedge E|_{\Gamma_0}).\ \bar{\tilde{E}}\, d\Gamma_0$$

in the Hilbert space (with the natural norm):

$$(260)\qquad V = V_{KL} = \{\tilde{E} \in H_{KL}(\text{curl}, \Omega),\ \text{with } \tilde{E}|_\Gamma = 0\}.$$

Thus problem (258) is equivalent to finding E in V so that:

$$(261)\qquad a(E,\tilde{E}) = \int_{\Gamma_0} f_I.\bar{\tilde{E}}\, d\Gamma_0\,,\qquad \forall \tilde{E} \in V.$$

Thanks to Proposition 10 on the Calderon operator C (that is (243) and that C is continuous in the natural trace space) and since the medium is lossy (thus with $\text{Im } \varepsilon_1 > 0$, $\text{Im } \mu_1 > 0$), we obtain that the sesquilinear form a is a V-coercive form, i.e., there exists a constant $C_0 > 0$ such that :

$$(262)\qquad \text{Re } a(E,E) \geq C_0 \|E\|_V^2,\qquad \forall E \in V.$$

Thus we can apply the Lax-Milgram lemma: therefore problems (261), (258) and also the scattering problem, have unique solution (E,H).
Now if we substitute free space for the lossy dielectric in Ω, we obtain a problem which depends on the Fredholm alternative (in a subspace of the Sobolev space $H^1(\Omega)^3$). Thus except for a countable set of singular values of ω, the scattering problem has one solution. This allows us to define a Calderon operator C_Γ for this profile, with the properties of C for the half-space (Proposition 10). For numerical applications, an integral method is preferred.

⊗

Example 4. "*A two-sided problem*". We consider the situation of Example 3 but the perfect conducting medium is replaced by free space (or by another dielectric medium). Now the scattering problem is in the whole space, with the electromagnetic field (E,H) propagating towards the negative axis x_3 under Ω. Then we can reduce this problem to the domain Ω with the Calderon operator C^i_Γ for the inferior domain (but generally we cannot use it for a numerical method), or to the domain $\Omega_\tau = R^2 \times (-\tau, 0)$ with the Calderon operator C_1 relative to the half-space $R^2 \times (-\infty, -\tau)$. In this case, we have to find (E,H) satisfying (258)i),ii),iv) and (257) and the condition (see (115)):

(263) $\qquad n \wedge H + C_1(n \wedge E) = 0 \quad \text{on } \Gamma_1 = R^2 \times \{-\tau\}, \text{ with } n = e_3.$

Now we define the sesquilinear form in $V = H_{KL}(\text{curl}, \Omega_\tau)$:

(264)
$$\left| \begin{array}{l} a(E,\tilde{E}) = \int_{\Omega_\tau} (-\frac{1}{i\omega\mu} \text{curl } E.\ \text{curl } \bar{\tilde{E}} - i\omega\varepsilon\, E.\ \bar{\tilde{E}})\, dx \\ \quad - \int_{\Gamma_0} C\,(e_3 \wedge E|_{\Gamma_0}).\ \bar{\tilde{E}}\, d\Gamma_0 - \int_{\Gamma_1} C_1\,(e_3 \wedge E|_{\Gamma_1}).\ \bar{\tilde{E}}\, d\Gamma_1, \end{array} \right.$$

with $\varepsilon = \varepsilon_1$ and $\mu = \mu_1$ in Ω, $\varepsilon = \varepsilon_0$, $\mu = \mu_0$ in free space.
This scattering problem is written in the variational form: find E in V such that

(265) $\qquad a(E,\tilde{E}) = \int_{\Gamma_0} f_I.\bar{\tilde{E}}\, d\Gamma_0, \qquad \forall \tilde{E} \in V.$

Since ε and μ in free space are positive real numbers, this new sesquilinear form is not V coercive. We will replace V by the following space

(266)
$$\left| \begin{array}{l} W_{KL}(\Omega_\tau) = \{E \in L^2_{KL}(\Omega_\tau)^3,\ E|_\Omega \in H^1_{KL}(\Omega)^3,\ E|_{\Omega'} \in H^1_{KL}(\Omega')^3, \\ \qquad\qquad [n \wedge E]_\Gamma = 0, \quad [n.\varepsilon_2 E]_\Gamma = 0\}, \end{array} \right.$$

where Ω' is the complementary set of Ω in Ω_τ. We note that the (transverse) traces of $W_{KL}(\Omega_\tau)$ on the boundaries Γ_0 and Γ_1 of Ω_τ are in $H^{1/2}_{KL}(\Gamma_0)^2$ and $H^{1/2}_{KL}(\Gamma_1)^2$, with natural inclusions $H^{1/2}_{KL}(\Gamma_0)^2 \subseteq H^{-1/2}_{KL}(\text{div},\Gamma_0) \subseteq H^{-1/2}_{KL}(\Gamma_0)^2$ and for Γ_1 also.

Thus the boundary terms in (264) which correspond to the Calderon operators C and C_1 are continuous in this new space. Since the natural mapping of $W_{KL}(\Omega_\tau)$ in $L^2_{KL}(\Omega_\tau)^3$ is compact, the problem (265) depends on the Fredholm alternative in $W_{KL}(\Omega_\tau)$. But uniqueness at most of the solution is an easy consequence of the dissipativity in the dielectric:
taking the real part of (265) with $\tilde{E} = E$ when $f_I = 0$ implies that $E = 0$ in Ω, we have $E = H = 0$ in Ω_τ; the scattering problem (265) has a unique solution in $W_{KL}(\Omega_\tau)$.⊗

Remark 18. *Integral method in periodic structures*. We first consider, like in section 1.3.2. chap. 3, the "standard problem" with given electric and magnetic currents J_Γ, M_Γ on a "regular" periodic surface Γ (contained in $R^2 \times (-\tau, 0[)$ in free space: find (E, H) of locally finite energy up to Γ such that

$$(267)\quad \left|\begin{array}{l} \text{i) } \operatorname{curl} H + i\omega\varepsilon E = J_\Gamma,\ J_\Gamma \in H^{-1/2}_{KL}(\operatorname{div},\Gamma), \\ \text{ii) } -\operatorname{curl} E + i\omega\mu H = M_\Gamma \text{ in } R^3,\quad M_\Gamma \in H^{-1/2}_{KL}(\operatorname{div},\Gamma); \\ \text{iii) E and H satisfy an "outgoing" wave condition above } \Gamma \text{ (see (233)).} \end{array}\right.$$

In order to simplify, we assume that the regularity condition (211) is satisfied (if not we have to assume that J_Γ, M_Γ satisfy supplementary conditions similar to (231) in the Helmholtz case, in order to satisfy the condition (251)). Then (267) has a solution (E, H) which has all the required properties and which is given by the convolution product (44) chap.3, but with the quasiperiodic elementary solution $\Phi^{(p)}$ given by (217) instead of Φ (and with the convolution product in $D'_{KL}(P \times R)$, which is allowed by the finite thickness of the grating). Then (E, H) is obtained thanks to (quasiperiodic) electric and magnetic layers $L_m^{(p)}$ and $P_m^{(p)}$ (like in (96) chap.3). By restriction to Γ, on each side of Γ, we obtain the operators $T^{(p)}$, $R^{(p)}$ (similar to T, R) see (100), (100)'chap.3) and thus the Calderon projectors $P_+^{(p)}$ and $P_-^{(p)}$ on each side of Γ and then the operator $S^{(p)}$

(with (104), (105), (106)) in the space $H^{-1/2}_{KL}(\operatorname{div},\Gamma) \times H^{-1/2}_{KL}(\operatorname{div},\Gamma)$.
Of course we should give this operator using Fourier expansions.
This allows us to apply the integral method in electromagnetism with various periodic situations: perfectly conducting gratings or inclusions or grid, or also with a dielectric medium (with constant permittivity and permeability).

⊗

Remark 19. *1D-Gratings*. Now we assume that Γ is a cylindrical periodic surface (with x_2 axis for instance). If the incident plane wave (E_I, H_I) given by (254) is such that its propagation direction α is orthogonal to the x_2 axis (i.e. $\alpha_2 = 0$), (E_I, H_I) is independent of x_2, then we can see: i) that the solution of the scattering problem is independent of x_2 also, ii) that this problem splits into two simple "scalar" problems (with Helmholtz equation) according to the polarization of the incident wave that is E_I parallel to x_2 (P polarization) or orthogonal (S polarization), see for instance Petit [1]. We can easily apply the theory developed above for 2D-gratings (changing double Fourier series into the usual Fourier series).

In more general cases, with an incident wave which depends on x_2, it may be useful to work with a trace space on a plane ($x_3 = 0$) which is in-between the plane situation (104) and the periodic situation (238) thanks to a Fourier expansion (with respect to x_1) and a Fourier transformation.

More precisely we can use definition (104), changing the Lebesgue measure $dx = d\xi_1 d\xi_2$ into the sum of measures $\nu_n = \delta(\xi_1 - \xi_{1n})d\xi_2$, and space $H_k^{1/2}(R^2)$ into the space (210) in Remark 14.

This (natural) trace space give the usual expected properties for an electromagnetic field with locally finite energy.

⊗

Remark 20. *On 2D-gratings with a small period.* When $K < |k|$ (which is the case for an incident plane wave), we always have the propagating mode $J = (0,0)$.

Since the propagating modes correspond to numbers J in Z^2 so that $|\xi_J| < k$, there is no other such mode (with notations (173)') when $|\xi_J^0| > 2k$, $J \neq (0,0)$. But $|\xi_J^0| > 2\pi L_M^{-1}$, with $L_M = \max(L_1, L_2)$. Thus when $2\pi L_M^{-1} > 2k$, i.e., $\lambda = 2\pi/k > 2L_M$ the unique propagating mode is for $J = (0,0)$.

This simple remark, useful for applications with composite materials, could have been made before.

⊗

3. CONICAL GEOMETRY

Here we consider stationary wave problems in domains Ω in R^n such that

$$\Omega = \Omega_{r_0,S} = \{x = (r,\alpha),\ r = |x|\ ,\ \alpha = x/r,\ r > r_0,\ \alpha \in S\}$$

with S a "regular" open subset of the sphere S^{n-1}, and $r_0 \in [0, \infty)$.

Thus in polar coordinates these domains are of the form $(r_0, \infty) \times S$. We essentially work on the Helmholtz equation (this will reduce Maxwell problems to problems on the boundary).

We use a method of separation of variables, which leads us to take into account Sommerfeld conditions, in a somewhat different way from the case of bounded obstacles. It will be interesting to compare the results of this powerful method with the usual "limiting absorption principle" (for which we must have sources of compact support) and to the other geometries. The fact that we have well-posed problems in the present framework implies the limiting absorption principle!

First we consider the Helmholtz equation in a domain Ω of the form $(0, \infty) \times S$:

(268) $$\Delta v + k^2 v = -g \text{ in } \Omega,$$

with given g in Ω, with a real wavenumber k, and with boundary conditions. Using polar coordinates, (268) is:

(269) $$r^{-(n-1)}\frac{\partial}{\partial r}r^{n-1}\frac{\partial V}{\partial r}+\frac{1}{r^2}\Delta_\alpha V+k^2V=-G \text{ in } R^+ \times S,$$

with Δ_α the Laplace-Beltrami operator on S, and $V(r,\alpha)=v(x)$, $G(r,\alpha)=g(x)$.
Now if we multiply (269) by r^2, and we change variables and the unknown function

(270) $$\rho = kr, \quad u(\rho,\alpha)=r^{(n-3)/2}V(r,\alpha), \qquad f(\rho,\alpha)=r^{(n+1)/2}G(r,\alpha),$$

we obtain the equation:

(271) $$\mathcal{A}u - \mathcal{B}u = f \text{ in } R^+ \times S,$$

with $\mathcal{A}$ and $\mathcal{B}$ given by (keeping notation r for ρ from now on):

(272) $$\mathcal{A}u = -(r^2u')' - r^2u - u/4 \text{ and } \mathcal{B} = \Delta_\alpha - ((n-2)/2)^2 I.$$

3.1. Properties of some unbounded operators associated to $\mathcal{A}$

First we study the main properties of the operator $\mathcal{A}$. Obviously the main difficulties are at infinity and at 0 since $\mathcal{A}$ is a "degenerated" operator at 0 (that is the coefficient of the derivative of higher degree is zero at the boundary) with weight r^2 at infinity. We choose the framework $L^2(R^+)$. This choice is not at all obvious since, through the change (270), it does not correspond to the usual $L^2(\Omega)$ space but to the space $L^2(\Omega, dx/r^2)$. But the properties obtained in such a space justify its use.
We first define unbounded operators in $L^2(R^+)$, with different domains associated to the differential operator $\mathcal{A}$. The first properties of these operators will be given with the usual notions on unbounded operators, see for instance Richtmyer [1], Dunford-Schwartz [1]. We define the "minimal" and "maximal" realizations of $\mathcal{A}$, denoted by A_m and A_M by their different domains:

(273) $$D(A_m) = \mathcal{D}((0,\infty)), \qquad D(A_M) = \{u \in L^2(R^+), \mathcal{A}u \in L^2(R^+)\}.$$

Proposition 11. *The minimal operator* A_m *is symmetric with deficiency indices* (1,1). *Its adjoint operator is the maximal operator* A_M *which has no boundary condition at* 0, *but two linearly independent boundary conditions at infinity* B_1 *and* B_2, *given by*

(274) $$B_1(u) = \lim_{r\to\infty} e^{2ir}(r\,u(r)\,e^{-ir})', \quad B_2(u) = \lim_{r\to\infty} e^{-2ir}(r\,u(r)\,e^{ir})',$$

In Weyl's terminology, we say that $\mathcal{A}$ is of limit point-type at 0, and $\mathcal{A}$ is of limit-circle type at ∞, see Dunford-Schwartz [1] p. 1306.

Definition 9. *The boundary condition* B_1 *(resp.* B_2*) is said to be the outgoing (resp. incoming) Sommerfeld condition (with respect to a time evolution* $e^{-i\omega t}$*).*

PROOF of Proposition 11. The minimal operator A_m is symmetric, i.e.,

(275) $(A u,v)=\int_0^\infty [r^2(u'\bar{v}' - u\bar{v}) - \frac{1}{4}u\bar{v}]\,dr=(u,Av), \quad \forall u, v \in D((0,\infty))$;

A_m has $(A_m)^* = A_M$ as adjoint; its deficiency indices are: $n_\pm = \dim\ker(A_M \pm iI)$. To determine n_+, we note that there are two independent solutions of the equation $(A + iI)\,u = 0$: $u(r) = J_\lambda(r)/\sqrt{r}$, $\lambda^2 = i$, thus $\lambda = \pm\exp(i\pi/4)$, with J_λ the Bessel function of order λ. Its behavior at 0, then at ∞ is:

(276) $|J_\nu(r)| \approx (r/2)^{\mathrm{Re}\,\nu}\dfrac{1}{|\Gamma(\nu+1)|}$ when $r \to 0$, and $|J_\nu(r)| \geq C\,r^{-1/2}$ when $r \to \infty$.

Thus $\int_0^\infty |J_\nu(r)|^2 dr/r < \infty$ only if $\mathrm{Re}\,\nu > 0$, therefore $n_+ = 1$. Similarly $n_- = 1$. Furthermore we obviously have $D(A_M) \subseteq H^2_{loc}(R^+)$, and thus for all u, v in $D(A_M)$:

(277) $<-Au,v>_{\varepsilon,R} = \int_\varepsilon^R\{(r^2u')' + r^2u + \frac{1}{4}u\}.\,\bar{v}\,dr = <u, -Av>_{\varepsilon,R} + [r^2(u'.\bar{v} - u.\bar{v}')]_\varepsilon^R$.

Then to obtain the limits of the last term in (277), we have to specify the behaviors of the elements of $D(A_M)$ at 0 and at infinity. We extensively use the Hardy inequality (see Dautray-Lions [1] chap. 8.2.7 lemma 3) for all T finite or not

(278) $\int_0^T|w|^2dr \leq 4\int_0^T|(rw)'|^2dr, \quad \forall w$ with $(rw)' \in L^2(0,T)$.

a) *Behavior at* 0 of u in $D(A_M)$. We have u and $(r^2u')'$ in $L^2(0,T)$ for all $T>0$. From (278) $r.u'$ is in $L^2(0,T)$, thus $(r.u)'$ is in $L^2(0,T)$, i.e., $w = r.u$ is in $H^1((0,T))$. Thus w(0) exists and since $u = w/r$ is in $L^2(0,T)$, this implies $w(0) = 0$. Furthermore thanks to Cauchy-Schwarz inequality,

(279) $|w(T)| = |\int_0^T w'(r)dr| \leq T^{1/2}(\int_0^T |w'(r)|^2dr)^{1/2} \leq T^{1/2}C_T$.

Arguing similarly with $\tilde{w} = r^2.u'$, we obtain:

(280) $u(r) = o(r^{-1/2})$, $u'(r) = o(r^{-3/2})$ when $r \to 0$,

and $u \in D(A_M)$ implies:

(281) $u \in L^2(0,T)$, $(r^2.u')' \in L^2(0,T)$, i.e., $u \in L^2(0,T)$, $r.u \in H^1(0,T)$, $r^2.u \in H^2(0,T)$.

Now in (277), $[r^2(u'.\bar{v} - u.\bar{v}')](\varepsilon) \to 0$ with ε, thus u *has no boundary condition at* 0.
b) *Behavior at infinity of* u in $D(A_M)$. Let:

(282) $w_u^1(r) = (e^{-ir}\,ru)'\,e^{2ir}$, $\quad w_u^2(r) = (e^{ir}\,ru)'\,e^{-2ir}$.

Then (for instance) $w(r) = w_u^2(r)$ satisfies: $rw'(r) = (-\mathcal{A}u - \frac{1}{4}u)\,e^{-ir} \in L^2(0,\infty)$. Therefore for all $T > 0$, $w' \in L^1(T,\infty)$ and $w(r)$ has a limit $l = B_2(u)$ when $r \to \infty$ with

(283) $|w(r) - l| = |\int_r^\infty w'(s)\,ds| \le (\int_r^\infty s^{-2}ds)^{1/2} (\int_r^\infty |sw'(s)|^2 ds)^{1/2} \le r^{-1/2} C_r$

with $C_r \to 0$ when $r \to \infty$. Then for all u and v in $D(A_M)$, we have:

(284) $a_r(u,v) \stackrel{\text{def}}{=} -i\,[ru'.r\bar{v} - ru.r\bar{v}'] = \frac{1}{2}[w_u^2 \bar{w}_v^2 - w_u^1 \bar{w}_v^1]$

and

(285)
$$\left|\begin{aligned} a(u,v) &\stackrel{\text{def}}{=} i\,[(A_M u,v) - (u,A_M v)] = -i \lim_{\varepsilon\to 0,\,R\to\infty} [r^2(u'\bar{v} - u\bar{v}')]_\varepsilon^R \\ &= \lim_{R\to\infty} a_R(u,v) = \frac{1}{2}[B_2(u)\overline{B_2(v)} - B_1(u)\overline{B_1(v)}]. \end{aligned}\right.$$

This implies that the two mappings $u \to B_\kappa(u)$, $\kappa = 1,2$ are continuous linear forms on $D(A_M)$ with the graph norm. Furthermore let u be in $D(A_M)$; using (283) with B_2 (and B_1) we have:

(286) $|(ru)' + iru - e^{ir}B_2(u)| = o(r^{-1/2})$, $\quad |(ru)' - iru - e^{-ir}B_1(u)| = o(r^{-1/2})$,

therefore giving when $r \to \infty$:

(286)' $|u - \frac{e^{ir}}{2ir}B_2(u) + \frac{e^{-ir}}{2ir}B_1(u)| = o(r^{-3/2})$, $\quad |u' - \frac{e^{ir}}{2r}B_2(u) - \frac{e^{-ir}}{2r}B_1(u)| = o(r^{-3/2})$.

⊗

Example 5. We take $u = J_\nu(r)/\sqrt{r}$, with Re $\nu > 0$. From the usual properties of Bessel functions (see Abramowitz-Stegun [1] p. 364), u is in $D(A_M)$, with

(287) $B_1(u) = \sqrt{\frac{2}{\pi}}\exp[i\frac{\pi}{2}(\nu - \frac{1}{2})]$, $\quad B_2(u) = \sqrt{\frac{2}{\pi}}\exp[-i\frac{\pi}{2}(\nu - \frac{1}{2})]$.

⊗

Definition 10. *We define the operators* A_κ, $\kappa = 1,2$ *by restriction of* A_M *to*

(288) $D(A_\kappa) \stackrel{\text{def}}{=} \{u \in D(A_M),\ B_\kappa(u) = 0\}$, $\kappa = 1, 2$.

Proposition 12. *The operators* A_1 *and* A_2 *are adjoint to each other. Moreover the operators* $-iA_1$ *and* iA_2 *are maximal dissipative operators and thus they are the infinitesimal generators of contraction semigroups of class* C^0.

PROOF. From relation (285), we easily see that $(A_1)^* = A_2$, $(A_2)^* = A_1$. Furthermore, taking $v = u$ in (285), we have

(289) $a(u,u) = i[(A_M u,u) - (u,A_M u)] = \frac{1}{2}[|B_2(u)|^2 - |B_1(u)|^2]$.

Thus

(290)
$$\left|\begin{array}{ll} \text{i) Re}\,(iA_1u,u) = \frac{1}{4}\,|B_2(u)|^2 \geq 0, & \forall\, u \text{ in } D(A_1)\,, \\ \text{ii) Re}\,(iA_2u,u) = -\frac{1}{4}\,|B_1(u)|^2 \leq 0, & \forall\, u \text{ in } D(A_2). \end{array}\right.$$

Thus the operators iA_1 and $(-iA_2)$ are accretive operators (and $-iA_1$ and iA_2 are dissipative operators). Since $(iA_1)^* = -iA_2$, Proposition 12 is a consequence of Dautray-Lions [1] Thm. 8 chap. 17A.3.

⊗

Then we define the operator A_θ when $0 < \theta < 2$ by:

(291)
$$\left|\begin{array}{l} D(A_\theta) = \{u \in D(A_M),\ BL_\theta u = 0\}, \quad A_\theta u = \mathcal{A}u, \\ BL_\theta u = \exp\,[i\frac{\pi}{2}(\frac{1}{2} - \theta)]\, B_1(u) - \exp\,[-i\frac{\pi}{2}(\frac{1}{2} - \theta)]\, B_2(u). \end{array}\right.$$

We can prove that it has the properties (see Dunford-Schwartz [1] 13, 9.15):

Proposition 13. *The operator* A_θ *is a self adjoint operator whose spectrum* $\sigma(A_\theta)$ *has a continuous spectrum* $\mathbb{R}^+$ *and a set of discrete simple eigenvalues*:

(292) $$\sigma(A_\theta) = \mathbb{R}^+ \cup \{-(2n+\theta)^2,\ n \in \mathbb{N}\},$$

with associated eigenvectors $J_{2n+\theta}(r)/\sqrt{r}$. *The resolvent* $R(-\lambda, A_\theta) = (\lambda I + A_\theta)^{-1}$ of A_θ *when* $-\lambda$ *is not in* $\sigma(A_\theta)$ *is given* (*with* Re $\sqrt{\lambda} > 0$) by:

(293) $$R(-\lambda, A_\theta)f(t) = \int_0^\infty K_\theta(t,s;\lambda)\, f(s)\, ds, \quad f \in L^2(\mathbb{R}^+),$$

with:

(294)
$$K_\theta(t,s;\lambda) = \left|\begin{array}{ll} \Gamma_{12}^\theta(\lambda)\,(J_{\sqrt{\lambda}}(s)/\sqrt{s})(J^\theta_{\sqrt{\lambda}}(t)/\sqrt{t})\,, & s < t\,, \\ \Gamma_{12}^\theta(\lambda)\,(J_{\sqrt{\lambda}}(t)/\sqrt{t})(J^\theta_{\sqrt{\lambda}}(s)/\sqrt{s})\,, & s > t\,, \end{array}\right.$$

(295) $$J_\lambda^\theta = \sin\frac{\pi}{2}(\lambda+\theta)\, J_\lambda + \sin\frac{\pi}{2}(\lambda-\theta)\, J_{-\lambda}\,, \quad \Gamma_{12}^\theta(\lambda) = \frac{\pi}{2}\,\frac{1}{\sin(\pi\sqrt{\lambda})\,\sin\,[\frac{\pi}{2}(\sqrt{\lambda}-\theta)]}.$$

PROOF. The fact that A_θ is a selfadjoint operator is a simple consequence of

(296) $$\overline{B_1(f)} = B_2(\bar{f}),\ \overline{B_2(f)} = B_1(\bar{f}), \quad \text{thus} \quad \overline{BL_\theta(f)} = -\,BL_\theta(\bar{f}).$$

Furthermore thanks to (287), we have:

(297) $$BL_\theta\,(J_\lambda(r)/\sqrt{r}) = 2i\,\sqrt{\frac{2}{\pi}}\,\sin\frac{\pi}{2}(\lambda-\theta).$$

Thus when λ is in $\mathbb{R}^+$, $J_{\sqrt{\lambda}}(r)/\sqrt{r}$ is an eigenvector of A_θ (for $-\lambda$) if:

(298) $\sin\frac{\pi}{2}(\sqrt{\lambda}-\theta)=0$, i.e., $\lambda=(\theta+2n)^2$, $n\in N$.

This gives the discrete spectrum. The continuous spectrum is obtained by finding the generalized eigenfunctions so that $(A+\lambda I)f=0$, with $BL_\theta f=0$.

We obtain: $f(r)=J^{\theta}_{\sqrt{\lambda}}(r)/\sqrt{r}$.
The general theory of differential operators (see Dunford-Schwartz [1] p. 1326) yields the resolvent of the operator A_θ as given in (293), (294).

⊗

Proposition 14. *The spectrum* $\sigma(A_\kappa)$ of A_κ *is continuous with* $\sigma(A_\kappa)=R^+$, $\kappa=1,2$.

The resolvent $R(-\lambda,A_\kappa)=(\lambda I+A_\kappa)^{-1}$ of A_κ when $\lambda\in C\backslash R^-$ *is an integral operator:*

$$R(-\lambda,A_\kappa)f(t)=\int_0^\infty K_\kappa(t,s,\lambda)\,f(s)\,ds,\quad \forall f\in L^2(R^+)$$

with kernel $K_\kappa(t,s,\lambda)$ *given by (when* $\mathrm{Re}(\sqrt{\lambda})>0$):

$$(299)\qquad K_\kappa(t,s,\lambda)=\left|\begin{array}{ll}(-1)^{\kappa-1}\frac{i\pi}{2}(J_{\sqrt{\lambda}}(s)/\sqrt{s})(H^{(\kappa)}_{\sqrt{\lambda}}(t)/\sqrt{t}) & \text{when } s<t,\\ (-1)^{\kappa-1}\frac{i\pi}{2}(J_{\sqrt{\lambda}}(t)/\sqrt{t})(H^{(\kappa)}_{\sqrt{\lambda}}(s)/\sqrt{s}) & \text{when } s>t,\end{array}\right.$$

where $H^{(1)}_\nu$, $H^{(1)}_\nu$ *are the usual Hankel functions of order* ν.

PROOF. The closure of $D(R^+)$ for the graph norm is the space $D(\bar{A}_m)$ defined by

(300) $D(\bar{A}_m)=\{u\in D(A_M),\ B_\kappa(u)=0,\ \kappa=1,2\}$.

Its range by $(\lambda I-A)$, $\lambda\in C\backslash R^+$, is closed of codimension 1. Let:

(301) $D_\pm\overset{\text{def}}{=}\ker(A_M\mp iI)$, with $\dim D_\pm=1$.

Then $D(A_M)$ is the orthogonal sum for the scalar product $\langle u,v\rangle=(Au,Av)+(u,v)$:

(302) $D(A_M)=D(\bar{A}_m)\oplus D_+\oplus D_-$,

and there are spaces D^κ, $\kappa=1,2$ of dimension 1 contained in $D_+\oplus D_-$ such that:

(303) $D(A_\kappa)=D(\bar{A}_m)\oplus D^\kappa$, $\kappa=1,2$.

In order to prove that $(\lambda I-A_\kappa)D(A_\kappa)=L^2(R^+)$, we only have to verify that:

(304) $(\lambda I-A_\kappa)u_\kappa=0$, $u_\kappa\in D(A_\kappa)$, implies $u=0$.

But the solutions of (304) are linear combinations of $(J_{\sqrt{\lambda}}(s)/\sqrt{s})$ and $(J_{-\sqrt{\lambda}}(s)/\sqrt{s})$ and the conditions u in $L^2(R^+)$, $B_\kappa u=0$ imply $u=0$. Thus we have proved that $\sigma(A_\kappa)\subseteq R^+$. Furthermore the asymptotic behavior of the Hankel functions gives:

$$B_\kappa((H_\nu^{(\kappa)}(s)/\sqrt{s}) = 0.$$

Thus for all λ in $C\backslash R^-$, the equation: $(A + \lambda I)\,\phi = 0$ has a unique (up to a multiplicative constant) generalized solution satisfying one of the conditions

i) ϕ is a square integrable function in a neighborhood of 0 only: $\phi(r,\lambda) = J_{\sqrt{\lambda}}(r)/\sqrt{r}$.

ii) ϕ satisfies the condition at infinity $B_1(\phi) = 0$: $\phi(r,\lambda) = H^{(1)}_{\sqrt{\lambda}}(r)/\sqrt{r}$.

This gives (299) according to the general theory, when $\kappa = 1$; $\kappa = 2$ is similar.

⊗

Proposition 15. *The resolvents* $R(-\lambda,A_\kappa)$, $\kappa = 1,2$ *and* $R(-\lambda,A_\theta)$ *are only different through a (degenerated) operator of rank* 1 *when* λ *is not in* $(-\infty,0]$ *and when* $\lambda \neq (2n+\theta)$, *with* n *an integer*:

(305) $$P_\lambda = R(-\lambda,A_1) - R(-\lambda, A_2), \qquad P_\lambda^\theta = R(-\lambda,A_1) - R(-\lambda,A_\theta),$$

with kernel $K(t,s,\lambda)$ and $K^\theta(t,s,\lambda)$:

(306) $$\left| \begin{array}{l} K(t,s,\lambda) = i\pi\,(J_{\sqrt{\lambda}}(s)/\sqrt{s}).(J_{\sqrt{\lambda}}(t)/\sqrt{t}), \\[2ex] K^\theta(t,s,\lambda) = \dfrac{-\exp[-i(\pi/2)(\sqrt{\lambda}-\theta)]}{\sin[(\pi/2)(\sqrt{\lambda}-\theta)]}\,K(t,s,\lambda). \end{array} \right.$$

The norms of the operators P_λ and P_λ^θ *are given with* $\sqrt{\lambda} = \mu_0 + i\mu_1$ by

(307) $$\|P_\lambda\| = \frac{\pi}{2\mu_0}\,\frac{\operatorname{sh}\mu_1\pi}{\mu_1\pi}, \quad \|P_\lambda^\theta\| = \|P_\lambda\|\exp(\mu_1\pi/2)\,[\operatorname{ch}^2(\tfrac{\pi}{2}\mu_1) - \cos^2(\tfrac{\pi}{2}(\mu_0-\theta))]^{-1/2};$$

thus when $\lambda = \lambda_0 \in R^+$, $\mu_0 = \sqrt{\lambda_0}$:

(308) $$\|P_{\lambda_0}\| = \frac{\pi}{2\mu_0}, \qquad \|P_{\lambda_0}^\theta\| = \frac{\pi}{2\mu_0}\,\frac{1}{|\sin[(\pi/2)(\mu_0-\theta)]|}.$$

Proposition 15 is a simple consequence of (299). It implies the basic result:

Proposition 16. *The resolvents of the operators* A_κ *satisfy:*

(309) $$\|R(-\lambda_0,A_\kappa)\| \le K/\sqrt{\lambda_0}, \qquad K = (\pi+1)/2, \quad \forall\,\lambda_0 > 0.$$

This result is the best in the sense that there cannot exist (K,α), $K > 0$, $\alpha > 1/2$ *so that*

(310) $$\|R(-\lambda_0,A_\kappa)\| \le K\,\lambda_0^{-\alpha}.$$

PROOF. From (305), (308) we have for instance for A_1:

(311) $$\|R(-\lambda_0,A_1)\| \le \|R_{\lambda_0}^\theta\| + \|P_{\lambda_0}^\theta\|.$$

Since $-A_\theta$ is a selfadjoint operator, the norm of its resolvent is given by:

(312) $\quad \|R^\theta_{\lambda_0}\| = 1/\text{dist}(\lambda_0, \sigma(-A_\theta)) = 1/\inf_{n\in N} |\lambda_0 - (2n+\theta)^2|.$

Let $[\mu_0]$ be the integer part of $\mu_0 = \sqrt{\lambda_0}$. When $[\mu_0]$ is even, taking $\theta = \mu_0 - [\mu_0] + 1$ gives $|\sin\frac{\pi}{2}(\mu_0 - \theta)| = 1$, $|\lambda_0 - (\theta + 2n)^2| \geq 2\mu_0$. When $[\mu_0]$ is odd, we obtain the same result choosing $\theta = \mu_0 - [\mu_0]$. Finally we cannot have inequality (310) since by difference, we would have $\|P_{\lambda_0}\| \leq 2K\lambda_0^{-\alpha}$, in contradiction with (308).

⊗

Remark 21. The so-called "Kantorovich-Lebedev" transformation (in the Hankel form) is used to diagonalize the operator A. But we can expect that it is not so easy, because the operators A_κ are not selfadjoint, nor normal, nor even spectral operators ! The definition of this transformation (with $\mu > 0$) is

(313) $\quad \hat{\phi}_\kappa(\mu) = \lim_{\eta\to 0} \int_\eta^\infty [H^{(\kappa)}_{i\mu}(s)/\sqrt{s}]\, \phi(s)\, ds, \quad \kappa = 1, 2,$

with $H^{(\kappa)}_\nu$ the Hankel function. The mapping $\phi \to (\hat{\phi}_1, \hat{\phi}_2)$ is an isomorphism from $L^2(R^+)$ onto $H_1^+ \times H_2^+$ with

(314) $\quad H^+_\kappa = \{g \in L^1_{loc}(R^+), \int_{R^+} |g(\mu)|^2 e^{\varepsilon\pi\mu} (\text{th}\frac{\pi\mu}{2})\, \mu\, d\mu < \infty\}, \quad \varepsilon = (-1)^\kappa,$

with (unusual) properties of diagonalization of A_κ. For developments on this transformation (which can be used to solve Helmholtz problems) see Cessenat [1], [2], [3].

⊗

3.2. Solution of Helmholtz problems

Now from section 3.1 we can solve Helmholtz problems. First in the domain $\Omega = R^+ \times S$, and in space $L^2(R^+ \times S)$, (i.e., in $L^2(\Omega, dx/r^2)$) we define the following framework:

(315) $\quad D_M(A - B) = \{u \in L^2(R^+ \times S), (A - B)u \in L^2(R^+ \times S)\}.$

This (maximal) space has traces on the boundary of Ω: let u be in $D_M(A - B)$, then:
i) $Bu \in L^2(R^+, H^{-2}(S))$, thus $Au \in L^2(R^+, H^{-2}(S))$ also, and thus $B_\kappa u$ exists in $H^{-2}(S)$.
ii) $Au \in L^2(S, H^{-2}_{loc}(\bar{R}^+))$, ($A$ maps $H^2_{oc}(R^+) = \{u \in H^2_0(R^+)$, supp u is bounded$\}$ into $L^2(R^+)$). Thus $Bu \in L^2(S, H^{-2}_{loc}(\bar{R}^+))$, which implies that the traces $\gamma_0 u$ and $\gamma_1 u$ of u on the boundary $R^+ \times \partial S$ make sense (at least in $H^s(\partial S, H^{-2}_{loc}(\bar{R}^+))$, for $s = -1/2$ and $s = -3/2$ respectively).

Definition 11. *We define the operators* $A_\kappa - B_D$, $A_\kappa - B_N$, $\kappa = 1,2$ *by restriction of the differential operator* $A - B$ *to the following domains*:

$$(316)\qquad \left|\begin{array}{l} D(A_\kappa - B_D) = \{u \in D_M(A - B),\ B_\kappa u = 0,\ \gamma_0 u = u|_{R^+ \times \partial S} = 0\}, \\ D(A_\kappa - B_N) = \{u \in D_M(A - B),\ B_\kappa u = 0,\ \gamma_1 u = \frac{\partial u}{\partial n}|_{R^+ \times \partial S} = 0\}, \end{array}\right.$$

that is with the Dirichlet or the Neumann boundary conditions and with the outgoing or the incoming Sommerfeld conditions.

Theorem 5. *The Dirichlet (or Neumann) problem with the outgoing (or incoming) Sommerfeld condition at infinity:*

$$(317)\qquad (A_\kappa - B_D)\,u = f \quad (\text{resp.}\ (A_\kappa - B_N)\,u = f), \quad f \in L^2(R^+ x S),$$

has a unique solution u *in* $D(A_\kappa - B_D)$ *(resp.* $D(A_\kappa - B_N)$*), except when* $n = 2$*, with the Neumann condition; in this case, the result is valid if we change (for* u *and* f*)* $L^2(R^+ \times S)$ *into*

$$(318)\qquad \tilde{L}^2(R^+ \times S) = \{f \in L^2(R^+ \times S),\ \int_S f(r,\theta)\, d\theta = 0\}.$$

We extensively use the following property: the Laplace-Beltrami operators $\Delta_{\alpha D}$ or $\Delta_{\alpha N}$ with the Dirichlet or Neumann condition are selfadjoint operators with compact resolvents, so that the operators B_D and B_N defined by (272) are strictly negative operators i.e. with pure point spectrum contained in $(-\infty, 0)$, except when $n = 2$ with the Neumann condition (in this case the supplementary condition (318) eliminates the eigenvalue 0, which correspond to constant eigenvectors). Thus the operators A_κ and B_D or B_N are commuting operators, with disjoint spectrum (thanks to Proposition 14). Under fairly general conditions (see Grisvard [3]), we can solve problems similar to (271), thanks to the Cauchy integral in the complex plane:

$$(319)\qquad u = \frac{-1}{2\pi i} \int_\gamma (A_\kappa + z)^{-1} (B + z)^{-1} f\, dz, \quad B = B_D \text{ or } B_N,$$

where γ is a contour in the complex plane including the eigenvalues of $-B$ (with horizontal asymptotes when Re z tends to $+\infty$); a priori the use of (319) is purely formal, since the estimate on the resolvent of A_κ at infinity (see (309)) is not sufficient to apply this formula. We will justify it below.

PROOF of Theorem 5. Let (λ_n, ϕ_n) be a spectral resolution of $-B$, (with $B = B_D$ or B_N) where the λ_n are the eigenvalues (accounting their multiplicity) and (ϕ_n) a corresponding orthonormal basis of eigenvectors, giving the decompositions

$$(320)\quad \left| \begin{array}{l} u(r\alpha) = \sum u_n(r)\phi_n(\alpha), \quad f(r\alpha) = \sum f_n(r)\phi_n(\alpha), \quad \alpha \in S,\ r>0, \\ \text{with: } \|u\|^2 = \sum \|u_n\|^2 < \infty, \quad \|f\|^2 = \sum \|f_n\|^2 < \infty, \end{array} \right.$$

so that (271) is reduced to find (u_n) satisfying

$$(321)\qquad (A_\kappa + \lambda_n I)\, u_n = f_n.$$

Then since $-\lambda_n$ is not in the spectrum of A_κ, (321) has a unique solution given by

$$(322)\qquad u_n = (A_\kappa + \lambda_n I)^{-1} f_n = R(-\lambda_n, A_\kappa)\, f_n.$$

Moreover inequality (309) implies:

$$(323)\qquad \|u_n\| \le K\|f_n\| / \sqrt{\lambda_n}.$$

Thus the condition $f \in L^2(R^+ \times S)$ implies $\sum \|u_n\|^2 < \infty$. The solution u of (317), is given by

$$(324)\qquad u(.,\alpha) = \sum R(-\lambda_n, A_\kappa) f_n\, \phi_n(\alpha),$$

and satisfies $u \in L^2(R^+ \times S)$. Thus $u \in D_M(\mathcal{A} - \mathcal{B})$ and also $u \in L^2(R^+, D(B^{1/2}))$ since

$$(325)\qquad \sum \lambda_n \|u_n\|^2 < \infty,$$

thanks to (323). Furthermore each u_n is in $D(A_\kappa)$ and thus we have $B_\kappa u = 0$ in a weak sense, but we don't have u in $L^2(S, D(A_\kappa))$. Note also that (323) implies:

$$(323)'\qquad \|u\| \le K \lambda_0^{-1/2} \|f\|, \quad \text{with } K = (\pi+1)/2,\ \lambda_0 = \inf \lambda_n.$$

⊗

The solution u of (317) is then given by the Cauchy integral:

$$(326)\qquad u = \sum_n u_n = \sum_n (A_\kappa + \lambda_n)^{-1} f_n = \frac{1}{2\pi i} \lim_{\nu\to\infty} \int_{\gamma_\nu} (A_\kappa + z)^{-1} (B+z)^{-1} f\, dz,$$

where γ_ν is a family of growing contours including the first ν eigenvalues in the complex plane. This is a consequence of the usual formulae (with Q_n the orthogonal projection on the eigenspace of $-B$ for λ_n and for large values of n)

$$(B+z)^{-1} = -\sum_n \frac{1}{\lambda_n - z} Q_n, \qquad (A_\kappa + \lambda_n)^{-1} f_n = \frac{1}{2\pi i} \int_{\gamma_\nu} \frac{1}{\lambda_n - z} (A_\kappa + z)^{-1} f_n\, dz.$$

Remark 22. *Rayleigh series.* Using formula (299) of the resolvent, the solution u of (317) (see (324)) is written in the form (for $\kappa = 1$):

$$(327)\quad \left| \begin{array}{l} u(r\alpha) = \frac{i\pi}{2} [\sum_n \phi_n(\alpha) (J_{\sqrt{\lambda_n}}(r)/\sqrt{r}) \int_r^\infty (H^{(1)}_{\sqrt{\lambda_n}}(s)/\sqrt{s})\, f_n(s)\, ds \\ \qquad + \sum_n \phi_n(\alpha) (H^{(1)}_{\sqrt{\lambda_n}}(r)/\sqrt{r}) \int_0^r (J_{\sqrt{\lambda_n}}(s)/\sqrt{s})\, f_n(s)\, ds\,], \end{array} \right.$$

with $f_n(r) = \int_S f(r\alpha)\,\phi_n(\alpha)\,d\alpha$. This gives the usual Rayleigh series, see (380) chap. 3:
i) when supp f is contained in the ball B_a, then for $r > a$,

(328) $$u(r\alpha) = \frac{i\pi}{2} \sum_n c_n \,\phi_n(\alpha)\,(H^{(1)}_{\sqrt{\lambda_n}}(r)/\sqrt{r}),$$

with $c_n = \int_0^\infty (J_{\sqrt{\lambda_n}}(s)/\sqrt{s})\, f_n(s)\, ds$;
ii) when supp f is contained in $R^n \setminus B_a$, then for $r < a$,

(328)' $$u(r\alpha) = \frac{i\pi}{2} \sum_n c_n \,\phi_n(\alpha)\,(J_{\sqrt{\lambda_n}}(r)/\sqrt{r}),$$

with $c_n = \int_0^\infty (H^{(1)}_{\sqrt{\lambda_n}}(s)/\sqrt{s})\, f_n(s)\, ds$.
In R^3, $\sqrt{\lambda_n} = n + \frac{1}{2}$, $n \in N$, with multiplicity equal to $2n+1$, the eigenvectors of B are the spherical harmonics.

⊗

Remark 23. In order to prove the following proposition, we note that the solution u in $D(A_1)$ of $(A_1 + \lambda I)u = f$, with f in $L^2(R^+)$, is obtained (with (305)) by: $u = R(-\lambda, A_1)f = R(-\lambda, A_2)f + P_\lambda f$; thus its traces at infinity are $B_1 u = 0$, and

(329) $$B_2 u = B_2 P_\lambda f = i\pi\, B_2(J_{\sqrt{\lambda}}(r)/\sqrt{r})\, L_{\sqrt{\lambda}} f, \quad \text{with } L_{\sqrt{\lambda}} f = \int_0^\infty (J_{\sqrt{\lambda}}(t)/\sqrt{t}) f(t)\, dt.$$

Hence from (287):

(330) $$B_2 u = i\sqrt{2\pi}\, \exp\left[-\frac{i\pi}{2}\left(\sqrt{\lambda} - \frac{1}{2}\right)\right] L_{\sqrt{\lambda}} f,$$

we have (thanks to the Cauchy-Schwarz inequality):

(331) $$|B_2 u|^2 \le 2\pi\, |L_{\sqrt{\lambda}} f|^2 \le \frac{\pi}{\sqrt{\lambda}} \|f\|^2,$$

using:

(331)' $$\int_{R^+} |J_\nu(t)|^2\, dt/t = 1/2\nu, \quad \nu > 0.$$

⊗

We note $Y = D(A_1 - B) + D(A_2 - B)$, $Y = Y_D$ when $B = B_D$, $Y = Y_N$ when $B = B_N$.
We recall that $D(B^{1/4}) = H^{1/2}_{00}(S)$ when $B = B_D$, $D(B^{1/4}) = H^{1/2}(S)$ when $B = B_N$.

Proposition 17. *Trace at infinity and Green formula. The mapping* $B_\kappa : u \longrightarrow B_\kappa u$ *is a continuous mapping from* Y *onto* $D(B^{1/4})$, $\kappa = 1,2$. *Furthermore all* u, v *in* Y *satisfying the Dirichlet or the Neumann condition, also satisfy the Green formula*

(332) $$a_B(u,v) \overset{\text{def}}{=} i\,[(A - B)u, v) - (u, (A - B)v) = \frac{1}{2} \int_S [B_2 u.\overline{B_2 v} - B_1 u.\overline{B_1 v}]\, d\alpha.$$

PROOF. First we prove that u in $D(A_1 - B)$ implies B_2u in $D(B^{1/4})$. Using the decomposition (320) with (322), we have with (331) and $f = (A_1 - B)u$:

$$\text{(333)} \qquad \|B_2u\|^2_{D(B^{1/4})} = \sum_n \sqrt{\lambda_n}\,|B_2u_n|^2 \le \pi \sum_n \|f_n\|^2 \le \pi\|f\|^2 < \infty.$$

Then given g in $D(B^{1/4})$, we construct now a lifting of g into $D(A_1 - B)$.
Let $g = \sum_n g_n\phi_n$ be an expansion of g with $\sum_n \sqrt{\lambda_n}\,|g_n|^2 < \infty$. Then the function u defined by:

$$\text{(334)} \qquad u = (A_1 - B)^{-1}F,$$

with

$$F = -\frac{2i}{\sqrt{2\pi}} \sum_n \sqrt{\lambda_n}\, g_n \exp\left[i\frac{\pi}{2}\left(\sqrt{\lambda_n} - \frac{1}{2}\right)\right] \left(J_{\sqrt{\lambda_n}}(r)/\sqrt{r}\right)\phi_n,$$

is such that F is in $L^2(R^+ \times S)$ (from (331)') and u satisfies (with (329) and (331)'):

$$B_2u = \sum_n 2\sqrt{\lambda_n}\, g_n\,\phi_n\, L_{\sqrt{\lambda_n}}\left(J_{\sqrt{\lambda_n}}(r)/\sqrt{r}\right) = \sum_n g_n\phi_n = g.$$

The first part of Proposition 17 follows easily. Now we prove the Green formula (332) using the spectral decomposition of B and (285), since we have:

$$\text{(335)} \qquad \left| \begin{aligned} a_B(u,v) &= i \sum_n \left[((A + \lambda_n)u_n, v_n) - (u_n, (A + \lambda_n)v_n)\right] \\ &= \sum_n a(u_n, v_n) = \frac{1}{2}\sum_n \left[B_2u_n\,\overline{B_2v_n} - B_1u_n\,\overline{B_1v_n}\right]. \end{aligned} \right.$$

⊗

Proposition 18. *The operators* $-i(A_1 - B_D)$, $i(A_2 - B_D)$ *are maximal dissipative operators, adjoint to each other, and so that*

$$\text{(336)}\ D((A_\kappa - B_D)^2) \subseteq L^2(S, D(A_\kappa)) \cap L^2(R^+, D(B_D)) \subseteq D(A_\kappa - B_D) \subseteq L^2(R^+, D(B_D^{1/2})).$$

The same properties are valid if we replace the Dirichlet condition by the Neumann condition, but for $n = 2$, *we have to substitute space* $L^2(R^+, L^2(S)_0)$ *for* $L^2(R^+ \times S)$, *where* $L^2(S)_0$ *is the orthogonal space to constants in* $L^2(S)$.

PROOF. The Green formula (332) implies that:

$$\text{(337)} \qquad \mathrm{Re}\left[\, i((A_1 - B_D)u, u)\right] \ge 0, \ \forall u \in D(A_1 - B_D).$$

Thus $-i(A_1 - B_D)$ is dissipative. Furthermore since this operator has a bounded inverse, it is a maximal dissipative operator (from Dautray-Lions [1] thm. 8 chap. 17A.4). We easily prove (336) and that $i(A_1 - B_D)$, $-i(A_2 - B_D)$ are operators which are adjoint to each other from Green formula (332). ⊗

We can prove (see Cessenat [1]) that if u is in $D(A_\kappa - B_D)$ (or in $D(A_\kappa - B_N)$), then

(338) $$\frac{\partial u}{\partial r} \in L^2_{loc}(R^+ \times S), \quad \text{thus } D(A_\kappa - B) \subseteq H^1_{loc}((0,\infty) \times \bar{S}).$$

With this result (and (323)'), we can obtain an estimate on the norm of the solution u of (317) in $H^1_{loc}((0,\infty) \times \bar{S})$, like the estimate (181) of the limiting absorption principle.

Remark 24. *Asymptotic behavior in* $D_M(A - B)$. *Sommerfeld conditions*.
We can specify the behavior of functions u in $D_M(A - B_D)$ at infinity like in the proof of Proposition 12. Using the decomposition (320), we define (where ' is used for derivation with respect to r):

(339) $$w_u^2(r,\alpha) = (e^{ir}\, ru(r,\alpha))' e^{-2ir}, \quad w_n^2(r) = (e^{ir}\, ru_n(r))' e^{-2ir},$$

thus $w_u^2(r,\alpha) = \sum_n w_n^2(r)\phi_n(\alpha)$. Then we can prove that the formulas (286) are also valid, the $o(r^{-3/2})$ refers *to the norm* $H^{-1}(S)$. This immediately implies:

if u is in $D(A_1 - B_D)$, i.e. $B_1 u = 0$, then $u(r,\alpha) = O(r^{-1})$, $u'(r,\alpha) - iu(r) = o(r^{-3/2})$,

that is, u *satisfies the outgoing Sommerfeld condition*.

We can also specify the behavior of functions u in $D_M(A - B_D)$ at the origin like in (280), using the spectral decomposition of $B = B_D$ (then the o refers to the $D(B^{-1/2})$ norm).

Remark 25. *Sources of compact support. Rellich lemma.*
i) If we have a source g (see (268)) with compact support and such that $f = r^2 g$ (see (270)) is in $L^2(R^+ \times S^2)$, when n = 3, then we know (see (15) Prop. 1 chap. 3) that the asymptotic behavior of the outgoing solution u = v is given by:

(340) $$u(r\alpha) = \frac{e^{ir}}{2ir} B_2 u(\alpha) + o(r^{-3/2}), \quad \text{with } B_2 u(\alpha) = \frac{i}{2\pi}\hat{g}(\alpha), \quad \alpha \in S^2,\ k = 1.$$

ii) Proof of the Rellich Lemma 1 (chap. 3). Let u satisfy outside a ball B_a

(341) $$(\Delta + k^2)u = 0, \text{ i.e., } (A - B)u = 0, \quad \text{and} \int_{S^2} |u(r\alpha)|^2\, d\alpha = o(r^{-2}),$$

so that $u(r\alpha) \in L^2(B_a')$. Furthermore (341) implies that $B_1 u = B_2 u = 0$. Let $u^a = u|_{S_a}$. Then using the spectral decomposition (320), we obtain that u_n satisfies:

(342) $$(A_\kappa + \lambda_n)\, u_n = 0, \quad r > a, \quad \kappa = 1,2, \quad u_n(a) = u_n^a.$$

Thus u_n is given by the two formulas (with $\kappa = 1$ *and* 2):

$$u_n(r) = u_n^a \,(H^{(\kappa)}_{\sqrt{\lambda_n}}(r)/\sqrt{r})\,(H^{(\kappa)}_{\sqrt{\lambda_n}}(a)/\sqrt{a})^{-1}, \quad \forall r \geq a,$$

which necessarily implies $u_n^a = u_n(r) = 0$, for all n, thus $u = 0$!
We emphasize that the Rellich lemma can be applied to any domain whose boundary is a conical surface when r is large enough. ⊗

Then we solve completely inhomogeneous problems thanks to Theorem 5 and a duality method. First we define the following trace spaces with $\kappa = 1,2$

(343)
$$\begin{cases} X_0^\kappa = \{u_0 = \gamma_0 u = u|_{R^+ \times \partial S},\ u \in D(A_\kappa - B_N)\} \\ X_1^\kappa = \{u_1 = \gamma_1 u = \frac{\partial u}{\partial n}|_{R^+ \times \partial S},\ u \in D(A_\kappa - B_D)\} \end{cases}$$

equipped with the quotient topology, and also with (315):

(344) $\tilde{X}_0 = \{u_0 = \gamma_0 u,\ u \in D_M(A - B)\}, \quad \tilde{X}_1 = \{u_1 = \gamma_1 u,\ u \in D_M(A - B)\}.$

We can prove that $X_0^1 = X_0^2$, $X_1^1 = X_1^2$ (thus denoted by X_0, X_1), and

(345) $\tilde{X}_0 = (X_1)', \tilde{X}_1 = (X_0)', \quad \text{with } X_0 \subseteq X_1 \subseteq (X_1)' \subseteq (X_0)'.$

Usual regularity results on elliptic problems in regular domains (see Gilbarg-Trudinger [1] for instance) imply:

$$X_0 \subset H^{3/2}_{loc}(R^+ \times \partial S),\ X_1 \subset H^{1/2}_{loc}(R^+ \times \partial S),\ \tilde{X}_0 \subset H^{-1/2}_{loc}(R^+ \times \partial S),\ \tilde{X}_1 \subset H^{-3/2}_{loc}(R^+ \times \partial S).$$

Theorem 6. *Let* (f, g, u_0), resp. (f, g, u_1), *be given such that*:

(346) $f \in L^2(R^+ x S), \quad g \in D(B^{1/4})', \quad u_0 \in \tilde{X}_0, \quad \text{resp. } u_1 \in \tilde{X}_1.$

Then there exists a unique solution u in $D_M(A - B)$ of:

(347)
$$\begin{cases} \text{i) } (A - B)\,u = f \quad \text{in } R^+ \times S, \\ \text{ii) } B_2 u = g \quad (\text{or } B_1 u = g) \\ \text{iii) } \gamma_0 u = u_0 \quad (\text{resp. } \gamma_1 u = u_1). \end{cases}$$

The only exception is the Neumann problem with n = 2. *In this case the theorem is valid if* (f,g,u_1) *satisfy the conditions (with* S *identified with* $(-\theta_0,\theta_0)$)

(348)
$$\begin{cases} \tilde{g} = 0, \quad \frac{1}{2\theta_0}(u_{1+} + u_{1-}) + \tilde{f} \in \operatorname{Im} A_1 \ (\text{or } \operatorname{Im} A_2), \text{ with} \\ u_{1+} = u_1(.,+\theta_0), \ u_{1-} = u_1(.,-\theta_0), \ \tilde{f} = \frac{1}{2\theta_0}\int_{-\theta_0}^{+\theta_0} f(.,\theta)\,d\theta, \ \tilde{g} = \frac{1}{2\theta_0}\int_{-\theta_0}^{+\theta_0} g(\theta)\,d\theta. \end{cases}$$

The proof is based on the weak form of the problem (347): find u in $L^2(R^+ \times S)$ such that

(349) $\quad (u, (A - B)v) = L_{f,g,u_0}(v), \quad \forall v \in D(A_1 - B_D),$

with

$$L_{f,g,u_0}(v) = (f,v) - \frac{1}{2}\langle g, B_2 v\rangle - \langle u_0, \gamma_1 v\rangle.$$

Thanks to Theorem 6, we can define the Calderon operator C relative to Ω: when $n \neq 2$, this is the isomorphism defined, with u the solution of (347) for $f = 0$, $g = 0$, by:

(350) $\quad u_0 \in \tilde{X}_0 (\text{resp. } X_0) \longrightarrow \gamma_1 u \in \tilde{X}_1 (\text{resp. } X_1).$

When $n = 2$, we have to replace spaces $\tilde{X}_1$ and X_1 by $\tilde{X}_1^+$ and X_1^+ with

(351) $\quad \tilde{X}_1^+ (\text{resp.} X_1^+) = \{u_1 = (u_{1+}, u_{1-}) \in \tilde{X}_1 (\text{resp. } X_1), \ u_{1+} + u_{1-} \in \text{Im } A_1\}.$

With the usual regularity properties of S (thus of $\Omega = R^+ \times S$) we can extend the functions u in $D^{\Omega}(A_\kappa - B_D)$ or $D^{\Omega}(A_\kappa - B_N)$ to R^n by symmetry and truncation.

This implies that the spaces of traces (343) and (344) are the same for Ω and its complement Ω'. This allows us to solve the "transmission" problem (or with given sources on the boundary Γ of a cone Ω): find u in $L^2(R^+ \times S^{n-1})$ such that

(352)
$$\left|\begin{array}{l} \text{i) } (A - B)\, u = 0 \quad \text{in } \Omega \text{ and } \Omega' = R^+ \times S', \quad S' = S^{n-1} \backslash \bar{S}, \\ \text{ii) } B_1(u|_{\Omega}) = 0, \quad B_1(u|_{\Omega'}) = 0 \ \text{(i.e., u is outgoing)}, \\ \text{iii) } [u]_\Gamma = \rho, \quad [\frac{\partial u}{\partial n}] = \rho', \end{array}\right.$$

with given jumps ρ and ρ' of u and its normal derivative across Γ.

Theorem 7. *Let* $\rho \in \tilde{X}_0$, $\rho' \in \tilde{X}_1$. *Then* (352) *has unique solution* u *such that*

(353) $\quad u|_{\Omega} \in D_M^{\Omega}(A - B), \quad u|_{\Omega'} \in D_M^{\Omega'}(A - B)$

(*if* $n = 2$ *the condition on* ρ' is: $\rho'_+ + \rho'_- \in \text{Im } A_1$ *with notations used in* (351)).
Furthermore we have the regularity result on the traces of u *and its normal derivative on each side of the boundary* Γ *of* Ω

(354) $\quad (\gamma_0^{\pm} u, \gamma_1^{\pm} u) \in X_0 \times X_1 \quad \text{when } (\rho, \rho') \in X_0 \times X_1$
(*when* $n = 2$, *we have to substitute* X_1^+ for X_1).

Let $P^{\pm}$ be the mappings $(\rho,\rho') \longrightarrow (\gamma_0^{\pm} u, \gamma_1^{\pm} u)$, u the solution of (352), we verify that $P^{\pm}$ are continuous projectors in spaces $\tilde{X}_0 \times \tilde{X}_1$ and in $X_0 \times X_1$ (or $X_0 \times X_1^+$ when $n = 2$).

These maps are the *Calderon projectors* in these spaces; they satisfy the usual relation: $P^+ + P^- = I$. These operators can be made explicit thanks to the spectral decomposition of the Laplace-Beltrami operator on S^{n-1}, and we can use them to solve scattering problems with an incident stationary field in free space on a cone occupied by a medium with wavenumber k_1.
We can generalize in many ways (see Cessenat [1]), for instance to domains $\Omega = (a, \infty) \times S$, $a > 0$, to dissipative media (associated with the limiting absorption principle).

Application to Maxwell equations. We can use the Calderon projectors above for Helmholtz equation to transform a boundary problem with Maxwell equations in a conical domain, into an (integrodifferential) problem on the boundary Γ of this domain. We can consider also the scattering of an incident wave by a dielectric cone. In order to do that, we have to transform relations between the electric field E and its normal derivative on Γ into relations between the tangential components of E and H on Γ, thanks to the differential formulas (310) in the Appendix. Using the same notations, the Maxwell equations at the boundary Γ are:

$$(355)\quad \left|\begin{array}{l} \text{i)}\ -\dfrac{\partial}{\partial n}(n \wedge E) + \overrightarrow{\mathrm{curl}}_\Gamma E^n - 2R_m\, n \wedge E + i\omega\mu\, H_\Gamma = 0, \\ \text{ii)}\ \dfrac{\partial}{\partial n}(n \wedge H) - \overrightarrow{\mathrm{curl}}_\Gamma H^n + 2R_m\, n \wedge H + i\omega\varepsilon\, E_\Gamma = 0, \\ \text{iii)}\ -\mathrm{curl}_\Gamma (E_\Gamma) + i\omega\mu\, H^n = 0, \quad \text{iv)}\ \mathrm{curl}_\Gamma (H_\Gamma) + i\omega\varepsilon\, E^n = 0, \end{array}\right.$$

with $E_\Gamma = t_\Gamma E$ the tangential component of E. We recall the divergence formulas:

$$(356)\qquad \mathrm{div}_\Gamma E_\Gamma + \frac{\partial E^n}{\partial n} + 2R_m E^n = 0,\ \ \mathrm{div}_\Gamma H_\Gamma + \frac{\partial H^n}{\partial n} + 2R_m H^n = 0.$$

We emphasize that $\frac{\partial}{\partial n}(n \wedge E)$ and $n \wedge \frac{\partial}{\partial n} E_\Gamma$ (for instance) are different!
Using an orthogonal system of coordinates (x^1, x^2, s_n), the Riemannian metric is: $g = ds^2 = g_1 dx_1^2 + g_2 dx_2^2 + ds_n^2$. If $X = X_1\partial_1 + X_2\partial_2$ is a tangent field to Γ, then:

$$(357)\qquad \frac{\partial}{\partial s_n}(n \wedge X) = n \wedge \frac{\partial X}{\partial s_n} + \tilde{\Gamma} X - 2R_m\, n \wedge X,$$

with:

$$\tilde{\Gamma} X = g_0[-X_2 \frac{\partial g_2}{\partial s_n}\partial_1 + X_1 \frac{\partial g_1}{\partial s_n}\partial_2],\quad g_0 = (g_1 g_2)^{-1/2}\ \text{ and }\ 2R_m = g_0 \frac{\partial}{\partial s_n} g_0^{-1}.$$

Of course, we would have to specify the functional framework for such Maxwell scattering problem in correspondence to the above scalar problems with Helmholtz equation. The functional spaces are based on weighted norms. We don't develop these (technical) spaces here, but results are especially interesting in the simplest case where the conical domain is the whole space! ⊗

CHAPTER 6

EVOLUTION PROBLEMS

Evolution problems in electromagnetism and more generally non-stationary wave problems are of many different types:

First we consider Cauchy problems, i.e., evolution problems with given initial conditions (or also mixed Cauchy problems, i.e., with boundary conditions). These problems are fairly easy to solve and quite "standard" in free space, or if we assume an "ideal" linear isotropic media with constitutive relations $D = \varepsilon E$, $B = \mu H$. In the general case the constitutive relations have to be written with a convolution in time, see (22) chap. 1. This implies that the usual Cauchy problem is ill-posed: we have to know all the past of the material to solve evolution problems, which are called Cauchy problems *with delay or with thick initial conditions*, see Dautray-Lions [1] chap. 18.

Here also scattering problems lead us to define *incoming and outgoing waves*, associated with Cauchy problems under fewer initial conditions.

Then we consider "*causal*" evolution problems, which are well-posed problems without any initial condition. Many evolution problems concerning waves and electromagnetism are of this type.

Maxwell equations are an example of symmetric systems of Friedrichs. We do not develop this general theory, but only refer to Friedrichs [1], Lax-Phillips [3], Chazarain-Piriou [1], Colton-Kress [2]. We do not either use microlocal analysis (and the trace theorem of Hörmander) in this book.

1. CAUCHY PROBLEMS

First we consider the following Cauchy problems. Let H be a Hilbert space; given U_0 and F, an H-valued function, find an H-valued function U such that:

$$(1)\qquad \left| \begin{array}{l} \text{i) } \dfrac{\partial U}{\partial t} + AU = F, \quad \text{with } t > 0 \text{ or with } t \in R, \\ \text{ii) } U(0) = U_0. \end{array} \right.$$

Problem (1) for $t > 0$ only, is called a *one-sided (or unilateral) problem*. Problem (1) for real t is called a *two-sided (or bilateral) Cauchy problem*.
First we assume that $-A$ is the infinitesimal generator of a semigroup $(G(t))$ of class C^0 in H. From usual theory (see Pazy [1], Dautray-Lions [1] chap. 17) we know that if $F \in C^0([0,\infty), H)$ (or if $F \in L^1([0,T), H)$ for all T) with $U_0 \in H$, the *one sided problem* (1) has a unique solution $U \in C^0([0,\infty), H)$ (in a weak sense) given by:

$$(2) \qquad U(t) = G(t)U_0 + \int_0^t G(t-s)F(s)\,ds.$$

Then assuming that $-A$ is the infinitesimal generator of a group $(G(t))$, and F is given for all real t, *the two-sided Cauchy problem* (1) has a unique solution with the same regularity properties.

1.1. Scalar wave Cauchy problems

We recall some well-known results on Cauchy problems for waves. As a reference on this subject see Hörmander [1], Chazarain-Piriou [1], Dautray-Lions [1]. The usual standard Cauchy problem for scalar waves is: given a (complex or real) Hilbert space H, find an H-valued function u so that:

$$(3) \quad \left| \begin{array}{l} \text{i) } \dfrac{\partial^2 u}{\partial t^2} + A\,u = f \quad \text{in } R_t, \text{ or } R_t^+, \\ \text{ii) } u(0) = u^0, \quad \dfrac{\partial u}{\partial t}(0) = u^1, \end{array} \right.$$

with given initial conditions u^0, u^1, and with a given source f. The operator A is a positive (or even bounded from below) selfadjoint operator defined using a variational framework $(V, H, a(u,v))$. The usual space with finite energy waves corresponds to $u^0 \in V$, $u^1 \in H$, and $f \in C^0(R_t, H)$ (or more generally $f \in L^1_{loc}(R_t, H)$). Then taking $H = V \times H$, and:

$$(4) \quad U_0 = \begin{pmatrix} u_0 \\ u_1 \end{pmatrix}, \; U(t) = \begin{pmatrix} u(t) \\ u'(t) \end{pmatrix}, \; F(t) = \begin{pmatrix} f(t) \\ 0 \end{pmatrix}, \; A = \begin{pmatrix} 0 & -I \\ A & 0 \end{pmatrix}, \text{ with } u' = \frac{\partial u}{\partial t},$$

problem (3) is written in the form (1), with A the infinitesimal generator of a group $(G(t))$, and thus (3) has a unique solution $U \in C^0(R, H)$, giving $u \in C^0(R, V)$, with $u' \in C^0(R, H)$; u is given by

$$(5) \quad u(t) = \cos(tA^{1/2})u^0 + A^{-1/2}\sin(tA^{1/2})u^1 + \int_0^t A^{-1/2}\sin((t-s)A^{1/2})\,f(s)\,ds.$$

We obtain very easily this formula of symbolic calculus, defining z and w by:

$$(6) \quad \left| \begin{array}{ll} z(t) = iA^{1/2}u(t) + u'(t), & w(t) = -iA^{1/2}u(t) + u'(t), \\ z_0 = iA^{1/2}u^0 + u^1, & w_0 = -\,iA^{1/2}u^0 + u^1. \end{array} \right.$$

Then z and w satisfy:

(7) $\quad z' = iA^{1/2} z(t) + f, \qquad w' = - iA^{1/2} w(t) + f,$

and the solution of (7) is given as in (2) by:

$$(8) \quad \begin{cases} z(t) = e^{iA^{1/2}t} z_0 + \int_0^t e^{iA^{1/2}(t-s)} f(s)\, ds, \\ w(t) = e^{-iA^{1/2}t} w_0 + \int_0^t e^{-iA^{1/2}(t-s)} f(s)\, ds. \end{cases}$$

We obtain (5) by taking the difference $z(t) - w(t)$. We note that the operator:

(9) $\quad M(t) = A^{-1/2} \sin(tA^{1/2})$

is a bounded operator, and that (5) is also given by:

(5)' $\quad u(t) = \frac{\partial}{\partial t} M(t)u^0 + M(t)u^1 + \int_0^t M(t-s)f(s)\, ds.$

Furthermore, multiplying (3)i) by u' (the time derivative of u) and integrating, we check that u satisfies the energy balance (for $f = 0$ this is the energy conservation law)

$$(10) \quad \frac{1}{2}[\|\frac{\partial u}{\partial t}(t)\|^2 + \|A^{1/2}u(t)\|^2] = \frac{1}{2}[\|u^1\|^2 + \|A^{1/2}u^0\|^2] + \mathrm{Re}\int_0^t (f, \frac{\partial u}{\partial s})\, ds.$$

The first usual example is the Cauchy problem in free space occupying R^n associated with the wave equation or *d'Alembert equation*:

$$(11) \quad \begin{cases} \text{i) } \Box u \overset{\text{def}}{=} \frac{1}{c^2}\frac{\partial^2 u}{\partial t^2} - \Delta u = f \quad \text{in } R^n \times R_t, \\ \text{ii) } u(0) = u^0, \ \frac{\partial u}{\partial t}(0) = u^1, \end{cases}$$

c being the velocity of light in free space, u^0, u^1 being given initial conditions, and f being a given source.

The usual framework is *the space of finite energy solutions*. Using the Beppo Levi space:

(12) $\quad W^1(R^n) = \{v \in \mathcal{D}'(R^n), \ v \in L^2_{loc}(R^n), \ \int_{R^n} (|\mathrm{grad}\, v|^2)\, dx < \infty\}$

(see also Appendix, Def. 7 section 5.2), we assume that the data satisfy:

(13) $\quad u^0 \in W^1(R^n), \ u^1 \in L^2(R^n), \ f \in \mathcal{C}^0(R_t, L^2(R^n)).$

Then taking $V = W^1(R^n)$, $H = L^2(R^n)$, $\mathcal{H} = W^1(R^n) \times L^2(R^n)$, $A = -c^2\Delta$ (but V is not contained in H), we see that problem (7) is of the form (3); thus it has a unique solution u obtained in the form (5), thanks to its Fourier transform (we take c = 1 in order to simplify the formulas):

$$(14) \qquad \hat{u}(\xi,t) = \cos(|\xi|t)\,\hat{u}^0(\xi) + \frac{\sin(|\xi|t)}{|\xi|}\hat{u}^1(\xi) + \int_0^t \frac{\sin(|\xi|(t-s))}{|\xi|}\hat{f}(\xi,s)\,ds.$$

We can write this formula using elementary solutions of the d'Alembert equation as will be seen below. We have used the semigroup theory to solve this Cauchy problem, but there are many other ways to solve it (see Dautray-Lions [1] for instance).

We can solve mixed Cauchy problems also, i.e., problems in a domain Ω with a Dirichlet or a Neumann condition. Let Ω be the exterior of a bounded domain with boundary Γ. For a Dirichlet condition, we have to substitute for $W^1(R^n)$ the Beppo Levi space (see Def. 8 in the Appendix) which is also defined in the regular case by:

$$(15) \qquad W_0^1(\Omega) = \{u \in \mathcal{D}'(\Omega),\ u \in L^2_{loc}(\Omega),\ \int_\Omega |\text{grad}\, u|^2 dx < \infty,\ u|_\Gamma = 0\}.$$

We first give a general property of waves satisfying (3) when t goes to infinity. We define the kinetic energy $E_c(t)$ and the potential energy $E_p(t)$ of u by:

$$E_c(t) = \frac{1}{2}\|\frac{\partial u}{\partial t}(t)\|^2, \quad E_p(t) = \frac{1}{2}\|A^{1/2}u(t)\|^2.$$

Then using the notion of spectral measure (see Kato [1]) and (5), we can prove the following property when there is no source (for instance when $A = -\Delta$ in the exterior of a bounded domain, see Goldstein [1], Lax-Phillips [1])

Proposition 1. *Equipartition of energy. If* 0 *is not an eigenvalue of* A, *then the solution* u *of* (3) *with* f = 0 *satisfies equipartition in the mean*:

$$\lim_{t\to\pm\infty} \frac{1}{t}\int_0^t E_c(s)ds = \lim_{t\to\pm\infty} \frac{1}{t}\int_0^t E_p(s)ds.$$

Moreover if the spectral measure $dE(\lambda)$ *is absolutely continuous with respect to the Lebesgue measure* $d\lambda$, u *satisfies*:

$$\lim_{t\to\pm\infty} E_c(t) = \lim_{t\to\pm\infty} E_p(t).$$

1.2. Cauchy problems in electromagnetism

1.2.1. Cauchy problems in free space R^3

We consider the Cauchy problem: find the electromagnetic field (E,B) (or (E,H)) in free space $\Omega = R^3$, satisfying the Maxwell equations (1) with (5) and (3) chap. 1, for positive t (one-sided problem), or for all real t (two-sided problem):

$$
(16)\quad \left|\begin{array}{l}
\text{i)}\ -\dfrac{\partial D}{\partial t}+\text{curl}\,H=J,\quad \dfrac{\partial B}{\partial t}+\text{curl}\,E=0,\\
\text{ii)}\ \text{div}\,D=\rho,\quad \text{div}\,B=0,\\
\text{iii)}\ D=\varepsilon_0 E,\quad B=\mu_0 H,\\
\text{iv)}\ E(0)=E_0,\ B(0)=B_0,\ \ (\text{or } H(0)=H_0,\ D(0)=D_0),
\end{array}\right.
$$

with given charge density ρ and current density J satisfying (3) chap.1, and with given initial conditions for E, B (or D, H). It is more common to work with (E,B) in free space than with (E,H) for evolution problems, so we will use (E,B) only. As indicated in chap.1, these equations are partially redundant. We assume that:

$$
(17)\qquad E_0 \in L^2(R^3)^3,\quad B_0 \in L^2(R^3)^3,\quad \text{and } J \in L^2_{loc}(R_t, L^2(R^3)^3.
$$

We define $H = L^2(R^3)^3 \times L^2(R^3)^3$, and with $\varepsilon_0\mu_0 = 1/c^2$,

$$
(18)\ U(t)=\begin{pmatrix} E(t)\\ B(t)\end{pmatrix},\ U_0=\begin{pmatrix} E_0\\ B_0\end{pmatrix},\ F(t)=\begin{pmatrix} -\frac{1}{\varepsilon_0}J(t)\\ 0\end{pmatrix},\ A=\begin{pmatrix} 0 & -c^2\,\text{curl}\\ \text{curl} & 0\end{pmatrix}.
$$

Then the Cauchy problem (16), without taking ii) into account, is written in the form of (1). Moreover we can prove easily by Fourier transformation that $-A$ is the infinitesimal generator of a unitary group (G(t)) in H, with domain $D(A) = H(\text{curl}, R^3) \times H(\text{curl}, R^3)$. Then we know that the two-sided Cauchy problem (16) with (17) has a unique solution (E,B) in $C^0(R_t, H)$ with finite energy (with conservation of energy when J = 0, see (7) chap.1):

$$
(19)\qquad W(t)=\tfrac{1}{2}\int_{R^3}(D(t).E(t)+B(t).H(t))\,dx = W(0)-\int_0^t ds\int_{R^3} E.J\,dx\,,
$$

but does not necessarily satisfy (16)ii). If the data satisfy, in addition to (3) chap.1:

$$
(20)\qquad \text{div}\,B_0 = 0 \text{ and } \text{div}\,E_0 = \rho/\varepsilon_0 \text{ with } \rho \in L^2_{loc}(R_t, L^2(R^3)),
$$

then we can prove that the solution (E,B) of the Cauchy problem also satisfies:

$$
\text{div}\,B(t) = 0, \text{ and } \text{div}\,E(t) = \rho(t)/\varepsilon_0 \text{ for all } t.
$$

1.2.2. Cauchy problems in a domain Ω

Now let Ω be a free space domain in R^3 bounded by a perfect conductor, with boundary Γ. We have to solve a mixed Cauchy problem: find (E,H) satisfying the Maxwell problem (16) in Ω, with the boundary conditions

$$
(21)\qquad n\wedge E|_{\Gamma\times R^+}=0,\qquad n.B|_{\Gamma\times R^+}=0.
$$

We assume that the data satisfy (17) with Ω instead of R^3. Then with $H = L^2(\Omega)^3 \times L^2(\Omega)^3$, we can prove that operator A, defined in (18), with domain $D(A) = H_0(\text{curl}, \Omega) \times H(\text{curl}, \Omega)$ is such that:

$$A^* = -A, \text{ with } D(A^*) = D(A)$$

(iA is a selfadjoint operator if we complexify the space). As a consequence of the Stone theorem, A is the infinitesimal generator of a unitary group (G(t)) in H, at least when Ω satisfies one of the following conditions:
i) Ω is a bounded domain, or the complement of a bounded domain,
ii) Ω is a cylinder (see Duvaut-Lions [1], Dautray-Lions [1] chap. 9A and 17B.4).
Of course we can take other frameworks which are subspaces of the above space H for these evolution problems, in order to take into account usual conditions of free divergence for B for instance, or regularity properties. This allows us to indicate in which sense the boundary conditions are satisfied (thanks to the notion of weak and strong solution). Thus we can use any variational framework of chap.2.11.1.
It is interesting to compare the Maxwell problem to the wave problem, for instance with respect to H (or B) only. We eliminate E by a usual trick to obtain:

$$(22)\quad \left|\begin{array}{l} \text{i)}\ \dfrac{1}{c^2}\dfrac{\partial^2 H}{\partial t^2} - \Delta H = \operatorname{curl} J, \qquad \operatorname{div} H = 0, \\ \text{ii)}\ H(0) = H_0, \qquad \dfrac{\partial H}{\partial t}(0) = -\dfrac{1}{\mu_0}\operatorname{curl} E_0, \\ \text{iii)}\ n.H|_{\Gamma \times R^+} = 0, \qquad n \wedge \operatorname{curl} H|_{\Gamma\times R^+} = 0. \end{array}\right.$$

The boundary conditions imply (see (147) chap. 2) that $(-\Delta)$ is the positive selfadjoint operator $(A^+ - I)$. For E, we obtain (in the regular case) a similar result but with the operator $(A^- - I)$.
Of course we can generalize these results with respect to the Cauchy problem in many ways: for instance, thanks to a variational method described in Duvaut-Lions [1], to the case of several linear (isotropic or not) media with different permittivities and permeabilities (but frequency independent in time Fourier transform).
Now we recall some properties of the wave equation and above evolution problems. These properties are common to hyperbolic equations: the wave equation is a typical hyperbolic equation.

1.3. Some hyperbolic properties of wave evolution

The main property of the d'Alembert equation is the *existence of an elementary solution* $\Phi(x,t)$ *(or* $\Phi_+(x,t)$*) with support in the future cone or forward light cone*:

$$(23)\qquad C_+ = \{(x,t) \in R^n \times R, \text{ with } |x| \le ct\}.$$

This solution which is thus characterized (taking c = 1) by:

$$(24)\qquad \Box\Phi \overset{\text{def}}{=} \frac{\partial^2\Phi}{\partial t^2} - \Delta\Phi = \delta(x,t) = \delta(x)\,\delta(t), \qquad \text{with supp}\,\Phi \subseteq C_+,$$

is *unique*: if Φ_0 is another elementary solution with the same property, then the convolution of Φ_0 and Φ is defined and we have

(25) $$\Box(\Phi * \Phi_0) = (\Box\Phi) * \Phi_0 = \Phi_0 = \Phi * (\Box\Phi_0) = \Phi.$$

This existence is *typical of hyperbolic equations* (see Dautray-Lions [1] chap. 5). There are many ways to calculate this elementary solution, each method having its own interest (one of them is the Radon transform method). We can also obtain simply by its space Fourier transform (compare to (14)):

(26) $$\hat{\Phi}(\xi,t) = Y(t)\,\hat{R}(\xi,t), \text{ with } \hat{R}(\xi,t) = \frac{\sin(|\xi|t)}{|\xi|},$$

where $Y(t)$ is the Heaviside function. Here we give the very simple result when $n = 3$:

(27) $$\Phi(x,t) = \frac{1}{4\pi}\frac{1}{t}\delta(r-t) = \frac{1}{2\pi}Y(t)\,\delta(r^2 - t^2), \quad r = |x|.$$

The operator $M(t)$ given by (9) in this case is simply given by the convolution:

(28) $$M(t)v = R(.,t) \underset{x}{*} v, \text{ with } R(x,t) = \frac{1}{2\pi}(\operatorname{sign} t)\,\delta(r^2 - t^2),\ t \neq 0.$$

In the general case, results differ essentially with the evenness of space: when n is even the support of Φ is the whole future cone, whereas when n is odd the support of Φ is the surface of this cone only.

There is also a unique *elementary solution* $\Phi_-(.,t)$ *with support in the past cone or backward light cone,* given by: $\Phi_-(.,t) = -Y(-t)\,R(.,t)$, and thus $R(.,t) = \Phi_+(.,t) - \Phi_-(.,t)$ is a solution of the homogeneous wave equation.

When Ω is the complement of a bounded domain Ω', we obtain as a consequence of the properties of Φ :

Proposition 2 (*Finite velocity of propagation*). *Let* $R > 0$ *be such that* Ω' *is contained in* B_R. *If the initial conditions* (u^0, u^1) *are zero outside* B_R, *the solution* u *of*:

(29) $$\begin{cases} \text{i)}\ \dfrac{1}{c^2}\dfrac{\partial^2 u}{\partial t^2} - \Delta u = 0 \quad \text{in } \Omega \times R_t^+, \\ \text{ii)}\ u(0) = u^0,\ \dfrac{\partial u}{\partial t}(0) = u^1, \quad (u^0,u^1) \in W_0^1(\Omega) \times L^2(\Omega), \\ \text{iii)}\ u|_{\Gamma\times R^+} = 0 \quad (\text{or } \dfrac{\partial u}{\partial n}|_{\Gamma\times R^+} = 0), \end{cases}$$

is zero outside the ball $B_{R+c|t|}$. *Furthermore if the initial conditions* (u^0, u^1) *are zero inside* the ball $B_R(x_0)$ *centered at* x_0 *then the solution* u *of* (29) *is zero inside the ball* $B_{R-ct}(x_0)$ *when* $t < R/c$.

Then we have the property of *local propagation of energy*: Let $E(u,t,\rho)$ be the energy of the wave u at time t in a ball B_ρ (B_ρ being arbitrary):

(30) $$E(u,t,\rho) = \frac{1}{2}\int_{\Omega \cap B_\rho} \left(\left|\frac{1}{c}\frac{\partial u}{\partial t}(x,t)\right|^2 + |\operatorname{grad} u(x,t)|^2\right) dx,$$

then we have

(31) $$E(u,T+t,\rho) \le E(u,T,\rho+ct) \text{ for all } T \in \mathbb{R} \text{ and } t > 0.$$

1.4. Radon transforms

The Radon transform method is a basic tool to obtain many properties in scattering and wave evolution. We briefly give the main definitions and properties, and refer to Gelfand-Graev-Vilenkin [1] and to Helgason [1] for more developments. First using regular functions we have:

Definition 1. *The Radon transform of a function* f *in* $S(R^n)$ *is the function on* $R \times S^{n-1}$ *defined by*:

(32) $$Rf(s,\alpha) = \tilde{f}(s,\alpha) = \int_{x.\alpha = s} f(x)\, d\tilde{x}, \qquad s \in R,\ \alpha \in S^{n-1},$$

$d\tilde{x}$ *being the surface element of the plane* $x.\alpha = s$, with $ds\, d\tilde{x} = dx$.

Note that we can take α in $R^n \backslash \{0\}$ instead of S^{n-1}. Then (32) would give a function satisfying the homogeneous condition:

(33) $$\tilde{f}(\lambda s, \lambda\alpha) = |\lambda|^{-1} \tilde{f}(s,\alpha), \quad \forall \lambda \neq 0,\ (s,\alpha) \in R x R^n_* .$$

We first have the basic property:

(34) $$\int_R e^{-irs} \tilde{f}(s,\alpha)\, ds = \int_R e^{-irs} ds \int_{x.\alpha = s} f(x)\, d\tilde{x} = \int_{R^n} e^{-ir\alpha.x} f(x)\, dx = \hat{f}(r\alpha),$$

where $\hat{f}$ is the Fourier transform of f. Thus if we define (like in chap.3, (396) (400)):

(35) $$\hat{F}(r,\alpha) = \hat{f}(r\alpha),\ \ r \ge 0, \ \text{ and } \ \hat{F}(r,\alpha) = \hat{f}((-r)(-\alpha)),\ \ r \le 0,\ \forall \alpha \in S^{n-1},$$

$\hat{F}(r,\alpha)$ is a function, defined on $R \times S^{n-1}$, which is in space $S_0(R \times S^{n-1})$ when f is in $S(R^n)$ with the notation 1 of chap. 3.11.3. Then we define the following space:

(36) $S^{\sim}(R \times S^{n-1})$ is the space of inverse Fourier transforms of $S_0(R \times S^{n-1})$.

Proposition 3. *The Radon transformation is an isomorphism from* $S(R^n)$ *onto* $S^{\sim}(R \times S^{n-1})$; *the Fourier transform of* f *and the Radon transform of* f *are related to each other through the Fourier transform* (34) *on* s.

The elements of $S^{\sim}(R \times S^{n-1})$ are characterized by

(37) $$\left| \begin{array}{ll} \text{i) } g(s,\alpha) = g(-s,-\alpha), & \forall (s,\alpha) \in R \times S^{n-1}, \\ \text{ii) } \int_R g(s,\alpha)\, s^k\, ds = P_k(\alpha) \text{ a spherical harmonic of order } k, & k \in N. \end{array} \right.$$

We note the properties of the Radon transformation with respect to translations, derivations and convolution:

$$(38)\quad \left|\begin{array}{l} \text{i) } \tilde{f}_a(s,\alpha) = \tilde{f}(s + a.\alpha,\alpha) \quad \text{with } f_a(x) = f(x+a), \\ \text{ii) } \left(\frac{\partial f}{\partial x_j}\right)^{\sim}(s,\alpha) = \alpha_j \frac{\partial \tilde{f}}{\partial s}(s,\alpha), \text{ and thus: } (\Delta f)^{\sim}(s,\alpha) = \frac{\partial^2 \tilde{f}}{\partial s^2}(s,\alpha), \\ \text{iii) } h = f * g \text{ with } f \text{ and } g \text{ in } S(R^n) \text{ implies: } \tilde{h}(s,\alpha) = \int_R \tilde{f}(s',\alpha)\,\tilde{g}(s-s',\alpha)\,ds'. \end{array}\right.$$

The inverse Radon transformation is obtained using the inverse Fourier transformation:

$$(39)\qquad f(x) = \frac{1}{(2\pi)^n}\int_{R^n} e^{ix.\xi}\,\hat{f}(\xi)d\xi = \frac{1}{(2\pi)^n}\int_{R^+ \times S^{n-1}} e^{irx.\alpha}\,\hat{f}(r\alpha)\,r^{n-1}dr d\alpha.$$

Then $g(r,\alpha) = e^{irx.\alpha}\,\hat{f}(r\alpha)$ has a natural extension to $R \times S^{n-1}$ by (36). Thus we have two cases according to the evenness of n:

i) n is *odd*:

$$(40)\qquad f(x) = \frac{1}{2(2\pi)^n}\int_{R \times S^{n-1}} e^{irx.\alpha}\,\hat{F}(r,\alpha)\,r^{n-1}dr d\alpha.$$

Then by the inverse Fourier transformation with respect to r, *with* $m = (n-1)/2$,

$$(41)\quad f(x) = \frac{(-1)^m}{2(2\pi)^{n-1}}\int_{S^{n-1}} \frac{\partial^{n-1}\tilde{f}}{\partial s^{n-1}}(x.\alpha\,,\,\alpha)\,d\alpha = \frac{1}{2(2\pi)^{n-1}}(-\Delta)^m \int_{S^{n-1}} \tilde{f}(x.\alpha\,,\,\alpha)\,d\alpha.$$

ii) n is *even*:

$$(42)\qquad f(x) = \frac{1}{2(2\pi)^n}\int_{R \times S^{n-1}} e^{irx.\alpha}\,\hat{F}(r,\alpha)\,(\text{sign } r)\,r^{n-1}dr d\alpha.$$

Then thanks to the Fourier transform of finite parts (see Schwartz [1] chap. 7.7):

$$(42)'\qquad F\left((-1)^p\frac{(2p-1)!}{\pi}\,\text{Pf}\,s^{-2p}\right) = (\text{sign } r)\,r^{2p-1},$$

we obtain

$$(43)\qquad f(x) = \frac{(-1)^{n/2}(n-1)!}{(2\pi)^n}\int_{R \times S^{n-1}} \text{Pf}\frac{\tilde{f}(s,\alpha)}{(x.\alpha - s)^n}\,ds\,d\alpha.$$

We can define *the Radon transformation on tempered distributions*, either by using a Fourier transformation, or by duality. Let g be in $S(R \times S^{n-1})$, and:

$$(44)\qquad \check{g}(x) = \int_{S^{n-1}} g(x.\alpha,\alpha)\,d\alpha.$$

Then $g \rightarrow \check{g}$ is a continuous mapping from $S(R \times S^{n-1})$ onto $S(R^n)$ so that

$$(44)'\qquad \langle f, \check{g}\rangle = \langle \tilde{f}, g\rangle, \quad \forall f \in S'(R^n).$$

That this mapping is onto is a consequence of inverse formulae (41) and (43). Then

$$\int_{R^n} f(x)\check{g}(x)\,dx = \int_{S^{n-1}} d\alpha \int_R ds \int_{\alpha.x=s} f(x)g(s,\alpha)\,d\tilde{x} = \int_{R\times S^{n-1}} \tilde{f}(s,\alpha)g(s,\alpha)\,ds d\alpha.$$

Thus (44)' defines the Radon transformation as a continuous mapping from $\mathcal{S}'(R^n)$ *into* $\mathcal{S}'(R\times S^{n-1})$.

Another *basic property of the Radon transformation* (and its advantage with respect to Fourier transformation) is that it *preserves the boundness of supports*. If supp f is bounded then it is obvious that supp $\mathcal{R}f$ is also bounded. For the converse we refer to Helgason [1]. We also have the obvious property for *odd* n

(45) $\tilde{f}(s,\alpha) = 0$ for $|s| \le a$, $a > 0$, $\forall \alpha \in S^{n-1}$, implies $f(x) = 0$ for $|x| \le a$.

Furthermore we have a "Plancherel" formula for Radon transformation; applying the Plancherel formula twice to Fourier transforms leads to:

i) when n is *odd* (and $m = (n-1)/2$):

$$(46) \qquad \int_{R^n} f(x)\bar{g}(x)\,dx = \frac{1}{2(2\pi)^{n-1}} \int_{R\times S^{n-1}} \frac{\partial^m \tilde{f}}{\partial s^m}(s,\alpha).\frac{\partial^m \bar{\tilde{g}}}{\partial s^m}(s,\alpha)\,ds\,d\alpha.$$

ii) when n is *even:*

$$(47) \int_{R^n} f(x)\bar{g}(x)\,dx = \frac{(-1)^{n/2}(n-1)!}{(2\pi)^n} \int_{R\times R\times S^{n-1}} \mathrm{Pf}\, \tilde{f}(s_1,\alpha)\,\bar{\tilde{g}}(s_2,\alpha)\,(s_1 - s_2)^{-n}\,ds_1 ds_2 d\alpha.$$

When n is *odd*, we define the following space:

(48) $L^2_m(R\times S^{n-1}) = \{v \in L^2(R\times S^{n-1}),\ v(-s,-\alpha) = (-1)^m v(s,\alpha), \text{ with } m = (n-1)/2\}$.

Then from the "Plancherel" formula we have:

Proposition 4. *The mapping*: $f \in \mathcal{S}(R^n) \to \frac{\partial^m \tilde{f}}{\partial s^m}$ *has a continuous unitary (up to a constant factor) extension from* $L^2(R^n)$ *onto the Hilbert space* $L^2_m(R\times S^{n-1})$.

Definition 2. *The unitary (up to a constant) extension of Proposition* 4 *(with* n *odd) is denoted by* $\mathcal{R}^d$ *and is called the "differentiated Radon transformation".*

Thus it satisfies, with $\mathcal{R}^d f = \frac{\partial^m \tilde{f}}{\partial s^m}$

$$(49) \qquad \int_{R^n} |f(x)|^2\,dx = \frac{1}{2(2\pi)^{n-1}} \int_{R\times S^{n-1}} |\mathcal{R}^d f(s,\alpha)|^2\,ds\,d\alpha.$$

Given two functions f_0 and f_1 in $L^2(R^n)$, we define the function $\hat{h}$ on $R \times S^{n-1}$ by:

(50) $\hat{h}(r,\alpha) = \hat{f}_0(r\alpha)$ if $r > 0$, $\hat{h}(r,\alpha) = \hat{f}_1((-r)(-\alpha))$ if $r < 0$.

Let $h = h(s,\alpha)$ be its inverse Fourier transform (with respect to r). Then we note

(51) $h = R_2(f_0,f_1)$.

Proposition 5. *The mapping*: $(f_0,f_1) \to \frac{\partial^m h}{\partial s^m}$ *is a unitary (up to a constant factor) mapping from* $L^2(R^n) x L^2(R^n)$ onto $L^2(R \times S^{n-1})$, *with* n *odd,* $m = (n-1)/2$:

(52) $$\int_{R^n} [|f_0(x)|^2 + |f_1(x)|^2] dx = \frac{1}{(2\pi)^{n-1}} \int_{R \times S^{n-1}} |\frac{\partial^m h}{\partial s^m}(s,\alpha)|^2 ds\, d\alpha.$$

Definition 3. *The unitary (up to a constant factor) mapping of Proposition 5 is denoted by* $R_2{}^d$ *and is called a "two-sided differentiated" Radon transformation.*

We see that $R_2{}^d$ and R_2 are natural extensions of R^d and R since $R_2{}^d(f,f) = R^d f$ and $R_2(f,f) = Rf$. Furthermore we have the following basic properties:

(53) $$\left| \begin{array}{l} R_2^d(f_0,f_1) = \frac{\partial^m}{\partial s^m} R_2(f_0,f_1), \quad R_2((-\Delta)^{1/2}f, -(-\Delta)^{1/2}f) = -i\frac{\partial \tilde{f}}{\partial s}, \\ R_2^d((-\Delta)^{1/2}f_0, -(-\Delta)^{1/2}f_1) = -i\frac{\partial}{\partial s} R_2^d(f_0,f_1) = -i\frac{\partial^{m+1}}{\partial s^{m+1}} R_2(f_0,f_1). \end{array} \right.$$

Definition 4. *We denote by* R^w *the mapping (called "wave Radon transformation") defined for* (f,g) *in* $W^1(R^n) \times L^2(R^n)$, *with* n *odd, by*

(54) $$R^w(f,g) = R_2^d(-i(-\Delta)^{1/2}f + g,\ i(-\Delta)^{1/2}f + g) = -\frac{\partial^{m+1}\tilde{f}}{\partial s^{m+1}} + \frac{\partial^m \tilde{g}}{\partial s^m}.$$

Proposition 6. *The mapping* R^w *is a unitary (up to a constant factor) transformation from* $W^1(R^n) \times L^2(R^n)$ *onto* $L^2(R \times S^{n-1})$:

(54)' $$\frac{1}{2(2\pi)^{n-1}} \int_{R \times S^{n-1}} |R^w(f,g)|^2 ds\,d\alpha = \|(-\Delta)^{1/2}f\|^2 + \|g\|^2 = \|f\|^2_{W^1(R^n)} + \|g\|^2_{L^2(R^n)},$$

and satisfies, with A given by (4), *and* $A = -\Delta$:

(55) $$R^w(A(f,g)) = \frac{\partial}{\partial s} R^w(f,g).$$

This Proposition is fairly easy to prove. Here we give its inverse only. Let h be in $L^2(R \times S^{n-1})$; we have to find f and g so that $h = R^w(f,g)$, that is with (54):

$$(56)\qquad -\frac{\partial^{m+1}\tilde{f}}{\partial s^{m+1}} + \frac{\partial^m \tilde{g}}{\partial s^m} = h.$$

Using the different behaviors of each term with respect to symmetry, we have:

$$\frac{\partial^m \tilde{g}}{\partial s^m}(s,\alpha) = \frac{1}{2}(h(s,\alpha) + (-1)^m h(-s,-\alpha)),\quad \frac{\partial^{m+1}\tilde{f}}{\partial s^{m+1}}(s,\alpha) = -\frac{1}{2}(h(s,\alpha) + (-1)^{m+1} h(-s,-\alpha)).$$

Then applying the inversion formula (41), we obtain:

$$(57)\ g(x) = \frac{(-1)^m}{2(2\pi)^{n-1}} \int_{S^{n-1}} \frac{\partial^m h}{\partial s^m}(x.\alpha,\alpha)\, d\alpha,\quad f(x) = \frac{(-1)^{m-1}}{2(2\pi)^{n-1}} \int_{S^{n-1}} \frac{\partial^{m-1} h}{\partial s^{m-1}}(x.\alpha,\alpha)\, d\alpha.$$

1.5. First applications of the Radon transform method

Now the (usual) Radon transform of the wave problem (11) with c = 1, f = 0, and with (38), is the one dimensional wave problem:

$$(58)\quad \left|\begin{array}{l} \text{i) } \dfrac{\partial^2 \tilde{u}}{\partial t^2}(s,\alpha,t) - \dfrac{\partial^2 \tilde{u}}{\partial s^2}(s,\alpha,t) = 0 \quad \text{in } R_s \times S^{n-1} \times R_t, \\ \text{ii) } \tilde{u}(s,\alpha,0) = \tilde{u}^0(s,\alpha),\ \dfrac{\partial \tilde{u}}{\partial t}(s,\alpha,0) = \tilde{u}^1(s,\alpha). \end{array}\right.$$

Its solution is given by the d'Alembert formula:

$$(59)\qquad \tilde{u}(s,\alpha,t) = \frac{1}{2}(\tilde{u}^0(s+t,\alpha) + \tilde{u}^0(s-t,\alpha)) + \frac{1}{2}\int_{s-t}^{s+t} \tilde{u}^1(s',\alpha)\, ds',$$

and we obtain the solution u of (11) with f = 0 applying the inverse Radon transformation (41) or (44). Now let $\chi_{[-a,+a]}$ be the characteristic function of the interval $[-a,+a]$ when $a>0$. When $a<0$, we define $\chi_{[-a,+a]} = -\chi_{[+a,-a]}$. Then the operator M(t) (see (9)) is easily obtained thanks to the convolution:

$$(60)\qquad M(t)v(x) = R(.,t) \underset{x}{*} v = \frac{1}{2} R^{-1}(\chi_{[-t,+t]} * \tilde{v}),$$

thus (using (38)):

$$(61)\qquad R(.,t) = \frac{1}{2} R^{-1}(\chi_{[-t,+t]}).$$

Thus to obtain the elementary solution Φ with support in the future light cone which is related to R by $\Phi(.,t) = Y(t)R(.,t)$ (see (26)), we have to calculate the inverse Radon transform of $\chi_{[-t,+t]}$ thanks to (41) or (44). We obtain the "Herglotz Petrovski" formulas for these elementary solutions:

$$(62)\quad \begin{cases} \Phi(x,t) = Y(t)\,(-1)^{m+1} \dfrac{1}{(2\pi)^{n-1}} \displaystyle\int_{S^{n-1}} \delta^{(n-2)}(x.\alpha - t)\, d\alpha, \text{ with odd } n,\ m = (n-1)/2 \\[2ex] \Phi(x,t) = Y(t)\,(-1)^{n/2} \dfrac{(n-2)!}{2\,(2\pi)^{n}} \displaystyle\int_{S^{n-1}} \mathrm{Pf}\dfrac{1}{(x.\alpha - t)^{n-1}}\, d\alpha, \text{ with even } n. \end{cases}$$

We can obtain R(t) when n is *odd* (at least formally): we have

$$(63)\qquad \check{\delta}_t(x) = \int_{S^{n-1}} \delta(x.\alpha - t)\, d\alpha = \sigma_{n-1} \int_{-\tau}^{\tau} \delta(|x|s - t)\,(1 - s^2)^{m-1}\, ds,$$

with $\tau = \inf(t/|x|, 1)$, $\sigma_{n-1} = |S^{n-2}|$ the area of the unit sphere S^{n-2}. Let:

$$J(x) = \frac{1}{|x|}\left(\frac{1}{|x|^2} - 1\right)^{m-1} \text{ if } |x| > 1, \quad J(x) = 0 \text{ if } |x| \leq 1.$$

Then we have

$$(64)\qquad \check{\delta}_t(x) = \sigma_{n-1} \frac{1}{t} J\left(\frac{x}{t}\right);$$

hence:

$$(65)\qquad R(t) = -\,c_n\,(-\Delta)^{m-1} \frac{\partial}{\partial t} \frac{1}{t} J\left(\frac{x}{t}\right), \quad \text{with } c_n = -\,\sigma_{n-1}/2(2\pi)^{n-1}.$$

We can also obtain usual formulas of the elementary solution (see Lax-Phillips [1] p. 125, Methee [1], Zieman [1]). Thus when n is odd, the support of this elementary solution is the surface of the future light cone. This implies:

Proposition 7. *Huyghens principle* (*with a velocity* c). *For odd dimension* n *of space* (n > 1), *if the supports of the initial data are contained in the ball* B_R, *then the support of the solution* u *of* (11) *with* f = 0 *is such that*:

$$(66)\quad \begin{cases} \operatorname{supp} u(t) = \{x \in R^n,\ \exists\, e \in S^{n-1},\ x + cte \in \operatorname{supp} u_0 \cup \operatorname{supp} u_1\} \subseteq K_{R,t}, \\ K_{R,t} = \{x \in R^n,\ c|t| - R \leq |x| \leq c|t| + R\} \ \text{ if } |t| > R/c, \\ K_{R,t} = \{x \in R^n,\ |x| \leq c|t| + R\} \ \text{ if } |t| \leq R/c. \end{cases}$$

PROOF. Thanks to (5)' and (60) the solution u of (11) (with c = 1) is given by:

$$(67)\qquad u(x,t) = \frac{\partial}{\partial t} R(t) \underset{x}{*} u^0 + R(t) \underset{x}{*} u^1,$$

and the property: $\operatorname{supp}(R(t) \underset{x}{*} v) \subseteq \operatorname{supp} R(t) + \operatorname{supp} v$, implies (66).
We can verify (66) on the Radon transform of the solution using again (45). ⊗

Furthermore we can prove that (for an odd dimension of space) when f = 0 and when the initial conditions are regular, *the solution* u of (11) *and all its derivatives converge to* 0 *at each point* x when t goes to infinity (see for instance Goldstein [1] for generalization to hyperbolic operators). Although the energy of the wave is globally constant in time, it decreases locally with time.

2. SCATTERING PROBLEMS - INCOMING AND OUTGOING WAVES

We now study problems which are relevant to scattering, following Lax-Phillips theory. We assume that *space dimension is odd*.

2.1. Another application of the Radon transformation

We consider the Cauchy problem: find u satisfying

$$(68)\quad \begin{cases} \text{i)}\ \dfrac{\partial^2 u}{\partial t^2} - \Delta u = 0 \quad \text{in } R^n \times R_t, \\ \text{ii)}\ u(0) = u^0,\ \dfrac{\partial u}{\partial t}(0) = u^1, \text{ with } (u^0, u^1) \text{ given in } H, \end{cases}$$

and we apply the wave Radon transformation R^w. Then with (6) let:

$$(69)\quad k = R^w(u^0, u^1), \qquad k(.,t) = R^w(u(t), v(t)) = R_2^d(w(t), z(t)), \quad \text{with } v(t) = \frac{\partial u}{\partial t}(t).$$

Using (54) we have:

$$(70)\quad \begin{cases} k(s,\alpha,t) = -\dfrac{\partial^{m+1}\tilde{u}}{\partial s^{m+1}}(s,\alpha,t) + \dfrac{\partial^m \tilde{v}}{\partial s^m}(s,\alpha,t), \\ k(s,\alpha) = -\dfrac{\partial^{m+1}\tilde{u}^0}{\partial s^{m+1}}(s,\alpha) + \dfrac{\partial^m \tilde{u}^1}{\partial s^m}(s,\alpha). \end{cases}$$

Thanks to (7) and (53), (or directly with (55) on (1)), (68) is transformed into:

$$(71)\quad \begin{cases} \text{i)}\ \dfrac{\partial k}{\partial t}(s,\alpha,t) = -\dfrac{\partial k}{\partial s}(s,\alpha,t), \\ \text{ii)}\ \dfrac{\partial k}{\partial t}(s,\alpha,0) = k(s,\alpha), \end{cases}$$

in $L^2(R \times S^{n-1})$. The solution is given by:

$$(72)\qquad k(s,\alpha,t) = k(s-t,\alpha).$$

Applying the inverse formulas (57), we obtain the solution of (68) by

$$(73)\quad \begin{cases} u(x,t) = \dfrac{(-1)^{m-1}}{2(2\pi)^{n-1}} \displaystyle\int_{S^{n-1}} \dfrac{\partial^{m-1} k}{\partial s^{m-1}}(x.\alpha - t, \alpha)\, d\alpha, \\ v(x,t) = \dfrac{\partial u}{\partial t}(x,t) = \dfrac{(-1)^m}{2(2\pi)^{n-1}} \displaystyle\int_{S^{n-1}} \dfrac{\partial^m k}{\partial s^m}(x.\alpha - t, \alpha)\, d\alpha, \end{cases}$$

and thus for $n = 3$:

$$(74)\quad u(x,t) = \frac{1}{8\pi^2}\int_{S^2} k(x.\alpha - t, \alpha)\, d\alpha, \qquad \frac{\partial u}{\partial t}(x,t) = -\frac{1}{8\pi^2}\int_{S^2} \frac{\partial k}{\partial s}(x.\alpha - t, \alpha)\, d\alpha.$$

Furthermore the conservation of energy is given by (10) and (55):

$$(75)\quad W(t) = \frac{1}{2}[\,\|v(t)\|^2 + \|(-\Delta)^{1/2}u(t)\|^2\,] = \frac{1}{4(2\pi)^{n-1}}\int_{R\times S^{n-1}} |k(s,\alpha)|^2\, ds\,d\alpha.$$

Thus R^w is a unitary transformation, which transforms the evolution group $(G(t))$ into translations; $R^w(u^0,u^1)$ is called *the translation representation* of (u^0,u^1) by Lax-Phillips [1].

Definition 5. *Outgoing waves, Incoming waves (with the velocity* c = 1*) . A wave* u *is called outgoing (resp. incoming) if it satisfies*

$$(76)\quad \left|\begin{array}{l} u(x,t) = 0 \text{ when } |x| < t, \text{ i.e., } u \textit{ is zero on the forward light cone } C_+ \\ (\text{resp. } u(x,t) = 0 \text{ when } |x| > t, \text{ i.e., } u \textit{ is zero on the backward light cone } C_-). \end{array}\right.$$

The corresponding initial conditions (u^0,u^1) *are called outgoing (resp. incoming), and we denote by* D^+ *(resp.* D^-*) the set of these initial conditions.*

Definition 6. *ρ-outgoing waves, ρ-incoming waves. Let* ρ *be a* real *number. A wave* u *which satisfies*

$$(77)\quad u(x,t) = 0,\ \ \forall (x,t) \text{ with } |x| < t + \rho \ (\text{resp. with } |x| < \rho - t)$$

is called a ρ-outgoing (resp. ρ-incoming) wave; the corresponding initial conditions are called ρ-outgoing (resp. ρ-incoming). The set of these initial conditions is denoted by D^+_ρ *(resp.* D^-_ρ*). Thus* $D^+_0 = D^+$, $D^-_0 = D^-$.
A wave u is called eventually outgoing (resp. eventually incoming) if there is a number ρ *such that* u *is* ρ*-outgoing (resp.* ρ*-incoming) and the corresponding initial conditions are called eventually outgoing (resp. eventually incoming).*

Note that changing t into − t changes outgoing waves into incoming waves and conversely. We can prove the equivalences (see Lax-Phillips [1] Thm. 2.3 chap. 4.2) with the translation representation:

$$(78)\quad \left|\begin{array}{l} (u^0,u^1) \in D^+ \ (\text{resp. } D^-) \Leftrightarrow k = R^w(u^0,u^1) \text{ satisfies } k(s,\alpha) = 0 \text{ if } s < 0 \ (\text{resp. } s > 0), \\ (u^0,u^1) \in D^+_\rho \ (\text{resp. } D^-_\rho) \Leftrightarrow k \text{ satisfies } k(s,\alpha) = 0 \text{ if } s < \rho \ (\text{resp. } s > -\rho). \end{array}\right.$$

Proposition 8. i) *The spaces* D^+, D^- *are orthogonal supplementary spaces in* H. ii) *The outgoing waves (resp. incoming) satisfy equipartition of energy for all* $t > 0$ *(resp.* $t < 0$*).* iii) *The evolution of outgoing waves for* $t > 0$ *is given by a semigroup* $(g^+(t))$ *in* $W^1(R^n)$*, and if* (u^0,u^1) *is an outgoing initial condition, then we have*

$$(79)\quad R^d u^1(s,\alpha) = -(\text{sign } s)\frac{\partial}{\partial s}R^d u^0(s,\alpha), \quad \text{or} \quad \tilde{u}^1(s,\alpha) = -(\text{sign } s)\frac{\partial}{\partial s}\tilde{u}^0(s,\alpha),$$

with a similar property for incoming waves (change − *into* + *in* (79)).

PROOF. i) is a straightforward consequence of the above equivalence (78).
ii) When (u^0,u^1) is in D^+, we have

(80) $(u^0,-u^1) \in D^-$ and for all $t>0$ $(u(t),u'(t)) \in D^+$ and $(u(t),-u'(t)) \in D^-$.

Since D^+ and D^- are orthogonal spaces we have:

(81) $$\frac{1}{2}\int_{R^n} |\text{grad}\, u(x,t)|^2\, dx - \frac{1}{2}\int_{R^n} |\frac{\partial u}{\partial t}(x,t)|^2\, dx = 0.$$

iii) From (70) (or from (56) with $h = k(.,t)$, $f = u(.,t)$, $g = v(.,t)$) we see that $k(s,\alpha,t) = 0$ when $t<0$ implies:

(82) $$\tilde{v}^{(m)}(s,\alpha,t) = \frac{\partial \tilde{u}^{(m)}}{\partial t}(s,\alpha,t) = -(\text{sign}\, s)\, \tilde{u}^{(m+1)}(s,\alpha,t),$$

that is (79) when $t = 0$. Then if we consider $L^2(R^+,L^2(S^{n-1}))$ as a subspace of $L^2(R,L^2(S^{n-1}))$, the translation group in $L^2(R,L^2(S^{n-1}))$ given by (72) induces on $L^2(R^+,L^2(S^{n-1}))$ an isometric semigroup for $t>0$ only, given by:

(83) $k(s,\alpha,t) = k(s-t,\alpha)$ if $s>t$, $k(s,\alpha,t) = 0$ if $s<t$,

whose infinitesimal generator is

(84) $$A_0 = -\frac{\partial}{\partial s}, \quad D(A_0) = H_0^1(R^+,L^2(S^{n-1})).$$

The translation semigoup gives, in $L^2_m(R \times S^{n-1})$ (see (48)), a semigroup $(\tilde{g}(t))$:

(85) $$\left|\begin{array}{l} \tilde{g}(t)v(s,\alpha) = v_-(s+t,\alpha) + v_+(s-t,\alpha) \text{ if } |s|>t, \quad \tilde{g}(t)v(s,\alpha) = 0 \text{ if } |s|<t, \\ \text{with } v_- = v|_{R^-},\ v_+ = v|_{R^+}, \end{array}\right.$$

with infinitesimal generator $\tilde{A} = -(\text{sign}\, s)\frac{\partial}{\partial s}$.
Then thanks to the inverse Radon transformation $(R^d)^{-1}$, we obtain the evolution semigroup $(g^+(t))$ (on u^0) in $W^1(R^n)$.

⊗

Thus we have proved that D^+ and D^- are graphs of continuous mappings C^+ and C^- from $W^1(R^n)$ into $L^2(R^n)$.

Definition 7. *The orthogonal projectors* P^+ *and* P^- *in* $H = W^1(R^n) \times L^2(R^n)$ *on* D^+ *and* D^- *are respectively called the outgoing and the incoming Calderon projectors for free waves. The operators* C^+ *and* C^- *are called the outgoing and the incoming Calderon operators for free waves, and they are defined also (thanks to* R^d*) by:*

(86) $$C^+ = -(R^d)^{-1}[(\text{sign}\, s)\frac{\partial}{\partial s}]\, R^d, \quad C^- = (R^d)^{-1}[(\text{sign}\, s)\frac{\partial}{\partial s}]\, R^d.$$

We can identify C^+ with the infinitesimal generator $\tilde{A}$ (thanks to R^d).

These Calderon operators are *unitary operators* from $W^1(R^n)$ onto $L^2(R^n)$ (this follows from equipartition of energy) *with*:

$$\mathrm{Re}\,(Cu,u) = 0,\ u \text{ in } W^1(R^n),\ C = C^+ \text{ or } C^-.$$

Proposition 9. *The problem: find an outgoing (resp. incoming) wave* u *with finite energy satisfying*

$$(87)\quad \left|\begin{array}{l} \text{i) } \dfrac{\partial^2 u}{\partial t^2} - \Delta u = 0 \quad \text{in } R^n \times R_t^+ \ (\text{resp. } R^n \times R_t^-), \\ \text{ii) } u(0) = u^0, \ \textit{with the only data } u^0 \text{ in } W^1(R^n), \end{array}\right.$$

has a unique solution, and thus is a well-posed Cauchy problem.

Now going back to the ρ-outgoing or ρ-incoming waves we have the properties:

$$(88)\qquad D_\rho^+ \subset D_{\rho'}^+ \quad \text{and} \quad D_{\rho'}^- \subset D_\rho^- \quad \text{if } \rho' < \rho.$$

Thus $(D_\rho^+)_{\rho \in R}$ is a decreasing family of spaces, whereas $(D_\rho^-)_{\rho \in R}$ is increasing. Spaces D_ρ^+ and $D_{-\rho}^-$ are orthogonal supplementary spaces: $H = D_\rho^+ \oplus D_{-\rho}^-$. Moreover ρ-outgoing (resp. ρ-incoming) waves satisfy equipartition of energy from time $t = -\rho$ (resp. ρ). The evolution of ρ-outgoing waves is given from this time thanks to an *isometric semigroup* $(g^+_\rho(t))$, $t \geq -\rho$, by: $g^+_\rho(t) = g^+(t+\rho)$. We have similar results for ρ-incoming waves but with $\rho \geq t$. The main properties of these spaces for scattering (which are very easy to verify thanks to the wave Radon transformation) are:

$$(89)\quad \left|\begin{array}{ll} \text{i) } G(t)D_\rho^+ \subset D_\rho^+ \ \text{ if } t > 0, & G(t)D_\rho^- \subset D_\rho^- \ \text{ if } t < 0, \\ \text{ii) } \bigcap_{t>0} G(t)D_\rho^+ = 0, & \bigcap_{t<0} G(t)D_\rho^- = 0, \\ \text{iii) } \overline{\bigcup_{t\in R} G(t)D_\rho^+} = \overline{\bigcup_{t\in R} G(t)D_\rho^-} = H, & \end{array}\right.$$

that is for iii): the set of eventually outgoing (resp. eventually incoming) initial conditions is dense in H.

2.2. Incoming and outgoing waves in electromagnetism

Now we consider the Maxwell Cauchy problem (16) in free space, without charges nor currents: find (E, B) satisfying

$$(100)\quad \left|\begin{array}{l} \text{i) } -\dfrac{1}{c^2}\dfrac{\partial E}{\partial t} + \mathrm{curl}\, B = 0, \quad \dfrac{\partial B}{\partial t} + \mathrm{curl}\, E = 0 \quad \text{in } R^3 \times R_t, \\ \text{ii) } \mathrm{div}\, E = 0, \quad \mathrm{div}\, B = 0, \\ \text{iii) } E(.,0) = E_0,\ B(.,0) = B_0, \ \text{(with free divergence) and } E_0,\ B_0 \text{ in } L^2(R^3)^3. \end{array}\right.$$

We write the Radon transform of this problem. First we easily have:

(101) $R(\text{curl } v)(s,\alpha) = \frac{\partial}{\partial s}(\alpha \wedge \tilde{v})(s,\alpha)$, $R(\text{div } v)(s,\alpha) = \frac{\partial}{\partial s}(\alpha.\tilde{v})(s,\alpha)$, $(s,\alpha) \in R \times S^2$.

Thus (100) becomes (with c = 1):

$$(100)' \quad \begin{cases} \text{i)} -\frac{\partial \tilde{E}}{\partial t} + \frac{\partial}{\partial s}(\alpha \wedge \tilde{B}) = 0, \quad \frac{\partial \tilde{B}}{\partial t} + \frac{\partial}{\partial s}(\alpha \wedge \tilde{E}) = 0 \text{ in } R_s \times S^2 \times R_t, \\ \text{ii)} \ \alpha.\tilde{E} = 0, \quad \alpha.\tilde{B} = 0, \\ \text{iii)} \ \tilde{E}(.,0) = \tilde{E}_0, \quad \tilde{B}(.,0) = \tilde{B}_0. \end{cases}$$

Let:

(102) $\eta(.,t) = -\tilde{E}(.,t) + \alpha \wedge \tilde{B}(.,t)$.

Thus η satisfies:

$$(103) \quad \begin{cases} \text{i)} \ \frac{\partial \eta}{\partial t} = -\frac{\partial \eta}{\partial s}, \quad \text{with } \alpha.\eta = 0, \\ \text{ii)} \ \eta(.,0) = \eta_0 \quad \text{with given } \eta_0 = -\tilde{E}_0 + \alpha \wedge \tilde{B}_0. \end{cases}$$

Note that for $c \neq 1$, then $\eta = -\frac{1}{c}\tilde{E} + \alpha \wedge \tilde{B}$ satisfies: $\frac{\partial \eta}{\partial t} = -c\frac{\partial \eta}{\partial s}$.
Here it is better to use the transformation R^d instead of R; with the notation $\tilde{v} = R^d v$, $\tilde{E}_0$ and $\tilde{B}_0$ are in $L_1^2(R \times S^2)$ (see (48)), thus η_0 is in $L^2(R \times S^2)$ but not in $L_1^2(R \times S^2)$:

(104) $\eta_0(-s,-\alpha) = -\tilde{E}_0(-s,-\alpha) + (-\alpha) \wedge \tilde{B}_0(-s,-\alpha) = (\tilde{E}_0(s,\alpha) + \alpha \wedge \tilde{B}_0(s,\alpha))$.

The evolution of η is given by a group of translations; if we define the outgoing electromagnetic waves as previously by $\eta(s,\alpha) = 0$ when $s < 0$, we obtain that an electromagnetic wave is outgoing if it satisfies the relation:

(105) $\tilde{E}(s,\alpha,t) = -(\text{sign } s)\, \alpha \wedge \tilde{B}(s,\alpha,t)$, for $t \geq 0$, $(s,\alpha) \in R \times S^2$,

and thus the initial conditions must satisfy:

(106) $\tilde{E}_0(s,\alpha) = -(\text{sign } s)\, \alpha \wedge \tilde{B}_0(s,\alpha)$, $(s,\alpha) \in R \times S^2$,

which is equivalent (with (101)ii) to:

(106)' $\tilde{B}_0(s,\alpha) = (\text{sign } s)\, \alpha \wedge \tilde{E}_0(s,\alpha)$.

Of course incoming electromagnetic waves will correspond to initial conditions with + instead of − in (106). Thus if (E, B) is an outgoing wave, then (E, −B) is an incoming wave for −t. Hence using orthogonality of these waves, we have:

Proposition 10. *Proposition 8 is valid in the electromagnetic case, by replacing only condition* (79) *by* (106), *concerning initial conditions of outgoing waves.*

We can also define *outgoing and incoming Calderon projectors* P^+ and P^- in the space H of free divergence fields of $L^2(R^3)^3 \times L^2(R^3)^3$ on the spaces D^+ and D^- of outgoing and incoming initial conditions, then *outgoing and incoming Calderon operators* C^+ and C^- in space of free divergence fields of $L^2(R^3)^3$) by: $B_0 = C^+E_0$ given by (106)' (for C^+). Furthermore C^+ and C^- are *unitary operators*, with:

$C^2 = -I$ ($C^2 = -c^{-2}I$ when $c \neq 1$) and $\mathrm{Re}\,(CE_0, E_0) = 0$, $C = C^+$ and C^-.

The Calderon projectors P^+ and P^- are also given by:

(107) $$P^+ = \frac{1}{2}(I - S), \quad P^- = \frac{1}{2}(I + S), \quad \text{with } S = \begin{pmatrix} 0 & C^+ \\ -C^+ & 0 \end{pmatrix}.$$

Like for scalar waves, we have:

Proposition 11. *The Cauchy problem: find an outgoing (resp. incoming) electromagnetic field* (E,B) *in* $R^3 \times R^+$ *(resp.* $R^3 \times R^-$*) with finite energy, satisfying* (100)i),ii) *and only the initial condition*:

(100)iii)' $$E(.,0) = E_0 \text{ with } \mathrm{div}\, E_0 = 0, \quad E_0 \in L^2(R^3)^3,$$

or with given initial condition on B *only, has a unique solution* (E,B) in $C(R^+, H)$ (resp. $C(R^-, H)$), *and thus it is a well posed problem.*

2.3. Incoming and outgoing waves with a bounded obstacle; Lax-Phillips theory

Let Ω be the open complement of a bounded obstacle Ω', and let B_ρ be a ball so that Ω' is contained in it. Let $(G_\Omega(t))$ be the evolution group of waves in $H_\Omega = W^1_0(\Omega) \times L^2(\Omega)$ with a Dirichlet boundary condition.

We first note that if u is a wave in $R^n \times R^+$ which satisfies (77), that is u is zero in the truncated forward (backward) cone C_ρ^+ (resp. C_ρ^-), then u(t) satisfies the Dirichlet boundary condition when $t > \rho$ (resp. $t < -\rho$). From this trivial remark we can consider D_ρ^+ (resp. D_ρ^-) as a ρ-outgoing (resp. ρ-incoming) space for the mixed problem. Then we apply the evolution group $(G_\Omega(t))$ to these spaces. We can prove (see Lax-Phillips[1]) that the essential properties (89) are also valid when substituting $G_\Omega(t)$ and H_Ω for G(t) and H. The only difficulty is (89)iii) which is proved to be equivalent to:

(108) $$\lim_{t \to \pm\infty} E(u,t,K) = 0, \text{ or only } \liminf_{t \to \pm\infty} E(u,t,K) = 0, \quad \forall K \text{ compact set, } K \subset \bar{\Omega}.$$

Now let $F = (f^0, f^1) \in \bigcup_{t \in R} G(t)D_\rho^+$. Then there is $T > 0$ so that $F \in G(-T)D_\rho^+$, and thus there is $F^+ \in H_\Omega$ so that $G_\Omega(t)F^+ = G(t)F$, $\forall t > T$.

This defines a mapping $W_+ : F \to F^+$ which can be extended continuously from H into H_Ω. We define a mapping W_- in a similar way.

Definition 8. *Wave operators. The mappings* W_+, W_- *from* H *into* H_Ω *defined by*:

(109) $W_\pm F = \lim_{t \to \pm\infty} G_\Omega(-t)G(t)F$, $\forall F$ in H, i.e., $W_\pm = \lim_{t \to \pm\infty} G_\Omega(-t)JG(t)$,

with J *the natural imbedding of* H_Ω *into* H, *are called the outgoing and the incoming wave operators.*

We can also define wave operators from $L^2(R^n)$ into $L^2(\Omega)$ using the groups generated by the operators $iA^{1/2}$, A being the Laplacian in R^n or in Ω, like in Wilcox [1]. Now these wave operators have the essential properties:

Proposition 12. *The wave operators are unitary operators from* H *onto* H_Ω, *with*

(110) $G_\Omega(t)W_\pm = W_\pm G(t)$, $\forall t \in R$.

From this we can define the "*scattering operator*" S in H by:

(111) $S = W_+^{-1} W_-$.

Then from Proposition 12, S is a *unitary operator in* H, such that $SG(t) = G(t)S$, for all real t.
These operators are a basic tool in scattering problems by a soft bounded obstacle. Obviously we can generalize to other boundary conditions, for example to hard obstacles with a Neumann condition. That the scattering operator S contains all the scattering information is seen from solving the inverse scattering problem (see Lax-Phillips [1] p. 173):
Let O_1 *and* O_2 *be two bounded soft obstacles; let* S_1 *and* S_2 *be the associated scattering operators in* H. *Then* $S_1 = S_2$ *implies* $O_1 = O_2$.
Furthermore we can easily deduce spectral properties from the wave operators on the selfadjoint operator $A = -\Delta$ with Dirichlet boundary condition: A has an absolutely continuous spectrum with no eigenvalue.
Here we are essentially interested in outgoing and incoming waves. Thanks to the wave Radon transform R^w we define mappings R^+ and R^- (called *outgoing and incoming representations*) by:

(112) $R^+ = R^w W_+^{-1}$, $R^- = R^w W_-^{-1}$.

They transform the evolution group in H_Ω into translations in $L^2(R, L^2(S^{n-1}))$, with:

$$R^+(D_\rho^+) = L^2((\rho, +\infty), L^2(S^{n-1})), \quad R^-(D_\rho^-) = L^2((-\infty, -\rho), L^2(S^{n-1})).$$

We can also define "*outgoing and incoming initial conditions*" D_Ω^+ and D_Ω^- as the inverse images of $L^2((0,\infty), L^2(S^{n-1}))$ and $L^2((-\infty, 0), L^2(S^{n-1}))$, respectively by R^+ and R^-, or $D_\Omega^+ = W_+D^+$ and $D_\Omega^- = W_-D^-$, and the Calderon projectors P_Ω^+, P_Ω^- on these spaces by:

$$P_\Omega^{\ +} = W_+ P^+ (W_+)^{-1}, \quad \text{and } P_\Omega^{\ -} = W_- P^- (W_-)^{-1},$$

but $D_\Omega^{\ +}$ and $D_\Omega^{\ -}$ are not orthogonal, nor necessarily graphs of operators !
Then the problem: *find an outgoing wave* u *with finite energy satisfying*:

$$(113)\quad \left|\begin{array}{l} \text{i)}\ \dfrac{\partial^2 u}{\partial t^2} - \Delta u = 0 \quad \text{in } \Omega \times R^+, \\ \text{ii)}\ u\big|_{\Gamma \times R^+} = 0, \\ \text{iii)}\ u(.,0) = u^0 \ \text{ with } u^0 \text{ given in } W_0^1(\Omega), \ \text{ and supp } u^0 \subset B'_\rho = R^n \backslash \bar{B}_\rho, \end{array}\right.$$

has a unique solution u in $C(R_t^+, W_0^1(\Omega))$, with $\frac{\partial u}{\partial t}$ in $C(R_t^+, L^2(\Omega))$.

Thus it is a well-posed problem (as a trivial consequence of Proposition 9), but we cannot eliminate the condition on the support of u^0!

Application to stationary waves

We can also define notions of outgoing or incoming waves in the stationary case in the following way. First we consider the stationary problem in a form similar to (1) in free space, in the whole space:

$$(114)\qquad \mathcal{A}W + \lambda W = F \quad \text{in } R^n,$$

where $\mathcal{A}$ is given by (4) with $A = -\ \Delta$, with given λ in C and $F = (f^1, f^2)$ given distributions in R^n with bounded supports. Using Definition 6, we have:

Definition 9. *A solution of* (114) *which is also an eventually outgoing (resp. incoming) initial condition for the evolution problem* (68) *is called a* λ*-outgoing (resp.* λ*-incoming) stationary wave. When* $F = (0,g)$, *then* $W = (w, \lambda w)$, *with* w *satisfying the Helmholtz equation*: $\lambda^2 w - \Delta w = g$, *and* w *is called a* λ*-outgoing (resp.* λ*-incoming) stationary wave when* W *is* λ*-outgoing (resp.* λ*-incoming).*

Note that a λ-outgoing stationary wave w is a $(-\lambda)$-incoming stationary wave also, and conversely. We can easily obtain results on stationary wave thanks to the wave Radon transformation R^W. We only give one example of application. Applying transformation R^W to (114) (this is possible up to now if F is in H, but we can generalize R^W to distributions, see Lax-Phillips) gives

$$(115)\qquad \frac{\partial k}{\partial s} + \lambda k = f \ \text{ in } R_s x S^{n-1},$$

with $k = R^W W$ and $f = R^W F$, f having a compact support. The solution with support bounded to the left (resp. to the right) which will give the outgoing (resp. incoming) solution of (114) by its inverse transform, is

$$(116)\quad k^{out}(s,\alpha) = \int_{-\infty}^{s} e^{\lambda(s'-s)} f(s',\alpha)\, ds', \quad k^{in}(s,\alpha) = -\int_{s}^{\infty} e^{\lambda(s'-s)} f(s',\alpha)\, ds'.$$

When F is in H, then f is in $L^2(R \times S^{n-1})$, and we see that if (Re λ) > 0, only k^{out} is in $L^2(R \times S^{n-1})$ whereas if (Re λ) < 0, only k^{in} is in $L^2(R \times S^{n-1})$, the other solution being not even tempered. (When λ is real, both solutions are tempered, but not in $L^2(R \times S^{n-1})$.)
Then applying this to $F = (0, -\delta)$ gives the elementary λ-outgoing (or λ-incoming) elementary solution of the Helmholtz equation Φ_λ. When n = 3 we have the very simple result:

$$(117) \qquad \Phi_\lambda(x) = \frac{1}{4\pi}\frac{e^{-\lambda r}}{r} \text{ with } r = |x|.$$

When $F = (0, -f)$, f being a distribution with bounded support, the λ-outgoing (resp. λ-incoming) stationary wave w is obtained by convolution of f with Φ_λ, and then we can verify that Definition 9 agrees with Definition 1 chap. 3 thanks to Theorem 1 chap. 3.

Now we consider ***a domain Ω exterior to a bounded soft obstacle***. Then we can define λ-outgoing (resp. λ-incoming) stationary waves as in Definition 9. Here we have to use a more general framework taking waves into account which are not of finite energy, thus not in $H_\Omega = W^1{}_0(\Omega) \times L^2(\Omega)$. So we define:

$$(118) \qquad H_{loc}(\bar{\Omega}) = \{U = (u^0, u^1), \text{ such that } \zeta U \in H_\Omega, \quad \forall \zeta \in D(R^n)\}.$$

Waves which are not of a finite energy may be of two very different types according to different situations:

i) ***scattering states***; they appear in ***usual stationary scattering problems*** as λ-outgoing (or λ-incoming) stationary waves. They are generalized eigenvectors, $w_+(x,\xi)$ (resp. $w_-(x,\xi)$) with ξ in R^n, i.e., satisfying:

$$(119) \quad \left| \begin{array}{l} \text{i) } (\Delta + \xi^2)\, w_\pm(x,\xi) = 0 \quad \text{in } \Omega, \\ \text{ii) } w_\pm(.,\xi) \in H^1_{0\,loc}(\bar{\Omega}), \text{ i.e., } \zeta w_\pm(.,\xi) \in H^1_0(\Omega), \ \forall \zeta \in D(R^n), \\ \text{iii) } w_{r+}(x,\xi) = w_+(x,\xi) - w_0(x,\xi) \text{ is an } i|\xi|\text{-outgoing wave,} \\ \text{(resp. } w_{r-}(x,\xi) = w_-(x,\xi) - w_0(x,\xi) \text{ is an } i|\xi|\text{-incoming wave),} \end{array} \right.$$

with $w_0(x,\xi) = e^{ix.\xi}$. Recall that iii) means that $w_{r\pm}(.,x)$ satisfies the outgoing (resp. incoming) Sommerfeld condition.

This problem (119) has a unique solution thanks to the Rellich lemma (Lemma 1 chap. 3), w_r being the usual reflected wave by the obstacle for the incident wave w_0. Then if we define for all f in $L^2(\Omega)$ with bounded support:

$$(120) \quad \hat{f}_+(\xi) = \int_\Omega f(x)\, \bar{w}_+(x,\xi)dx, \qquad \text{resp. } \hat{f}_-(\xi) = \int_\Omega f(x)\, \bar{w}_-(x,\xi)dx,$$

the mapping $f \to \hat{f}_+$ (resp. $\hat{f}_-$) has a continuous extension F^+ (resp. F^-) from $L^2(\Omega)$

onto $L^2(R^n)$. Furthermore $\mathcal{F}^+$ and $\mathcal{F}^-$ are unitary (up to a constant), and are natural generalizations of the Fourier transformation which diagonalize the Laplacian Δ^D with Dirichlet conditions. They are called the *incoming (resp. outgoing) spectral representation of* Δ^D. The operators $\mathcal{F}^{-1}\mathcal{F}^+$ and $\mathcal{F}^{-1}\mathcal{F}^-$ with $\mathcal{F}$ the Fourier transformation, correspond to (inverse) wave operators W_- and W_+ in L^2, see Wilcox [1]. Note that $w_+(x,\xi)$, $w_-(x,\xi)$ are tempered distributions.

ii) *resonance states. Let* w *be a non trivial solution with locally finite energy of*:

$$(121)\quad \begin{cases} \text{i) } -\Delta w + \lambda^2 w = 0 \quad \text{in } \Omega, \text{ with } \lambda \in C, \\ \text{ii) } w|_\Gamma = 0, \quad \text{thus } w \in H^1_{0\,loc}(\bar{\Omega}), \\ \text{iii) } w \text{ is a } \lambda\text{-}\textit{outgoing stationary wave}, \text{ i. e. } w \textit{ has the asymptotic behavior} \\ w(x) = \frac{e^{-\lambda r}}{r}\theta(\alpha) + O(\frac{e^{-\lambda r}}{r^2}) \text{ for } r = |x| \to \infty,\ \theta(\alpha) \textit{ a regular function of } \alpha = x/r. \end{cases}$$

Definition 10. *A wave* w *satisfying* (121) *is called a resonance state and* λ *a resonance.*

(We emphasize that this terminology is not universal: sometimes, see Sanchez Palencia [1], resonances are called scattering states.) From Rellich lemma, a resonance cannot be real. Furthermore it must satisfy Re $\lambda < 0$ (see Lax-Phillips [1] p.161). Thus a resonance state cannot be tempered; its physical nature is quite different from that of a scattering state. There is at most a countable set of resonances, as a consequence of the Rellich compactness result:

Proposition 13. *The unit ball of* D(A) in H_Ω, $\{F \in \mathcal{H}_\Omega,\ \|\mathcal{A}F\| + \|F\| \le 1\}$ *is a precompact set in* $H_{loc}(\bar{\Omega})$ *equipped with the family of semi-norms for all compact set* K *in* R^n

$$(122)\qquad \|U\|_K = \|(u_1,u_2)\|_K = [\tfrac{1}{2}\textstyle\int_{K\cap\Omega} [|\text{grad } u_1|^2 + |u_2|^2]\, dx]^{1/2}.$$

A complementary result on resonance states is the following (Lax-Phillips [1])

Proposition 14. *The problem: find* w *with locally finite energy satisfying*

$$(123)\quad \begin{cases} \text{i) } \lambda^2 w - \Delta w = f, \quad \text{with } f \in L^2(\Omega) \text{ with compact support}, \\ \text{ii) } w|_\Gamma = 0, \quad \text{thus } w \in H^1_{0\,loc}(\bar{\Omega}), \\ \text{iii) } w \text{ (or } W = (w, \lambda w)) \text{ is a } \lambda\text{-}\textit{outgoing stationary wave}, \end{cases}$$

has a unique solution when λ *is not a resonance.*

In order to study resonance we define the following spaces and projectors:

(124) $D_\rho = D_\rho^+ \oplus D_\rho^-$, K_ρ the orthogonal of D_ρ in $\mathcal{H}_\Omega$, thus $\mathcal{H}_\Omega = D_\rho \oplus K_\rho$,

$Q^+_\rho = I - P^+_\rho$ (resp. $Q^-_\rho = I - P^-_\rho$), the orthogonal projector in H_Ω on the orthogonal space to D^+_ρ (resp. D^-_ρ), and $Q_\rho = Q^+_\rho Q^-_\rho = I - P_\rho$, the orthogonal projector on K_ρ.

A basic tool to study resonance is the family:

(125) $$Z_\rho(t) = Q^+_\rho G_\Omega(t) Q^-_\rho, \quad t \geq 0.$$

We can prove that $(Z_\rho(t))$ has the properties:

i) $D_\rho \subset \ker Z_\rho(t)$, $\forall t \geq 0$, $Z_\rho(t) K_\rho \subset K_\rho$, $\forall t \geq 0$,

ii) $(Z_\rho(t))$ is a contraction semigroup of class C^0 in K_ρ, with:

$$Z_\rho(0) = Q_\rho, \text{ and } Z_\rho(t) U_0 \longrightarrow 0 \text{ when } t \longrightarrow +\infty.$$

Let B^ρ be its infinitesimal generator in K_ρ. Then the main point is:

Proposition 15 (Lax-Phillips). *The operator* $Z_\rho(2\rho)(\lambda I - B^\rho)^{-1}$ *is compact in* K_ρ, *for all* $\lambda > 0$.

This implies that the spectrum of B^ρ is a sequence of eigenvalues in half plane $\mathrm{Re}\,\lambda < 0$, with finite multiplicity without accumulation point at finite distance, and *that the set of eigenvalues of* B^ρ *is the set of resonances*.

Moreover if W^ρ_j is an eigenvector of B^ρ for the eigenvalue λ_j then we obtain a resonance state W_j by: $W_j(x) = \lim W^\rho_j(x)$ when $\rho \longrightarrow \infty$, x fixed in Ω.

Then resonances are useful to study evolution of waves when t goes to infinity:

Proposition 16. *Limiting Amplitude Principle. Let* g *be in* $L^2(\Omega)$ *with bounded support, and* λ *a complex number which is not a resonance. Then the solution* u *with finite energy of the problem*:

(126) $$\begin{cases} \text{i) } \dfrac{\partial^2 u}{\partial t^2} - \Delta u = e^{\lambda t} g & \text{in } \Omega \times R_t, \\ \text{ii) } u(.,t)|_\Gamma = 0, & \forall t \in R, \\ \text{iii) } u(.,0) = u^0, & \dfrac{\partial u}{\partial t}(.,0) = u^1, \end{cases}$$

with given (u^0, u^1) in $W^1_0(\Omega) \times L^2(\Omega)$, *satisfies:*

(127) $U(t) = (u(t), \frac{\partial u}{\partial t}(t))$ *converge in* $H_{loc}(\bar{\Omega})$ to $V(t) = (e^{\lambda t} v, \lambda e^{\lambda t} v)$ *when* $t \longrightarrow +\infty$,

where v *is the* λ*-outgoing stationary wave solution of:*

(128) $$\lambda^2 v - \Delta v = g \text{ in } \Omega.$$

Another result is the local decrease with time of wave with respect to energy. The first result is due to Morawetz [1]:

Proposition 17. *Let* Ω' *be a bounded star-shaped domain. Then the solution* u *with finite energy of*:

$$(129)\quad \begin{cases} \text{i) } \dfrac{\partial^2 u}{\partial t^2} - \Delta u = 0 \quad \text{in } \Omega \times R_t, \quad \Omega = R^n \setminus \bar{\Omega}', \\ \text{ii) } u(.,t)|_{\Gamma} = 0, \quad \forall t \in R, \\ \text{iii) } u(.,0) = u^0, \ \dfrac{\partial u}{\partial t}(.,0) = u^1, \text{ with given } U_0 = (u^0, u^1) \text{ in } W_0^1(\Omega) x L^2(\Omega), \end{cases}$$

satisfies (*with* C a *constant and with* $t \geq 2\rho$):

$$(130)\quad E(u,t,B_\rho \cap \Omega) = \frac{1}{2} \int_{B_\rho \cap \Omega} [|\text{grad } u(t)|^2 + |\frac{\partial u}{\partial t}(t)|^2]\, dx \leq \frac{C}{t^2}\, \|U_0\|^2_{H_\Omega}.$$

When the obstacle is star-shaped, and even when there is no captive ray (for this notion, see for instance Bardos [1]), the energy of the wave is exponentially decreasing on every compact set when t goes to infinity (for an odd dimension of space). Moreover we have (Lax-Phillips [2]):

Proposition 17'. *Let* Ω' *be a bounded star-shaped domain. Then for every* U_0 *given in* H_Ω *with bounded support, there are resonances* (λ_j), $j = 1,\dots,N$, *with associated resonance states* w_j, *constant numbers* C_j, $j = 1,\dots,N$, *so that the solution* u *of* (129) *satisfies for any compact set* K:

$$(131)\quad E((u(t) - \sum_{j=1 \text{ to } N} C_j e^{\lambda_j t} w_j), t, K) \leq C(K)\, e^{(\text{Re}\,\lambda_{N+1})t}.$$

Thus resonance states allow us to have an asymptotic behavior of the wave when t goes to infinity. This formula would be generalized to more general domains. But we know (see Ralston [1]) that if there are arbitrarily long rays following optical geometry, contained in a ball of finite radius, energy may be arbitrarily slowly decreasing in compact sets. For a soft obstacle which has a cavity with a tiny hole (of diameter δ) resonances converge to the eigenvalues of the (closed) cavity when δ goes to 0 (see Beale [1]). There are other similar examples with waves due to a drum in Sanchez Palencia [1].
Obviously all these results can be generalized, to hard obstacles with a Neumann condition, and to systems of equations, in elasticity and in electromagnetism. For Maxwell equations we can use potentials (scalar and vector potentials, and Debye potentials also) to obtain results thanks to those for the wave equation. We shall not develop these questions here, and we refer to Schmidt [1], Dassios [1], Bardos [1].

3. CAUSAL PROBLEMS

A great number of evolution problems in scalar waves and in electromagnetism have no initial condition but are of the type:

Causal problem with a support condition. Given a distribution f in $R^n \times R_t$ with support bounded in the past:

(132) $\quad \operatorname{supp} f \subset Q_a^+ = \{(x,t)\,,\ x \in R^n,\ t \geq a\},$

find a distribution u in $R^n \times R_t$ satisfying:

(133) $\quad \left| \begin{array}{l} \text{i) } \square u = f \ \text{ in } \mathcal{D}'(R^n \times R_t), \\ \text{ii) } \operatorname{supp} u \subset Q_a^+ \text{, (or } \operatorname{supp} u \subset \operatorname{supp} f + C_+ \text{ with (23)),} \\ \text{i.e., with support bounded in the past, with the same bound a.} \end{array} \right.$

Thanks to the elementary solution Φ with support in the future light cone (see section 1.3) we have:

Theorem 1. *The causal problem* (133) *has a unique solution* u, *which is given by convolution*:

(134) $\quad u = \Phi \underset{x,t}{*} f.$

PROOF. We have to prove that the convolution product has a sense with the above hypotheses on the supports of f and Φ, that is their supports satisfy:

(135) $\quad A \cap (K - B)$ is a bounded set, for all compact set K in $R^n \times R_t$,

with $A = \operatorname{supp} f$, $B = C_+ = \operatorname{supp} \Phi$. Let (x,t) be in A, (y,τ) in C_+, and (z,λ) in K, with $t = \lambda - \tau$, $x = z - y$. We have $a \leq t$, thus $\tau \leq \lambda - a$, and $|y| \leq \tau \leq \lambda - a$, thus x and t are bounded, giving (135).
Now if w is a solution of (133)i) with support bounded in the past only, then applying the convolution by Φ to (133)i) gives:

$$u = \Phi * f = \Phi * (\square w) = (\square \Phi) * w = \delta * w = w\,,$$

which proves uniqueness of the solution with a weaker hypothesis on the support. ⊗

Remark 1. Obviously we can transform a one-sided Cauchy problem (11) (for $t > 0$ only), into a causal problem, by usual extension by 0 for $t < 0$, so that it becomes: find U a distribution in $R^n \times R_t$ with support in $Q^+ = R^n \times R^+_t$ satisfying:

(136) $\quad \square U = F + u^1\, \delta(t) + u^0\, \delta'(t),$

F being the extension of f by 0 in the past. Equivalence between the two problems is realized if we assume usual regularity conditions on the data.

⊗

Remark 2. For $n = 3$, the solution u of (133) is given at least formally by the "retarded potential":

$$(137) \qquad u(x,t) = \frac{1}{4\pi}\int_{R^3} \frac{1}{|x - x'|} f(x', t - \frac{|x - x'|}{c})\, dx'.$$

⊗

We have solved the causal problem (133) by convolution. Under similar hypotheses we can apply a Laplace transformation method which gives also Theorem 1. For instance, let X be a Banach space of functions on R^n, let $\mathcal{S}'(R,X)$ be the space of tempered X-valued distributions. We define

$$(138) \qquad \begin{cases} L_+(R_t,X) = \{f \in \mathcal{D}'(R_t,X),\ \exists\, \xi \in R,\ e^{-\xi t} f \in \mathcal{S}'(R_t,X)\}, \\ \mathcal{D}'_+(X) = \{f \in \mathcal{D}'(R_t,X) \text{ with } \mathrm{supp}_t\, f \text{ bounded in the past}\}, \\ L_+^0(R_t,X) = L_+(R_t,X) \cap \mathcal{D}'_+(X). \end{cases}$$

Taking for instance f in $L_+^0(R_t,X)$ with $X = L^2(R^n)$, then we prove that (133) has a unique solution in the same space (see Dautray-Lions [1] chap. 16.1) which is obtained thanks to the resolvent of the operator $A = -\Delta$.
Furthermore the Laplace method allows us to define *well-posed causal problems without condition on supports*: we substitute for the condition of bounded support in the past a L^2-condition with a weight in time, such as $L^2_\gamma(X)$ (also denoted by $e^{\gamma t}L^2(R,X)$), with $\gamma > 0$, defined by:

$$(139) \qquad L^2_\gamma(X) = \{f : R_t \to X \text{ measurable with } \|f\|^2_{\gamma,X} = \int_R e^{-2\gamma t}\|f\|^2_X\, dt < \infty\}.$$

A typical example (which is relevant for wave equation and for Maxwell equations) is the following: let A be an operator in a Hilbert space X, which is the infinitesimal generator of a unitary group (G(t)) in X. Let f be a given function in $L^2_\gamma(X)$, with $\gamma > 0$; find u in $L^2_\gamma(X)$ satisfying:

$$(140) \qquad \frac{\partial u}{\partial t} + Au = f.$$

Proposition 19. *The problem* (140) *has a unique solution* u *in* $L^2_\gamma(X)$.

PROOF. The Laplace transform of a function ϕ in $L^2_\gamma(X)$ being defined by:

$$(141) \qquad \hat{\phi}(p) = \int_R e^{-pt}\phi(t)dt, \qquad p = \xi + i\eta,\ \xi \geq \gamma,$$

the Laplace transform of (140) is:

$$(142) \qquad (pI + A)\hat{u}(p) = \hat{f}(p).$$

When $\xi \geq \gamma > 0$, this equation has a unique solution, given by:

(143) $\hat{u}(p) = R(p)\hat{f}(p) = (pI + A)^{-1}\hat{f}(p)$.

Thanks to the Stone theorem, we have:

(144) $\|R(p)\| \leq \frac{1}{\xi}$,

and thus:

(145) $\int_R \|\hat{u}(p)\|_X^2 \, d\eta = \int_R \|R(p)\hat{f}(p)\|_X^2 \, d\eta \leq \frac{1}{\gamma^2} \|f\|^2_{L^2_\gamma(X)}$,

which implies the Proposition.

⊗

Furthermore if f has a bounded support in the past, we obtain that u also has this property, with the same bound (thanks to a Paley-Wiener theorem).

3.1. Some examples with the wave equation

3.1.1. A typical example of causal problem

We first consider the simple problem which will be very useful for solving other problems with the d'Alembert equation: find a function v on R satisfying, with a given function f on R, and real $\lambda > 0$

(146)i) $v'' + \lambda^2 v = f$, where $v'' = \frac{\partial^2 v}{\partial t^2}$,

with either:

(146)ii) $\text{supp}\, v \subset \text{supp}\, f + R^+$, if supp f is bounded in the past,

or

(146)ii)' $v \in L^2_\gamma(R) = e^{\gamma t} L^2(R)$, if $f \in L^2_\gamma(R)$.

In both cases, the solution for $f = 0$, $v_0(t) = A\, e^{i\lambda t} + B\, e^{-i\lambda t}$ must be zero.
Thus (146) has the unique solution:

(147) $v(t) = \int_{-\infty}^{t} \frac{\sin \lambda(t-s)}{\lambda} f(s)\, ds = (G_\lambda * f)(t)$,

with $G_\lambda(t) = Y(t)\frac{\sin \lambda t}{\lambda}$, Y(t) being the Heaviside function, and also:

(147)' $e^{-\gamma t} v(t) = \int_{-\infty}^{t} e^{-\gamma(t-s)} . \frac{\sin \lambda(t-s)}{\lambda} e^{-\gamma s} f(s)\, ds = G_{\gamma,\lambda} * (e^{-\gamma t} f(t))$,

with: $G_{\gamma,\lambda}(t) = Y(t)\, e^{-\gamma t} \frac{\sin \lambda t}{\lambda}$; thus $G_{\gamma,\lambda} \in L^1(R)$, with $\int_R |G_{\gamma,\lambda}(t)|\, dt \leq \frac{1}{\gamma\lambda}$.
Therefore when $f \in L^2_\gamma(R)$, we have $v \in L^2_\gamma(R)$ and:

(148) $\|v\|_\gamma \leq \frac{1}{\gamma\lambda} \|f\|_\gamma$ with $\|v\|_\gamma = \int_R e^{-2\gamma t} |v|^2\, dt$.

3.1.2. An example with boundary conditions

Now we consider the evolution problem: given a (regular) bounded domain Ω in R^n, and given f in $L^2_\gamma(X)$, find u in $L^2_\gamma(X)$, $X = L^2(\Omega)$, so that

$$(149)\qquad \begin{vmatrix} \text{i)}\ \dfrac{\partial^2 u}{\partial t^2} - \Delta u = f \quad \text{in } \Omega \times R_t, \\ \text{ii)}\ u|_\Sigma = 0, \quad \Sigma = \Gamma \times R. \end{vmatrix}$$

Using the spectral decomposition (λ^2_j, ϕ_j) of the Laplacian with Dirichlet condition in $L^2(\Omega)$, we have:

$$(150)\qquad u(t) = \sum_j u_j(t)\phi_j, \quad \text{and } f(t) = \sum_j f_j(t)\phi_j \quad \text{with } \sum_j \int_R e^{-2\gamma t} |f_j(t)|^2 dt < \infty,$$

and u_j has to satisfy: $u_j \in L^2_\gamma(R)$, with:

$$(151)\qquad \frac{\partial^2 u_j}{\partial t^2} + \lambda_j^2 u_j = f_j, \quad j \in N.$$

From section 3.1.1, we know that this problem has the unique solution:

$$(152)\qquad u_j(t) = \int_{-\infty}^{t} \frac{\sin \lambda_j(t-s)}{\lambda_j} f_j(s)\, ds = (G_{\lambda_j} * f)(t),$$

and u_j satisfies (thanks to (148)):

$$(153)\qquad \|u_j\|_\gamma \leq \frac{1}{\gamma \lambda_j} \|f_j\|_\gamma .$$

This implies that u is in $L^2_\gamma(X)$, $X = L^2(\Omega)$, with:

$$(154)\qquad \|u\|_\gamma \leq C \|f\|_\gamma, \quad C = \frac{1}{\gamma \lambda_0} .$$

Moreover since $((-\Delta_D)^{1/2}u)_j = \lambda_j u_j$, we obtain with $Y = H^1_0(\Omega)$, and $\tilde{C} = \frac{1}{\gamma} \max\left(\frac{1}{\lambda_0}, 1\right)$

$$(155)\qquad \|u\|_{L^2_\gamma(Y)} \leq \tilde{C} \|f\|_\gamma .$$

Furthermore differentiating (152) with respect to t, we have:

$$(156)\qquad u'_j(t) = (Y(t) \cos(\lambda_j t)) \underset{t}{*} f_j$$

and thus: $\|u'_j\|_\gamma \leq \frac{1}{\gamma} \|f_j\|_\gamma$, giving:

$$(157)\qquad \|u'\|_\gamma \leq \frac{1}{\gamma} \|f\|_\gamma .$$

Thus we have obtained:

(158) $u \in L^2_\gamma(Y)$ with $Y = H^1_0(\Omega)$, $u' \in L^2_\gamma(X)$ with $X = L^2(\Omega)$, thus $e^{-\gamma t}u \in H^1{}_0(\Omega \times R)$.

Then we can prove also the trace result: $e^{-\gamma t}\frac{\partial u}{\partial n}\big|_\Sigma \in H^{-1/2}(\Sigma)$ on $\Sigma = \Gamma \times R$.

But this trace result is not optimal. Using the proof of Lions [3] p. 40, 41 in the framework of $L^2{}_\gamma(X)$ functions, we get:

(159) $\frac{\partial u}{\partial n}\big|_\Sigma \in e^{\gamma t} L^2(\Sigma)$, i.e., $e^{-\gamma t}\frac{\partial u}{\partial n}\big|_\Sigma \in L^2(\Sigma)$.

Remark 3. *An a priori estimate.* We multiply (149)i) by $e^{-2\gamma t}u'$ (the time derivative of u) and integrate by parts. We get:

(160) $$\gamma \int_{\Omega\times R} e^{-2\gamma t}\,[\,|u'|^2 + |\mathrm{grad}\,u|^2]\,dx\,dt = \int_{\Omega\times R} e^{-2\gamma t} f\,u'\,dx\,dt \geq 0.$$

Then using the Cauchy-Schwarz inequality, we obtain the estimate where $\|v\|_\gamma$ denotes the $L^2_\gamma(L^2(\Omega))$ norm of v:

(161) $$\gamma\,[\tfrac{1}{2}\|u'\|^2_\gamma + \|\mathrm{grad}\,u\|^2_\gamma] \leq \tfrac{1}{2\gamma}\|f\|^2_\gamma.$$

This also implies (158), thanks to the Poincaré inequality.

⊗

Remark 4. We can solve in the same way the evolution problem (149) but with a (homogeneous) Neumann condition. The only new difficulty is with the eigenvalue 0. We obtain finally the same results (158) for the solution.

Usual trace results give $u|_\Sigma \in e^{\gamma t} H^{1/2}(\Sigma)$, but this is not an optimal result. If the uniform Lopatinski condition (see Chazarain-Piriou [1] p. 362) were valid, we could hope to have $u|_\Sigma \in e^{\gamma t} H^1(\Sigma)$. But this is not the case.

From results of Lasiecka-Triggiani [1], we have $u|_\Sigma \in e^{\gamma t} H^{2/3}(\Sigma)$, and when Ω is a parallelepiped $u|_\Sigma \in e^{\gamma t} H^{(3/4)-\varepsilon}(\Sigma)$, $\varepsilon > 0$.

⊗

3.1.3. An example with transmission conditions, Integral methods

Let Ω be a regular bounded in R^n with boundary Γ. We consider the problem: find u with given jumps (ρ, ρ') across $\Sigma = \Gamma \times R$, so that

(162) $$\left|\begin{array}{l} \text{i)}\ \dfrac{\partial^2 u}{\partial t^2} - \Delta u = 0 \quad \text{in } (R^n\backslash\Gamma)\times R_t\,, \\ \text{ii)}\ [u]_\Sigma = \rho\,,\ \ [\frac{\partial u}{\partial n}]_\Sigma = \rho',\ \ (\rho,\rho') \in L^0_+(R_t, Y),\ \ Y = H^{1/2}(\Gamma)\times H^{-1/2}(\Gamma), \\ \text{iii)}\ \mathrm{supp}\,u \subset \mathrm{supp}\,(\rho,\rho') + C_+, \end{array}\right.$$

where $[v]_\Sigma$ is the jump of v across $\Sigma = \Gamma \times R$. Writing (162)i) in $D'(R^n x R_t)$, we see that u must satisfy:

$$(163) \qquad \frac{\partial^2 u}{\partial t^2} - \Delta u = f, \quad \text{with } f = -(\rho'\delta_\Sigma + \operatorname{div}(\rho n \delta_\Sigma)),$$

and thus we obtain the solution of (162) in the form:

$$(164) \qquad u = \Phi \underset{x,t}{*} f.$$

But we have to specify the functional space, especially with respect to the boundary conditions. So it is better to use a Laplace transformation.

Then problem (162) becomes: find $\hat{u} = \hat{u}(p)$ (with Re p $\geq \xi_0$), satisfying

$$(165) \qquad \left| \begin{array}{l} \text{i) } (p^2 - \Delta)\,\hat{u} = 0 \ \text{ in } R^n \backslash \Gamma, \\ \text{ii) } [\hat{u}]_\Gamma = \hat{\rho}, \ [\frac{\partial \hat{u}}{\partial n}]_\Gamma = \hat{\rho}'. \end{array} \right.$$

This problem has for each p a unique solution in:

$$(166) \qquad X = H^1(\Delta, R^n \backslash \Gamma) = \{u, \ u|_\Omega \in H^1(\Delta,\Omega), \ u|_{\Omega'} \in H^1(\Delta,\Omega')\}.$$

Then we have a continuous lifting of the jump conditions into $L^0_+(R_t,X)$ so that we are reduced to solving the problem: find U in $L^0_+(R_t, H^2(R^n))$ satisfying

$$(167) \qquad \frac{\partial^2 U}{\partial t^2} - \Delta U = F, \qquad F \text{ given in } L^0_+(R_t, L^2(R^n)).$$

This problem has a unique solution. Thus we obtain a unique solution u of (162), in $L^0_+(R_t,X)$ with X given by (166).

Thus we can define *causal Calderon projectors* P^c_i and P^c_e in $L^0_+(R_t, Y)$ by:

$$(168) \qquad P^c_i(\rho,\rho') = (u|_{\Gamma_i}, \frac{\partial u}{\partial n}|_{\Gamma_i}), \qquad P^c_e(\rho,\rho') = (u|_{\Gamma_e}, \frac{\partial u}{\partial n}|_{\Gamma_e}).$$

with the usual property: $P^c_i + P^c_e = I$.

We get u in R^3 x R from (164) and the retarded potentials by the Kirchhoff formula (see Bamberger-Ha Duong [1]):

$$(169) \qquad \left| \begin{array}{l} u = \mathcal{L}^c\rho' + \mathcal{P}^c\rho, \text{ with:} \\ \mathcal{L}^c\rho'(x,t) = \dfrac{1}{4\pi}\displaystyle\int_\Gamma \frac{1}{|x-y|}\,\rho'(y,\tau)\,d\Gamma_y, \qquad \tau = t - |x-y|, \\ \mathcal{P}^c\rho(x,t) = -\dfrac{1}{4\pi}\displaystyle\int_\Gamma \frac{(x-y).n_y}{|x-y|^2}\,(\frac{\rho(y,\tau)}{|x-y|} + \frac{\partial \rho}{\partial t}(y,\tau))\,d\Gamma_y. \end{array} \right.$$

From these definitions, there is no jump of the trace of $v = \mathcal{L}^c\rho'$ across Σ, nor of the normal derivative of $w = \mathcal{P}^c\rho$ across Σ. From $\mathcal{L}^c$ and $\mathcal{P}^c$, we can define "usual retarded" integral operators on the surface Σ, and an operator S^c such that

(170) $$P_i^c = \frac{1}{2}(I + S^c), \quad P_e^c = \frac{1}{2}(I - S^c), \quad \text{with } S^c = \begin{pmatrix} K^c & 2L^c \\ 2R^c & J^c \end{pmatrix},$$

and:

(171) $$\left| \begin{array}{ll} L^c\rho' = \mathcal{L}^c\rho'|_{\Sigma}, & K^c\rho = \mathcal{P}^c\rho|_{\Sigma_e} + \mathcal{P}^c\rho|_{\Sigma_i}, \\ J^c\rho' = \frac{\partial}{\partial n}\mathcal{L}^c\rho'|_{\Sigma_e} + \frac{\partial}{\partial n}\mathcal{L}^c\rho'|_{\Sigma_i}, & R^c\rho = \frac{\partial}{\partial n}\mathcal{P}^c\rho|_{\Sigma}. \end{array} \right.$$

Of course it is interesting to specify these operators in a $L^2_\gamma(X)$ framework. We will do it, but optimal regularity results are missing.

a) A regular framework. We define for real s and $\gamma > 0$

(172) $$e^{\gamma t}H^s_\gamma(R^n x R) \stackrel{def}{=} \{f,\ e^{-\gamma t}f \in \mathcal{S}'(R^{n+1}),\ \|f\|^2_{s,\gamma} = \int_{R^n x R} (|p|^2 + \xi^2)^s |\hat{f}(\xi,p)|^2 d\xi d\eta < \infty\}$$

with $p = \gamma + i\eta$, $\hat{f}(\xi,p)$ being the x-Fourier transform and t-Laplace transform of f, Then we define for Ω a regular open set in R^n, by restriction, the space $e^{\gamma t}H^s_\gamma(\Omega \times R)$, and (thanks to an atlas) $e^{\gamma t}H^s_\gamma(\Gamma \times R)$, see Chazarain-Piriou [1] p. 412. Thus we have:

(172)' $$e^{\gamma t}H^1_\gamma(\Omega \times R) \stackrel{def}{=} \{u \in L^2_\gamma(L^2(\Omega)),\ \frac{\partial u}{\partial t}, \frac{\partial u}{\partial x_j} \in L^2_\gamma(L^2(\Omega))\}.$$

Now if we successively consider the wave problem (149) with Dirichlet boundary conditions, and Neumann conditions, we define the trace spaces:

(173) $$\left| \begin{array}{l} \text{i) } X^i_\gamma = \{u^1 = \frac{\partial u}{\partial n}|_{\Sigma},\ u \in e^{\gamma t}H^1_\gamma(\Omega \times R) \text{ the solution of (149)}\}, \\ \text{ii) } Y^i_\gamma = \{u^0 = v|_{\Sigma},\ v \in e^{\gamma t}H^1_\gamma(\Omega \times R) \text{ the solution of (149)i) with } \frac{\partial v}{\partial n}|_{\Sigma} = 0\}. \end{array} \right.$$

We recall from the regularity results (Remark 4 and (159)) that:

(174) $$X^i_\gamma \subset e^{\gamma t}H^{2/3}(\Sigma), \quad Y^i_\gamma \subset e^{\gamma t}L^2(\Sigma).$$

The elements of $X_\gamma \times Y_\gamma$ are traces at the boundary of waves u such that:

$$u \in D_\gamma(\mathcal{B}) \stackrel{def}{=} D_\gamma(\mathcal{B}_D) + D_\gamma(\mathcal{B}_N),$$

with:

$$\left| \begin{array}{l} D_\gamma(\mathcal{B}_D) = \{u \in e^{\gamma t}H^1_\gamma(\Omega x R),\ \Box u \in L^2_\gamma(X),\ X = L^2(\Omega),\ u|_{\Sigma} = 0\}, \\ D_\gamma(\mathcal{B}_N) = \{u \in e^{\gamma t}H^1_\gamma(\Omega x R),\ \Box u \in L^2_\gamma(X),\ X = L^2(\Omega),\ \frac{\partial u}{\partial n}|_{\Sigma} = 0\}. \end{array} \right.$$

Let f be given in $L^2_\gamma(X)$, $X = L^2(\Omega)$. Let u_f (resp. v_f) be the solution of the Dirichlet (Neumann) problem for f. Let $U = U_f = v_f - u_f$. Then U satisfies:

$$(175)\qquad \left|\begin{array}{l} \text{i) } \Box U = 0, \\ \text{ii) } U|_\Sigma = v|_\Sigma = u^0, \quad \frac{\partial U}{\partial n}\Big|_\Sigma = -\frac{\partial u}{\partial n}\Big|_\Sigma. \end{array}\right.$$

Let θ (resp. θ_1) be the mapping $f \in L^2_\gamma(X) \longrightarrow (v|_\Sigma, -\frac{\partial u}{\partial n}|_\Sigma) \in X^i_\gamma \times Y^i_\gamma$) (resp. $U_f \in D^0_\gamma(\mathcal{B})$) with:

$$D^0_\gamma(\mathcal{B}) = \{U \in D_\gamma(\mathcal{B}),\ \Box U = 0\}.$$

We can prove that $\ker\theta = \ker\theta_1$. Then taking the quotient space $L_\gamma(X)/\ker\theta$, we obtain an isomorphism from $L_\gamma(X)/\ker\theta$ onto $D^0{}_\gamma(\mathcal{B})$ and onto a subspace G_i of $X^i_\gamma x Y^i_\gamma$. Then we define an isomorphism $C^i = C^i_c$ from X^i_γ onto Y^i_γ by:

$$(176)\qquad C^i(u^0) = \frac{\partial U}{\partial n}\Big|_\Sigma \quad \text{(with (175))}.$$

This operator C^i is the ***interior causal Calderon operator***.

Now we can operate in a similar way for the exterior domain. If the trace space $X^i{}_\gamma$ (resp., $Y^i{}_\gamma$) is the same, for the interior domain as well as the exterior domain (which seems to be a very natural conjecture for regular domains), that is $X^i{}_\gamma = X^e{}_\gamma = X_\gamma$ (resp. $Y^i{}_\gamma = Y^e{}_\gamma = Y_\gamma$), we also define a subspace G^e of $X_\gamma \times Y_\gamma$ and an operator C^e for the exterior domain which is the ***exterior causal Calderon operator*** with $G(C^e) = G^e$; C^e is also an isomorphism from X_γ onto Y_γ.

Note that G^i and G^e are closed subspaces in $X_\gamma x Y_\gamma$ with intersection reduced to $\{0\}$. A natural conjecture is that $X_\gamma \times Y_\gamma = G(C^i) \oplus G(C^e)$.

b) A more general framework of "finite energy". Now we define:

$$(177)\qquad \left|\begin{array}{l} W^\Omega_\gamma = \{u \in e^{\gamma t}H^1_\gamma(\Omega \times R),\ \Box u = 0 \text{ in } \Omega \times R\}, \\ \\ W_\gamma = \{u,\ u|_{\Omega \times R} \in W^\Omega_\gamma,\ u|_{\Omega' \times R} \in W^{\Omega'}_\gamma,\ \Omega' = R^n \backslash \bar{\Omega}\}. \end{array}\right.$$

We can consider the space W_γ as a natural space of waves with finite energy on both sides of $\Sigma = \Gamma \times R$. The corresponding space of sources concentrated on Σ is

$$(178)\qquad Z_\gamma = \{(\rho,\rho') = ([u]_\Sigma, [\tfrac{\partial u}{\partial n}]_\Sigma),\ u \in W_\gamma\}.$$

The mapping $u \longrightarrow (\rho,\rho')$ is an isomorphism from W_γ onto Z_γ, with inverse:

$$(179)\qquad u = \Phi \underset{x,t}{*} (\rho'\delta_\Sigma + \operatorname{div}(\rho n \delta_\Sigma)).$$

Then Z_γ is a Hilbert space, which is the direct sum of two closed subspaces G_i, G_e with intersection reduced to {0}, which are the set of boundary values of elements in W_γ^{Ω} and $W_\gamma^{\Omega'}$: $Z_\gamma = G_i \oplus G_e$. The projectors on G_i and G_e (along G_e and G_i) are the *causal Calderon projectors* in Z_γ. Furthermore we easily see that G_i and G_e are the graphs of two operators, the *causal Calderon operators*, which are extensions of those defined in the preceding section.

A reasonable conjecture is that $Z_\gamma = \tilde{X}_\gamma \times \tilde{Y}_\gamma$ with the spaces of traces

(180) $$\tilde{X}_\gamma = \{v = u|_\Gamma,\ u \in W_\gamma^{\Omega}\},\ \tilde{Y}_\gamma = \{v = \frac{\partial u}{\partial n}|_\Gamma,\ u \in W_\gamma^{\Omega}\},$$

and that these spaces of traces for Ω are equal to the spaces of traces for Ω'.
We refer to Bamberger-Ha Duong [1] for numerical implementations, and for results with the space $H_\gamma^s(R,X) = \{u, \int_R |p|^{2s} \|\hat{u}(p)\|_X^2 d\eta < \infty\}$, with $s = 3/2$, and $X = H^{1/2}(\Gamma)$.

We note that proceeding as in Remark 3, we obtain the positivity property:

(181) $$\gamma \int_{\Omega x R} e^{-2\gamma t} ([|u'|^2 + |\text{grad}\, u|^2])\, dx\, dt = \int_\Sigma e^{-2\gamma t} u' \frac{\partial u}{\partial n} d\Gamma\, dt > 0.$$

3.1.4. An example with a waveguide

Let $\Omega_+ = \Omega \times R^+_z$ (with $z = x_3$) be a semi-infinite cylinder (modelling a soft waveguide) with a regular bounded cross-section Ω (with boundary Γ) in R^2. Then we consider the wave problem with a given value u^0 at the end of Ω_+:

(182)
$$\begin{cases} \text{i)}\ \dfrac{\partial^2 u}{\partial t^2} - (\dfrac{\partial^2 u}{\partial z^2} + \Delta_x u) = 0 \text{ in } \Omega_+ \times R_t, \text{ with } x = (x_1, x_2) \in \Omega, \\ \text{ii)}\ u(x,0,t) = u^0(x,t) \text{ on } \Omega x R_t,\ u^0 \in L^2_\gamma(X),\ X = L^2(\Omega),\ \gamma > 0, \\ \text{iii)}\ u(x,z,t) = 0 \quad \text{on } \Sigma_1 \times R_t = \Gamma \times R_z^+ \times R_t, \\ \text{iv)}\ u \in L^2_\gamma(L^2(\Omega_+)). \end{cases}$$

We can solve this problem due to a spectral decomposition of the Laplacian Δ_x with a Dirichlet condition. Let (λ^2_j, ϕ_j) be eigenvalues and eigenmodes so that we can decompose u and u^0 into:

(183) $$u(.,z,t) = \sum_j u_j(z,t)\, \phi_j, \qquad u^0 = \sum_j u_j^0\, \phi_j.$$

Then (182) is reduced to finding u_j satisfying:

(184)
$$\begin{cases} \text{i)}\ \dfrac{\partial^2 u_j}{\partial z^2} - \dfrac{\partial^2 u_j}{\partial t^2} - \lambda_j^2 u_j = 0 \quad \text{in } R_z^+ \times R_t, \\ \text{ii)}\ u_j(0,t) = u_j^0(t), \quad t \in R. \end{cases}$$

Now using the (time) Laplace transformation, we obtain:

$$(185)\qquad \left| \begin{array}{l} \text{i) } \dfrac{\partial^2 \hat{u}_j}{\partial z^2} - (p^2 + \lambda_j^2)\,\hat{u}_j = 0 \quad \text{in } R_z^+, \\ \text{ii) } \hat{u}_j(0) = \hat{u}_j^0. \end{array} \right.$$

The solutions of (185)i) are given with two constants A_j, B_j by:

$$(186)\qquad \hat{u}_j(z) = A_j\, e^{\theta_j z} + B_j\, e^{-\theta_j z}, \quad \text{with } \theta_j^2 = p^2 + \lambda_j^2.$$

Now in order to have a solution in $L^2_\gamma(X)$, we have to take $A_j = 0$, with Re $\theta_J \geq \gamma$. Thus the solution $\hat{u}_j$ is given by:

$$(187)\qquad \hat{u}_j(z) = e^{-\theta_j z}\, \hat{u}_j^0.$$

The image of the half plane Re $p \geq \gamma$ by the map $p \rightarrow p^2$ is the set:

$\{(x,y) \in C,\ (y/2\gamma)^2 + x - \gamma^2 \geq 0\}$,

i.e., the exterior of a parabola P_γ, which contains the set $\{\theta_j^2 = p^2 + \lambda_j^2,\ \text{Re } p \geq \gamma\}$. We have the inequalities:

$$(188)\qquad |\hat{u}_j(z)| \leq |\hat{u}_j^0|, \quad \int_{R^+} |\hat{u}_j(z)|^2\, dz = \frac{1}{2\,\text{Re}\,\theta_j}\, |\hat{u}_j^0|^2 \leq \frac{1}{2\gamma}\, |\hat{u}_j^0|^2.$$

This implies that (184) (and thus (182)) has a unique solution which is given thanks to a contraction semigroup $(G(z))$ (of class C^0) in $L^2_\gamma(X)$ by $u(.,z,.) = G(z)u^0$. The infinitesimal generator A of $(G(z))$ is defined thanks to the spectral decomposition of the (transverse) Laplacian with Dirichlet condition and thanks to the time Laplace transformation by:

$$(189)\qquad \hat{A}\hat{u}_j(p) = -\,\theta_j \hat{u}_j(p) = (p^2 + \lambda_j^2)^{1/2} \hat{u}_j(p), \quad \text{with Re } p \geq \gamma,\ \text{Re } \theta_j \geq \gamma,$$

and its domain is:

$$(189)'\qquad D(A) = \{v \in L^2_\gamma(X),\ \sum_j \int_R |\theta_j|^2\, |\hat{v}_j(p)|^2\, d\eta < \infty,\ \text{with } p = \gamma + i\eta\}.$$

Since $|p| \leq |\theta_j|$, we have $D(A) \subset e^{\gamma t} H^1_\gamma(\Omega \times R_t)$.

Thus the problem (182) is equivalent to finding u in $L^2_\gamma(X)$, $X = L^2(\Omega)$, satisfying:

$$(190)\qquad \left| \begin{array}{l} \text{i) } \dfrac{\partial u}{\partial z} - Au = 0, \quad z > 0, \\ \text{ii) } u(0) = u^0 \quad \text{on } \Omega \times R_t. \end{array} \right.$$

The operator A is the *causal Calderon operator of the waveguide*.

3.2. Some examples with Maxwell equations

3.2.1. Causal Maxwell problems in the whole space

We first consider the basic problem: let J and M be given electric and magnetic currents in free space, which are distributions in the whole space $R^3 \times R_t$, with supports bounded in the past. Like in the stationary case, we assume that there exists magnetic currents, which will be useful to consider in order to solve other problems in electromagnetism. We also assume the existence of related electric and magnetic charges ρ and ζ. Let ε_0, μ_0 the permittivity and the permeability of the free space resp.. We have to find the electromagnetic field (E, B) satisfying

$$(191)\quad \left|\begin{array}{l} \text{i)}\ \frac{\partial E}{\partial t} - c^2 \,\text{curl}\, B = -\frac{1}{\varepsilon_0} J,\quad \text{ii)}\ \frac{\partial B}{\partial t} + \text{curl}\, E = -M \quad \text{in } R^3_x \times R_t, \\ \text{iii)}\ \text{div}\, E = \frac{1}{\varepsilon_0}\rho\,,\ \text{and div}\, B = \zeta, \\ \text{iv)}\ \text{supp}\,(E,B) \subset \text{supp}\,(J,M) + C_+\,, \end{array}\right.$$

with:

$$(192)\quad \frac{\partial \rho}{\partial t} + \text{div}\, J = 0, \quad \frac{\partial \zeta}{\partial t} + \text{div}\, M = 0.$$

We first note that the equations (191) iii) are consequences of i), ii), iv) (and (192)). By the usual tricks, we eliminate B, then E, to obtain the d'Alembert equations (with (11)):

$$(193)\quad \square E = j\,, \quad \square B = \mu_0 m,$$

with (j, m) given by:

$$(194)\quad -j = \frac{1}{\varepsilon_0}\left(\frac{1}{c^2}\frac{\partial J}{\partial t} + \text{grad}\,\rho\right) + \text{curl}\, M, \qquad -m = \frac{1}{\mu_0}\left(\frac{1}{c^2}\frac{\partial M}{\partial t} + \text{grad}\,\zeta\right) - \text{curl}\, J,$$

and thus we get the solution (E, B) of (191) thanks to the convolution product with the elementary solution Φ (see (24) and (135)) (with $c = 1$ to simplify):

$$(195)\quad E = \Phi * j\,, \quad B = \Phi * (\mu_0 m).$$

We only note that taking the divergence of (195) implies (191) iii):

$$(196)\quad \text{div}\, E = \Phi * \text{div}\, j = \Phi * \frac{1}{\varepsilon_0}\square \rho = \frac{\rho}{\varepsilon_0}, \quad \text{div}\, B = \Phi * \mu_0 \,\text{div}\, m = \Phi * \square \zeta = \zeta\,.$$

Theorem 2. *Let* (J, M) *be given currents with support bounded in the past. The causal problem* (191) *has a unique solution* (E, B) *in* $D'(R^3 \times R_t)^3 \times D'(R^3 \times R_t)^3$ *given by* (195) *with* (194).

Remark 5. *Causal vector and scalar potentials.* When $M = 0$, we define the causal vector and scalar potentials: $A = \mu_0 \Phi * J$, $\phi = \frac{1}{\varepsilon_0}\Phi * \rho$; they satisfy: $\frac{1}{c^2}\frac{\partial \phi}{\partial t} + \text{div}\, A = 0$; and from (195) (E, B) is given by: $B = \text{curl}\, A$, $E = -\frac{\partial A}{\partial t} - \text{grad}\,\phi$. ⊗

3.2.2. Causal problems with currents on a surface. Transmission problems
Now let (J_Γ, M_Γ) be given electric and magnetic currents concentrated on a regular surface Γ in R^3 which is the boundary of a bounded open set Ω. Furthermore we assume that they satisfy, with notations (138):

(197) $\quad J_\Gamma,\ M_\Gamma \in L^0_+(R_t, X) \quad \text{with } X = H^{-1/2}(\text{div}, \Gamma).$

Then the electromagnetic field (E, B) due to these currents must satisfy (191). Theorem 2 implies its existence and uniqueness at least in the sense of distributions. Thanks to the hypothesis (197) we can prove (for instance thanks to a lifting) that the field (E, B) has finite energy, that is:

(198) $\ (E,B) \in L^0_+(R_t, Y) \times L^0_+(R_t, Y),\ Y = L^2(R^3)^3, \text{ then } E,\ B \in L^0_+(R_t, H(\text{curl}, R^3 \backslash \Gamma)).$

Thus E and B have traces $(n \wedge E|_\Sigma,\ n \wedge B|_\Sigma)$ on each side of the boundary Σ of $\Omega \times R$. Furthermore they satisfy (197). Using jumps formulas (25) chap. 2 we obtain that (E, B) satisfy

(199) $$\left| \begin{array}{l} \dfrac{\partial E}{\partial t} - c^2 \,\text{curl}\, B = c^2 [n \wedge B]_\Sigma\, \delta_\Sigma, \quad \dfrac{\partial B}{\partial t} + \text{curl}\, E = -[n \wedge B]_\Sigma\, \delta_\Sigma, \quad \text{in } R^3_x \times R_t, \\ \text{div}\, E = -[n.E]_\Sigma, \quad \text{div}\, B = -[n.B]_\Sigma, \end{array} \right.$$

so that if we compare to (191), we obtain:

(200) $$J_\Sigma = -\frac{1}{\mu_0}[n \wedge B]_\Sigma, \qquad M_\Sigma = [n \wedge E]_\Sigma.$$

Moreover applying the divergence to (199), we obtain

(201) $$-\frac{\partial}{\partial t}[n.E]_\Sigma = c^2\, \text{div}\,([n \wedge B]_\Sigma\, \delta_\Sigma), \qquad \frac{\partial}{\partial t}[n.B]_\Sigma = \text{div}\,([n \wedge E]_\Sigma\, \delta_\Sigma).$$

These basic relations correspond to those of the stationary case (see (84) chap.3). They are compatibility relations for traces, and they give the normal components of E and B thanks to the tangential components of B and E.
Thus we have obtained with hypotheses (197), results similar to those of the stationary case (see Theorem 3 chap.3), but here (E, B) satisfies (198) and thus is of finite energy in the whole space. Furthermore (E, B) is obtained by the formulas (195) (with retarded potentials), with (j, m) given by (194), and with:

(202) $$\rho_\Sigma = -\varepsilon_0 [n.E]_\Sigma, \quad \zeta_\Sigma = -[n.B]_\Sigma.$$

These formulas (195) with surface currents are the "Stratton-Chu" formulas in evolution case (see (87) chap. 3 in the stationary case), which are simply written in a form similar to (96) chap. 3. We define like in (95) chap. 3:

(203) $$\mathcal{L}^c_m J = c\left(\frac{1}{c^2}\frac{\partial J}{\partial t} + \text{grad}\, \rho\right) * \Phi, \qquad \mathcal{P}^c_m J = -(\text{curl}\, J) * \Phi,$$

with $\frac{\partial \rho}{\partial t} + \operatorname{div} J = 0$, thus $\rho = - Y \underset{t}{*} \operatorname{div} J$, Y being the Heaviside function.

Thus keeping the magnetic field as unknown, we have

(203)' $\quad E = \mathcal{P}^c_m M_\Gamma + Z\, \mathcal{L}^c_m J_\Gamma, \qquad H = Z^{-1} \mathcal{L}^c_m M_\Gamma - \mathcal{P}^c_m J_\Gamma.$

Now we define the *causal Calderon projectors for the electromagnetic field* (E,H), P^c_i and P^c_e in $L^0_+(R,X) \times L^0_+(R,X)$ (see (197)) by:

(204) $P^c_i(M,J) = (n \wedge E|_{\Sigma_i}, - n \wedge H|_{\Sigma_i}), \qquad P^c_e(M,J) = -(n \wedge E|_{\Sigma_e}, - n \wedge H|_{\Sigma_e}),$

like in (93) chap. 3. These operators satisfy the usual relations (104), (105), (106) chap. 3 with an operator S^c given by:

(205) $$S^c = \begin{pmatrix} -T^c & ZR^c \\ -\frac{1}{Z}R^c & -T^c \end{pmatrix},$$

with

$$T^c J = -(\gamma_i + \gamma_e)\, \mathcal{P}^c_m J, \qquad R^c J = 2\, n \wedge \mathcal{L}^c_m J|_\Sigma.$$

We have to specify these operators in spaces with a L^2_γ time behavior. Here also regularity results are missing. Like in the case of scalar waves we can define spaces with more or less regularity properties: here we define a general space with "finite energy" (but always with the time decreasing weight $e^{-\gamma t}$)

(206) $$\left| \begin{array}{l} W^\Omega_\gamma = \{(E,H) \in L^2_\gamma(Y_\Omega), \ \text{with } Y_\Omega = L^2(\Omega)^3 \times L^2(\Omega)^3, \\ \qquad \varepsilon_0 \frac{\partial E}{\partial t} - \operatorname{curl} H = 0, \quad \mu_0 \frac{\partial H}{\partial t} + \operatorname{curl} E = 0, \text{ in } \Omega \times R\}, \\ W_\gamma = \{(E,H),\ (E,H)|_{\Omega \times R} \in W^\Omega_\gamma,\ \ (E,H)|_{\Omega' \times R} \in W^{\Omega'}_\gamma\}. \end{array} \right.$$

Then we define the space of jumps across $\Sigma = \Gamma \times R$:

(207) $\quad Z_\gamma = \{(M_\Sigma, J_\Sigma),\ M_\Sigma = [n \wedge E]_\Sigma,\ J_\Sigma = -[n \wedge H]_\Sigma, \text{ with } (E,H) \in W_\gamma\}.$

The mapping $(E,H) \to (M_\Sigma, J_\Sigma)$ is an isomorphism from $W\gamma$ onto $Z\gamma$ with inverse given by (195) with (194). Like in the scalar case Z_γ is the direct sum of two closed subspaces G_i, G_e (with intersection reduced to {0}) corresponding to the set of boundary values of elements in W_γ^{Ω} and in $W_\gamma^{\Omega'}$: $Z_\gamma = G_i \oplus G_e$.

The projectors on G_i and G_e (along G_e and G_i) are the *causal Calderon projectors* P^c_i and P^c_e in Z_γ (see (204)). Furthermore we easily see that G_i and G_e are the graphs of two operators, $-C^c_i$ and $-C^c_e$, i.e., up to the sign, the *causal Calderon operators*, C^c_i and C^c_e. Then let:

(208) $\quad X^m_\gamma \overset{\text{def}}{=} \{n\wedge E|_\Sigma,\ (E,H)\in W^\Omega_\gamma\}, \qquad Y^m_\gamma \overset{\text{def}}{=} \{n\wedge H|_\Sigma,\ (E,H)\in W^\Omega_\gamma\}.$

We easily see that $X^m_\gamma = Y^m_\gamma$. Here also a reasonable conjecture is that $Z_\gamma = X^m_\gamma \times X^m_\gamma$ (we thus admit that the spaces of trace are the same for the interior as for the exterior). Then the Calderon operators C^c_i and C^c_e are isomorphisms in X^m_γ. As usual, these operators are directly related to interior and exterior boundary problems, since C^c_i and C^c_e are defined by:

(209) $\quad C^c_i(n\wedge E^0) = n\wedge H|_\Sigma$, (resp. C^c_e for the exterior).

where (E,H) is the unique solution in $L^2_\gamma(Y_\Omega)$ (resp. $Y_{\Omega'}$) satisfying:

(210)
$$\left|\begin{array}{l} \text{i) } \varepsilon_0 \dfrac{\partial E}{\partial t} - \text{curl } H = 0, \\ \text{ii) } \mu_0 \dfrac{\partial H}{\partial t} + \text{curl } E = 0, \text{ in } \Omega\times R \text{ (resp. } \Omega'\times R), \\ \text{iii) } n\wedge E|_\Sigma = n\wedge E^0 \text{ with given } n\wedge E^0 \text{ in } X^m_\gamma. \end{array}\right.$$

Furthermore multiplying i) by $E\, e^{-2\gamma t}$ and ii) by $H\, e^{-2\gamma t}$ then integrating, we get

(211) $$\gamma \int_{\Omega\times R} (\varepsilon_0 E^2 + \mu_0 H^2)\, e^{-2\gamma t}\, dx\, dt = -\int_\Sigma n\wedge E \,.\, H\, e^{-2\gamma t}\, d\Sigma \geq 0$$

for the interior, and obviously a similar result for the exterior. This gives the usual positivity properties (see (201), (212) chap. 3):

(212)
$$\left|\begin{array}{ll} (C^c_i m,\ n\wedge m) = \int_\Sigma (n\wedge E).H\, e^{-2\gamma t}\, d\Sigma \leq 0 & (\text{with } m = n\wedge E|_\Sigma), \\ (C^c_e m,\ n\wedge m) = \int_\Sigma (n\wedge E).H\, e^{-2\gamma t}\, d\Sigma \geq 0. & \end{array}\right.$$

For a numerical implementation of these time integral methods, see for instance Pujols [1].

3.2.3. Causal Maxwell problems with boundary conditions

i) First we recall that the evolution problem in free space (for a regular domain Ω bounded or not): find (E,H) in $L^2_\gamma(Y_\Omega)$, $Y_\Omega = L^2(\Omega)^3 \times L^2(\Omega)^3$, such that

(213)
$$\left|\begin{array}{l} \text{i) } \varepsilon_0 \dfrac{\partial E}{\partial t} - \text{curl } H = J, \quad \mu_0 \dfrac{\partial H}{\partial t} + \text{curl } E = M \text{ in } \Omega\times R, \\ \text{ii) } n\wedge E|_\Sigma = 0, \end{array}\right.$$

with given electric and magnetic currents (J,M) in $L^2_\gamma(Y_\Omega)$, has a unique solution thanks to Proposition 19. Furthemore it satisfies the *energy relation*:

(214) $\gamma \int_{\Omega \times R} (\varepsilon_0 E^2 + \mu_0 H^2)\, e^{-2\gamma t}\, dx\, dt = - \int_{\Omega \times R} (J.E + M.H)\, e^{-2\gamma t}\, dx\, dt.$

ii) Now we solve a more general evolution problem: we assume that the boundary Γ is coated with a dielectric medium, in a domain Ω_1, and the complementary domain in Ω, Ω_0 is occupied by free space. The dielectric medium is linear isotropic, its constitutive relations are given by (22), (or (23)) chap.1, with hypotheses H5) ((27),(28)) and H14) in chap.1.

Thus with $p = \gamma + i\eta = -i\omega$, $\omega = ip = -\eta + i\gamma$, $\eta \in R$, $\gamma > 0$ fixed, we have:

(215) $\alpha_\omega \overset{\text{def}}{=} \mathrm{Re}(-i\omega\hat{\varepsilon}) > 0, \qquad \beta_\omega \overset{\text{def}}{=} \mathrm{Re}(-\frac{1}{i\omega\hat{\mu}}) > 0,$

and there exists a constant $C > 0$ so that:

(216) $\alpha_\omega \geq C\varepsilon_0\gamma, \quad \beta_\omega \geq C\gamma/(\mu_0 |\omega|^2).$

Note that (216) is also true in free space, with $C = 1$, and even with equalities. Now given an electric current J in $L^2_\gamma(X_\Omega)$, $X_\Omega = L^2(\Omega)^3$, we consider the evolution problem: find the electromagnetic field E, B, D, H in $L^2_\gamma(X_\Omega)$, such that

(217) $\left| \begin{array}{l} \text{i)} \dfrac{\partial D}{\partial t} - \mathrm{curl}\, H = J, \quad \dfrac{\partial B}{\partial t} + \mathrm{curl}\, E = 0 \ \text{ in } \Omega \times R, \\ \text{ii)} D = \varepsilon \underset{t}{*} E, \quad B = \mu \underset{t}{*} H \ \text{ in } \Omega \times R \ \ (\text{with } \varepsilon = \varepsilon_0\,\delta(t),\ \mu = \mu_0\,\delta(t) \text{ in } \Omega_0 \times R), \\ \text{iii)} n \wedge E|_\Sigma = 0. \end{array} \right.$

This is an evolution problem with delay. Using a time Fourier-Laplace transform method like in chap.1, with $\gamma = \mathrm{Re}\, p = \mathrm{Im}\,\omega > 0$, problem (217) becomes:

(218) $\left| \begin{array}{l} \text{i)} i\omega\hat{\varepsilon}\, \hat{E} + \mathrm{curl}\, \hat{H} = \hat{J}, \quad - i\omega\hat{\mu}\, \hat{H} + \mathrm{curl}\, \hat{E} = 0 \ \text{ in } \Omega, \\ \text{ii)} n \wedge \hat{E}|_\Sigma = 0. \end{array} \right.$

We can solve this problem where ω is a *complex* parameter, with $\mathrm{Im}\,\omega > 0$, using a variational method. We define the sesquilinear form:

(219) $a(\phi,\psi) = \int_\Omega (-\frac{1}{i\omega\hat{\mu}} \mathrm{curl}\,\phi.\ \mathrm{curl}\,\bar{\psi} - i\omega\hat{\varepsilon}\,\phi.\,\bar{\psi})\, dx, \qquad \phi, \psi \in V = H_0(\mathrm{curl},\Omega),$

and then the problem (218) is equivalent to finding $\hat{E}(\omega)$ in V with:

(220) $a(\hat{E}(\omega),\psi) = - \int_\Omega \hat{J}(\omega).\, \bar{\psi}\, dx, \quad \forall\, \psi \in V.$

The sesquilinear form $a(\phi,\psi)$ is coercive on V for all ω with $\mathrm{Im}\,\omega \neq 0$, thus (220) has a unique solution for all ω with $\mathrm{Im}\,\omega \neq 0$.

Now we integrate with respect to η, and take the real part of (220) with $\varphi = \hat{E}$; using (216) in Ω, and the Cauchy-Schwarz inequality, we obtain with $C_1 = C\varepsilon_0\gamma$:

(221) $C_1 \int_{\Omega x R} |\hat{E}(\omega)|^2 \, dx \, d\eta \le \int_{\Omega x R} [\beta_\omega \, |\text{curl } \hat{E}(\omega)|^2 + \alpha_\omega \, |\hat{E}(\omega)|^2] \, dx \, d\eta \le \|\hat{J}\| \, \|\hat{E}\|$

with $\|\hat{J}\|^2 = \int_{\Omega x R} |\hat{J}(\omega)|^2 \, dx \, d\eta$. Thus we have:

(222) $\hat{E}(\omega) \in L^2(\Omega \, x \, R_\eta)^3$, and $\frac{1}{i\omega} \text{curl } \hat{E}(\omega) \in L^2(\Omega \, x \, R_\eta)^3$, with $\omega = -\eta + i\gamma$,
that is:

(223) $E \in L^2_\gamma(X)$, $X = L^2(\Omega)^3$, and $E(t) \overset{\text{def}}{=} \int_{-\infty}^{t} E(s)ds$ satisfies: $\text{curl } E \in L^2_\gamma(X)$.

This implies:

(224) $\hat{H}(\omega) \in L^2(\Omega \, x \, R_\eta)^3$, and $\frac{1}{i\omega} \text{curl } \hat{H}(\omega) \in L^2(\Omega \, x \, R_\eta)^3$,
thus:

(225) $H \in L^2_\gamma(X)$, and $H(t) \overset{\text{def}}{=} \int_{-\infty}^{t} H(s) \, ds$ satisfies: $\text{curl } H \in L^2_\gamma(X)$.

Furthermore since $\hat{E}(\omega) = -\frac{1}{i\omega} \hat{E}(\omega)$ and $\hat{H}(\omega) = -\frac{1}{i\omega} \hat{H}(\omega)$, (with $|\omega| \ge \gamma$), thus $\hat{E}(\omega)$ and $\hat{H}(\omega)$ are in $L^2(\Omega \, x \, R_\eta)^3$, and since $n \wedge \hat{E}(\omega)|_\Gamma = 0$, we have:

(226)
$$\left| \begin{array}{l} \hat{E} \in L^2(R_\eta, H_0(\text{curl},\Omega)), \quad \hat{H} \in L^2(R_\eta, H(\text{curl},\Omega)), \\ \text{thus: } E \in L^2_\gamma(H_0(\text{curl},\Omega)), \quad H \in L^2_\gamma(H(\text{curl},\Omega)). \end{array} \right.$$

Proposition 20. *The causal evolution problem* (217) *with delay, that is in a domain* Ω *occupied in part by free space and in part by a linear anisotropic medium with permittivity and permeability* (ε,μ) *satisfying* **H5)** *and* **H14)** chap.1, *has a unique solution* (E, D, B, H) *in* $L^2_\gamma(X_\Omega)$, *with* $X_\Omega = L^2(\Omega)^3$, *for a given electric current* J *in* $L^2_\gamma(X_\Omega)$, *and this solution satisfies* (226).

3.2.4. A causal waveguide problem

Like in section 3.1.4, let $\Omega_+ = \Omega \, x \, R_z^+$ be a free space domain, which is a semi-infinite cylinder bounded on its lateral boundary by a perfect conductor (modelling a waveguide), with a regular bounded cross-section Ω (with boundary Γ) in R^2. We assume that a tangential electric field is given at the end of the waveguide. Then we consider the evolution problem:

find (E, H) in the waveguide, with $(E,H) \in L^2_\gamma(X^3 \, x \, X^3)$, $X = L^2(\Omega_+)$, satisfying:

(227)
$$\left| \begin{array}{l} \text{i)} -\varepsilon_0 \dfrac{\partial E}{\partial t} + \text{curl } H = 0, \quad \mu_0 \dfrac{\partial H}{\partial t} + \text{curl } E = 0 \ \text{ in } Q_+ = \Omega_+ \, x \, R_t, \\ \text{ii)} \ n \wedge E|_{\Sigma_1} = 0 \ \text{ on } \Sigma_1 = \Gamma \, x \, R^+_z \, x \, R_t, \\ \text{iii)} \ n \wedge E|_{\Sigma_0} = n \wedge E^0 \ \text{ on } \Sigma_0 = \Omega \, x \, R_t, \quad E^0_T \in L^2_\gamma(H_0(\text{curl}_T,\Omega)). \end{array} \right.$$

We first write Maxwell equations like in (30) chap. 4, using the same notations:

$$(228)\quad \begin{cases} \text{i) } \partial_3 H_T - \text{grad}_T H_3 - \varepsilon_0 \partial_t S E_T = 0, \\ \text{ii) } \partial_3 E_T - \text{grad}_T E_3 + \mu_0 \partial_t S H_T = 0 \quad \text{in } \Omega_+ x R_t, \end{cases}$$

(where $\partial_t = \frac{\partial}{\partial t}$) with:

$$(229)\quad \text{i) } \text{curl}_T H_T - \varepsilon_0 \partial_t E_3 = 0, \qquad \text{ii) } \text{curl}_T E_T + \mu_0 \partial_t H_3 = 0 \text{ in } \Omega_+ x R_t,$$

and with the boundary conditions:

$$(230)\quad \text{i) } n_T \wedge E_T|_{\Sigma_1} = 0, \quad E_3|_{\Sigma_1} = 0, \qquad \text{ii) } E_T|_{\Sigma_0} = E_T^0.$$

From now on we drop the subscript T for differential operators. The determination of the electromagnetic field will be made in several steps.

Determination of the transverse electric field E_T. In the same way as in chap. 4, we first obtain that E_T satisfies:

$$(231)\quad \begin{cases} \text{i) } \partial_3^2 E_T + \Delta E_T - c^{-2} \partial_t^2 E_T = 0, \quad \text{in } \Omega_+ x R_t, \\ \text{ii) } n_T \wedge E_T|_{\Sigma_1} = 0, \quad (\text{div } E_T)|_{\Sigma_1} = 0, \\ \text{iii) } E_T(0) = E_T^0 \text{ on } \Sigma_0, \\ \text{iv) } E_T \in L_\gamma^2(X^2),\ X = L^2(\Omega_+); \end{cases}$$

the boundary condition on the divergence is a consequence, with (230)i), of:

$$(232)\quad \text{div}_{x,x_3} E = \partial_3 E_3 + \text{div}_T E_T = 0.$$

Let A_0 be the positive selfadjoint operator in $L^2(\Omega)^2$ defined by: $A_0 u = -\Delta u$, with

$$(233)\quad D(A_0) = \{u \in H^1(\Omega)^2,\ \text{curl } u \text{ and } \text{div } u \in H^1(\Omega),\ n \wedge u|_\Gamma = 0,\ (\text{div } u)|_\Gamma = 0\}$$

(see (147) chap. 2 with n = 3). Then using the spectral decomposition (λ^2_j, ϕ_j) of A_0, with $c = 1$, we define the operator A_T from its Laplace transform

$$(234)\quad \begin{cases} \text{i) } \hat{A}_T \hat{v}_j(p) = -\theta_j \hat{v}_j(p),\ \theta_j = (p^2 + \lambda_j^2)^{1/2}, \quad \text{with Re } \theta_j > \gamma,\ v = \sum_j v_j \phi_j \text{ in } L_\gamma^2(X_T), \\ \text{ii) } D_\gamma(A_T) = \{v \in L_\gamma^2(X_T),\ X_T = L^2(\Omega)^2,\ \sum_j \int_R |\theta_j|^2 |\hat{v}_j|^2 d\eta < +\infty,\ p = \gamma + i\eta\}. \end{cases}$$

We can prove that A_T is the infinitesimal generator of a contraction semigroup $(G_T(x_3))$ in $L^2_\gamma(X_T)$, $X_T = L^2(\Omega)^2$, of class C^0. Thus like in section 3.1.4, the problem (231) has a unique solution E_T which is given thanks to $(G_T(x_3))$ by:

(235) $\quad E_T(x_3) = G_T(x_3)E_T^0.$

Then we can verify thanks to the Laplace transformation that $E_T \in L^2_\gamma(X^2)$.

Determination of E_3. First assuming that E_3 is given at $x_3 = 0$, we have to find E_3 so that

$$
(236)\qquad
\left|
\begin{array}{l}
\text{i) } \partial_3^2 E_3 + \Delta E_3 - c^{-2}\partial_t^2 E_3 = 0 \quad \text{in } \Omega_+ \times R_t,\\
\text{ii) } E_3|_{\Sigma_1} = 0,\\
\text{iii) } E_3|_{\Sigma_0} = E_3^0, \text{ with given } E_3^0 \text{ in } L^2{}_\gamma(L^2(\Omega)),\\
\text{iv) } E_3 \in L^2_\gamma(X),\ X = L^2(\Omega_+).
\end{array}
\right.
$$

Then E_3 is obtained thanks to the semigroup $(G(x_3))$ of section 3.1.4 (with generator A defined by (189),(189)') by:

(237) $\quad E_3(x_3) = G(x_3)\, E_3^0.$

Then we have to find E_3^0. It is obtained (like in chap.4) from (232) by:

(238) $\quad \partial_3 E_3(0) = AE_3^0 = -\operatorname{div} E_T^0.$

We have $\operatorname{div} E_T^0 \in L^2_\gamma(H^{-1}(\Omega)) \subset (D(A))'$, and thus, since A is an isomorphism from D(A) onto $L^2_\gamma(L^2(\Omega))$

(239) $\quad E_3^0 = -A^{-1}(\operatorname{div} E_T^0) \in L^2_\gamma(L^2(\Omega))$, thus $E_3 \in L^2_\gamma(X)$.

Determination of H_3. We get H_3 from $\varphi = \operatorname{curl}_T E_T$ by time integration of (229)ii). Applying curl_T to (231)i), then by taking the scalar product of (228)i) with the normal n_T, we get φ satisfying

$$
(240)\qquad
\left|
\begin{array}{l}
\text{i) } \partial_3^2 \varphi + \Delta \varphi - c^{-2}\partial_t^2 \varphi = 0, \quad \text{in } \Omega_+ \times R_t,\\
\text{ii) } \dfrac{\partial \varphi}{\partial n_T}\Big|_{\Sigma_1} = 0,\\
\text{iii) } \varphi(0) = \varphi^0 \in L^2_\gamma(L^2(\Omega)),
\end{array}
\right.
$$

and $\varphi(x_3)$ is obtained by the semigroup $(G_N(x_3))$ with the Neumann condition:

(241) $\quad \varphi(x_3) = G_N(x_3)\varphi^0.$

This gives $H_3^0 \in L^2_\gamma(L^2(\Omega))$ and $H_3 \in L^2_\gamma(X)$.

Determination of H_T. Now we can obtain H_T by time integration of (228)ii); we have to verify that H_T is in $L_\gamma^2(X^2)$. We can prove this result like in the stationary case, using the Hodge decomposition for H_T, see (68) chap.4 (we can also use an inverse Laplace transformation on the formulas of chap. 4 taken with complex ω).

Causal Calderon operator of the waveguide. Now we are especially interested in having the boundary value of H_T at the end of the waveguide, i.e., for $x_3 = 0$. This is also obtained thanks to (228)ii) with $x_3 = 0$:

(242) $$\partial_t S H_T(.,0,t) = A_T E_T(.,0,t) + \text{grad}\, A^{-1} \text{div}\, E_T^0(.,0,t).$$

We get a formula similar to (61) chap.4 (with (29) and (38) chap. 4):

(243) $$H_T(.,0,t) = - \int_{-\infty}^{t} [S A_T E_T^0(.,0,s) + \overrightarrow{\text{curl}}_T A^{-1} \text{div}_T E_T^0(.,0,s)]\, ds.$$

We can also obtain a form similar to (68), chap.4, more useful for spaces.
The operator

$$C: E_T^0 \rightarrow H_T(.,0,.) \quad (\text{or } C: SE_T^0 \rightarrow SH_T(.,0,.))$$

is the *causal Calderon operator of the waveguide.*

Of course we have to specify its domain. We can do so thanks to the operators A_T and A; we do not detail this question here. The interest of this operator is (like all previous Calderon operators) that it contains all informations at hand for the domain exterior to the waveguide. Furthermore we can prove that the causal Calderon operator satisfies the positivity properties:

(244) $$(CSE_T^0, E_T^0) = -(CE_T^0, SE_T^0) \geq 0$$

which we get from

(245) $$\int_{\Omega_+ \times R_t} (\varepsilon_0 \frac{\partial E}{\partial t} . E + \mu_0 \frac{\partial H}{\partial t} . H)\, e^{-2\gamma t}\, dx\, dt = \int_{\Omega_+ \times R_t} \text{div}\,(H \wedge E)\, e^{-2\gamma t}\, dx\, dt,$$

and integrating by parts:

(246) $$\gamma \int_{\Omega_+ \times R_t} (\varepsilon_0 E^2 + \mu_0 H^2)\, e^{-2\gamma t}\, dx\, dt = \int_{\Sigma_0 \times R_t} n \wedge H . E\, e^{-2\gamma t}\, d\Sigma_0\, dt.$$

3.2.5. Uniqueness at most

In all these causal problems we have uniqueness (at most) results for the solution of the wave equation like the Maxwell equation (this is to compare to the stationary case, chap. 2.13; these results are given for a bounded domain Ω, to simplify):

Theorem 3. *Uniqueness at most. Given a (regular bounded) domain* Ω *in* R^n, *with boundary* Γ, *the wave problem: find* u *so that*

$$(247)\quad \left|\begin{array}{l} \text{i) } \Box u = f, \quad f \text{ given in } \mathcal{D}'_+(X), \text{ in } \Omega \times R_t, \\ \text{ii) } u|_\Sigma = u^0, \ \frac{\partial u}{\partial n}|_\Sigma = u^1, \text{ with } \Sigma = \Gamma \times R_t, \\ \text{with } u^0 \in \mathcal{D}'_+(H^{-1/2}(\Gamma)),\ u^1 \in \mathcal{D}'_+(H^{-3/2}(\Gamma)) \\ \text{iii) } u \in \mathcal{D}'_+(X),\ X = L^2(\Omega), \end{array}\right.$$

has at most one solution.

Theorem 4. *Uniqueness at most. The Maxwell problem: find* (E, H) *so that*

$$(248)\quad \left|\begin{array}{l} \text{i) } \varepsilon_0 \frac{\partial E}{\partial t} - \text{curl } H = J, \quad \mu_0 \frac{\partial H}{\partial t} + \text{curl } E = M \text{ in } \Omega \times R, \\ \text{with M and J given in } \mathcal{D}'_+(X),\ X = L^2(\Omega)^3 \\ \text{ii) } n \wedge E|_\Sigma = M^0, \ n \wedge H|_\Sigma = J^0 \ \text{ with } M^0 \text{ and } J^0 \in \mathcal{D}'_+(H^{-1/2}(\text{div}, \Gamma)) \\ \text{iii) } (E, H) \in \mathcal{D}'_+(X),\ X = L^2(\Omega)^3 x L^2(\Omega)^3, \end{array}\right.$$

has at most one solution.

The essential difference with the stationary case is that these uniqueness results are not local in time: we can substitute a part $\Sigma_0 = \Gamma_0 \times R$, with Γ_0 a "regular" part of Γ, for Σ. However we cannot replace R_t by finite time intervals, except for bounded Ω, in which case these time intervals have to be large enough. This is a consequence of the finite velocity of propagation. Finding the optimal time interval for which we have uniqueness is of great interest in control problems (see for instance Lions [3]). These uniqueness results imply that the causal Calderon operators and the causal Calderon projectors have the same importance as in the stationary case: these operators on the boundary Γ of a domain Ω contains all the informations on the wave inside the domain Ω that we consider. Of course these operators are not very easy to handle especially for numerical implementations. So they are often replaced by approximate boundary conditions. However they are very useful in obtaining a priori estimates.

⊗

CONCLUSION We have solved many evolution problems, but we would have to deal with many other linear interesting problems, which are not a simple application of the above theory, for example with a moving obstacle, and a moving antenna... Furthermore there are very often parameters giving different scales, so that an asymptotic analysis is necessary, with singular perturbations.

Appendix

DIFFERENTIAL GEOMETRY FOR ELECTROMAGNETISM

Using differential geometry in electromagnetism is quite natural, at first in the modelling of Maxwell equations by differential forms, according to the right transformation laws when changing a coordinate system into another one with a different orientation (this explains the notion of "polar vector" often used in electromagnetism).

The notion of cohomology allows a deep understanding of many points, notably in the study of the differential operators grad, curl, div with their kernels and their images, in an open domain Ω; and then in the trace spaces on the boundary Γ of Ω for electromagnetic fields of finite energy. We have to use the manifold structure of Γ, and the differential operators grad_Γ, curl_Γ, div_Γ on Γ, with the technical difficulty of having enough regularity in order to apply usual tools generally developed for very regular manifolds.

Differential geometry is also useful in numerical methods in electromagnetism with Whitney forms (see Bossavit [1]).

We suppose only elementary prerequisites on finite dimensional manifolds. We review some simple notions in order to set notations, following Morrey [1], Arnold [1], Abraham-Marsden-Ratiu [1]. We also need some elementary notions on variational framework (the notion of a V-coercive bilinear form) and we often use the Peetre lemma (see chapter 2).

We emphasize that this appendix is not a course on differential geometry; we refer to Malliavin [1] or to Kobayashi-Nomizu [1] for instance for that.

1. INTRODUCTION. MATHEMATICAL FRAMEWORK

1.1. Manifold with boundary

The notion of *manifold* is supposed to be known, with the notions of *charts*, of *mutually $C^{k,\alpha}$-compatible charts, of atlases and of $C^{k,\alpha}$-compatible atlases* (we refer to Dieudonné [1], Bourbaki [1]).

We recall the standard definition of an n-dimensional *manifold M with boundary* Γ of class $C^{k,\alpha}$, $k \in \mathbb{N}$, $0 \leq \alpha \leq 1$ (C^{∞}, analytic).

Each point of M is contained in some open set V of M, which is the homeomorphic image either of the unit ball in R^n, or of the part of it for which $x_n \leq 0$, the points where $x_n = 0$ correspond to the boundary Γ of M; any two coordinate systems are related by a transformation of class $C^{k,\alpha}$ (C^{∞}, analytic).

Let Ω be an open subset of R^n; $\bar{\Omega}$ may be considered as an n-dimensional manifold with boundary Γ; following Grisvard [1], we then define:

Definition 1. *We say that $\bar{\Omega}$ is an n-dimensional continuous (resp Lipschitz, continuously differentiable, of class* $C^{k,\alpha}$, $k \in N$, $0 \leq \alpha \leq 1$) *submanifold with boundary in* R^n, *if for every* $x \in \Gamma$, *there are a neighborhood V of x in* R^n *and an injective map ϕ from V into* R^n *so that:*

i) $\Omega \cap V = \{ y \in \Omega, \ \phi_n(y) < 0 \}$, *where $\phi_n(y)$ denotes the n-th component of $\phi(y)$;*

ii) ϕ *together* ϕ^{-1} *defined on* $\phi(V)$ *is continuous (resp. Lipschitz, continuously differentiable, of class* $C^{k,\alpha}$).

The boundary Γ of $M = \bar{\Omega}$ is then locally defined by $\phi_n(y) = 0$. The interior $\overset{o}{M} = \Omega$ is a C^{∞} (even analytic) manifold without boundary! The boundary Γ of an n-dimensional continuously differentiable (resp. C^k, $C^{k,\alpha}$) manifold M, has an induced structure of (n − 1)-dimensional continuously differentiable (resp. C^k, $C^{k,\alpha}$) manifold, according to the implicit function theorem. In the case of Lipschitz regularity, this is not true (see Grisvard [1]).

⊗

Example 1. Polyhedrons, which are in common use for numerical applications, are only Lipschitz manifold, so we often have to make direct proofs of the main properties, like for the Stokes theorem!

⊗

Let M be a regular (differentiable) manifold, we will use the following notations in this appendix.

T_xM the *tangent space* at a point x of M (the space of tangent vectors at x),

$TM = \bigcup_{x \in M} T_xM$ the *tangent bundle* of M;

T_x^*M the *cotangent space* at $x \in M$ (i.e., the dual space of T_xM, or the space of *covectors* at x),

$T^*M = \bigcup_{x \in M} T_x^*M$ the *cotangent bundle* of M (or the *phase space*);

$T^r_{s\,x}M$ the *tensor product* $\overset{r}{\otimes} T_xM \otimes \overset{s}{\otimes} T_x^*M$ (the space of r-*contravariant* s-*covariant* tensors),

T^r_sM the *tensor bundle of type* (r,s) over M, $T^r_sM = \bigcup_{x \in M} T^r_{s\,x}M$;

$\overset{r}{\Lambda} T_x^*M$ and $\overset{r}{\Lambda} T_xM$ the spaces of r-*covectors* and of r-*vectors*,

$\overset{r}{\Lambda} T^*M = \bigcup_{x \in M} \overset{r}{\Lambda} T_x^*M$ the r-*exterior bundle* over M (with $\wedge$ the *exterior or wedge product*).

A *vector field* (resp. a (r,s) *tensor field*) is a *section* of TM (resp. of T^r_sM), that is, a mapping v from M into TM (resp. T^r_sM), such that $\pi \circ v = 1$, with π the *canonical projection* from TM (resp. T^r_sM) onto M; *a differential form* is a section of T^*M, i.e., a mapping ω: $M \to T^*M$ such that $\pi^* \circ \omega = 1$, with π^* the canonical projection from T^*M onto M.

A (differential) r-*form* $\overset{r}{\omega}$ is a section of $\overset{r}{\Lambda} T^*M$, and an r-*vector field* is a section of $\overset{r}{\Lambda} TM$, the regularity of these mappings being so far not specified.

In the domain of any coordinate system $(x^1,\dots,x^n)$ we denote by $\frac{\partial}{\partial x^1},\dots,\frac{\partial}{\partial x^n}$ (or simply by $\partial_1,\dots,\partial_n$) the *respectively associated* vector fields, by $dx^1,\dots,dx^n$ the associated 1-forms, so that $(\frac{\partial}{\partial x^1})_x,\dots,(\frac{\partial}{\partial x^n})_x$ is a basis of T_xM, $(dx^1)_x,\dots,(dx^n)_x$ is the dual basis of T_x^*M, with $(dx^i)_x \cdot \partial_j = \delta_{ij}$, $\forall$ i,j.

An r-form ω may be represented as follows:

$$\omega = \sum_{1 \le i_1 < \dots < i_r \le n} \omega_{i_1 \dots i_r} \, dx^{i_1} \wedge \dots \wedge dx^{i_r}, \tag{1}$$

where $(\omega_{i_1 \dots i_r})$ are the components of ω in that coordinate system and $\wedge$ denotes the exterior product. Let I be a sequence $I = \{i_1,\dots,i_r\}$ with $1 \le i_1 < \dots < i_r \le n$. We often abbreviate notations by:

$$\omega = \sum_I \omega_I \, dx_I. \tag{1)'}$$

We assume that M is an n-dimensional *orientable manifold* (i.e., there is a continuous n-form ω on M such that $\omega(x) \neq 0$, $\forall x \in M$). Then two orientations of M may be chosen M_+ and M_- (so that M becomes an orientated manifold). Let σ be the symmetry exchanging the orientations. Then we can define in two equivalent ways even and odd r-forms (see Schwartz [1], Bourbaki [2], Morrey [1]. The first one is intrinsic, the second uses local charts). *Even* r-forms are r-forms which are invariant by σ; *odd* r-forms are r-forms which are changed into their opposite by σ. (We can also define even and odd r-fields on M.)

If two coordinate systems with coordinates (x) and $(\tilde{x})$ overlap, the components of ω in each system are related by:

$$\omega_{i_1 \ldots i_p}(\tilde{x}) = \varepsilon \sum \omega_{j_1 \ldots j_p}(x(\tilde{x})) \frac{\partial (x^{j_1}, \ldots, x^{j_p})}{\partial (\tilde{x}^{i_1}, \ldots, \tilde{x}^{i_p})}, \tag{2}$$

with $\varepsilon = 1$ for even r-forms, $J/|J|$ for odd r-forms, $J = \dfrac{\partial(x^1, \ldots, x^n)}{\partial(\tilde{x}^1, \ldots, \tilde{x}^n)}$.

An odd form is usually (see Schwartz [1]) denoted by $\underline{\omega}$; we also note $\pm\omega$ or $\varepsilon\omega$.

If ω is a (regular) r-form on M, its *exterior derivative* is the (r + 1)-form dω:

$$d\omega = \sum_{I,\alpha} \frac{\partial \omega_I}{\partial x^\alpha} dx^\alpha \wedge dx^I, \tag{3}$$

or

$$\left| \begin{array}{l} d\omega = \sum\limits_J (d\omega)_J \, dx^J, \; (d\omega)_J = \sum\limits_{\nu=1}^{r+1} (-1)^{\nu-1} \dfrac{\partial \omega_{\tilde{J}_\nu}}{\partial x^{j_\nu}}, \qquad \tilde{J}_\nu = J \setminus \{j_\nu\}, \\ d\omega_x(v_0, \ldots, v_r) = \sum\limits_i (-1)^i (D\omega_x . v_i)(v_0, \ldots, \hat{v}_i, \ldots, v_r), \qquad v_i \in T_x M, \end{array} \right. \tag{3'}$$

where $\hat{v}_i$ denotes that v_i is missing, and $D\omega_x$ is the derivative of ω in charts. Note also

$$\left| \begin{array}{l} d^2 = 0, \text{ i.e., } d^2\omega = 0, \quad \forall \omega, \\ d(\alpha \wedge \beta) = d\alpha \wedge \beta + (-1)^r \alpha \wedge d\beta, \quad \text{for all r-forms } \alpha \text{ and all p-forms } \beta. \end{array} \right. \tag{4}$$

1.2. Riemannian manifold M, with or without boundary

A *Riemannian structure (or metric)* on a manifold M is a (covariant) tensor field of order 2, i.e., a mapping $g: M \to T^*M \times T^*M$ which is a positive definite bilinear form on T_xM

$$g_x(v,v) > 0, \qquad \forall v \in T_xM, \; v \neq 0. \tag{5}$$

In a local coordinate system $x^1, \ldots, x^n$, g (which is often written as ds^2) is given with $(g_{ij}) = (g(\partial_i, \partial_j))$ a positive definite matrix, by:

$$g = \sum g_{ij} \, dx^i \otimes dx^j. \tag{5'}$$

A manifold M with a Riemannian stucture is called a ***Riemannian manifold***. If M is of class $C^{k,\alpha}$, $g_{ij}(x)$ is of class $C^{k-1,\alpha}$. For all x in M, g_x is an inner product on T_xM; we note:

(6) $$g_x(h,k) = (h,k)_{g_x} = \sum_{i,j} g_{ij}(x)h^i k^j,$$

for all tangent vectors h, k of T_xM, given in a local coordinate system $x^1,\ldots,x^n$ by $h = \Sigma h^i \partial_i$, $k = \Sigma k^i \partial_i$. Recall that (∂_i) is the dual basis of (dx^i), $i = 1,\ldots,n$.

The *length* of the tangent vector $h \in T_xM$ is defined by:

(7) $$|h| = (h, h)^{1/2}_{g_x}.$$

A tangent vector is said to be *unitary* if $|h| = 1$ (we say that h is a unit vector). For all $x \in M$, the metric g_x assigns to each tangent vector $h \in T_xM$ the covector $\omega_h = G_x(h) \in T^*_xM$ by:

(8) $$\omega_h(k) = G_x(h)(k) = g_x(h,k), \qquad \forall\, h,k \in T_xM,$$

and G_x is an isomorphism from T_xM onto $T^*{}_xM$. The cotangent space $T^*{}_xM$ is therefore also equipped with a positive definite bilinear form (thus an inner product, corresponding to a tensor field g^* on M):

(9) $$(\omega_1, \omega_2)_{g^*_x} = g^*_x(\omega_1, \omega_2) = g_x(\overset{-1}{G}_x\omega_1, \overset{-1}{G}_x\omega_2), \quad \forall\, \omega_1, \omega_2 \in T^*_xM.$$

Thus using notations (8):

(10) $$(\omega_h, \omega_k)_{g^*_x} = g^*_x(\omega_h, \omega_k) = g_x(h,k) = (h,k)_{g_x}, \qquad \forall\, h, k \in T_xM$$

(i.e., G_x carries the metric of the tangent space onto the cotangent space), or in a local coordinate system:

(11) $$g^* = \Sigma\, g^{ij}\, \partial_i \otimes \partial_j, \quad (g^{ij}) \text{ inverse matrix of } (g_{ij}).$$

This can be extended more generally to exterior algebras: on the space $\overset{r}{\Lambda} T_xM$ of tangent r-vectors at x, and on the space $\overset{r}{\Lambda} T^*xM$ of r-covectors at x, we define, for all h_i, k_j in T_xM, and all ω_i, $\tilde{\omega}_j$ in T^*_xM

(12) $$\left|\begin{aligned} &(h_1 \wedge \ldots \wedge h_r, k_1 \wedge \ldots \wedge k_r)_{g_x} = \det\left((h_i, h_j)_{g_x}\right), \\ &(\omega_1 \wedge \ldots \wedge \omega_r, \tilde{\omega}_1 \wedge \ldots \wedge \tilde{\omega}_r)_{g^*_x} = \det\left((\omega_i, \tilde{\omega}_j)_{g^*_x}\right); \end{aligned}\right.$$

G_x defines an isomorphism (again denoted by G_x) from $\overset{r}{\Lambda} T_xM$ onto $\overset{r}{\Lambda} T^*_xM$ by:

(13) $$G_x(h_1 \wedge \ldots \wedge h_r) \overset{\text{def}}{=} G_x(h_1) \wedge \ldots \wedge G_x(h_r) = \omega_{h_1} \wedge \ldots \wedge \omega_{h_r}.$$

We will often denote by $(,)_x$ the inner products (6), or (9), or (12).

For r = n, we can define an odd n-form $v = v^g$ which is positive (for the chosen orientation of the orientable manifold M):

(14) $$v^g = \pm\, \omega_1 \wedge \ldots \wedge \omega_n \cdot \frac{1}{|\omega_1 \wedge \ldots \wedge \omega_n|} = \pm\, \omega_1 \wedge \ldots \wedge \omega_n \,(\det g^*(\omega_i, \omega_j))^{-1/2},$$

for a family $\omega_1,\ldots,\omega_n$ of n independent 1-forms, where $|\omega_1 \wedge \ldots \wedge \omega_n|$ is a function of x defined according to (12): $|\omega_1 \wedge \ldots \wedge \omega_n|_x = (\omega_1 \wedge \ldots \wedge \omega_n, \omega_1 \wedge \ldots \wedge \omega_n)_x^{1/2}$.

This definition is independent of the chosen family. Thus (14) defines a Lebesgue measure on M. In a coordinate system $x^1,\ldots,x^n$, v is given by:

(14)' $$v = v^g = \pm\, dx^1 \wedge \ldots \wedge dx^n (\det(g^*(dx^i, dx^j)))^{-1/2} = \pm\, dx^1 \wedge \ldots \wedge dx^n \,(\det g_{ij})^{1/2}.$$

We often denote by dM the measure defined by v^g (and also by g the determinant of the matrix $(g_{i,j})$). But the notation dM may be misleading, since in general it is not the exterior derivative of any form.

We then define *a scalar product* on the space of (smooth) r-forms (with compact support if M is only locally compact) by:

(15) $$(\omega,\eta) = (\omega,\eta)_g \overset{\text{def}}{=} \int_M \omega_x(\overset{-1}{G}_x \eta_x)\, dM,$$

with G_x defined by (8) and (13). We also have, from the symmetry of g:

(16) $$\omega\,(\overset{-1}{G}\eta) = \eta\,(\overset{-1}{G}\omega), \qquad (\omega,\eta) = (\eta,\omega).$$

In a coordinate system, with r-forms ω, η,

$$\omega = \sum_I \omega_I\, dx^I, \qquad \eta = \sum_K \eta_K\, dx^K,$$

we have:

(17) $$\omega_x(\overset{-1}{G}_x \eta_x) = \sum_{IK} g^{IK}(x)\omega_I(x)\eta_K(x),$$

with:

$$g^{IK} = \det \begin{vmatrix} g^{i_1k_1} & \ldots & g^{i_1k_r} \\ \cdot & \ldots & \cdot \\ g^{i_rk_1} & \ldots & g^{i_rk_r} \end{vmatrix}, \qquad dM = (\det(g_{ij}))^{1/2}\, dx^1 \ldots dx^n.$$

The scalar product (15) is independent of the system of local charts. Thus we can define a Hilbert space $L^2_r(M)$ as the closure of the space of smooth differential r-forms with compact support in M. It is also possible to define it as the space of r-forms with square integrable coefficients in every coordinate system. The elements of the space

(18) $L^2_r(M) = \{\omega \text{ "generalized r-forms"}, (\omega,\omega) < +\infty\}$

are called *square integrable r-forms*. We often do not specify whether ω is an even or odd r-form. But if necessary, we shall use the notations $L^2_{r,e}(M)$ and $L^2_{r,o}(M)$. Note that from the definition (14), we have

(19) $(v^g_x, v^g_x)_x = 1, \quad \forall x \in M,$

and thus

(19)' $(v^g, v^g)_g = \int_M v^g = |M|$ (finite or not).

For all n-forms ω on M, we have

(20) $\omega = (\omega_x, v^g_x)_x v^g, \qquad \int_M \omega = (\omega, v^g)_g.$

Then we define by duality to the exterior product the *inner (or interior) product* $\lrcorner$ of a p-covector u by an r-vector k, $r \leq p$, as the $(p - r)$-covector $i_k u = k \lrcorner u$ such that:

(21) $<h, i_k u> = <h, k \lrcorner u> = <k \wedge h, u>, \qquad \forall\ (p - r)\text{-vector } h,$

using the duality bracket between p-vectors, and p-covectors $<,>$. We also define the inner (or interior) product $k^* \lrcorner_g u$ of a p-covector u by an r-covector k^*, $r \leq p$, by

(21)' $(k^* \wedge h^*, u)_x = (h^*, k^* \lrcorner_g u)_x, \qquad \forall\ (p - r)\text{-covector } h^*,$

i.e., we also have:

(22) $<G_x^{-1}(k^* \wedge h^*), u> = <G_x^{-1}h^*, G_x^{-1}k^* \lrcorner u>.$

Thus

(23) $k^* \lrcorner_g u = G_x^{-1}k^* \lrcorner u.$

We then define the *Hodge transform* (or $*$ transform) of an r-form η by:

(24) $(*\eta)_x = \eta_x \lrcorner_g v^g_x = (G^{-1}\eta_x) \lrcorner v^g_x.$

Therefore the Hodge transform of an even (resp. odd) r-form is an odd (resp. even) $(n - r)$-form, with:

(25) $(\eta \wedge \alpha, v^g)_g = (\alpha, *\eta)_g, \qquad \forall \alpha \in L^2_{p-r}(M), \eta \in L^2_r(M).$

Note that the Hodge transformation has the following properties:

(26) $*v^g = 1, \qquad *1 = v^g.$

One of the main properties of the Hodge transformation is:

(27) $\omega \wedge *\eta = (\omega, \eta)_x v^g = <\omega, G^{-1}\eta>_x v^g,$

giving for all ω, η in $L^2{}_r(M)$, even or odd:

(28) $$\int_M \omega \wedge *\eta = (\omega, \eta)_g.$$

This is expressed in a coordinate system, with

(29) $$\omega = \sum \omega_I \, dx^I, \quad \eta = \sum \eta_K dx^K, \qquad \text{card } I = \text{card } K = r,$$
by:

(30) $$\int_M \omega \wedge *\eta = (\omega,\eta)_g = \sum_{I,K} \int_M g^{IK} \omega_I \eta_K \, g^{1/2} \, dx^1 \ldots dx^n.$$

Taking the inner product of (27) with v^g_x, and using (19) and the commutativity relation:

(31) $$\omega \wedge *\eta = (-1)^{r(n-r)} *\eta \wedge \omega, \qquad \forall \text{ r-forms } \omega, \eta,$$

we obtain for all r-forms ω, η

(32) $$(\omega,\eta)_x = (\omega \wedge *\eta, v^g)_x = (-1)^{r(n-r)}(*\eta \wedge \omega, v^g)_x = (-1)^{r(n-r)}(\omega, **\eta)_x,$$

hence for all r-form η:

(33) $$**\eta = (-1)^{r(n-r)}\eta.$$

Besides, we have:

(34) $$\omega \wedge *\eta = (-1)^{r(n-r)}(*\omega) \wedge \eta = \eta \wedge *\omega$$

(which may be verified by inner product with v^g_x), thus

(35) $$(\omega,\eta)_g = (*\eta, *\omega)_g = (*\omega, *\eta)_g, \qquad \forall \omega, \eta \in L^2_r(M).$$

The Hodge transformation is an isometry from $L^2_r(M)$ onto $L^2_{n-r}(M)$, changing even into odd differential forms, and conversely.

Remark 1. Expression of the Hodge transformation in a coordinate system. For each sequence $K \subset \{1,\ldots,n\}$, with card $K = r$, there are numbers (α_{KJ}), with card $J = n - r$, so that

(36) $$*dx^K = \varepsilon \sum_J \alpha_{KJ} dx^J,$$

with $\varepsilon = +1$ if $(x^1,\ldots,x^n)$ has the chosen orientation, $\varepsilon = -1$ if not. To specify (36), we define with the usual notations (see Bourbaki [1] A III 79-87) $I' = \{1,\ldots,n\} \setminus I$

(37) $$\rho_{II'} = (-1)^\nu, \quad \nu = \text{number of couples } (i,j) \in I \times I' \text{ such that } i > j,$$

(38) $$\left|\begin{array}{l} \rho_{JK} = 0 \quad \text{if } J \cap K \neq \varnothing, \quad \rho_{JK} = (-1)^\nu \quad \text{if } J \cap K = \varnothing, \\ \nu = \text{number of couples } (\lambda,\mu) \in J \times K \text{ such that } \lambda > \mu, \end{array}\right.$$

we have for all r-forms ω, η:

(39) $$\omega \wedge {}_*\eta = \sum_{IK} \omega_I \eta_K \, dx^I \wedge {}_*dx^K,$$

(40) $$dx^I \wedge {}_*dx^K = \varepsilon\, \alpha_{KI'}\, dx^I \wedge dx^{I'}, \qquad dx^I \wedge dx^{I'} = \rho_{I\,I'}\, dx^1 \wedge \ldots \wedge dx^n.$$

Comparing with (30), we find:

(41) $$\alpha_{KI'} = g^{IK} \rho_{I\,I'}\, g^{1/2}, \qquad \text{card}\, K = r, \quad \text{card}\, I' = n - r,$$

(42) $${}_*dx^K = \varepsilon \sum_I g^{IK} \rho_{I\,I'}\, g^{1/2} dx^{I'}, \qquad \text{card}\, I = \text{card}\, K = r.$$

When $r = n$, let $N = \{1,\ldots,n\}$; we have $I' = \varnothing$, $dx^{\varnothing} = 1$, $\rho_{N\varnothing} = 1$, $\alpha_{N\varnothing} = g^{-1/2}$, thus ${}_*dx^N = \varepsilon\, g^{-1/2}$.

When $r = 1$, ${}_*dx^k = \varepsilon \sum_i g^{ik}(-1)^{i-1} g^{1/2} dx^1 \wedge \ldots \wedge \widehat{dx^i} \wedge \ldots \wedge dx^n$, where $\widehat{dx^i}$ indicates that dx^i is missing; in a Cartesian system, we get

$$ {}_*dx^k = \varepsilon (-1)^{k-1} dx^1 \wedge \ldots \wedge \widehat{dx^k} \wedge \ldots \wedge dx^n. $$

⊗

1.3. Definition of the codifferential

For any (regular) r-form ω^r, the *codifferential* $\delta\omega^r$ is an $(r-1)$-form defined with the exterior derivative d by:

(43) $$\delta\omega^r = (-1)^{n(r+1)+1}\, {}_*d\, {}_*\omega^r.$$

We can easily show the following properties:

(44) $$\left|\begin{array}{l} \delta\,\delta = 0, \quad \delta 1 = 0, \quad \delta f = 0 \ \text{ for all functions (0-forms) } f, \\ {}_*\delta\, d = d\,\delta\,{}_*, \quad {}_*d\,\delta = \delta\, d\,{}_*, \quad \delta\,{}_*d = d\,{}_*\delta = 0, \end{array}\right.$$

(45) $${}_*(\delta\,\omega^r) = (-1)^r d({}_*\omega^r), \qquad \delta({}_*\omega^r) = (-1)^{r-1}\, {}_*d\omega^r.$$

From Stokes formula (see (81) below), δ appears as the adjoint of the exterior derivative:

(46) $$(d\alpha, \beta) = (\alpha, \delta\beta),$$

for all (smooth) $(r-1)$-forms α and r-forms β with compact support in $\overset{o}{M}$, because:

(46)' $$\left|\begin{array}{l} (d\alpha, \beta) - (\alpha, \delta\beta) = \int_M d\alpha \wedge {}_*\beta - \alpha \wedge {}_*\delta\beta = \int_M d\alpha \wedge {}_*\beta + (-1)^{r-1} \alpha \wedge d\,{}_*\beta \\ \qquad = \int_M d(\alpha \wedge {}_*\beta) = 0. \end{array}\right.$$

We will show the generalizations of (46) below.

1.4. The gradient, divergence and Laplace-Beltrami operator

Let f be a (smooth) function on M. Then its *gradient*, denoted by grad f, is the vector field defined by:

(47) $\operatorname{grad} f = \overset{-1}{G}(df)$, i.e., $df = G(\operatorname{grad} f) = \omega_{\operatorname{grad} f}$.

Let X be a (smooth) vector field on M. Then its *divergence* denoted by div X is the function defined by:

(48) $\operatorname{div} X = -\delta G X = -\delta\omega_X$.

Then we define the *Laplace-Beltrami operator* (also called the *Laplace-de Rham operator*) by:

(49) $\Delta = -(d\delta + \delta d)$.

Applying Δ to a (smooth) function f, we obtain:

(49)' $\Delta f = -\delta df = -\delta G\overset{-1}{G} df = -\delta G \operatorname{grad} f = \operatorname{div} \operatorname{grad} f$;

the Laplace-Beltrami operator applied to the functions is identical to the Laplacian. Note also that for all smooth r-forms β with compact support in $\overset{\circ}{M}$

$$(\Delta\alpha, \beta) = -(\delta\alpha, \delta\beta) - (d\alpha, d\beta) = (\alpha, \Delta\beta).$$

Expression in a coordinate system $(x_1,\ldots,x_n)$. We have $df = \sum_i \frac{\partial f}{\partial x^i} dx^i$, thus:

(50) $\operatorname{grad} f = \overset{-1}{G} df = \sum_i \frac{\partial f}{\partial x^i} \overset{-1}{G} dx^i = \sum_{ij} \frac{\partial f}{\partial x^i} g^{ij} \partial_j$,

and for $X = \sum X^i \partial_i$, we have:

(51) $\omega_X = G(X) = \sum_{ij} X^i g_{ij} dx^j$,

(52) $*\omega_X = \pm \sum_j (-1)^{j-1} X^j g^{1/2} dx^1 \wedge \ldots \wedge \widehat{dx^j} \wedge \ldots \wedge dx^n$.

We verify that

$$\omega_X \wedge *\omega_X = \pm (X,X)_{g_x} g^{1/2} dx^1 \wedge \ldots \wedge dx^n = \pm \sum g_{ij} X^i X^j g^{1/2} dx^1 \wedge \ldots \wedge dx^n,$$

and we have

$$\delta\omega_X = - * d * \omega_X = - * \pm \sum \frac{\partial}{\partial x^j} (X^j g^{1/2})\, dx^1 \wedge \ldots \wedge dx^n,$$

thus:

(53) $\operatorname{div} X = g^{-1/2} \sum_j \frac{\partial}{\partial x^j} (X^j g^{1/2})$.

Furthermore the Laplace-Beltrami operator is given by

(54) $\Delta f = g^{-1/2} \sum_{jk} \frac{\partial}{\partial x^j} (g^{jk} g^{1/2} \frac{\partial f}{\partial x^k})$.

1.5. Decomposition of the space of tangent p-vectors and tangent p-covectors into tangential and normal parts on the boundary

Under the usual regularity hypotheses on the Riemannian manifold M, on its boundary Γ and on g, the tangent space T_xM of M at a point x of Γ has an orthogonal decomposition with respect to g_x into tangent and normal part to Γ

(55) $T_xM = T_x\Gamma \oplus R_n$;

similarly the cotangent space T_x^*M of M at x is orthogonally decomposed (with respect to g_x^*) into:

(55)* $T_x^*M = T_x^*\Gamma \oplus R_n^*$,

and we write the decomposition of elements with n a unit normal to T_xM, $\omega_n^1 = G_x n$,

(55)' $\left| \begin{array}{ll} h = th + (nh)\, n, & \forall h \in T_xM, \\ h^* = th^* + \langle h^*, n \rangle \omega_n^1, & \forall h^* \in T_x^*M. \end{array} \right.$

There are corresponding decompositions of the spaces of tangent r-vectors and r-covectors:

(56) $\left| \begin{array}{l} \overset{r}{\Lambda} T_xM = \overset{r}{\Lambda} (T_x\Gamma \oplus R_n) = \overset{r}{\Lambda} T_x\Gamma \oplus (\overset{r-1}{\Lambda} T_x\Gamma \overset{g}{\otimes} R_n) \\ \overset{r}{\Lambda} T_x^*M = \overset{r}{\Lambda} (T_x^*\Gamma \oplus R_n^*) = \overset{r}{\Lambda} T_x^*\Gamma \oplus (\overset{r-1}{\Lambda} T_x^*\Gamma \overset{g}{\otimes} R_n^*), \quad \forall x \in \Gamma, \end{array} \right.$

and we write for all $\omega_x \in \overset{r}{\Lambda} T_x^*M$

(57) $\omega_x = t\omega_x + n\omega_x \wedge \omega_n^1 = t\omega_x + \bar{n}\omega_x$.

Then G_x (see (30)) transforms the decomposition (56) of $\overset{r}{\Lambda} T_xM$ into that of $\overset{r}{\Lambda} T_x^*M$

(58) $GtA = tGA, \quad GnA = nGA, \quad \forall A \in \overset{r}{\Lambda} T_xM$.

These decompositions involve decompositions of the restrictions to Γ of (smooth) r-fields and r-forms on an oriented manifold M

(59) $\beta = t\beta + n\beta \wedge \omega_n^1 = t\beta + \bar{n}\beta$,

$t\beta$ (resp. $n\beta$) is called the *tangential part* (*resp. normal part*) of β.

Admissible boundary coordinate system; adapted chart.
The decomposition (59) may be extended to a neighborhood of Γ, using an "admissible boundary coordinate system" or a chart "adapted to the unit outgoing normal vector field". We recall the definition (Morrey [1]).

An *admissible boundary coordinate system* on a manifold M with boundary Γ of class $C^{k,\mu}$ is a coordinate system (of class $C^{k,\mu}$) which maps its domain $G \cup \sigma \subset R^n_-$ (with σ the part of ∂G on $x^n = 0$, $\sigma \neq \emptyset$) onto a boundary neighborhood N, in such a way that σ is mapped onto $N \cap \Gamma$ and the metric is given on σ by:

$$(60) \qquad d\,s^2 = \sum_{i,j=1}^{n-1} g_{ij}(x', 0)\, dx^i \otimes dx^j + (dx^n)^2, \quad x = (x', x^n).$$

[Also, if (W, φ) is a chart of Γ with $y_o \in W$, there are $\alpha > 0$ and a diffeomorphism: $(t, y) \in [0, \alpha) \times W \to z = v_y(t) \in U \subset M$, such that $U \cap \Gamma = W$, and:

$$(61) \qquad v_y(0) = y, \quad v'_y(0) = n, \text{ the unit normal to } \Gamma.$$

Thus we can define a chart (U, χ) of M on U by:

$$(62) \qquad \chi_j(z) = \varphi_j(y), \; j = 1, \dots, n-1, \;\; \chi_n(z) = t \;\text{ with } z = v_y(t) \in U.$$

Then $(\frac{\partial}{\partial \chi_j}|_y)_{j=1,\dots,n}$ can be identified with the normal n.]

If M is a manifold $\bar{\Omega}$, Ω being an open set of R^n, we can use the Euclidean distance of $z \in \Omega$ to the boundary Γ, $d(z,\Gamma)$: $t = s_n = \varphi_n(z) = d(z, \Gamma)$, and: $\omega^1_n = ds_n = d\varphi_n$. This allows us to extend the decomposition (59) in a neighborhood V of Γ. ⊗

Let j be the *canonical mapping* $\Gamma \to M$. If β is a C^k r-form on M, the *pull-back* $j^*\beta$ of β by j is the C^{k-1} r-form on Γ defined by (see Schwartz [1], Bourbaki [2])

$$(63) \qquad (j^*\beta)_x(v_1, \dots, v_r) = \beta_x(v_1, \dots, v_r), \quad \forall\; x \in \Gamma,\; v_1, \dots, v_r \in T_x\Gamma.$$

This r-form $j^*\beta$ can be identified with the restriction of the r-form $t\beta$ to Γ, and we write

$$(64) \qquad t_\Gamma\beta = j^*\beta = t\beta|_\Gamma.$$

Using notations (21), (23) we define an (r − 1)-form $n_\Gamma\beta$ on Γ for all r-form β on M:

$$(65) \quad n_\Gamma\beta = (-1)^{r-1}(\omega^1_n \lrcorner_g \beta)|_\Gamma = (-1)^{r-1}\omega^1_n \lrcorner_g (n\beta \wedge \omega^1_n)|_\Gamma = n\beta|_\Gamma = (-1)^{r-1}\, i_n\beta|_\Gamma.$$

Indeed for all (r − 1)-forms α on M, we have:

$$(66) \quad \left|\begin{aligned} &(\alpha, \omega^1_n \lrcorner_g (n\beta \wedge \omega^1_n)) = (\omega^1_n \wedge \alpha, n\beta \wedge \omega^1_n) = (-1)^{r-1}(\alpha \wedge \omega^1_n, n\beta \wedge \omega^1_n) \\ &\quad = (-1)^{r-1}(t\alpha \wedge \omega^1_n, n\beta \wedge \omega^1_n) = (-1)^{r-1}(t\alpha, n\beta)(\omega^1_n, \omega^1_n) \\ &\quad = (-1)^{r-1}(t\alpha, n\beta) = (-1)^{r-1}(\alpha, n\beta). \end{aligned}\right.$$

Let g_Γ denote the Riemannian metric induced by g on Γ, and $v_\Gamma = v^{g_\Gamma}$ the (n − 1)-form on Γ (corresponding to the Lebesgue measure $d\Gamma$ on Γ, see (14)); then

(67) $$v_\Gamma = \omega_n^1 \underset{g}{\lrcorner} v^g|_\Gamma = (-1)^{n-1} n_\Gamma v^g \quad (\text{or also } v_\Gamma = t_\Gamma * \omega_n^1).$$

It is thus possible to define the Hodge transformation on Γ as in (25).

Lemma 1. *For all (smooth) r-forms β on M, we have*

(68) $$\begin{cases} \text{i) } t_\Gamma * \beta = (-1)^{r-1} \underset{\Gamma}{*} n_\Gamma \beta, \\ \text{ii) } n_\Gamma * \beta = (-1)^{n-1} \underset{\Gamma}{*} t_\Gamma \beta. \end{cases}$$

PROOF. Applying the Hodge transformation to (59), we obtain

(69) $$*\beta = *t\beta + *(n\beta \wedge \omega_n^1) = *t\beta + *\bar{n}\beta.$$

We have

(70) $$t*\beta = *\bar{n}\beta, \quad \bar{n}*\beta = *t\beta, \quad t_\Gamma * \beta = t_\Gamma * (n\beta \wedge \omega_n^1).$$

Using definitions (25), (24), we have for all (n − r)-forms α on a neighborhood V of Γ

(71) $$\begin{aligned} &(\alpha, *(n\beta \wedge \omega_n^1))_g = (\alpha, (n\beta \wedge \omega_n^1) \underset{g}{\lrcorner} v)_g = (n\beta \wedge \omega_n^1 \wedge \alpha, v)_g \\ &= (-1)^{r-1}(\omega_n^1 \wedge n\beta \wedge \alpha, v)_g = (-1)^{r-1}(n\beta \wedge \alpha, \omega_n^1 \underset{g}{\lrcorner} v)_g. \end{aligned}$$

But using (67), we obtain:

(72) $$(\alpha, *(n\beta \wedge \omega_n^1))_g = (-1)^{r-1}(n\beta \wedge \alpha, v_\Gamma)_g = (-1)^{r-1}(\alpha, n\beta \lrcorner v_\Gamma)_g;$$

therefore we deduce the identity on Γ:

(73) $$t_\Gamma * (n\beta \wedge \omega_n^1) = (-1)^{r-1} n_\Gamma \beta \underset{g}{\lrcorner} v_\Gamma = (-1)^{r-1} \underset{\Gamma}{*} n_\Gamma \beta,$$

giving (68)i) thanks to (70). Applying the $\underset{\Gamma}{*}$ transformation to (68)i), we find:

(74) $$\underset{\Gamma}{*} t_\Gamma * \beta = (-1)^{r-1} \underset{\Gamma}{*}\underset{\Gamma}{*} n_\Gamma \beta = (-1)^{(r-1)(n-r-1)} n_\Gamma \beta.$$

Then taking $\beta = *\alpha$, with α an (n − r)-form on M, we have:

(75) $$n_\Gamma * \alpha = (-1)^{n-1} \underset{\Gamma}{*} t_\Gamma \alpha,$$

with (33), thus (68).

⊗

Remark 2. We also define t_Γ and n_Γ on 0-forms by:

$$t_\Gamma 1 = 1, \quad t_\Gamma f = f|_\Gamma, \quad n_\Gamma 1 = 0, \quad n_\Gamma f = 0.$$

On n-forms, we have: $t_\Gamma \omega = 0$, $n_\Gamma \omega = (-1)^{n-1} \omega_n^1 \underset{g}{\lrcorner} \omega = (-1)^{n-1} f|_\Gamma v_\Gamma$, for $\omega = fv^g$.

⊗

Remark 3. From the properties of the exterior derivative, we have for all smooth r-forms ω on M, using (63):

(76) $t_\Gamma d\omega = j^* d\omega = d(j^*\omega) = dt_\Gamma\omega,$

thus:

(77) $t_\Gamma d\omega = dt_\Gamma\omega.$

We also have, in a similar manner for all smooth r-forms β

(78) $\delta(n_\Gamma\beta) = n_\Gamma\delta\beta.$

[We prove this, taking the exterior derivative of (68)i) and using (45):

$$d\, t_\Gamma * \beta = (-1)^{r-1} d \underset{\Gamma}{*} n_\Gamma \beta;$$

thus with (77):

(79)
$$\left|\begin{array}{l} d\, t_\Gamma * \beta = t_\Gamma d * \beta = (-1)^r t_\Gamma * \delta\beta = \underset{\Gamma}{*} n_\Gamma \delta\beta, \\ d \underset{\Gamma}{*} n_\Gamma \beta = (-1)^{r-1} * \delta n_\Gamma \beta\ .] \end{array}\right.$$

We remark that in an admissible boundary coordinate system, $t_\Gamma u$ and $n_\Gamma u$ (with $u = \sum_J u_J dx^J$, $J \subseteq \{1,\ldots,n-1\}$, card $J = r$) are given by:

(80)
$$\left|\begin{array}{ll} t_\Gamma u = \sum_J u_J(x', 0)\, dx'^J, & J \subset \{1,\ldots,n-1\}, \quad \text{card } J = r \\ n_\Gamma u = \sum_{J'} u_{J'n}(x',0) dx'^{J'}, & J' \subset \{1,\ldots,n-1\}, \quad \text{card } J' = r-1. \end{array}\right.$$

Using the tangential part of a (smooth) $(n-1)$-form ω on M, the *Stokes theorem* (see Bourbaki [2] and Arnold [1]) may be written:

(81) $\int_M d\omega = \int_\Gamma t_\Gamma\omega,$

where the (bounded) boundary Γ has its usual orientation induced by that of M. All this is true for smooth differential forms. We shall generalize it below.

We give here another application of the Stokes formula. Let α and β be respectively a smooth $(r-1)$-form and an r-form, with the same parity. From (45), we have:

(82) $d\alpha \wedge *\beta - \alpha \wedge *\delta\beta = d(\alpha \wedge *\beta),$

then integrating on M, and using (81):

(83) $(d\alpha,\beta) - (\alpha,\delta\beta) = \int_\Gamma t_\Gamma(\alpha \wedge *\beta).$

But we have from (64):

(84) $t_\Gamma(\alpha\wedge *\beta)=j^*(\alpha\wedge *\beta)=j^*\alpha\wedge j^*(*\beta)=t_\Gamma\alpha\wedge t_\Gamma(*\beta).$

Then with (68), we obtain Stokes formula for all (smooth) (r − 1)-forms α, and all (smooth) r-forms β

(85) $(d\alpha\,,\beta)-(\alpha\,,\delta\beta)=(-1)^{r-1}\int_\Gamma t_\Gamma\alpha\wedge \underset{\Gamma}{*}\, n_\Gamma\beta=(-1)^{r-1}(t_\Gamma\alpha\,,n_\Gamma\beta).$

⊗

Example 2. We take $\alpha=f$, $\beta=\omega_X$, X a vector field; thus with (47), (48)

(86) $$\left|\begin{array}{l}(df\,,\omega_X)=(\omega_{\mathrm{grad}\,f}\,,\omega_X)=(\mathrm{grad}\,f\,,X),\\ (f\,,\delta\omega_X)=(f\,,\mathrm{div}\,X),\qquad n_\Gamma\,\omega_X=n.X\,,\end{array}\right.$$

and (85) gives the usual Stokes formula:

(87) $(\mathrm{grad}\,f\,,X)+(f\,,\mathrm{div}\,X)=(f|_\Gamma\,,n.X)=\int_\Gamma f\,n.X\,d\Gamma.$

⊗

From (85), we can deduce other useful formulas for the Laplace-Beltrami operator. We first take $\beta=d\gamma$ in (85), then we apply (85) replacing α by δγ and β by α, to get

(88) $$\left|\begin{array}{l}(d\alpha\,,d\gamma)-(\alpha\,,\delta d\gamma)=(-1)^{r-1}(t_\Gamma\alpha\,,n_\Gamma d\gamma)\\ (d\delta\gamma\,,\alpha)-(\delta\gamma\,,\delta\alpha)=(-1)^{r}(t_\Gamma\delta\gamma\,,n_\Gamma\alpha).\end{array}\right.$$

Then by difference, we have with $\Delta=-(d\delta+\delta d)$

(89) $(d\alpha\,,d\gamma)+(\delta\alpha,\delta\gamma)+(\alpha\,,\Delta\gamma)=(-1)^{r-1}[(t_\Gamma\alpha\,,n_\Gamma d\gamma)+(n_\Gamma\alpha\,,t_\Gamma\delta\gamma)]$

for all (smooth) (r − 1)-forms α, γ. Since we also have

(89)' $(d\gamma\,,d\alpha)+(\delta\gamma\,,\delta\alpha)+(\Delta\alpha\,,\gamma)=(-1)^{r-1}[(t_\Gamma\gamma\,,n_\Gamma d\alpha)+(n_\Gamma\gamma\,,t_\Gamma\delta\alpha)],$

we get by difference:

(90) $(\Delta\alpha,\gamma)-(\alpha,\Delta\gamma)=(-1)^{r-1}[(t_\Gamma\gamma\,,n_\Gamma d\alpha)-(t_\Gamma\alpha\,,n_\Gamma d\gamma)+(t_\Gamma\delta\alpha\,,n_\Gamma\gamma)-(n_\Gamma\alpha\,,t_\Gamma\delta\gamma)].$

1.6. Currents; generalized r-forms (or distribution r-forms)

Let $D_r(M)$ (resp. $\underline{D}_r(M)$) be the space of C^∞ even (resp. odd) r-forms with compact support in the interior $\overset{o}{M}$ of a C^∞-manifold M, these spaces being equipped with the usual inductive topology ("Schwartz topology") generalizing the scalar case. Note that it would be better to use notation $D_r(\overset{o}{M})$ instead of $D_r(M)$, but usual notations are as if M were an open set.

The dual space is called the *space of odd (resp. even)* (n – p)*-currents,* and often denoted by $\underline{D}'_r(M)$ (resp. $D'_r(M)$), thus: $\underline{D}'_r(M) = (D_{n-r}(M))'$, $D'_r(M) = (\underline{D}_{n-r}(M))'$.
We identify *currents with generalized (or distribution) even or odd r-forms*: for all ω in $D_r(M)$, we define the odd current T_ω by

(91) $$<T_\omega, \phi> = \int_M \omega \wedge *\phi = \int_M \phi \wedge *\omega, \quad \forall \phi \in D_r(M).$$

The mapping T: $\omega \longrightarrow T_\omega$ extends naturally (continuously) from the closure of $D_r(M)$ onto the space of currents $\underline{D}'_r(M)$. We denote also by $D'_r(M)$ this closure, and we call it the space of generalized (or distribution) even r-forms. Similarly we have generalized odd r-forms, and we denote by $\underline{D}'_r(M)$ their space; (91) is then written

(92) $$< T_\omega, \phi> = < \omega \wedge *\phi, 1 >, \quad \forall \phi \in D_r(M).$$

[The exterior product between smooth (n – p)-forms and generalized p-forms is naturally defined, and then $\omega \wedge *\phi$ is a generalized n-form with compact support.]

The Hodge transformation has a natural extension from $D'_r(M)$ onto $\underline{D}'_{n-r}(M)$ (and from $\underline{D}'_r(M)$ onto $D'_{n-r}(M)$).

The exterior derivative of generalized r-forms is also defined by

(93) $$< dT_\omega, \phi > = < T_\omega, \delta\phi >, \quad < \delta T_\omega, \phi > = < T_\omega, d\phi>, \quad \forall \phi \in D_r(M),$$
with
$$d^2 T_\omega = 0, \quad \delta^2 T_\omega = 0.$$

1.7. Application: Jump formula of exterior derivative for a p-form which is discontinuous across a surface Γ

Let Γ be a regular n – 1 surface of class C^1 in R^n (for example Γ is the boundary of an open set Ω in R^n). Let ω be an r-form which is continuous on each side of Γ, which is naturally oriented; ω has a jump σ across Γ: $\sigma = \omega_1 - \omega_2 = [\omega]_\Gamma$. The exterior derivative of ω in the generalized sense is given by (see Schwartz [1])

(94) $$d\omega = (d\omega) - \Gamma \wedge \sigma = (d\omega) - \delta_\Gamma\, \omega_n^1 \wedge t_\Gamma\sigma, \qquad \sigma = [\omega]_\Gamma,$$

where $(d\omega)$ is the "usual" derivative of ω (this is an (r + 1)-form with continuous coefficients), and where the last term is an (r + 1)-form defined (using (65) with δ_Γ the Dirac distribution on Γ) by

(95) $$(\Gamma \wedge \sigma, \beta) = (\delta_\Gamma\, \omega_n^1 \wedge t_\Gamma\sigma, \beta) = (\omega_n^1 \wedge t_\Gamma\sigma, \beta|_\Gamma) = (-1)^r (t_\Gamma\sigma, n_\Gamma\beta),$$

for all smooth (r + 1)-forms β on R^n. In fact, (94) is a direct consequence of the Stokes formula (81) written for the interior part Γ_1 and the exterior part Γ_2, with a smooth (r + 1)-form on R^n. Naturally we also have (using (85)):

(96) $$\delta\omega = (\delta\omega) + (-1)^{r-1}\,\delta_\Gamma n_\Gamma \sigma\,,\ \omega \text{ a r-form},\ \ \sigma = [\omega]_\Gamma.$$

Remark 4. We assume that ω is of class C^2, with limits, on each side of Γ. We can then take the exterior derivative of (94), and we get, using (93):

(97) $$0 = \delta_\Gamma\,\omega_n^1 \wedge t_\Gamma\,[d\omega]_\Gamma + d\,(\delta_\Gamma\,\omega_n^1 \wedge t_\Gamma\,\sigma),$$

thus:

(97)' $$d\,(\delta_\Gamma\,\omega_n^1 \wedge t_\Gamma\,\sigma) = -\,\delta_\Gamma\,\omega_n^1 \wedge t_\Gamma\,[d\omega]_\Gamma.$$

⊗

2. SOBOLEV SPACES OF r-FORMS; MAXIMA SPACES

We introduce the following space of r-forms of Sobolev type:

(98) $$H_r^1(M) = \{\,\omega \in L_r^2(M) \text{ with all its components } \omega_I \text{ in a coordinate system in the Sobolev space } H^1\},$$

and more generally, for all real s:

(99) $$H_r^s(M) = \{\,\omega \in L_r^2(M) \text{ with all its components } \omega_I \text{ in a coordinate system in the Sobolev space } H^s\}.$$

We write $\underline{H}_r^s$ or $H_{r,o}^s$ for odd r-forms and $H_{r,e}^s$ for even r-forms, if we have to specify these even or odd r-forms.

Let $U = (U_1,\dots,U_Q)$ be a finite open covering of M by coordinate patches θ_k with domains G_q and ranges U_q. If ω and η are elements of $H_r^1(M)$, we define:

(100) $$((\omega,\eta))_U = \sum_{q=1}^{Q} \int_{G_q} \sum_I \Big(\omega_I^{(q)}\,\eta_I^{(q)} + \sum_{\alpha=1}^{n} \frac{\partial\omega_I^{(q)}}{\partial x_\alpha}\frac{\partial\eta_I^{(q)}}{\partial x_\alpha}\Big)\,dx$$

where $\omega_I^{(q)}$ and $\eta_I^{(q)}$ are the components of ω and η with respect to θ_q; we could also use (15) for that definition.

We can show that this space of r-forms is a real Hilbert space with inner product given by (100), which depends on the chosen coordinate covering U. Any two such inner products are topologically equivalent; we say that the space $H^1{}_r(M)$ is Hilbertible, see Marsden [1] p. 54.

We also define the two following "*maxima*" *spaces*:

$$(101)\quad \left| \begin{array}{l} H_r(d,M) = \{\, \omega \in L^2_r(M),\ \ d\omega \in L^2_{r+1}(M)\}, \\ H_r(\delta,M) = \{\, \omega \in L^2_r(M),\ \ \delta\omega \in L^2_{r-1}(M)\}. \end{array} \right.$$

We can equip these two spaces with the following inner products:

$$(102)\quad \left| \begin{array}{l} (\omega,\eta)_{H_r(d,M)} = (\omega,\eta) + (d\omega,d\eta), \\ (\omega,\eta)_{H_r(\delta,M)} = (\omega,\eta) + (\delta\omega,\delta\eta), \end{array} \right.$$

with (,) on the left hand side given by (15). These are Hilbert spaces with the natural norm.

Remark 5. Since the coefficients of an r-form in $H^s_r(M)$ are in the Sobolev space $H^s(U)$ for any local chart U of a C^∞ manifold, then the Sobolev spaces of r-forms inherit from the usual Sobolev spaces the principal properties of the Sobolev spaces, essentially Sobolev inclusions and compacity properties:

$H^s_r(M) \to H^{s'}_r(M)$, $s > s'$, with compact inclusion for M a compact manifold.

We also have trace properties, such as:

The trace map: $\omega \to t_\Gamma\omega$ (resp. $n_\Gamma\omega$) is continuous from $H^1_r(M)$ onto $H^{1/2}_r(\Gamma)$ (resp. $H^{1/2}_{r-1}(\Gamma)$), and from $H^s_r(M)$ onto $H^{s-1/2}_r(\Gamma)$ (resp. $H^{s-1/2}_{r-1}(\Gamma)$), for any $s > 1/2$. [Using a partition of the unit (α_i), $\Sigma\alpha_i = 1$, with regular α_i with support in the local chart U_i, we can prove that the mapping is onto.]

But the imbeddings: $H_r(d,M) \to L^2_r(M)$ and $H_r(\delta,M) \to L^2_r(M)$ are not compact and there are no "Sobolev" imbeddings: $H_r(d,M)$ (or $H_r(\delta,M)$) $\to L^p_r(M)$, $p \neq 2$. That will be specified by the Hodge decomposition: $H_r(d,M)$ contains the space $dH^1_r(M)$ which is closed in $H_r(d,M)$ and in $L^2_r(M)$, and is of infinite dimension (and $H_r(\delta,M)$ contains the space $\delta H^1_r(M)$ which is closed in $H_r(\delta,M)$ and in $L^2_r(M)$, and is of infinite dimension). Also the imbedding:

$$H_r(d,\delta,M) = H_r(d,M) \cap H_r(\delta,M) \to L^2_r(M)$$

is not compact since: $\mathcal{H}_r(M) = \{\, \alpha \in L^2_r(M),\ d\alpha = 0\ ,\ \delta\alpha = 0\,\}$ is a closed subspace of $H_r(d,\delta,M)$ and of $L^2_r(M)$ and has infinite dimension. ⊗

We also define the (Hilbertible) spaces, for real s

$$(103)\quad H^s_r(d,M) = \{\, \omega \in H^s_r(M),\ \ d\omega \in H^s_{r+1}\},\quad H^s_r(\delta,M) = \{\, \omega \in H^s_r(M),\ \ \delta\omega \in H^s_{r-1}\}.$$

Then for Γ a smooth manifold, we have the following basic theorems (see Paquet [1] pp. 113, 116):

Theorem 1 (*Trace theorem*). *The map* $t_\Gamma: \omega \to t_\Gamma\omega = j^*(\omega)$ *which is naturally defined from the space of smooth r-forms on* M *into the space of smooth r-forms on its boundary* Γ

$$C^\infty(\overset{r}{\Lambda}T^*\bar{M}) = D_r(\bar{M}) \to C^\infty(\overset{r}{\Lambda}T^*\Gamma) = D_r(\Gamma),$$

extends continuously into a (continuous) linear map from $H_r(d,M)$ *onto* $H_r^{-1/2}(d,\Gamma)$. *Furthermore the space* $C_{00}^\infty(\overset{r}{\Lambda}T^*M) = D_r(M)$ *is dense in*:

(104) $\quad H_{r,0}(d,M) = \ker t_\Gamma.$

Theorem 2 (*Trace theorem*). *The map* $n_\Gamma: \omega \to n_\Gamma\omega$ *which is naturally defined from* $D_r(\bar{M})$ into $D_{r-1}(\Gamma)$ *extends continuously into a (continuous) linear map from* $H_r(\delta,M)$ onto $H_{r-1}^{-1/2}(\delta,\Gamma)$. *Furthermore the space* $D_r(M)$ *is dense in*:

(105) $\quad H_{r,0}(\delta,M) = \ker n_\Gamma.$

We also denote by $t_\Gamma\omega$ and $n_\Gamma\omega$ the extended trace of ω for ω in $H_r(d,\Omega)$ or in $H_r(\delta,\Omega)$.

Theorem 3. *The spaces* $H_r^{-1/2}(d,\Gamma)$ *and* $H_r^{-1/2}(\delta,\Gamma)$ *are dual spaces.*

Remark 6. The Hodge transformation $*$ is continuous from $H_r(d,M)$ onto $H_{n-r}(\delta,M)$ and from $H_r(\delta,M)$ onto $H_{n-r}(d,M)$ (with opposite parity).
The Hodge transformation $\underset{\Gamma}{*}$ extends into a continuous transformation from: $H_r^{-1/2}(d,\Gamma)$ onto $H_{n-r}^{-1/2}(\delta,\Gamma)$ (therefore also from $H_r^{-1/2}(\delta,\Gamma)$ onto $H_{n-r}^{-1/2}(d,\Gamma)$). Thus by the Hodge transformation, Theorems 1 and 2 can be deduced one from the other. ⊗

We can prove Theorem 1 (or 2) using local charts in order to work on the half-space (see Paquet [1]).

Remark 7. i) Using Stokes formula (85) with $\alpha \in H_r(d,M)$ and $\beta \in H_{r+1}^1(M)$ or with $\beta \in H_{r+1}(\delta,M)$ and $\alpha \in H_r^1(M)$, we obtain:

(106) $\quad t_\Gamma\alpha \in H_r^{-1/2}(\Gamma)$ for $\alpha \in H_r(d,M)$, $\quad n_\Gamma\beta \in H_r^{-1/2}(\Gamma)$ for $\beta \in H_{r+1}(\delta,M)$.

ii) But if $\alpha \in H_r(d,M)$ (resp. $\beta \in H_{r+1}^1(\delta,M)$), then $d\alpha \in H_{r+1}(d,M)$, since we have: $d\alpha \in L_{r+1}^2(M)$ and $d(d\alpha) = d^2\alpha = 0$ (resp. $\delta\beta \in H_r(\delta,M)$ since $\delta\beta \in L_r^2(M)$, and also $\delta(\delta\beta) = \delta^2\beta = 0$), then from (106) applied to $d\alpha$ and $\delta\beta$:

(107) $\quad t_\Gamma\, d\alpha \in H^{-1/2}_{r+1}(\Gamma), \quad n_\Gamma\, \delta\beta \in H^{-1/2}_{r-1}(\Gamma).$

Thus using relations (77) and (78) (generalized to this case) we have:

(108) $\quad t_\Gamma \alpha \in H^{-1/2}_r(d,\Gamma), \quad n_\Gamma \beta \in H^{-1/2}_r(\delta,\Gamma).$

iii) Therefore we only have to prove that the mappings t_Γ and n_Γ are onto. Using local charts, we are reduced to prove this result on the half-space (see Paquet [1] p. 114). ⊗

3. THE HODGE DECOMPOSITION (FOR A COMPACT MANIFOLD WITH BOUNDARY)

We remark that:

i) d maps $H_r(d,M)$ into $H_{r+1}(d,M)$, and δ maps $H_r(\delta,M)$ into $H_{r-1}(\delta,M)$.

ii) d maps $H_{r,0}(d,M)$ into $H_{r+1,0}(d,M)$, δ maps $H_{r,0}(\delta,M)$ into $H_{r-1,0}(\delta,M)$ with (see (104), (105)):

(109)
$$\left|\begin{array}{l} H_{r,0}(d,M) = \{\,\omega \in H_r(d,M),\ t_\Gamma\omega = 0\,\} \\ H_{r,0}(\delta,M) = \{\,\omega \in H_r(\delta,M),\ n_\Gamma\omega = 0\,\}. \end{array}\right.$$

This results from (77). Thus we have the following sequences:

i) for the "maxima spaces":

$$\left|\begin{array}{l} H_0(d,M) \xrightarrow{d} H_1(d,M) \to \ldots \to H_{n-1}(d,M) \xrightarrow{d} H_n(d,M) \\ H_0(\delta,M) \xleftarrow{\delta} H_1(\delta,M) \leftarrow \ldots \leftarrow H_{n-1}(\delta,M) \xleftarrow{\delta} H_n(\delta,M), \end{array}\right.$$

ii) for the "minima spaces":

$$\left|\begin{array}{l} H_{0,0}(d,M) \xrightarrow{d} H_{1,0}(d,M) \to \ldots \to H_{n-1,0}(d,M) \xrightarrow{d} H_n(d,M) \\ H_{0,0}(\delta,M) \xleftarrow{\delta} H_{1,0}(\delta,M) \leftarrow \ldots \leftarrow H_{n-1,0}(\delta,M) \xleftarrow{\delta} H_{n,0}(\delta,M), \end{array}\right.$$

each image of d (or δ) being contained in the kernel of d (or δ) in the following step, that is:

(110)
$$\left|\begin{array}{l} \mathrm{Im}\, d^{(r)} = d\, H_r(d,M) \subseteq \ker d^{(r+1)} = \{\,\omega \in H_{r+1}(d,M),\ d\omega = 0\,\} \\ \mathrm{Im}\, \delta^{(r)} = \delta\, H_{r+1}(\delta,M) \subseteq \ker \delta^{(r-1)} = \{\,\omega \in H_r(\delta,M),\ \delta\omega = 0\} \end{array}\right.$$

with finite codimension.

The operators d and δ may be viewed either as unbounded operators in $L^2_r(M)$, or as bounded operators from one space into another.

i) First we define various realizations of operators d and δ in $L^2_r(M)$, denoted $d_0 = d_m$, d_M; $\delta_0 = \delta_m$, δ_M, by their domains:

(111) $$\begin{cases} D(d_o) = D(d_m) = H_{r,o}(d,M); \quad D(\delta_o) = D(\delta_m) = H_{r,o}(\delta,M) \\ D(d_M) = H_r(d,M); \quad D(\delta_M) = H_r(\delta,M) \end{cases}$$

(with $d_o\omega = d_m\omega = d\omega$, $d_M\omega = d\omega$, $\delta_o\omega = \delta_m\omega = \delta\omega$, $\delta_M\omega = \delta\omega$, for ω in the relevant domain); d_m, d_M (resp. δ_m, δ_M) are called the minimal and maximal realizations of operator d (resp. δ) in L_r^2. From the Stokes formula, we have

(112) $$\begin{cases} (d_m^{(r)}\alpha,\beta) = (\alpha,\delta_M^{(r)}\beta), \quad \forall\, \alpha \in D(d_m),\ \beta \in D(\delta_M), \\ (d_M^{(r)}\alpha,\beta) = (\alpha,\delta_m^{(r)}\beta), \quad \forall\, \alpha \in D(d_M),\ \beta \in D(\delta_m), \end{cases}$$

thus $d_m^{(r)}$, $\delta_M^{(r)}$ (resp. $d_M^{(r)}$, $\delta_m^{(r)}$) are adjoint operators

(113) $$(d_m^{(r)})^* = \delta_M^{(r)}, \quad (\delta_m^{(r)})^* = d_M^{(r)}, \quad (\delta_M^{(r)})^* = \delta_m^{(r)}, \quad (d_M^{(r)})^* = \delta_m^{(r)}.$$

From the Stokes formula (or also from (113)), we have the following relations

(114) $$\begin{cases} (\mathrm{Im}\,\delta_M)^\perp = \ker d_o, \quad (\mathrm{Im}\,d_M)^\perp = \ker \delta_o, \\ (\mathrm{Im}\,\delta_o)^\perp = \ker d, \quad (\mathrm{Im}\,d_o)^\perp = \ker \delta, \end{cases}$$

(where $V^\perp$ is the orthogonal of space V) and:

(114)' $$\begin{cases} \overline{\mathrm{Im}\,\delta_M} = (\ker d_o)^\perp, \quad \overline{\mathrm{Im}\,d_M} = (\ker \delta_o)^\perp, \\ \overline{\mathrm{Im}\,\delta_o} = (\ker d)^\perp, \quad \overline{\mathrm{Im}\,d_o} = (\ker \delta)^\perp. \end{cases}$$

(Actually, we will prove that spaces $\mathrm{Im}\,\delta_o$, $\mathrm{Im}\,\delta_M$, $\mathrm{Im}\,d_o$, $\mathrm{Im}\,d_M$ are closed.)

ii) If we consider $\delta = \delta_m$ as a continuous operator from $H_{r,o}(\delta,M)$ into $L_{r-1}^2(M)$ and $d = d_m$ as a continuous operator from $H_{r-1,o}(d,M)$ into $L_r^2(M)$, then we can identify their dual mappings, i.e., the continuous mapping from L_{r-1}^2 into $(H_{r,o}(\delta,M))'$ (resp. from L_r^2 into $(H_{r-1,o}(d,M))'$) with the operator d (resp. δ); we denote

(115) $$H_r^{-1}(d,M) = (H_{r,o}(\delta,M))', \quad H_r^{-1}(\delta,M) = (H_{r,o}(d,M))',$$

these spaces being current spaces (or generalized r-forms) since $D_r(M)$ is dense in $H_{r,o}(d,M)$ and $H_{r,o}(\delta,M)$.

Then we define the spaces of harmonic r-forms by

(116) $$\begin{cases} \mathcal{H}_r^0(M) = \{\alpha \in L_r^2(M),\ d\alpha = 0,\ \delta\alpha = 0\} \\ \mathcal{H}_r^1(M) = \{\alpha \in H_r^1(M),\ d\alpha = 0,\ \delta\alpha = 0\}, \end{cases}$$

and also the so-called *cohomology spaces*

(117) $$\begin{cases} \mathcal{H}_r^+ = \mathcal{H}_r^+(M) = \{ \alpha \in \mathcal{H}_r^0(M),\ n_\Gamma\alpha = 0 \} \\ \mathcal{H}_r^- = \mathcal{H}_r^-(M) = \{ \alpha \in \mathcal{H}_r^0(M),\ t_\Gamma\alpha = 0 \}. \end{cases}$$

Note that $*\mathcal{H}_r^+ = \underline{\mathcal{H}}_{n-r}^-$, $*\mathcal{H}_r^- = \underline{\mathcal{H}}_{n-r}^+$, hence for an orientable manifold M, spaces $\mathcal{H}_r^+$, $\underline{\mathcal{H}}_{n-r}^-$ and $\mathcal{H}_{n-r}^-$ (resp. $\mathcal{H}_r^-$, $\underline{\mathcal{H}}_{n-r}^+$ and $\mathcal{H}_{n-r}^+$) are isomorphic, thus:

$$\dim \mathcal{H}_r^+ = \dim \mathcal{H}_{n-r}^-,\ \mathcal{H}_r^- = \dim \mathcal{H}_{n-r}^+.$$

We have $\mathcal{H}_r^\pm(M) \subseteq H_r^s(M)$ for all s, for M of class C^∞, thus $\mathcal{H}_r^\pm(M) \subseteq C^\infty$.

In order to study operators d, δ, dδ, δd with the decomposition of relevant spaces and boundary problems, we have two variational framework levels in L^2.

1st level: the usual one, associated to the Laplace-Beltrami operator, with the bilinear form

(118) $$a_\lambda(u,v) = (du,du) + (\delta u,\delta u) + \lambda(u,v),$$

(with $\lambda > 0$), in a subspace of $H_r^1(M)$.

2nd level: linked to the operator d or δ separately, with the bilinear form

(119) $$a_\lambda^+(u,v) = (du,du) + \lambda(u,v),$$

or:

(119)' $$a_\lambda^-(u,v) = (\delta u,\delta u) + \lambda(u,v),$$

in a subspace of $H_r(d,M)$ or $H_r(\delta,M)$ only.
Each variational framework gives interesting specific results.

3.1. Variational frameworks for the Laplace-Beltrami operator
We define the following spaces

(120) $$H_{ro}^1(M) = \{ \alpha \in H_r^1(M),\ \alpha|_\Gamma = 0 \} = V_r^+ \cap V_r^-,$$

(this is also the closure of $\mathcal{D}_r(M)$ in $H_r^1(M)$), with:

(121) $$\begin{cases} V_r^+ = H_{r,t}^1(M) = \{ \omega \in H_r^1(M),\ n_\Gamma\omega = 0 \} \\ V_r^- = H_{r,n}^1(M) = \{ \omega \in H_r^1(M),\ t_\Gamma\omega = 0 \} \end{cases}$$

(122) $$\begin{cases} H_{r,n}(d,\delta,M) = H_{r,o}(d,M) \cap H_r(\delta,M) \\ \quad = \{ \omega \in L_r^2(M),\ d\omega \in L_{r+1}^2(M),\ \delta\omega \in L_{r-1}^2(M),\ t_\Gamma\omega = 0 \}, \\ H_{r,t}(d,\delta,M) = H_{r,o}(\delta,M) \cap H_r(d,M) \\ \quad = \{ \omega \in L_r^2(M),\ d\omega \in L_{r+1}^2(M),\ \delta\omega \in L_{r-1}^2(M),\ n_\Gamma\omega = 0 \}. \end{cases}$$

Later, we will show (see (177)) that (122) and (121) are equivalent. A fundamental property relative to these spaces is given by the following theorem

Theorem 4 (Morrey [1]). *For each finite system U of admissible coordinate systems whose range covers M, there are constants λ_U and $C_U > 0$ such that*

$$(123) \qquad a(\omega,\omega) + \lambda_U(\omega,\omega) \geq C_U((\omega,\omega))_U, \qquad \forall \omega \in V_r^+ \text{ (resp. } V_r^-)$$

with $((\omega,\omega))_U$ defined by (100), and

$$(124) \qquad a(\omega,\omega) = (d\omega, d\omega) + (\delta\omega, \delta\omega).$$

Thus the bilinear form $a(\omega, \tilde{\omega})$ is coercive on V_r^+ (resp. V_r^-) relatively to $L_r^2(M)$.

Formula (123) is called the *Gaffney inequality*.

Consequences. The map: $\omega \to (d\omega, \delta\omega, \omega)$ being continuous from $H_r^1(M)$ into the space $L_{r+1}^2(M) \times L_{r-1}^2(M) \times L_r^2(M)$, there is also a constant C' so that

$$(125) \qquad a(\omega,\omega) + \lambda(\omega,\omega) \leq C'((\omega,\omega))_U, \qquad \forall \omega \in V_r^{\pm}.$$

Thus the norms $(a(\omega,\omega) + \lambda(\omega,\omega))^{1/2}$ and $((\omega,\omega))_U^{1/2}$ are equivalent, and we can identify the spaces (for a proof see below) $V_r^+ = H_{r,t}^1(M)$ with $H_{r,t}(d, \delta, M)$, and $V_r^- = H_{r,n}^1(M)$ with $H_{r,n}(d, \delta, M)$, equipped with the norm $(a(\omega,\omega) + \lambda(\omega,\omega))^{1/2}$.

Corollary 1. *We assume* M *to be compact and regular (of class $C^{1,1}$). Then the natural mappings* $\omega \in V_r^- \to \omega \in L_r^2(M)$ and $\omega \in V_r^+ \to \omega \in L_r^2(M)$ *are compact.*

Corollary 2. *The trace mapping* $\omega \to n_\Gamma\omega$ (resp. $t_\Gamma\omega$) *is continuous from* V_r^- (resp. V_r^+) *into* $H_{r-1}^{1/2}(\Gamma)$ (resp. $H_r^{1/2}(\Gamma)$).

From the Peetre lemma, we deduce that the cohomology spaces $\mathcal{H}_r^{\pm}$ have finite dimension. Then we denote by $\overset{\circ}{V}{}_r^-$ and $\overset{\circ}{V}{}_r^+$ the spaces in V_r^- and V_r^+ which are L^2-orthogonal to $\mathcal{H}_r^-$ and $\mathcal{H}_r^+$ respectively; we have the orthogonal decomposition:

$$(126) \qquad V_r^- = H_{r,n}^1(M) = \mathcal{H}_r^- \oplus \overset{\circ}{V}{}_r^-, \quad V_r^+ = H_{r,t}^1(M) = \mathcal{H}_r^+ \oplus \overset{\circ}{V}{}_r^+.$$

Theorem 5 (Morrey). *The bilinear form* $a(\omega, \tilde{\omega})$ is $\overset{\circ}{V}{}_r^+$ *(resp.* $\overset{\circ}{V}{}_r^-$*)-coercive, that is, there is a positive constant* C^+ *(resp.* C^-*) so that*

$$(127) \quad a(\omega,\omega) \geq C^+((\omega,\omega))_U, \quad \forall \omega \in \overset{\circ}{V}{}_r^+, \quad (\text{resp. } a(\omega,\omega) \geq C^-((\omega,\omega))_U, \quad \forall \omega \in \overset{\circ}{V}{}_r^-).$$

We can also obtain this theorem as a consequence of Theorem 4 and the Peetre lemma. Note that (127) is a generalization of Poincaré inequality.
We use the following new spaces:

(128) $\quad V^-_{r\delta} = V^-_r \cap \ker\delta = \mathcal{H}^-_r \oplus \overset{o}{V}{}^-_{r\delta}, \quad V^+_{r\delta} = V^+_r \cap \ker\delta = \mathcal{H}^+_r \oplus \overset{o}{V}{}^+_{r\delta},$

with notation $W_\delta = W \cap \ker\delta$, for $W = V^-_r$, $\overset{o}{V}{}^-_r$, V^+_r, $\overset{o}{V}{}^+_r$.

Lemma 2. *With definitions* (111), (109), *spaces* $\mathrm{Im}\, d_M$, $\mathrm{Im}\, d_0$, $\mathrm{Im}\,\delta_M$, $\mathrm{Im}\,\delta_0$ *are closed in* $L^2(M)$.

PROOF. We first prove that $\mathrm{Im}\, d_M \subseteq \ker d$, from which we deduce for the orthogonal spaces (in L^2)

(129) $\quad (\mathrm{Im}\, d_M)^\perp \supseteq (\ker d)^\perp.$

But for all $\alpha \in H_r(d, M)$, there are $\beta \in \ker d$, $\gamma \in (\ker d)^\perp \cap H_r(d, M)$, such that $\alpha = \beta + \gamma$, thus $d\alpha = d\gamma$, and then, from (129):

(130) $\quad \mathrm{Im}\, d_M = d\,((\ker d)^\perp \cap H_r(d, M)) = d((\mathrm{Im}\, d_M)^\perp \cap H_r(d, M)).$

Using (114), we have (with (128)):

(131) $\quad \mathrm{Im}\, d_M = d\,(\ker\delta_0 \cap H_r(d, M))\, d\, V^+_{r\delta} = d\, \overset{o}{V}{}^+_{r\delta}.$

Now, from inequality (127), we have:

(132) $\quad \|d\omega\|^2 \geq C^+ \|\omega\|^2_{H^1_r(M)}, \quad \forall\, \omega \in \overset{o}{V}{}^+_{r\delta},$

which proves that:

(133) $\mathrm{Im}\, d_M = d\, \overset{o}{V}{}^+_{r\delta}$ is closed in L^2_{r+1} and d is an isomorphism from $\overset{o}{V}{}^+_{r\delta}$ onto Im d.

In a similar manner, we can prove that

(134) $\quad \mathrm{Im}\, d_0 \overset{\mathrm{def}}{=} d\, H_{r0}(d, M) = d\,((\mathrm{Im}\, d_0)^\perp \cap H_{r0}(d, M)) = d\,(\ker\delta \cap H_{r0}(d, M)).$

Thus with spaces defined by (128),

$$\mathrm{Im}\, d_0 = d\, V^-_{r\delta} = d\, \overset{o}{V}{}^-_{r\delta},$$

and we have also inequality (132) on $\overset{o}{V}{}^-_{r\delta}$ from which we deduce:

(135) $\mathrm{Im}\, d_0 = d\, \overset{o}{V}{}^-_{r\delta}$ is closed in L^2_{r+1}, and d is an isomorphism from $\overset{o}{V}{}^-_{r\delta}$ onto $\mathrm{Im}\, d_0$.

⊗

Remark 8. Anticipating on the Hodge decomposition (see below), we have:

$$(136)\quad \begin{vmatrix} \overset{\circ}{V}{}^{+}_{r\delta} = \mathrm{Im}\,\delta_o \cap H_r(d,M) = \mathrm{Im}\,\delta_o \cap V_r^+ , \\ \overset{\circ}{V}{}^{-}_{r\delta} = \mathrm{Im}\,\delta \cap H_{ro}(d,M) = \mathrm{Im}\,\delta \cap V_r^- . \end{vmatrix}$$

Hence we have proved:

$$(137)\quad \begin{vmatrix} \mathrm{Im}\,d_M = d\,H_r(d,M) = dV_r^+ = d\,(\mathrm{Im}\,\delta_o \cap V_r^+), \\ \mathrm{Im}\,d_o = d\,H_{ro}(d,M) = dV_r^- = d\,(\mathrm{Im}\,\delta \cap V_r^-), \end{vmatrix}$$

$$(138)\quad \begin{vmatrix} \mathrm{Im}\,\delta_M = \delta\,H_r(\delta,M) = \delta V_r^- = \delta\,(\mathrm{Im}\,d_o \cap V_r^-), \\ \mathrm{Im}\,\delta_o = \delta\,H_{ro}(\delta,M) = \delta V_r^+ = \delta\,(\mathrm{Im}\,d \cap V_r^+), \end{vmatrix}$$

and:

d is an isomorphism from $\mathrm{Im}\,\delta_o \cap V_r^+$ (resp. $\mathrm{Im}\,\delta \cap V_r^-$) onto Im d (resp. Im d_o),
δ is an isomorphism from $\mathrm{Im}\,d_o \cap V_r^-$ (resp. $\mathrm{Im}\,d \cap V_r^+$) onto Im δ (resp. Im δ_o).

⊗

Remark 9. Let $V_{r\Delta}^+$ (resp. $V_{r\Delta}^-$) be the space which is orthogonal to $H^1_{ro}(M)$ in V_r^+ (resp. V_r^-), for the scalar product $((\alpha,\phi)) = (d\alpha,d\phi) + (\delta\alpha,\delta\phi)$.
These spaces are also characterized by:

$$(139)\quad V_{r\Delta}^+ = \{\alpha \in V_r^+ ,\ (\delta d + d\delta)\alpha = 0\}, \qquad V_{r\Delta}^- = \{\beta \in V_r^- ,\ (\delta d + d\delta)\beta = 0\}.$$

Note that the intersection with harmonic r-forms (see (116)) is:

$$(140)\quad V_{r\Delta}^+ \cap \mathcal{H}_r(M) = \mathcal{H}_r^+ , \qquad V_{r\Delta}^- \cap \mathcal{H}_r(M) = \mathcal{H}_r^- .$$

Thus relation $(d\delta + \delta d)\,\alpha = 0$ does not imply $d\alpha = 0$, $\delta\alpha = 0$: the kernel of the Laplace-Beltrami operator is not the set of harmonic r-forms (in the general case). ⊗

We deduce fairly easily from Theorem 4 and from Lemma 2 (and the Stokes formula) the following theorem of decomposition of space $L^2_r(M)$ for a *compact manifold M with boundary*.

Theorem 6 *(The Hodge decomposition). Using the notations above (see (117)), we have the orthogonal decompositions of the kernel spaces*

$$(141)\quad \begin{vmatrix} \ker d = \mathrm{Im}\,d \oplus \mathcal{H}_r^+ , & \ker d_o = \mathrm{Im}\,d_o \oplus \mathcal{H}_r^- , \\ \ker \delta = \mathrm{Im}\,\delta \oplus \mathcal{H}_r^- , & \ker \delta_o = \mathrm{Im}\,\delta_o \oplus \mathcal{H}_r^+ , \end{vmatrix}$$

and also

$$(142)\quad \ker d = \mathrm{Im}\,d_o \oplus \mathcal{H}_r , \qquad \ker \delta = \mathrm{Im}\,\delta_o \oplus \mathcal{H}_r ,$$

with

(143) $\quad H_r = H_r(M) = H_r^0(M) = \ker d \cap \ker \delta = H_r^+ \oplus (\mathrm{Im}\, d \cap \mathrm{Im}\, \delta) \oplus H_r^-$,

and also the orthogonal decomposition of the space $L_r^2(M)$:

(144) $\quad L_r^2(M) = \mathrm{Im}\, d \oplus \ker \delta_0 = \ker d_0 \oplus \mathrm{Im}\, \delta = \mathrm{Im}\, d_0 \oplus H_r^0(M) \oplus \mathrm{Im}\, \delta_0$.

We will sum up these results in a diagram at the end of this chapter.

Definition 2. An r-form ω ***is said to be closed (resp. coclosed) if*** $d\omega = 0$ ***(resp.*** $\delta\omega = 0$) ***and to be exact if there is an*** $(r-1)$-form θ ***so that*** $\omega = d\theta$ ***(resp. coexact if there is an*** $(r+1)$-form θ ***so that*** $\omega = \delta\theta$).

(Of course, we have to specify spaces in this definition.) An exact r-form is closed and the converse is locally true (it is the Poincaré lemma) but globally wrong as we can see it from (141).

We can also deduce from Theorem 2 another coerciveness result. Let W be the subspace of $H^1{}_r(M)$ which is L^2-orthogonal to $H_r(M)$, thus (with (116)):

(145) $\quad H_r^1(M) = W \oplus H_r^1(M)$.

Note that we have:

(145)' $\quad W = (\mathrm{Im}\, d_0 \oplus \mathrm{Im}\, \delta) \cap H_r^1(M) = (\mathrm{Im}\, d_0 \cap H_r^1(M)) \oplus (\mathrm{Im}\, \delta_0 \cap H_r^1(M))$.

We also have with notation (124):

Theorem 7 (Morrey). ***The bilinear form*** $a(\omega, \bar{\omega})$ ***is coercive on*** W: ***there is a constant*** $C > 0$ ***so that***

(146) $\quad a(\omega,\omega) \geq C\,((\omega,\omega))_U, \quad \forall\, \omega \in W.$

Let V_r^0 be the intersection of the two spaces $\overset{o}{V}{}_r^-$ and $\overset{o}{V}{}_r^+$ (see (126)):

(147) $\quad V_r^0 = \overset{o}{V}{}_r^- \cap \overset{o}{V}{}_r^+ = \{\, \omega \in H_{r0}^1(M),\ \omega \text{ orthogonal to } H_r^- \oplus H_r^+\}$.

Then from Theorem 2, we also have:

Theorem 7'. ***The bilinear form*** $a(\omega, \bar{\omega})$ ***is coercive on*** V_r^0 : ***there is a constant*** $C > 0$ ***so that***

(148) $\quad a(\omega,\omega) \geq C\,((\omega,\omega))_U, \quad \forall\, \omega \in V_r^0.$

Remark 10. The orthogonality of H_r^+ and H_r^- is easily seen from the fact that H_r^+ and H_r^- are contained respectively in $\mathrm{Im}\, \delta$ and $\mathrm{Im}\, d$. ⊗

We give the particular cases $r = 0$ ***and*** n.

i) In the $r = 0$ case, we have:

(149) $\quad L_0^2(M) = L^2(M) = \mathrm{Im}\, \delta = H_0^+ \oplus \mathrm{Im}\, \delta_0$,

and:

(150) $\mathcal{H}_0^- = H_0(M) = \{0\}, \qquad \mathcal{H}_0^+ = \ker d = \{f \in L^2(M),\ \mathrm{grad}\, f = 0\}$

thus $\mathcal{H}_0^+$ is the space of constant functions on each connected component of M, the dimension of $\mathcal{H}_0^+$ is equal to the number of connected components of M, and:

$$\mathrm{Im}\,\delta_0 = \delta H_{1,0}(\delta, M) = \{f \in L^2(M),\ f = \delta\omega_X = \mathrm{div}\, X,\ \omega_X \in H_{1,0}(\delta, M)\}.$$

ii) In the r = n case, we have:

(151) $L_n^2(M) = \mathrm{Im}\, d = \mathcal{H}_n^- \oplus \mathrm{Im}\, d_0,$

(152) $\mathcal{H}_n^+ = H_n(M) = \{0\}, \quad \text{and} \quad \mathcal{H}_n^- = \ker \delta.$

Thus $\mathcal{H}_n^- = \{ f v^g$, f constant on each connected component of M $\}$, the dimension of $\mathcal{H}_n^-$ is also equal to the number of connected components of M.

⊗

Application to some boundary problems. Potentials.

Proposition 1. *Let* ω *be a given* r*-form in* $L^2(M)$, *orthogonal to* $\mathcal{H}_r^+$ (resp. $\mathcal{H}_r^-$). *Then there is a unique* r*-form* ω^+ *(resp.* ω^-*) in* $\overset{o}{V}{}_r^+$ (resp. $\overset{o}{V}{}_r^-$) *such that (with* $\alpha = \omega^+$, resp. ω^-):

(153) $(d\alpha, d\zeta) + (\delta\alpha, \delta\zeta) = (\omega, \zeta), \qquad \forall \zeta \in \overset{o}{V}{}_r^+ \text{ (resp. } \overset{o}{V}{}_r^-).$

Furthermore, for M of class $C^{1,1}$, *we have*:

(154) $d\omega^+ \in V_{r+1}^+, \quad \delta\omega^+ \in V_{r-1}^+ \quad (\text{resp. } d\omega^- \in V_{r+1}^-, \quad \delta\omega^- \in V_{r-1}^-).$

Note that we also have: $d\omega^+ \in \overset{o}{V}{}_{r+1}^+$, $\delta\omega^+ \in \overset{o}{V}{}_{r-1}^+$.

Definition 3. *We call* ω^+ *(resp.* ω^-*) the plus potential (resp. minus) potential of* ω.

We can prove that ω^+ (resp. ω^-) is the solution of the problem:

(155)$_+$ $\quad \left| \begin{array}{l} (\delta d + \delta d)\,\omega^+ = \omega, \\ n_\Gamma \omega^+ = 0, \quad n_\Gamma d\omega^+ = 0 \end{array} \right.$

(resp.(155)$_-$ $\quad \left| \begin{array}{l} (\delta d + \delta d)\,\omega^- = \omega, \\ t_\Gamma \omega^- = 0, \quad t_\Gamma \delta\omega^- = 0 \,). \end{array} \right.$

PROOF OF PROPOSITION 1. The first part of Proposition 1 follows directly from Theorem 2 via the Lax-Milgram lemma. The second part is a regularity result, for which we refer to Morrey [1]. For other regularity results, see Morrey [1]. ⊗

Then we define the Laplace-Beltrami operators Δ^+ and Δ^- as selfadjoint negative operators in $L^2_r(M)$ thanks to the variational framework: $(V_r^{\pm}, L^2_r(M), a(\omega, \tilde{\omega}))$

$$(156)\quad \left| \begin{array}{l} D(\Delta^{\pm}) = \{\omega \in V_r^{\pm}, \text{ such that the map } \zeta \longrightarrow a(\omega,\zeta) \text{ is continuous on } \\ \qquad\qquad V^{\pm} \text{ equipped with the } L^2\text{-topology}\}, \\ \quad = \{\omega \in V_r^{\pm},\ \Delta\omega \in L^2_r(M),\ a(\omega,\zeta) = -(\Delta\omega,\zeta),\quad \forall \zeta \in V_r^{\pm}\}, \end{array} \right.$$

and

$$\Delta^{\pm}\omega = \Delta\omega = -(d\delta + \delta d)\,\omega, \qquad \forall\, \omega \in D(\Delta^{\pm}).$$

Thanks to Proposition 1, we have the following characterization of their domains:

$$(156)' \qquad D(\Delta^{\pm}) = \{\omega \in V_r^{\pm},\ d\omega \in V_{r+1}^{\pm},\ \delta\omega \in V_{r-1}^{\pm}\},$$

or also:

$$(156)'' \qquad D(\Delta^+)(\text{resp. } \Delta^-) = \{\omega \in H^1_r(M),\ d\omega \in H^1_{r+1}(M),\ \delta\omega \in H^1_{r-1}, \\ n_\Gamma\omega = 0,\ n_\Gamma d\omega = 0 \ (\text{resp. } t_\Gamma\omega = 0,\ t_\Gamma\delta\omega = 0)\}.$$

Since

$$(157) \qquad V_r^+ \subseteq H_{ro}(\delta, M) = D(\delta_m), \qquad V_r^- \subseteq H_{ro}(d, M) = D(d_m),$$

using notations (111), we can write

$$(158) \qquad -\Delta^+ = d_M\delta_m + \delta_m d_M, \quad -\Delta^- = d_m\delta_M + \delta_M d_m.$$

Remark 11. The operators $(d+\delta)_{\pm}$ in $\oplus_r L^2_r(M)$ defined by:

$$(159) \qquad D((d+\delta)_{\pm}) = \oplus_r V_r^{\pm}, \qquad (d+\delta)_{\pm}\,\omega = d\omega + \delta\omega, \qquad \forall\, \omega \in D((d+\delta)_{\pm}),$$

are selfadjoint and elliptic (see Gilkey p.243), and we have:

$$(160) \qquad ((d+\delta)_{\pm})^2 = \Delta^{\pm} \quad \text{in } L^2_r(M),\ 0 \le r \le n.$$

⊗

We also have:

Theorem 8. *Let* $\omega \in H_r(d, M)$ (*resp.* $H_r(\delta, M)$), *be orthogonal to* $\mathcal{H}_r^+(M)$ (resp. $\mathcal{H}_r^-(M)$), ω^+ (resp. ω^-) *the plus (resp. minus) potential of* ω ($\omega^+ \in V_r^+$, $\omega^- \in V_r^-$). *Then*

$$(161) \qquad d\omega = d\,\delta\,d\,\omega^+ \ (\text{resp. } \delta\omega = \delta\,d\,\delta\,\omega^-),$$

and we have:

$$(162) \qquad d\omega^+ = (d\omega)^+ \quad (\text{resp. } \delta\omega^- = (\delta\omega)^-)$$

i.e., $d\omega^+$ (resp. $\delta\omega^-$) *is the plus (resp. minus) potential of* $d\omega$.

PROOF. Let $\omega \in H_r(d, M)$ be orthogonal to $\mathcal{H}_r^+(M)$. Since $d\omega \in L^2_{r+1}$ and $d\omega \in \operatorname{Im} d$, we have $d\omega$ orthogonal to $\mathcal{H}_r^+$. Since $\delta\,d\,(d\omega^+) = 0$ and $d\omega^+ \in V_r^+$, then from (161), $d\omega^+$ is the plus potential of $d\omega$, therefore: $\delta\,d\omega^+ \in V_r^+$ from Proposition 1.

We would prove the other part of the theorem in a similar way. ⊗

This theorem is a simple generalization of theorem 7.7.5 of Morrey [1], for ω in the space $H^1{}_r(M)$. Theorem 8 leads to many consequences.

Proposition 2 (Morrey). *Let $\omega \in L^2_r(M)$ be orthogonal to $H_r(M)$. Then there is a unique r-form $\omega_N \in H^1_r(M)$, L^2-orthogonal to $H_r(M)$ such that:*

(163) $$(d\omega_N, d\zeta) + (\delta\omega_N, \delta\zeta) = (\omega, \zeta), \quad \forall \zeta \in H^1_r(M).$$

Moreover:

(164) $$d\omega_N = d\omega^+, \quad \delta\omega_N = \delta\omega^-,$$

with the above notations and then: $d\omega_N \in H^1_{r+1}(M)$, $\delta\omega_N \in H^1_{r-1}(M)$.

Proposition 3. *Let $\omega \in L^2_r(M)$ be orthogonal to $H^-_r \oplus H^+_r$. Then there is a unique r-form $\omega_D \in H^1_{ro}$ orthogonal to $H^-_r \oplus H^+_r$ such that:*

(165) $$(d\omega_D, d\zeta) + (\delta\omega_D, \delta\zeta) = (\omega, \zeta), \quad \forall \zeta \in H^1_{ro}(M).$$

Moreover we have:

(166) $$d\omega_D = d\omega^-, \quad \delta\omega_D = \delta\omega^+.$$

and then: $d\omega_D \in H^1_{r+1}(M)$, $\delta\omega_D \in H^1_{r-1}$.

PROOF OF PROPOSITIONS 2 AND 3. The first part is a mere application of the Gaffney inequality (Theorem 4) and of the Lax-Milgram lemma. The second part follows from Theorem 8. ⊗

Definition 4. *We call ω_N the Neumann potential of ω and ω_D the Dirichlet potential of ω.*

We can prove that ω_N, ω_D are the unique solutions to the following problems: find ω_N in $H^1_r(M)$, ω_N orthogonal to $H_r(M)$, with

(167) $$\left| \begin{array}{l} (\delta\, d + d\, \delta)\, \omega_N = \omega, \quad \omega \text{ given in } L^2_r(M) \text{ orthogonal to } H_r(M) \\ n_\Gamma\, d\omega_N = 0, \quad t_\Gamma \delta\omega_N = 0, \end{array} \right.$$

find ω_D in $H^1_r(M)$, and orthogonal to $H^+_r(M)$ and $H^-_r(M)$, with

(168) $$\left| \begin{array}{l} (\delta\, d + d\, \delta)\, \omega_D = \omega, \quad \omega \text{ given in } L^2_r(M) \text{ orthogonal to } H^+_r(M) \oplus H^-_r(M), \\ t_\Gamma\, \omega_D = 0, \quad n_\Gamma \omega_D = 0. \end{array} \right.$$

Then using the variational framework $H^1_{ro}(M)$ and $H^1_r(M)$ in $L^2_r(M)$ with the bilinear form $a(\omega,\zeta)$, we can define the Dirichlet Laplace-Beltrami operator Δ^D and the Neumann Laplace-Beltrami operator Δ^N as (unbounded) selfadjoint negative operators in $L^2_r(M)$, by:

(169)D
$$\left| \begin{array}{l} D(\Delta^D) = \{ \omega \in H^1_{ro}(M) \text{, such that } \zeta \to a(\omega,\zeta) \text{ is continuous on } \\ \qquad\qquad H^1_{ro} \text{ equipped with the } L^2\text{-topology} \} \\ = \{ \omega \in H^1_{ro}(M)\ \Delta\omega \in L^2_r(M),\ a(\omega,\zeta) = (-\Delta\omega,\zeta),\quad \forall \zeta \in H^1_{ro} \} \end{array} \right.$$

(169)N
$$D(\Delta^N) = \{ \omega \in H^1_r(M),\ \Delta\omega \in L^2_r(M),\ a(\omega,\zeta) = (-\Delta\omega,\zeta),\quad \forall \zeta \in H^1_r \}.$$

We also have the characterizations:

(170)
$$\left| \begin{array}{l} D(\Delta^D) = \{ \omega \in H^1_{ro}(M),\ d\omega \in H^1_{r+1}(M),\quad \delta\omega \in H^1_{r-1} \} \\ D(\Delta^N) = \{ \omega \in H^1_r(M),\ d\omega \in H^1_{r+1}(M),\quad \delta\omega \in H^1_{r-1}(M),\ n_\Gamma d\omega = 0,\ t_\Gamma \delta\omega = 0 \}, \end{array} \right.$$

so that we can write with notations (111):

(171)
$$-\Delta^D = d_M \delta_m + \delta_M d_m,\quad -\Delta^N = d_m \delta_M + \delta_m d_M.$$

CONSEQUENCES OF THEOREM 8: *The Hodge decomposition of* $H_r(d,M)$, $H_r(\delta,M)$
From Theorem 8, we easily deduce the following Hodge decomposition of the spaces $H_r(d,M)$ and $H_r(\delta,M)$ into orthogonal subspaces:

(172)
$$\left| \begin{array}{l} H_r(d,M) = dH^1_{r-1}(M) \oplus (\operatorname{Im} \delta_0 \cap H^1_r(M)), \\ H_r(\delta,M) = \delta H^1_{r+1}(M) \oplus (\operatorname{Im} d_0 \cap H^1_r(M)). \end{array} \right.$$

An easy consequence of these decompositions is the following theorem due to Tartar and Murat:

Theorem 9. "*Compensated compactness*" *theorem. Let* (ω_n), $(\tilde{\omega}_n)$ *be two sequences with* $\omega_n \in H_r(d,M)$, $\tilde{\omega}_n \in H_r(\delta,M)$ *so that* (ω_n) (resp. $(\tilde{\omega}_n)$) *converges weakly to* ω^0 (resp. $\tilde{\omega}^0$) in $H_r(d,M)$ (resp. $H_r(\delta,M)$). *Then*:

(173)
$$\omega_n \wedge * \tilde{\omega}_n \to \omega^0 \wedge \tilde{\omega}^0 \quad \text{in } D'_n(M),$$

or equivalently:

(173)'
$$(\phi\omega_n, \tilde{\omega}_n) \to (\phi\omega^0, \tilde{\omega}^0),\quad \forall \phi \in D_0(M).$$

PROOF. From (172), there are sequences (α_n), (β_n), $(\tilde{\alpha}_n)$, $(\tilde{\beta}_n)$ such that:

(174)
$$\left| \begin{array}{l} \omega_n = d\alpha_n + \beta_n,\quad \alpha_n \in H^1_{r-1}(M),\quad \beta_n \in \operatorname{Im} \delta_0 \cap H^1_r(M), \\ \tilde{\omega}_n = \delta\tilde{\alpha}_n + \tilde{\beta}_n,\quad \tilde{\alpha}_n \in H^1_{r+1}(M),\quad \tilde{\beta}_n \in \operatorname{Im} d_0 \cap H^1_r(M), \end{array} \right.$$

are weakly convergent in $H^1(M)$, therefore strongly convergent in $L^2(M)$. Thus:

(175) $(\phi\omega_n, \widetilde{\omega}_n) = (\phi d\alpha_n + \phi\,\beta_n, \delta\,\widetilde{\alpha}_n + \widetilde{\beta}_n)$.

Using the formula:

(176) $(\phi d\alpha_n, \delta\,\widetilde{\alpha}_n) = (d\phi \wedge d\alpha_n, \widetilde{\alpha}_n)$

we immediately see that this sequence converges as well as each of the term of the development of $(\phi\omega_n, \widetilde{\omega}_n)$. ⊗

We now prove the equality between the spaces (see (121),(122)):

(177) $H_{r,t}(d,\delta,M) = V_r^+, \quad H_{r,n}(d,\delta,M) = V_r^-$.

We first prove that if $\omega \in H_{r,t}(d,\delta,M)$, then $\omega \in V_r^+$.
Let h^+ be its projection onto $\mathcal{H}_r^+(M)$, ω_+ the plus potential of $\omega - h^+$, thus:

(178) $\omega - h^+ = (d\,\delta + \delta\,d)\,\omega_+, \quad \omega_+ \in V_r^+ \quad (\text{with } d\omega_+ \in V_{r+1}^+, \ \delta\omega_+ \in V_{r-1}^+)$

then:

(179) $d\omega = d\,\delta\,(d\omega_+) \in L_{r+1}^2, \quad \delta\omega = \delta\,d\,(\delta\omega_+) \in L_{r-1}^2,$

$d\omega_+$ is the plus potential of $d\omega$ from Theorem 8, thus $\delta\,d\,\omega_+ \in V_r^+$ from Proposition 1. Besides, we have $\delta\omega_+ \in V_{r-1}^+ \cap \mathrm{Im}\ \delta_0$ from (138), thus $\delta\omega_+$ is orthogonal to $\mathcal{H}_r^+$ (see diagram), thus $\delta\omega_+ \in \overset{o}{V}{}_r^+$, then $\delta\omega_+$ is the plus potential of $\delta\omega$, and Proposition 1 implies that $d\delta\omega_+ \in V_r^+$; thus we have $\omega \in V_r^+$. ⊗

Remark 12. From the proof above, it also follows that if $\omega \in \overset{o}{V}{}_r^+$, then its plus potential ω^+ satisfies $\beta^+ = \delta\,\omega^+ = (\delta\omega)^+$, i.e., is the plus potential of $\delta\omega$. Also if $\omega \in \overset{o}{V}{}_r^-$ then its minus potential ω^- is such that $\alpha^- = d\omega^- = (d\omega)^-$, i.e., is the minus potential of $d\omega$. ⊗

REGULARITY RESULTS (FOR REGULAR MANIFOLDS WITH BOUNDARY):
The Hodge decomposition of (regular) Sobolev spaces. We give here, without proof, the following decomposition (see Marsden [1] p. 62, see also Dautray-Lions [1]), for all positive integer s, with usual notations (see (121) for s = 1):

(180)
$$\left| \begin{array}{l} H_r^s(M) = d\,H_{r-1,t}^{s+1}(M) \oplus \delta\,H_{r+1,n}^{s+1}(M) \oplus \mathcal{H}_r^s(M), \\ H_r^s(M) = d\,H_{r-1,t}^{s+1}(M) \oplus (\ker \delta_0 \cap H_r^s(M)) = (\ker d_0 \cap H_r^s(M)) \oplus \delta\,H_{r+1}^{s+1}(M), \end{array} \right.$$

with

$$\mathcal{H}_r^s(M) = \{\alpha \in H_r^s(M),\ d\alpha = 0,\ \delta\alpha = 0\},$$

(181) $H_{r,t}^s(M) = \{\omega \in H_r^s(M),\ n_\Gamma\omega = 0\}, \quad H_{r,n}^s(M) = \{\omega \in H_r^s(M),\ t_\Gamma\omega = 0\}$.

[This follows from regularity results on elliptic operators, see Paquet [1], Morrey [1]. For a regular manifold M (C^∞), the hypotheses:

$$\omega \in H^s_r(M),\ d\omega \in H^s_{r+1}(M),\ \delta\omega \in H^s_{r-1}(M),\ t_\Gamma\omega = 0 \text{ (or } n_\Gamma\omega = 0),$$

implies $\omega \in H^{s+1}_r(M)$. Or also if ω satisfies: $\Delta\omega = (d\,\delta + \delta\, d)\,\omega = f \in H^{s-1}_r(M)$, with $t_\Gamma\omega = 0$ (or $n_\Gamma\omega = 0$), then $\omega \in H^{s+1}_r(M)$.]

3.2. Variational frameworks for the operators dδ and δd

For many boundary problems, it is interesting to use other variational frameworks which do not a priori suppose H^1 regularity. They are linked to the $d\delta$ or the δd operator instead of the Laplace-Beltrami operator.

3.2.1. Inhomogeneous problems

a) We first consider the inhomogeneous problem:

Let $f \in L^2_r(M)$ be given, with $0 \le r < n$, find $u \in H_r(d, M)$ so that:

$$(182)\qquad \left|\begin{array}{l} \delta\, d\, u + u = f, \\ t_\Gamma u = 0. \end{array}\right.$$

We define in the space $V = H_{ro}(d, M)$, the bilinear form:

$$(183)\qquad a(u,v) = (du, dv) + (u,v), \qquad u,\ v \in V.$$

Then the problem (182) can be rewritten as: find u in V such that

$$(184)\qquad a(u,v) = (f,v)\,, \qquad \forall\, v \in V.$$

From Theorem 1, V is a closed subspace of $H_r(d,M)$, and the bilinear form $a(u,v)$ is V-coercive. Therefore, there is a unique solution u of (184) (then of (182)) in V.

Note that (V, $L^2{}_r(M)$, a(u,v)) is the variational framework relative to the selfadjoint operator:

$$(185)\qquad D^+_o + I, \text{ with } D^+_o = \delta\, d\,, \text{ and } D(D^+_o) = \{ u \in H_{ro}(d,M),\ \delta\, d\, u \in L^2_r(M) \}.$$

b) In a similar manner, we can solve the problem in $H_r(\delta, M)$, $0 < r \le n$:

$$(186)\qquad \left|\begin{array}{l} d\,\delta\, u + u = f\,,\ \ f \text{ given in } L^2_r(M) \\ n_\Gamma\, u = 0, \end{array}\right.$$

with the variational framework: $\tilde{V} = {}_* V = H_{ro}(\delta, M)$, and

$$(187)\qquad \tilde{a}\,(u,v) = (\delta u, \delta v) + (u,v), \quad u,\ v \in \tilde{V}\,;$$

then (186) is equivalent to:

(188) $\tilde{a}(u,v) = (f,v), \quad \forall v \in \tilde{V}$,

which has a unique solution in $\tilde{V}$. We define the associated selfadjoint operator

(189) $D_0^- + I$, with $D_0^- = d\,\delta$, and $D(D_0^-) = \{u \in H_{ro}(\delta,M),\ d\,\delta\,u \in L_r^2(M)\}$.

Note that it is possible to solve (182) with $f \in V' = (H_{ro}(d,M))'$ (resp. (186) with: $f \in \tilde{V}' = H_{ro}(\delta,M)'$), since $\mathcal{D}'_r(M)$ is dense in H_{ro} and $H_{ro}(\delta,M)$.

c) If we substitute $H_r(d,M)$ for the space V in (184) (resp. $H_r(\delta,M)$ for $\tilde{V}$ in (188)), we solve the following problem:

(190) $\left| \begin{array}{l} \delta\,d\,u + u = f \\ n_\Gamma d\,u = 0, \end{array} \right.$ resp. $\left| \begin{array}{l} d\,\delta\,u + u = f \\ t_\Gamma \delta\,u = 0. \end{array} \right.$

Indeed the variational framework ($H_r(d,M)$, $L_r^2(M)$, $a(u,v)$) (resp. $H_r(\delta,M)$, $L_r^2(M)$, $\tilde{a}(u,v)$) defines the positive selfadjoint operator $D^+ + I$ (resp. $D^- + I$), with:

(191) $D(D^+) = \{ u \in H_r(d,M),\ \delta\,d\,u \in L_r^2(M),\ a(u,v) = (\delta\,du + u, v),\ \ \forall v \in H_r(d,M) \}$,

that is, using Stokes formula:

(192) $\left| \begin{array}{l} D(D^+) = \{ u \in H_r(d,M),\ \delta\,d\,u \in L_r^2(M),\ n_\Gamma\,du = 0\} \\ D^+u = \delta\,d\,u, \quad \forall u \in D(D^+), \end{array} \right.$

and also for D^-

(193) $\left| \begin{array}{l} D(D^-) = \{ u \in H_r(\delta,M),\ d\,\delta\,u \in L_r^2(M),\ t_\Gamma\,du = 0\} \\ D^-u = d\,\delta\,u, \quad \forall u \in D(D^-). \end{array} \right.$

d) Then we consider the inhomogeneous problem:

(194) $\delta\,d\,u = f,\ \ t_\Gamma u = 0, \quad f$ given in $L_r^2(M)$,

or in variational form with $a_0(u,v) = (du,dv)$

(195) $a_0(u,v) = (f,v), \quad \forall v \in V = H_{ro}(d,M)$.

In order to have a solution to this problem, it is required that f be orthogonal to the space ker d_0, therefore that $f \in \operatorname{Im}\delta$, thus $\delta f = 0$ with f orthogonal to $\mathcal{H}_r^-(M)$.
Then using the L^2-decomposition (with (136)):

(196) $\quad H_{ro}(d,M) = \ker d_0 \oplus (\operatorname{Im}\delta \cap H_{ro}(d,M)) = \ker d_o \oplus \overset{o}{V}{}^{-}_{r\delta}$,

we have from the proof of Lemma 2

(197) $\quad a_0(u,u) = (du,du) \geq C^+ \|u\|^2_{H^1_r(M)} \geq C^+ (u,u)_{L^2}, \quad \forall u \in \overset{o}{V}{}^{-}_{r\delta}$.

This is a (generalized) Poincaré inequality. Thus $a_0(u,v)$ is $\overset{o}{V}{}^{-}_{r\delta}$-coercive and from Lax-Milgram lemma, the problem (195) has a unique solution in the quotient space $H_{ro}(d,M)/\ker d_0$, or also in space $\overset{o}{V}{}^{-}_{r\delta}$. We have thus solved the problem: find u in $H_r(d,M)$ satisfying: $\delta d\, u = f$, $\delta u = 0$, $t_\Gamma u = 0$, u orthogonal to $\mathcal{H}^-_r(M)$.

We can also solve (195) with $V = H_r(d,M)$; it corresponds to the problem:

(198) $\quad \delta\, d\, u = f, \; n_\Gamma\, du = 0, \quad f$ given in $L^2_r(M)$,

with f orthogonal to ker d, thus $f \in \operatorname{Im}\delta_0$, i.e., $\delta f = 0$, $n_\Gamma f = 0$, f orthogonal to $\mathcal{H}^+_r(M)$. Then using the L^2-orthogonal decomposition (with (136)):

(199) $\quad H_r(d,M) = \ker d \oplus (\operatorname{Im}\delta_0 \cap H_r(d,M)) = \ker d \oplus \overset{o}{V}{}^{+}_{r\delta}$,

and using inequality (197), we obtain a unique solution u of (198) in the quotient space $H_r(d,M)/\ker d$ identified with $\overset{o}{V}{}^{+}_{r\delta}$, thus corresponding to:

$$\delta\, d\, u = f, \;\; \delta\, u = 0, \;\; n_\Gamma du = 0, \;\; n_\Gamma u = 0, \text{ u orthogonal to } \mathcal{H}^+_r(M).$$

In a similar manner, we can solve the problem

(200) $\quad d\,\delta\, u = f, \;\; n_\Gamma u = 0, \;\; f$ given in $L^2_r(M)$,

with f orthogonal to $\ker \delta_o$, then (200) has a unique solution u in $H_{ro}(\delta,M)/\ker\delta_0$, thus in $\operatorname{Im} d \cap H_{ro}(\delta,M)$ (that is with $du = 0$, u orthogonal to $\mathcal{H}^+_r(M)$); or

(201) $\quad d\,\delta\, u = f, \;\; t_\Gamma\delta\, u = 0, \;\; f$ given in $L^2_r(M)$,

with f orthogonal to ker δ; then (201) has a unique solution u in $H_r(\delta,M)/\ker\delta$ identified with $\operatorname{Im} d \cap H_{ro}(d,M)$, that is with $d\,u = 0$, u orthogonal to $\mathcal{H}^-_r(M)$.

3.2.2. Homogeneous problems

We have the following results for boundary value problems (see Paquet [1]):

Proposition 4. *i) Let* $u_o \in H^{-1/2}_r(d,\Gamma)$, $r < n$; *the problem: find* u in $H_r(d,M)$ *with*

(202) $\quad \delta\, d\, u + u = 0, \quad t_\Gamma u = u_o$,

has a unique solution.

ii) Let $u_1 \in H_r^{-1/2}(\delta,\Gamma)$, $0 < r < n$; *the problem: find* u in $H_{r+1}(\delta, M)$ *so that*

(203) $$d\,\delta\,u + u = 0, \quad n_\Gamma u = u_1,$$

has a unique solution.

PROOF. Using a lifting of the boundary conditions (thanks to Theorems 1 and 2), we are reduced to resolve a problem of type (182) or (186). We can also prove directly, following Paquet, that u is the projection of 0 onto the convex set:

(204) $$E_{u_0} = \{ v \in H_r(d,M),\ t_\Gamma v = u_0 \},\ \text{resp.}\ \widetilde{E}_{u_1} = \{ v \in H_r(\delta, M),\ n_\Gamma v = u_1 \}.$$

Remark 13. Note that the orthogonal space in $H_r(d,M)$ of $H_{ro}(d,M)$ is

(205) $$Z = \{ \alpha \in H_r(d,M),\ (d\alpha, d\beta) + (\alpha,\beta) = 0,\ \ \forall \beta \in H_{ro}(d,M) \},$$

therefore:

(205)' $$Z = \{ \alpha \in H_r(d,M),\ \delta\, d\alpha + \alpha = 0 \}.$$

In a similar manner the orthogonal space of $H_{ro}(\delta, M)$ in $H_r(\delta, M)$ is given by:

(206) $$\widetilde{Z} = \{ \alpha \in H_r(\delta,M),\ (\delta\alpha,\delta\beta) + (\alpha,\beta) = 0 \ \ \forall \beta \in H_{ro}(\delta, M) \} = \{ \alpha \in H_r(\delta, M),\ d\,\delta\,\alpha + \alpha = 0 \}.$$

Thus we see that the solution u of (202) (resp. (203)) belongs to Z (resp. to $\widetilde{Z}$). ⊗

Proposition 5. *i) Let* ϕ *be in* $H_r^{-1/2}(\delta,\Gamma)$, $0 \le r < n$; *the following problem*:

(207) $$\delta\, d\, u + u = 0 \ \text{in } M, \quad n_\Gamma du = \phi,$$

or in a variational form:

(207)' $$a(u,v) = (du, dv) + (u,v) = (-1)^r(\phi, t_\Gamma v), \quad \forall v \in H_r(d,M),$$

has a unique solution u in $H_r(d,M)$,

ii) Let $\varphi \in H_r^{-1/2}(d,\Gamma)$, $0 < r < n$; *the following problem*:

(208) $$d\,\delta\, u + u = 0 \ \text{in } M, \quad t_\Gamma \delta u = \varphi,$$

or in a variational form:

(208)' $$\tilde{a}(u,v) = (\delta u, \delta v) + (u,v) = (-1)^r(\varphi, n_\Gamma v), \quad \forall v \in H_{r+1}(\delta, M),$$

has a unique solution u in $H_{r+1}(\delta, M)$.

Let $v = du$ (resp. $w = \delta u$) in (207) (resp.(208)), applying d to (207) (resp. δ to (208)), we see that v is the solution of (203) for $u_1 = \phi$, and that w is the solution of (202) for $u_0 = \varphi$.

3.2.3. Capacity or Calderon operator relative to $(-\Delta + I)$ and to M
Using Propositions 4 and 5, we can give:

Definition 5. *The mapping*:

$$C_r^+ : u_0 \in H_r^{-1/2}(d,\Gamma) \to (-1)^r n_\Gamma du \in H_r^{-1/2}(\delta,\Gamma), \quad 0 \le r < n,$$

with u *the unique solution of* (202) *is called the capacity operator* (*or the Calderon operator*) *relative to* $(-\Delta + I)$ *in r-forms. We also define*

$$C_r^- : u_1 \in H_r^{-1/2}(\delta,\Gamma) \to (-1)^{r-1} t_\Gamma \delta u \in H_r^{-1/2}(d,\Gamma), \quad 0 < r < n,$$

with u *the unique solution of* (203).

Proposition 6. *The mapping* C_r^+ (*resp.* C_r^-) *is an isomorphism from* $H_r^{-1/2}(d,\Gamma)$ onto $H_r^{-1/2}(\delta,\Gamma)$ *(resp. from* $H_r^{-1/2}(\delta,\Gamma)$ onto $H_r^{-1/2}(d,\Gamma)$), *with*:

(209) $\quad C_r^- C_r^+ = I, \quad C_r^+ C_r^- = I, \quad 0 < r < n.$

Remark 14. *A regularity result on Propositions* 4 *and* 5. We assume that $u_0 \in H_r^{1/2}(\Gamma)$ (resp. u_1, ϕ, $\varphi \in H_r^{1/2}(\Gamma)$), then the solution u of (202) (resp. (203), (207), (208)) satisfies $u \in H_r^1(M)$.

PROOF (in the case of problem (202)). Let $U_0 \in H_r^1(M)$ with $t_\Gamma U_0 = u_0$. Then with $u = U_0 + U$, we have to find U solution of:

(202)' $\quad \delta d U + U = -(\delta d U_0 + U_0) \in (H_0^1(d,M))', \quad t_\Gamma U = 0.$

Using the variational form of problem (202)', we obtain a unique solution U of (202)' in $H_0^1(d,M)$, with: $\delta U = -\delta U_0 \in L_r^2(M)$, thus $U \in H_r^1(M)$, and $u = U + U_0 \in H_r^1(M)$. Furthermore, if we suppose that Γ is C^∞, with $u_0 \in C_r^\infty(\Gamma)$ (resp. u_1, ϕ, φ), then the solution u of (202) (resp. (203), (207), (208)) satisfies $u \in C_r^\infty(M)$. ⊗

Some consequences.
Then the capacity operator is also a continuous mapping from $H_r^{1/2}(\Gamma)$ onto $H_r^{1/2}(\Gamma)$. Using interpolation between $H_r^{-1/2}(d,\Gamma)$ and $H_r^{1/2}(\Gamma)$, it is thus also a continuous mapping from $X_0 = [H^{1/2}(\Gamma), H_r^{-1/2}(d,\Gamma)]_{1/2}$ onto $Y_0 = [H_r^{1/2}(\Gamma), H_r^{-1/2}(\delta,\Gamma)]_{1/2}$ (with u the solution of (202) in $[H_r^1(M), H_r(d,M)]_{1/2}$ for u_0 in X_0.

This also implies regularity results for problem (202)). This allows us to consider the capacity operator in the $L^2{}_r(\Gamma)$-framework if we have:

(210) $X_o = \{ u_o \in L^2_r(\Gamma),\ d u_o \in H^{-1/2}_{r+1}(\Gamma) \},\ Y_o = \{ u_1 \in L^2_r(\Gamma),\ \delta u_1 \in H^{-1/2}_{r+1}(\Gamma) \}.$

Finally if Γ is regular (C^∞), then C^+_r and C^-_r are isomorphisms in $C^\infty(\Gamma)$.

⊗

Remark 15. Applying the Hodge transformation to relation:

(211) $(-1)^r n_\Gamma d\, u = C^+_r u_o\,,\ \ u_o = t_\Gamma u$ (with $u \in H_r(d,M)$, $u_o \in H_r(d,\Gamma)$),

we obtain, for all v_1 in $H^{-1/2}_{n-r-1}(\delta,\Gamma)$, with $v = {*}\, u$, $v_1 = (-1)^{n-1} \underset{\Gamma}{*}\, u_0$, $0 \le r < n-1$,

(212) $$C^-_{n-r-1} v_1 = (-1)^{n-r-2} t_\Gamma \delta {*}\, u = (-1)^{n-1}\, t_\Gamma \underset{\Gamma}{*}\, du$$
$$= (-1)^{n-r-1} \underset{\Gamma}{*}\, n_\Gamma\, du = (-1)^{n-1} \underset{\Gamma}{*}\, C^+_r u_o,$$

thus

(213) $$C^-_{n-r-1} \underset{\Gamma}{*} = \underset{\Gamma}{*}\, C^+_r\,, \quad C^-_{n-r-1} = (-1)^m \underset{\Gamma}{*}\, C^+_r \underset{\Gamma}{*}\,.$$

The capacity operator has many other properties.

Let us equip the spaces $H^{-1/2}_r(d,\Gamma)$ and $H^{-1/2}_r(\delta,\Gamma)$ with the quotient norm:

(214) $$\begin{cases} \|u_o\|_{H^{-1/2}_r(d,\Gamma)} = \inf\limits_{v \in H_r(d,M),\ t_\Gamma v = u_o} ((dv,dv) + (v,v))^{1/2} \\ \|\phi\|_{H^{-1/2}_r(\delta,\Gamma)} = \inf\limits_{v \in H_{r+1}(\delta,M),\ n_\Gamma v = \phi} ((\delta v,\delta v) + (v,v))^{1/2}\,. \end{cases}$$

The infimum is at $v = u$, the solution of (202) for the first formula, and of (203) for the second; thus:

(215) $$\begin{cases} \|u_o\|_{H^{-1/2}_r(d,\Gamma)} = ((du,du) + (u,u))^{1/2} \\ \|\phi\|_{H^{-1/2}_r(\delta,\Gamma)} = \|n_\Gamma du\| = [(\delta\, d\, u, \delta\, d\, u) + (du,du)]^{1/2} = ((du,du) + (u,u))^{1/2}. \end{cases}$$

Therefore we have proved that the capacity operator C^+_r (or C^-_r) is unitary:

(216) $$\|C^+_r u_o\|_{H^{-1/2}_r(\delta,\Gamma)} = \|u_o\|_{H^{-1/2}_r(d,\Gamma)}\,.$$

We also have

Proposition 7. *The capacity operator is a positive symmetric operator in the following sense*

(217) $(u_0, C_r^+ u_0) = \|u_0\|_{H_r^{-1/2}(d,\Gamma)} > 0, \quad \forall u_0 \in H_r^{-1/2}(d,\Gamma),\ u_0 \neq 0,$

(218) $(u_0, C_r^+ \tilde{u}_0) = (C_r^+ u_0, \tilde{u}_0), \quad \forall u_0,\ \tilde{u}_0 \in H_r^{-1/2}(d,\Gamma).$

PROOF. Using Stokes formula (85) with $\alpha = u$ the solution of (202) and $\beta = d\tilde{u}$ with $\tilde{u}$ the solution of (202) for $\tilde{u}_0$, we have:

(219) $(du, d\tilde{u}) - (u, \delta d\tilde{u}) = (-1)^r \langle t_\Gamma u, n_\Gamma d\tilde{u}\rangle,$

thus

(220) $(du, d\tilde{u}) + (u, \tilde{u}) = \langle u_o, C_r^+ \tilde{u}_o\rangle = \langle C_r^+ u_o, \tilde{u}_o\rangle,$

from which we deduce Proposition 7. ⊗

The capacity operator is therefore also a positive symmetric operator in $L_r^2(\Gamma)$, with C_r^+ as an unbounded operator with

(221) $D(C_r^+) = \{u_0 \in L_r^2(\Gamma),\quad C_r^+ u_0 \in L_r^2(\Gamma)\}.$

Note that we can also define a capacity operator for $-\Delta$ (instead of $-\Delta + I$) as in definition 4, here using the unique solution in $H_r(d,\delta,M)$ of:

$$\delta d u = 0,\ \delta u = 0,\ u \text{ orthogonal to } \mathcal{H}_r^-(M),\ t_\Gamma u = u_0$$

$$(\text{resp. } d\delta u = 0,\ d u = 0,\ u \text{ orthogonal to } \mathcal{H}_r^+(M),\ n_\Gamma u = u_1),$$

but since $n_\Gamma(\delta du) = \delta(n_\Gamma du) = 0$, the capacity operator is not an isomorphism from $H^{-1/2}(d,\Gamma)$ onto $H^{-1/2}(\delta,\Gamma)$.

4. THE HODGE DECOMPOSITION FOR A COMPACT MANIFOLD WITHOUT BOUNDARY

For a compact manifold Γ without boundary, results are much simpler. Let n be the dimension of Γ. First using a variational point of view with the bilinear symmetric form $a(\omega,\omega)$ defined by (124) on $H^1{}_r(\Gamma)$, we have:

Theorem 10 (Morrey). *For* $r = 0$ *to* n, *and for every coordinate covering* U *of* Γ, *there are constants* $C_U > 0$ *and* λ_U *such that:*

(222) $a(\omega,\omega) + \lambda_U(\omega,\omega) \geq C_U((\omega,\omega))_U, \quad \forall \omega \in H_r^1(\Gamma),$

with $((\omega,\omega))_U$ *defined by* (100).

Thus the bilinear form $a(\omega, \tilde{\omega})$ is coercive on $H^1_r(\Gamma)$ with respect to $L^2_r(\Gamma)$.
Since the canonical map $H^1_r(\Gamma) \to L^2_r(\Gamma)$ is compact, we can apply the Peetre lemma with the maps:

$$\omega \in H^1_r(\Gamma) \xrightarrow{A_1} (d\omega, \delta\omega) \in L^2_{r+1}(\Gamma) \times L^2_{r-1}(\Gamma), \quad \omega \in H^1_r(\Gamma) \xrightarrow{A_2} \omega \in L^2_r(\Gamma).$$

Theorem 11. *For* $r = 0,\dots,n$ *the space:*

(223) $\quad \mathcal{H}_r(\Gamma) = \ker A_1 = \{\omega \in H^1_r(\Gamma),\ d\omega = 0,\ \delta\omega = 0\}$

is a finite dimensional vector space. The image space

(224) $\quad \operatorname{Im} A_1 = \{(d\omega, \delta\omega) \in L^2_{r+1}(\Gamma) \times L^2_{r-1}(\Gamma),\ \forall \omega \in H^1_r(\Gamma)\}$

is closed, and there is a positive constant C *so that:*

(225) $\quad a(\omega,\omega) \geq C((\omega,\omega))_U, \quad \forall \omega \in V_r,$

with:

(226) $\quad V_r = \{\omega \in H^1_r(\Gamma),\ \omega$ is L^2*-orthogonal to* $\mathcal{H}_r(\Gamma)\}$.

Thus $a(\omega,\omega)$ is V_r-coercive.
Using Definition (101), (102), we derive that we can identify the space $H^1_r(\Gamma)$ with

(227) $\quad H_r(d,\delta,\Gamma) = H_r(d,\Gamma) \cap H_r(\delta,\Gamma) = \{\omega \in L^2_r(\Gamma),\ d\,\omega \in L^2_{r+1}(\Gamma),\ \delta\omega \in L^2_{r-1}\}$.

Then we have:

Lemma 3. *The spaces* $\operatorname{Im} d = d\, H_r(d,\Gamma)$ *and* $\operatorname{Im} \delta = \delta\, H_r(\delta,\Gamma)$ *are closed in* $L^2_r(\Gamma)$.

PROOF. As for Lemma 2, we prove:

(228) $\quad \operatorname{Im} d = d\,((\ker d)^{\perp} \cap H_r(d,\Gamma))$.

From Stokes formula, we have (as in (114)):

(229) $\quad (\delta\alpha, \beta) = (\alpha, d\beta), \qquad \forall \alpha \in H_r(\delta, M),\ \forall \beta \in H_{r-1}(d, M),$

and $(\delta\alpha, d\beta) = 0$, therefore:

(230) $$\left| \begin{array}{l} (\operatorname{Im} \delta)^{\perp} = \ker d, \quad (\operatorname{Im} d)^{\perp} = \ker \delta, \\ \overline{\operatorname{Im} \delta} = (\ker d)^{\perp}, \quad \overline{\operatorname{Im} d} = (\ker \delta)^{\perp}. \end{array} \right.$$

Since we have:

(231) $\operatorname{Im} d \subseteq \ker d\,,\quad \operatorname{Im}\delta \subseteq \ker\delta,$

thus:

(232) $(\operatorname{Im} d)^{\perp} \supseteq (\ker d)^{\perp}, \quad (\operatorname{Im}\delta)^{\perp} \supseteq (\ker\delta)^{\perp},$

so that, with (228) and (230):

(233) $\operatorname{Im} d = d\,((\operatorname{Im} d)^{\perp} \cap H_r(d,\Gamma)) = d\,(\ker\delta \cap \dot{H}_r(d,\Gamma)).$

But $V_0 \overset{\text{def}}{=} \ker\delta \cap H_r(d,\Gamma)$ is a closed subspace of $H^1_r(\Gamma)$ which contains $H_r(\Gamma)$. Thus it is decomposed into:

(234) $V_0 = H_r(\Gamma) \oplus V_{00}\,,\quad V_{00} \overset{\text{def}}{=} (H_r(\Gamma))^{\perp} \cap V_0$ a closed subspace of V_r,

and we also have:

(235) $\operatorname{Im} d = dV_0 = dV_{00}\,.$

Now from inequality (225), we have:

(236) $\|d\omega\|^2 \geq C\,\|\omega\|^2_U, \quad \forall\,\omega \in V_{00}\,,$

which proves that $\operatorname{Im} d = dV_{00}$ is closed in $L^2_{r+1}(\Gamma)$, and that d is an isomorphism from V_{00} onto Im d. ⊗

We then derive the theorem:

Theorem 12. (*Hodge decomposition for a compact manifold without boundary*). *We have the* L^2*-decomposition of the kernel spaces:*

(237) $\ker d = \operatorname{Im} d \oplus H_r(\Gamma), \quad \ker\delta = \operatorname{Im}\delta \oplus H_r(\Gamma),$

with

(238) $H_r(\Gamma) = \ker d \cap \ker\delta$

and also the L^2*-orthogonal decomposition of space* $L^2_r(\Gamma)$:

(239) $L^2_r(\Gamma) = \operatorname{Im} d \oplus \ker\delta = \ker d \oplus \operatorname{Im}\delta = \operatorname{Im} d \oplus H_r(\Gamma) \oplus \operatorname{Im}\delta.$

We shall sum up these results in a diagram at the end of the chapter.

Definition 6. *The numbers* $b_r = b_r(\Gamma) = \dim H_r(\Gamma)$ *and* $\chi_\Gamma = \sum_{r=0}^{n} (-1)^r b_r(\Gamma)$ *are respectively called the* r*-Betti number and the Euler-Poincaré characteristic.*

A compact orientable manifold Γ *without boundary*, of dimension n, satisfies:

i) $\chi_\Gamma = 0$ if it is odd-dimensional, ii) $b_r(\Gamma) = b_{n-r}(\Gamma)$ if it is connected,

iii) $H_{n-1}(\Gamma) = \{0\}$, $H_1(\Gamma) = \{0\}$ if it is simply-connected.

If Γ is *contractible to a point*, then $H_r(\Gamma) = \{0\}$ for all r different from 0 and n. Note that $H_0(\Gamma)$ is the space of functions on Γ which are constant on each connected component of Γ, therefore: dim $H_0(\Gamma)$ is equal to the number of components of Γ.

The variational framework given by $H^1{}_r(\Gamma)$ and the bilinear form $a(\omega,\omega)$ allows us also to give the following proposition:

Proposition 8. *Let* ω *be a given r-form in* L^2 *orthogonal to* $H_r(\Gamma)$. *Then there exists a unique r-form* ω_0 *in* V_r (i.e. in $H^1_r(\Gamma)$, *orthogonal to* $H_r(\Gamma)$), *such that*

$$(240) \qquad (d\omega_0, d\zeta) + (\delta\omega_0, \delta\zeta) = (\omega, \zeta), \qquad \forall \zeta \in H^1_r(\Gamma).$$

We can write (at least formally, or in $(H^1_r(\Gamma))'$)

$$(241) \qquad \omega = (d\delta + \delta d)\omega_0 = \Delta\omega_0.$$

Definition 7. *The r-form* ω_0 *is called the potential of* ω.

Note that Proposition 8 is a simple consequence of Theorem 11 (inequality (225)) and of the Lax-Milgram lemma. There are also regularity results of ω_0 with respect to ω (see Morrey [1] p. 296).

Proposition 8 allows us to give a Hodge decomposition of $L^2{}_r(\Gamma)$ in a more precise form:

Theorem 13. *Let* $\omega \in L^2_r(\Gamma)$; *then there are unique forms* h, α, β *with* $h \in H_r(\Gamma)$, $\alpha \in H^1_{r+1}(\Gamma)$, $\beta \in H^1_{r-1}(\Gamma)$ *such that*

$$(242) \qquad \omega = h + \delta\alpha + d\beta\,,\ d\alpha = 0,\ \delta\beta = 0,\ \alpha = d\omega_0\,,\ \beta = \delta\omega_0\,,$$

where ω_0 *is the potential of* $\omega - h$; h, $\delta\alpha$ and $d\beta$ *are mutually orthogonal in* $L^2_r(\Gamma)$. *Thus*

$$(243) \qquad L^2_r(\Gamma) = dH^1_{r-1}(\Gamma) \oplus \delta H^1_{r+1}(\Gamma) \oplus H_r(\Gamma).$$

Furthermore if $\omega \in H^1_r(\Gamma)$, *then we also have* $\delta\alpha$ *and* $d\beta$ in $H^1_r(\Gamma)$.

For a regular (C^∞) compact manifold without boundary, we have a similar Hodge decomposition of spaces $H^s_r(\Gamma)$, $s \in \mathbb{N}$, given by (see Marsden [1] p.58)

$$(244) \qquad H^s_r(\Gamma) = dH^{s+1}_{r-1}(\Gamma) \oplus \delta H^{s+1}_{r+1}(\Gamma) \oplus H_r(\Gamma).$$

The set of harmonic forms is contained in the space $C^\infty_r(\Gamma)$, and

$$(245) \qquad H_r(\Gamma) = \{\, \alpha \in C^\infty_r(\Gamma) \ (\text{or } H^s_r(\Gamma)),\ \Delta\alpha = 0 \,\}.$$

[Note that here the relation $\Delta\alpha = 0$ implies $d\alpha = 0$ and $\delta\alpha = 0$.] Actually we can prove that $\omega \in H_r(\Gamma)$ is of class $C^{k-1,\mu}$ if Γ is of class $C^{k,\mu}$ (see Morrey [1] p. 296) then that $\omega \in H_r(\Gamma)$ is of class C^∞ (or even analytic) if Γ is so. Thus $H_r(\Gamma)$ is independent of s. Using duality, we prove that the decomposition (244) is true for *any real* s.

Remark 16. Let us apply the trace map t_Γ (or also n_Γ) to the "sequences" of $H_r(d,M)$ (and to the Hodge decomposition) for M a compact Riemannian manifold with boundary Γ. We thus obtain the diagrams:

$$\begin{array}{ccccccc}
H_0(d,M) & \xrightarrow{d} & H_1(d,M) & \xrightarrow{d} & H_2(d,M) & \xrightarrow{d} \\
\downarrow t_\Gamma & & \downarrow t_\Gamma & & \downarrow t_\Gamma & \\
H_0^{-1/2}(d_\Gamma,\Gamma) & \xrightarrow{d_\Gamma} & H_1^{-1/2}(d_\Gamma,\Gamma) & \xrightarrow{d_\Gamma} & H_2^{-1/2}(d_\Gamma,\Gamma) & \xrightarrow{d_\Gamma}
\end{array}$$

$$\begin{array}{ccc}
\mathrm{Im}\ d & \xrightarrow{d} & \ker\ d \\
\downarrow t_\Gamma & & \downarrow t_\Gamma \\
\mathrm{Im}_{-1/2} d_\Gamma & \xrightarrow{d_\Gamma} & \ker_{-1/2} d_\Gamma
\end{array}$$

Diagram 1

The image of the spaces $H_r(d,M)$ by t_Γ is in the framework of the Sobolev spaces $H_r^s(\Gamma)$, $s = -1/2$; if $\omega \in H_r(M)$, then $t_\Gamma\omega$ is not necessary in the space $H_r(\Gamma)$.
We have (see below):

$$\dim t_\Gamma H_r^+ \geq \dim H_r(\Gamma), \qquad \dim n_\Gamma H_{r+1}^- \geq \dim H_r(\Gamma).$$

For a comparison of the Euler-Poincaré characteristics of M and Γ, see Gilkey [1] p. 246.

5. THE HODGE DECOMPOSITION IN L^2 FOR UNBOUNDED MANIFOLDS

We are essentially interested in the Hodge decomposition for $L^2_r(M)$ when M is the complement of a bounded open set Ω in R^n. The main difficulty lies in the new spaces to introduce. First the study for M being the whole space R^n is interesting by itself and will allow us to treat the general case in view.

5.1. The Hodge decomposition for $L^2_r(R^n)$

We will extensively use the Fourier transform of r-forms in $L^2_r(R^n)$ or in $S'_r(R^n)$, that is, with coefficients (in a Cartesian system of coordinates) in $L^2(R^n)^r$ or in $S'(R^n)$, the temperate distributions.

We define $F\omega(\xi) = \hat{\omega}(\xi) = \sum_I \hat{\omega}_I(\xi)\, dx_I$ as the Fourier transform of ω.

Lemma 4. *The Fourier transforms of $d\omega$ and $\delta\omega$ are:*

(246) $$F(d\omega) = i\,\xi \wedge \hat{\omega}\,, \quad F(\delta\omega) = -\,i\,i_\xi\hat{\omega}\,,$$

i.e., the action of d (resp. δ) is (up to a coefficient i or − i) simply the exterior (resp. interior) product by ξ.

PROOF. The first equality is a direct consequence of the definition (formula (3)) of d, the second equality is obtained by duality: from (46)' and (21)' and the Parseval equality, we have:

(247) $$(F(dv), Fu) = (Fv, F\delta v) = (i\xi \wedge \hat{v}, \hat{u}) = (\hat{v}, -\,i\,i_\xi\hat{u}).$$ ⊗

The operators $\sigma(d) = i\xi \wedge .$ and $\sigma(\delta) = -\,i\,i_\xi$ are called the *symbols* of the operators d and δ (see Gilkey [1]). From the properties of the interior product, we have

(248) $$i_\xi(\xi \wedge \hat{v}) = (i_\xi\xi)\,\hat{v} - \xi \wedge i_\xi\hat{v} = \xi^2\hat{v} - \xi \wedge i_\xi\hat{v}\,,$$

thus, corresponding to (49):

(249) $$\xi \wedge i_\xi\hat{v} + i_\xi\,\xi \wedge \hat{v} = F\,((d\delta + \delta d)\,v) = \xi^2\hat{v} = F(-\,\Delta v).$$

⊗

Then we have to introduce some spaces (for $n > 2$):

Definition 8. The Beppo Levi spaces on R^n. *Let $W^1(R^n)$ be the closure of the space of smooth functions, with compact support or rapidly decreasing at infinity, for the norm:*

(250) $$\|u\|_{W^1(R^n)} = \left(\int_{R^n} \sum_i \left|\frac{\partial u}{\partial x_i}\right|^2 dx\right)^{1/2}.$$

Let $W^1_r(R^n)$ be the space of r-forms with coefficients in $W^1(R^n)$.

It is also the closure of the space of smooth r-forms (in $D_r(R^n)$ or in $S_r(R^n)$) for the norm:

(251) $$\|\omega\|_{W^1_r(R^n)} = \left(\|d\omega\|^2_{L^2_{r+1}} + \|\delta\omega\|^2_{L^2_{r-1}}\right)^{1/2}.$$

This is a direct consequence of (3), since from Parseval equality:

(252) $$(2\pi)^n\|\,\omega\|^2_{W^1_r(R^n)} = \|\xi \wedge \hat{\omega}\|^2 + \|i_\xi\hat{\omega}\|^2 = (i_\xi(\xi \wedge \hat{\omega}), \hat{\omega}) + (\xi \wedge i_\xi\hat{\omega}, \hat{\omega}) = (\xi^2\hat{\omega}, \hat{\omega}).$$

We can also characterize the Beppo Levi space $W^1(R^n)$ by (see Lions [2]):

(253) $$W^1(R^n) = \{u \in L^q(R^n),\ q = \frac{2n}{n-2},\ \frac{\partial u}{\partial x_i} \in L^2(R^n),\ i = 1,\dots,n\},\ n > 2,$$

or also using a weight function (see Nedelec[1]), for example in the n = 3 case:

$$W^1(R^3) = \{u,\ \frac{u}{(1+r^2)^{1/2}} \in L^2(R^3),\ r = |x|,\ \frac{\partial u}{\partial x_i} \in L^2(R^3),\ i = 1,\dots,n\}.$$

We then define spaces $W_r(d,R^n)$, $W_r(\delta,R^n)$ and $W_r(d,\delta,R^n)$ by:

(254) $$\begin{cases} W_r(d,R^n) = \{\omega \in L^q(R^n),\ q \text{ given by (253)},\ d\omega \in L^2(R^n)\} \\ W_r(\delta,R^n) = \{\omega \in L^q(R^n),\ q \text{ given by (253)},\ \delta\omega \in L^2(R^n)\}, \\ W_r(d,\delta,R^n) = W_r(d,R^n) \cap W_r(\delta,R^n). \end{cases}$$

Thus $\omega \in W_r(d,R^n)$ or $W_r(\delta,R^n)$ implies: $\omega \in L^2_{loc,r}(R^n) \cap \mathcal{S}'(R^n)$.

We could also use a weight function instead of q, in (254), as before for $W^1(R^3)$. Remark that $W^1(R^n)$ for example is strictly contained in the space (which contains constant functions):

$$\{u \in L^2_{loc}(R^n) \cap \mathcal{S}'(R^n),\quad \frac{\partial u}{\partial x_i} \in L^2(R^n),\quad i = 1,\dots,n\}.$$

From (246) and the definition 7, we see that d maps $W^1_r(R^n)$ into $L^2_{r+1}(R^n)$, δ maps $W^1_r(R^n)$ into $L^2_{r-1}(R^n)$, and by duality d maps $L^2_r(R^n)$ into $W^{-1}_{r+1}(R^n) = (W^1_{r+1}(R^n))'$ and δ maps $L^2_r(R^n)$ into $W^{-1}_{r-1}(R^n) = (W^1_{r-1}(R^n))'$.

Note that we can also define Beppo Levi space $W^2(R^n)$ to order 2 as the closure of smooth functions (with compact support) for the norm:

(255) $$\|u\|_{W^2} = \left(\sum_{i,j} \int_{R^n} \left|\frac{\partial^2 u}{\partial x_i \partial x_j}\right|^2 dx\right)^{1/2},$$

and then we define spaces $W^2_r(R^n)$, and by duality $W^{-2}_r(R^n)$.

Let P and Q be the following operators in $L^2_r(R^n)$:

(256) $$\mathcal{F}Pu(\xi) = \frac{1}{\xi^2} i_\xi(\xi \wedge \hat{u}), \quad \mathcal{F}Qu(\xi) = \frac{1}{\xi^2} \xi \wedge (i_\xi \hat{u}).$$

To be more precise (and not to use space $W^2_r(R^n)$ nor its dual) we have to write:

(256)' $$\mathcal{F}Pu(\xi) = i_\xi \frac{1}{\xi^2}(\xi \wedge \hat{u}), \quad \mathcal{F}Qu(\xi) = \xi \wedge \frac{1}{\xi^2}(i_\xi \hat{u}),$$

since $-\Delta$ is an isomorphism from W^1_r onto W^{-1}_r. Thus we have

Theorem 14. *(The Hodge decomposition in* $L^2(R^n)$*). The operators* P *and* Q *are hermitian orthogonal projectors with* P + Q = I, *and correspond to the orthogonal Hodge decomposition:*

$$L_r^2(R^n) = \ker d \oplus \ker \delta = \operatorname{Im} d \oplus \operatorname{Im} \delta = dW_{r-1}^1 \oplus \delta W_{r+1}^1. \tag{257}$$

PROOF. From the definition and the properties of d and δ, operators P and Q are continuous in $L^2(R^n)$. Using (256), we have: $P^2 = P$, $Q^2 = Q$, and (from (249)) $P + Q = I$. Therefore P and Q are supplementary projectors, and they are also hermitian since:

$$(FPu, Fv) = (i_\xi \frac{1}{\xi^2} \xi \wedge \hat{u}, \hat{v}) = (\frac{1}{\xi^2} \xi \wedge \hat{u}, \xi \wedge \hat{v}) = (\hat{u}, i_\xi \frac{1}{\xi^2} \xi \wedge \hat{v}) = (Fu, FPv).$$

Then, using (256), we have:

$$\ker P = \operatorname{Im} Q = dW_{r-1}^1(R^n), \ \ker Q = \operatorname{Im} P = \delta W_{r+1}^1(R^n). \tag{258}$$

⊗

5.2. The Hodge decomposition of $L^2{}_r(M)$, M the complement of a bounded open set in R^n

We also define Beppo Levi spaces in this case.

Definition 9. *(Beppo Levi spaces for unbounded domains). Let* Ω *be an open regular (Lipschitz) bounded set in* R^n, *and let* M *be the complement of* Ω *in* R^n. *Then* $W^1(M)$ *(resp.* $W^1{}_0(M)$*) is the closure of the space of smooth functions,* i.e.,

$D(\bar{\Omega}')$ *(resp.* $D(\Omega')$*) with* $\Omega' = R^n \setminus \bar{\Omega}$, *relatively to the norm*

$$\|u\|_{W^1} = \left(\int_{\Omega'} \sum_i \left| \frac{\partial u}{\partial x_i} \right|^2 dx \right)^{1/2}. \tag{259}$$

Let $W_r^1(M)$(resp. $W_{ro}^1(M)$) *be the space of r-forms on* $M = \bar{\Omega}'$, *with coefficients (for a Cartesian system of coordinates) in* $W^1(M)$ (resp. $W^1{}_0(M)$).

Note that $W^1(M)$ is contained in the space (which contains constant functions)

$$\{u \in L^2_{loc}(M), \ \frac{\partial u}{\partial x_i} \in L^2(M), \ \forall i\}, \text{ with } M = \bar{\Omega}',$$

We can also define spaces $W^1(M)$ and $W_r^1(M)$ by restriction to M of the elements of $W^1(R^n)$ and $W_r^1(R^n)$. Then we define the spaces $W_r(d, M)$ and $W_r(\delta, M)$ as those of the restrictions to M of elements of $W^1(R^n)$ and $W_r^1(R^n)$, and then $W_{r0}(d, M)$ and $W_{r0}(\delta, M)$ as the subspaces of $W^1(R^n)$ and $W_r^1(R^n)$ with vanishing tangential (resp. normal) trace.

Thus spaces $W^1(M)$, $W^1{}_r(M)$, $W_r(d,M)$, $W_r(\delta,M)$ have the same trace properties and the same trace spaces for $M = \Omega$ bounded open set or its complement.
From these properties, we fairly easily obtain (using extensions of r-forms to R^n) that spaces $dW_{r-1}(d,M)$, $\delta W_{r+1}(\delta,M)$, $dW_{r-1,o}(d,M)$, $\delta W_{r+1,o}(\delta,M)$ are closed in $L^2{}_r(M)$.
Using all these spaces we can prove that the Hodge decomposition of $L^2{}_r(M)$ remains valid as in the case of a compact bounded manifold, see theorem 6, formula (141), (143), (144), with finite dimensional spaces of cohomology defined as above by:

$$\left| \begin{array}{l} \mathcal{H}_r^+(M) = \{\omega \in L_r^2(M),\ d\omega = 0,\ \delta\omega = 0,\ n_\Gamma\omega = 0\} \\ \mathcal{H}_r^-(M) = \{\omega \in L_r^2(M),\ d\omega = 0,\ \delta\omega = 0,\ t_\Gamma\omega = 0\} \end{array} \right.$$

giving then: $\ker d = \operatorname{Im} d \oplus \mathcal{H}_r^+$, $\ker \delta = \operatorname{Im} \delta \oplus \mathcal{H}_r^-$.

Remark 17. As in the bounded case (see (177)), we have:

(260) $\quad W_{ro}(d,M) \cap W_r(\delta,M) = W_{rn}^1(M),\ \ W_{ro}(\delta,M) \cap W_r(d,M) = W_{rt}^1(M).$

Indeed using a smooth function ϕ with compact support in R^n, $\phi = 1$ in a neighbourhood of the boundary, we easily prove that for any u in the first left hand side of (260), ϕu is in the Sobolev space $H_r^1(M)$ thus is also in $W_{rn}^1(M)$, then we have $(1 - \phi)u$ in $W^1(R^n)$ thus is also in $W_{rn}^1(M)$.
Note that as in the bounded case:

(261) $\quad W_r(d,M) \cap W_r(\delta,M) \supset W_r^1(M)$ strictly.

We can also prove the trace property:

$$t_\Gamma W_r(d,M) = t_\Gamma W_r(d,\delta,M) = t_\Gamma(W_r(d,\delta,M) \cap \ker \delta), \quad \text{similarly for } \delta,$$

but note that the mappings $u \to (t_\Gamma u, n_\Gamma u)$ from $W_r(d,\delta,M)$ and $H_r(d,\delta,M)$ into $H_r^{-1/2}(d,\Gamma) \times H_{r-1}^{-1/2}(d,\Gamma)$ cannot be both surjective: if not, every element u in space $W_r(d,\delta,M)$ or in $H_r(d,\delta,M)$ will have an extension U to R^n in $W_r(d,\delta,R^n) = W^1(R^n)$, thus giving $U|_M = u \in W_r^1(R^n)$, in contradiction to (261). ⊗

Finally remark that the cohomology spaces $\mathcal{H}_r^-$, $\mathcal{H}_r^+$ are contained respectively in $W_{rt}^1(M)$ and in $W_{rn}^1(M)$, since they are also in the Sobolev space $H_r^1(M)$, and thus we have an orthogonal decomposition similar to (126):

(262) $\quad W_{rt}^1(M) = \mathcal{H}_r^-(M) \oplus \overset{o}{W}{}_r^-, \quad W_{rn}^1(M) = \mathcal{H}_r^+(M) \oplus \overset{o}{W}{}_r^+.$

Let $a(\omega,\omega')$ be as in the bounded case, the bilinear form:

(263) $a(\omega,\omega') = (d\omega, d\omega') + (\delta\omega, \delta\omega')$.

Then the main tool for proving results similar to the bounded case is the following theorem, analogous to Theorem 5 for Beppo Levi spaces.

Theorem 15. *The bilinear form* $a(\omega,\omega')$ *is coercive on space* $\overset{\circ}{W}{}^{-}_{r}$ *(resp.* $\overset{\circ}{W}{}^{+}_{r}$*): there exists a positive constant* C *such that*:

(264) $a(\omega,\omega') \geq C\,((\omega,\omega'))_U, \quad \forall \omega \in \overset{\circ}{W}{}^{+}_{r} \text{ (resp. } \omega \in \overset{\circ}{W}{}^{-}_{r}),$

with the notation (used for (100)):

(265) $((\omega,\omega))_U = \sum_{q=0,\ldots,Q} \int_{G_q} \sum_{I} \sum_{j=1,\ldots,n} \left| \frac{\partial \omega_I^{(q)}}{\partial x_j} \right|^2 dx$

where the chart $q = 0$ *is unbounded, the others being bounded.*

PROOF. Using a partition of unity (ϕ_q), $q = 0,\ldots,Q$, we decompose all elements ω in $W^1_r(M)$ into $\omega = \sum_{q=0,\ldots,Q} \omega_q$, $\omega_q = \phi_q \omega$, with support in the chart U_q.

Then we have two cases: first on each bounded chart, we apply the Gaffney inequality (123) of Theorem 4, then on the unbounded chart U_0 we have:

$$a(\omega_0,\omega_0) = ((\omega_0,\omega_0))_{W^1(R^n)}.$$

Then summing (and applying Peetre Lemma to eliminate the L^2 norm relative to terms in the bounded charts) we obtain the theorem.

⊗

5.3. Comparison between the cohomology of M = Ω, its complement and its boundary

Let M be (the closure of) a bounded open (regular) set of R^n, or its complement. At first we need the following lemma:

Lemma 5. *The Hodge decompositions of the trace spaces* $H_r^{-1/2}(d,\Gamma)$ of $H_r(d,M)$, $H_r^{-1/2}(\delta,\Gamma)$ of $H_r(d,M)$ *are:*

(266)
$$\left| \begin{array}{l} H_r^{-1/2}(d,\Gamma) = dH^{1/2}_{r-1}(\Gamma) \oplus \delta H^{3/2}_{r+1}(\Gamma) \oplus \mathcal{H}_r(\Gamma), \\ H_r^{-1/2}(\delta,\Gamma) = \delta H^{1/2}_{r+1}(\Gamma) \oplus dH^{3/2}_{r-1}(\Gamma) \oplus \mathcal{H}_r(\Gamma). \end{array} \right.$$

PROOF. Let ω be in $H_r^{-1/2}(d,\Gamma)$, then using the potential decomposition, there exists ω_0 and a harmonic r-form h such that $\omega = (d\delta + \delta d)\omega_0 + h = -\Delta\omega + h$. Then we have $d\omega = -\Delta\, d\omega_0 \in H^{-1/2}_{r+1}(\Gamma)$, therefore $d\omega_o \in H^{3/2}_{r+1}(\Gamma)$, $\delta\omega = -\Delta\,\delta\omega_o \in H^{-3/2}_{r-1}(\Gamma)$, and thus $\delta\omega_o \in H^{1/2}_{r-1}(\Gamma)$, from which we obtain the lemma.

⊗

Then we have (with Definition 2):

Proposition 9. *Let* $\omega \in L^2_r(M)$ *be a closed form, with* $t_\Gamma\omega$ *an exact form. Then* ω *is an exact form.*

PROOF. For M bounded or its complement, let $\omega \in L^2_r(M)$, $d\omega = 0$, $t_\Gamma\omega = d_\Gamma\theta$, $\theta \in H^{1/2}_{r+1}(\Gamma)$ from Lemma 5. Then if M' is the complement of M, there exists $\widetilde{\theta} \in H^1_r(M')$, with $t_\Gamma \widetilde{\theta} = \theta$. Let $\widetilde{\omega} = \omega$ on M, $d\widetilde{\theta}$ on M'. Then we have $d\widetilde{\omega} = 0$, with $\widetilde{\omega} \in L^2_r(R^n)$, then $\widetilde{\omega}$ is exact ($\widetilde{\omega} = d\phi$ with ϕ in W^1_{r+1}) and thus ω the restriction of $\widetilde{\omega}$ to M is exact. ⊗

Remark 18. Let ω_1 and ω_2 be two closed r-forms on M and M' (respectively) in L^2, with $t_\Gamma\omega_1 = t_\Gamma\omega_2$. Then $\widetilde{\omega} = \omega_1$ on M, ω_2 on M' satisfies $d\widetilde{\omega} = 0$, with $\widetilde{\omega}$ in $L^2_r(R^n)$ and thus $\widetilde{\omega}$ is exact, therefore ω_1 and ω_2 are exact.

⊗

Some consequences. With notations d_i and d_e for d in M (bounded) and M' respectively, we have:

(267) $$t_\Gamma \ker d_i \cap t_\Gamma \ker d_e = \operatorname{Im} d_\Gamma,$$

and there are finite dimensional spaces $\mathcal{H}^i_r(\Gamma)$ and $\mathcal{H}^e_r(\Gamma)$ so that:

(268) $$\left| \begin{array}{ll} t_\Gamma \ker d_i = \operatorname{Im} d_\Gamma \oplus \mathcal{H}^i_r(\Gamma), & \mathcal{H}^i_r(\Gamma) \subseteq \mathcal{H}_r(\Gamma), \\ t_\Gamma \ker d_e = \operatorname{Im} d_\Gamma \oplus \mathcal{H}^e_r(\Gamma), & \mathcal{H}^e_r(\Gamma) \subseteq \mathcal{H}_r(\Gamma). \end{array} \right.$$

Thus if P_Γ denotes the orthogonal projection (in $H^{-1/2}_r(d,\Gamma)$ on $\mathcal{H}_r(\Gamma)$), we have:

(269) $$P_\Gamma t_\Gamma \mathcal{H}^+_r(M) = \mathcal{H}^i_r(\Gamma), \qquad P_\Gamma t_\Gamma \mathcal{H}^+_r(M') = \mathcal{H}^e_r(\Gamma),$$

and $P_\Gamma t_\Gamma$ is an isomorphism from $\mathcal{H}^+_r(M)$ and $\mathcal{H}^+_r(M')$ onto $\mathcal{H}^i_r(\Gamma)$ and $\mathcal{H}^e_r(\Gamma)$.

Proposition 10. *The cohomology space of the boundary* Γ *of* M *and* M' *is decomposed into:*

(270) $$\mathcal{H}_r(\Gamma) = \mathcal{H}^i_r(\Gamma) \oplus \mathcal{H}^e_r(\Gamma),$$

and thus the Betti numbers:

$$b^+_r(M) \stackrel{\text{def}}{=} \dim \mathcal{H}^+_r(M), \quad b^+_r(M') \stackrel{\text{def}}{=} \dim \mathcal{H}^+_r(M'), \quad b_r(\Gamma) \stackrel{\text{def}}{=} \dim \mathcal{H}_r(\Gamma),$$

and the Euler-Poincaré characteristics of M, M' *and* Γ *satisfy:*

(271) $$b_r(\Gamma) = b^+_r(M) + b^+_r(M'), \qquad \chi(\Gamma) = \chi(M) + \chi(M').$$

PROOF. We only have to prove that every σ in $H_r(\Gamma)$ is decomposed according to (270). Thus we have to solve with notation (94):

$$d\omega = -\delta_\Gamma \omega_n^1 \wedge \sigma, \quad \delta\omega = 0 \ \text{ in } R^n, \quad \text{with } \omega \text{ in } H_r^1(M) \text{ and } W_r^1(M'),$$

therefore ω must satisfy:

$$\Delta\omega = \delta(\delta_\Gamma \omega_n^1 \wedge \sigma) \ \text{ in } R^n, \tag{272}$$

and thus ω is obtained by convolution of the usual elementary solution of the Laplacian with the right member of (272). We can see (by Fourier transformation) that ω has the required properties, with a jump of its traces on each side of Γ given by σ. We use in fact the properties of the simple layer potential, see for example Dautray-Lions [1].

⊗

6. APPLICATION TO 3-DIMENSIONAL AND 2-DIMENSIONAL CASES

6.1. The 3-dimensional case

Let M be a (regular at least for its interior) oriented Riemannian dimensional manifold: for the applications in view, M is the closure of an open set in R^3. There are special features of differential geometry in 3 dimensions due to the identification by the Hodge transformation of even 2-covectors (or 2-forms) with odd covectors (or 1-forms), and also (using the Riemannian G-transform) of even 2-vectors with odd vectors (often called polar vectors in physics). Remark that in 3 dimensions we always have $**=1$. Let us define the vector product of two vectors X_1 and X_2 on M, (here denoted by $X_1 \underline{\wedge} X_2$, to differentiate with the exterior product), using the exterior product and Hodge transformation by:

$$X_1 \underline{\wedge} X_2 = G^{-1} * (\omega_{X_1} \wedge \omega_{X_2}) = G^{-1} * (GX_1 \wedge GX_2) = G^{-1} * G(X_1 \wedge X_2). \tag{273}$$

We also have, with usual notations (see (24)):

$$X_1 \underline{\wedge} X_2 = X_1 \wedge X_2 \lrcorner_g G^{-1} v_g . \tag{274}$$

Indeed we have for all vectors X:

$$\left| \begin{aligned} &(X, X_1 \underline{\wedge} X_2)_g = (X, G^{-1}(X_1 \wedge X_2 \lrcorner v_g))_g = \langle X, X_1 \wedge X_2 \lrcorner v_g \rangle \\ &= \langle X_1 \wedge X_2 \wedge X, v_g \rangle = (X_1 \wedge X_2 \wedge X, G^{-1} v_g)_g = (X, X_1 \wedge X_2 \lrcorner_g G^{-1} v_g)_g , \end{aligned} \right. \tag{275}$$

which proves (274). In a coordinate system (x^1, x^2, x^3), we notably have for $X_1 = \partial_1$, $X_2 = \partial_2$, $X_3 = \partial_3$ (since $\langle \partial_1 \wedge \partial_2 \wedge \partial_3, v_g \rangle = g^{1/2}$ from (275)):

$$\partial_1 \underline{\wedge} \partial_2 = g^{1/2} G^{-1} dx^3 = \sum_k g^{1/2} g^{3k} \partial_k . \tag{276}$$

Differential operators grad, curl, div (Representatives of d and δ.)
We successively define for all (smooth) functions f and vector fields A, B:

(277) $\quad df = \omega_{\mathrm{grad}\, f}$, or $\mathrm{grad}\, f = G^{-1} df$,

(278) $\quad d\omega_A = {}_*\, \omega_{\mathrm{curl}\, A}$, or $\mathrm{curl}\, A = G^{-1}{}_*\, d\omega_A$,

(279) $\quad d\, {}_*\, \omega_B = (\mathrm{div}\, B)\, v^g$, or $\mathrm{div}\, B = {}_*\, d\, {}_*\, \omega_B = -\, \delta\omega_B$.

Definitions of grad and div are that of the general case (see (47),(48)). For the curl operator we can also write (with (45)):

(280) $\quad {}_*\, d\omega_A = \delta\, {}_*\, \omega_A = \omega_{\mathrm{curl}\, A}$, or $\mathrm{curl}\, A = G^{-1} \delta\, {}_*\, \omega_A$.

$$
\begin{array}{ccccccc}
H^1(\Omega) & \xrightarrow{\mathrm{grad}} & H(\mathrm{curl},\Omega) & \xrightarrow{\mathrm{curl}} & H(\mathrm{div},\Omega) & \xrightarrow{\mathrm{div}} & L^2(\Omega) \\
\downarrow & & \downarrow G & & \downarrow {}_* G & & \downarrow {}_* \\
H_0(d,\Omega) & \xrightarrow{d} & H_1(d,\Omega) & \xrightarrow{d} & H_2(d,\Omega) & \xrightarrow{d} & H_3(d,\Omega) = L^2(\Omega) \\
\downarrow * & & \downarrow * & & \downarrow * & & \downarrow * \\
H_3(\delta,\Omega) & \xrightarrow{(-\delta)} & H_2(\delta,\Omega) & \xrightarrow{\delta} & H_1(\delta,\Omega) & \xrightarrow{(-\delta)} & H_0(\delta,\Omega) = L^2(\Omega).
\end{array}
$$

Diagram 2

Usual relations are mere expressions of the fundamental relations: $d^2 = 0$, and $\delta^2 = 0$. The Laplace-Beltrami operator may be identified with the usual Laplacian (on functions, see (49)'):

$$\Delta f = -\, \delta df = \mathrm{div}\, \mathrm{grad}\, f, \quad \Delta\omega_A = \omega_{\mathrm{grad}\, \mathrm{div}\, A\, -\, \mathrm{curl}\, \mathrm{curl}\, A} = \omega_{\Delta A}\, ,$$

giving the "usual" relation on vector fields:

(281) $\quad \Delta\, A = \mathrm{grad}\, \mathrm{div}\, A - \mathrm{curl}\, \mathrm{curl}\, A$.

In a coordinate system, we have the expressions of grad f and div X given by (50), (53). Let us give that of curl A; from formula (51) for A:

(282) $\quad \omega_A = \sum_{ij} A^i g_{ij} dx^j$, for $A = \sum A^i \partial_i$, $\quad d\omega_A = \sum_{ijk} \dfrac{\partial (A^i g_{ij})}{\partial x^k}\, dx^k \wedge dx^j$,

we have:

(283) $\quad \mathrm{curl}\, A = \sum_{ijk} \dfrac{\partial (A^i g_{ij})}{\partial x^k}\, G^{-1}{}_*\, (dx^k \wedge dx^j) = \sum_{ijk} \dfrac{\partial (A^i g_{ij})}{\partial x^k}\, G^{-1}\, dx^k \underline{\wedge}\, G^{-1}\, dx^j$.

Applying (275) with $X_1 = G^{-1} dx^k$, $X_2 = G^{-1} dx^j$, $X = G^{-1} dx^i$, we get:

(284) $$(G^{-1}dx^i, G^{-1}dx^k \underline{\wedge} G^{-1}dx^j)_g = \langle G^{-1}(dx^i \wedge dx^k \wedge dx^j), v_g\rangle = g^{-1/2}\varepsilon^{ikj},$$

ε^{ikj} the symbol of Levi Civita, = 1 if ikj is an even permutation of 123, − 1 if it is odd, 0 otherwise. Thus:

(285) $$G^{-1}dx^k \underline{\wedge} G^{-1}dx^j = g^{-1/2}\varepsilon^{mkj}\partial_m$$

(286) $$\operatorname{curl} A = \sum_{ijkm} g^{-1/2}\varepsilon^{mkj}\frac{\partial(A^i g_{ij})}{\partial x^k}\partial_m .$$

For an orthogonal coordinate system, with Ω an open set of R^n, the Riemannian metric is:

(287) $$ds^2 = \sum_i g_i\, dx^i \otimes dx^i, \text{ (also denoted as } ds^2 = \sum_i g_i (dx^i)^2 \text{ with } g_{ij} = g_i\delta_{ij}).$$

With:

(288) $$\operatorname{curl} A = \sum_j (\operatorname{curl} A)^j \partial_j, \quad \operatorname{grad} f = \sum_j (\operatorname{grad} f)^j \partial_j,$$

we thus obtain, for A (or B) given by $A = \sum A^i\partial_i$:

(289) $$\begin{cases} (\operatorname{grad} f)^j = \dfrac{1}{g_j}\dfrac{\partial f}{\partial x^j}, \quad \operatorname{div} B = \dfrac{1}{\sqrt{g_1 g_2 g_3}}\sum_j \dfrac{\partial}{\partial x^j}(B^j\sqrt{g_1 g_2 g_3}), \\ (\operatorname{curl} A)^j = \sum_{\ell k}\dfrac{1}{\sqrt{g_1 g_2 g_3}}\varepsilon_{jkm}\dfrac{\partial(A^m g_m)}{\partial x^k}, \end{cases}$$

and thus:

(290) $$\Delta f = \operatorname{div}\operatorname{grad} f = \frac{1}{\sqrt{g_1 g_2 g_3}}\sum_j \frac{\partial}{\partial x^j}\left(\frac{\sqrt{g_1 g_2 g_3}}{g_j}\frac{\partial f}{\partial x^j}\right).$$

Some applications. In a Cartesian coordinate system, formulas (289) give the usual expressions for grad, curl, div. Then we give in two other coordinate systems, the Riemannian metric.

In a *cylindrical coordinate system*, we have:

(291) $$\begin{cases} x_1 = \rho\cos\phi,\ x_2 = \rho\sin\phi,\ x_3 = z; \\ ds^2 = d\rho^2 + \rho^2 d\phi^2 + dz^2, \text{ thus } g_\rho = 1,\ g_\phi = \rho^2,\ g_z = 1. \end{cases}$$

In a *spherical (Euler) coordinate system*, we have:

(292) $$\begin{cases} x_1 = r\sin\theta\cos\phi,\ x_2 = r\sin\theta\sin\phi,\ x_3 = r\cos\theta, \\ ds^2 = dr^2 + r^2\sin^2\theta\, d\phi^2 + r^2 d\theta^2, \text{ thus } g_r = 1,\ g_\phi = r^2\sin^2\theta,\ g_\theta = r^2, \end{cases}$$

and also $dx_1 \wedge dx_2 \wedge dx_3 = \rho d\rho \wedge d\phi \wedge dz = r^2\sin\theta\, dr \wedge d\theta \wedge d\phi$.

Due to (291), (292), we obtain the grad, div, rot operators in these coordinate systems by (289) and (290). Note that the orientation of (r, θ, ϕ) is the same as (x_1, x_2, x_3).
The cylindrical (and the spherical) coordinate system given by (291) (resp. (291)') does not define a chart on the whole space, but only on a part of it: the whole space without the half-plane $x_2 = 0$, $x_1 \geq 0$, onto the *open set*:

$$(0,+\infty) \times (0,2\pi) \times R \text{ for } (\rho,\phi,z), \quad \text{resp. } (0,+\infty) \times (0,2\pi) \times (0,\pi) \text{ for } (r,\phi,\theta).$$

Note that relations (289) *are often written in the "orthonormal" system:*

$$e_i = (g_i)^{-1/2}\, \partial_i \quad (\text{recall that } \|\partial_i\|_g^2 = g(\partial_i, \partial_i) = g_i).$$

The advantage of this system rests on the possible identification with the dual basis $\varepsilon_i = (g_i)^{1/2} dx^i$, and the very simple formulation of the Hodge transformation:

$$*\,\varepsilon_1 = \varepsilon_2 \wedge \varepsilon_3, \quad *\,\varepsilon_2 = \varepsilon_3 \wedge \varepsilon_1, \quad *\,\varepsilon_3 = \varepsilon_1 \wedge \varepsilon_2,$$

therefore for $A = \sum_i A^i e_i$, $\omega_A = \sum_i A^i \varepsilon_i$, we have (see Arnold [1]):

(293) $$\omega_A^2 = *\,\omega_A = A^1 \varepsilon_2 \wedge \varepsilon_3 + A^2 \varepsilon_3 \wedge \varepsilon_1 + A^3 \varepsilon_1 \wedge \varepsilon_2.$$

6.2. The 2-dimensional case

Let Γ be an oriented Riemannian 2-manifold. For the applications we have in mind, Γ will be the boundary of a ("regular") open set Ω in R^3. This situation seems fairly simple, but it is interesting to specify the differential geometric properties of Γ with respect to that of Ω. At first we prove that the Hodge transformation on Γ is directly related to the vector product with the normal to Γ, more precisely:

(294) $$\underset{\Gamma}{*}\,\omega_X = t_\Gamma G(n \underline{\wedge} X) = t_\Gamma\, \omega_{n \underline{\wedge} X}, \quad \text{thus} \quad G^{-1} \underset{\Gamma}{*} G(X) = n \underline{\wedge} X$$

for all vectors X tangent to Γ.

PROOF. Using the definition (273), $t_\Gamma G(n \underline{\wedge} X)$ is given by:

(295) $$t_\Gamma G(G^{-1} * (Gn \wedge GX)) = t_\Gamma * (Gn \wedge GX) = -\underset{\Gamma}{*}\, n_\Gamma (Gn \wedge GX) = -\underset{\Gamma}{*}\,(\omega_n^1 \wedge GX)$$

(see (60)). Since (see (58))

(296) $$n_\Gamma(\omega_n^1 \wedge GX) = -\,GX,$$

we obtain (294) from (295). Note that $\underset{\Gamma}{*}\underset{\Gamma}{*}\,\omega = -\,\omega$ for all 1-forms ω on Γ.

Differential operators on a surface representatives of d *and* δ

With usual notations (see (8)) adapted to the present case: $\omega_X = G_\Gamma X$, G_Γ being the canonical isomorphism due to the Riemannian metric g_Γ induced by g on Γ, we successively define for all (smooth) function f and vector fields A, B on Γ:

(297) $df = \omega_{\mathrm{grad}_\Gamma f}, \quad d\omega_A = \mathrm{curl}_\Gamma A\, v_\Gamma,$

or:

(297)' $\mathrm{grad}_\Gamma f = G_\Gamma^{-1}\, df, \quad \mathrm{curl}_\Gamma A = \underset{\Gamma}{*}\, d\,\omega_A,$

then

(298) $-\delta(f v_\Gamma) = \omega_{\overrightarrow{\mathrm{curl}}_\Gamma f}, \quad \delta\,\omega_B = -\,\mathrm{div}_\Gamma B,$

or

(298)' $\overrightarrow{\mathrm{curl}}_\Gamma f = -\,G^{-1}\,\delta(f v_\Gamma), \quad d * \omega_B = \mathrm{div}_\Gamma B\, v_\Gamma .$

These definitions are that of the general case (47), (48). Applying the Hodge transformation to (297), we obtain:

(299) $\underset{\Gamma}{*}\, df = -\,\delta \underset{\Gamma}{*} f = -\,\delta f v_\Gamma = \underset{\Gamma}{*}\,\omega_{\mathrm{grad}_\Gamma f} = \omega_{\overrightarrow{\mathrm{curl}}_\Gamma f}, \quad \underset{\Gamma}{*}\, d\omega_A = \mathrm{curl}_A = \delta \underset{\Gamma}{*}\,\omega_A,$

then using (294), we have:

(300) $\overrightarrow{\mathrm{curl}}_\Gamma f = \underline{n \wedge}\, \mathrm{grad}_\Gamma f, \quad \mathrm{curl}_\Gamma A = -\,\mathrm{div}_\Gamma (\underline{n \wedge} A).$

$$\begin{array}{ccccc}
H^1(\Gamma) & \xrightarrow{\mathrm{grad}_\Gamma} & H(\mathrm{curl}_\Gamma,\Gamma) & \xrightarrow{\mathrm{curl}_\Gamma} & L^2(\Gamma) \\
\downarrow & & \downarrow G & & \downarrow * \\
H_0(d,\Gamma) & \xrightarrow{d} & H_1(d,\Gamma) & \xrightarrow{d} & H_2(d,\Gamma) = L^2(\Gamma) \\
\downarrow * & & \downarrow * & & \downarrow * \\
H_2(\delta,\Gamma) & \xrightarrow{(-\delta)} & H_1(\delta,\Gamma) & \xrightarrow{\delta} & H_0(\delta,\Gamma) = L^2(\Gamma)
\end{array}$$

Diagram 3

$$\begin{array}{ccccc}
H^1(\Gamma) & \xrightarrow{\overrightarrow{\mathrm{curl}}_\Gamma} & H(\mathrm{div}_\Gamma,\Gamma) & \xrightarrow{\mathrm{div}_\Gamma} & L^2(\Gamma) \\
\downarrow & & \downarrow * G & & \downarrow * \\
H_0(d,\Gamma) & \xrightarrow{(-d)} & H_1(d,\Gamma) & \xrightarrow{d} & H_2(d,\Gamma) = L^2(\Gamma) \\
\downarrow * & & \downarrow * & & \downarrow * \\
H_2(\delta,\Gamma) & \xrightarrow{\delta} & H_1(\delta,\Gamma) & \xrightarrow{\delta} & H_0(\delta,\Gamma) = L^2(\Gamma)
\end{array}$$

Diagram 3'

We can easily verify that relations (297) may be obtained from (277), (278), (279), relative to Ω by taking the trace on Γ, that is by applying t_Γ to relations (277), (278), (279). Then we obtain (298) with the Hodge transformation. Thus Diagrams 3 and 3' are deduced from Diagram 2, but we have to change spaces $H(d,\Gamma)$ into $H^{-1/2}(d,\Gamma)$, and $H(\delta,\Gamma)$ into $H^{-1/2}(\delta,\Gamma)$ for all r. ⊗

The Laplace-Beltrami operator Δ_Γ on Γ is given by the formulas below (using (297) to (299)):

(301) $\Delta_\Gamma f = -\,(d\delta + \delta d)\, f = -\,\delta d\, f = \mathrm{div}_\Gamma\, \mathrm{grad}_\Gamma f, \ \Delta_\Gamma(f v_\Gamma) = -\,d\delta\,(f v_\Gamma) = (\mathrm{curl}_\Gamma\, \overrightarrow{\mathrm{curl}}_\Gamma f) v_\Gamma$

giving: $\Delta_\Gamma = \mathrm{div}_\Gamma\, \mathrm{grad}_\Gamma = \mathrm{curl}_\Gamma\, \overrightarrow{\mathrm{curl}}_\Gamma$. On 1-forms we have
$$\Delta_\Gamma \omega_A = -(d\delta + \delta d)\,\omega_A = d(\mathrm{div}_\Gamma A) - \delta(\mathrm{curl}_\Gamma A\, v_\Gamma) = \omega_{\Delta_\Gamma A},$$

with $\Delta_\Gamma A = \mathrm{grad}_\Gamma\, \mathrm{div}_\Gamma A + \overrightarrow{\mathrm{curl}}_\Gamma\, \mathrm{curl}_\Gamma A$.

⊗

Remark 19. Since Γ is a manifold without boundary, the operators d and δ are adjoint (this follows from Stokes formula).
Thus the operators grad_Γ and $-\mathrm{div}_\Gamma$ are adjoint, as well as $\overrightarrow{\mathrm{curl}}_\Gamma$ and $-\mathrm{curl}_\Gamma$.
This follows from relations:

(302) $\quad (df, \omega_A) = (f, \delta\omega_A), \quad (d\omega_A, f v_\Gamma) = (\omega_A, \delta f v_\Gamma)$,
giving:

(302)' $\quad (\mathrm{grad}_\Gamma f, A) = (f, -\mathrm{div}_\Gamma A), \quad (\mathrm{curl}_\Gamma A, f) = -(A, \overrightarrow{\mathrm{curl}}_\Gamma f)$.

⊗

In a coordinate system (x^1, x^2) we have the following expressions of $\mathrm{curl}_\Gamma A$ and $\overrightarrow{\mathrm{curl}}_\Gamma f$, for $A = \sum A^i \partial_i$:

(303) $$\mathrm{curl}_\Gamma A = \underset{\Gamma}{*}\, d\,\omega_A = g^{-1/2}\left(\frac{\partial(A^i g_{i2})}{\partial x_1} - \frac{\partial(A^i g_{i1})}{\partial x_2}\right),$$

using $\underset{\Gamma}{*}(dx^1 \wedge dx^2) = g^{-1/2}\, \underset{\Gamma}{*} v_\Gamma = g^{-1/2}$. We recall that:

(304) $$\mathrm{div}_\Gamma B = g^{-1/2} \sum_i \frac{\partial}{\partial x^i}(B^i g^{1/2}).$$

From (294), (also from (300) and (304)) we have, with $(\underline{n} \wedge A) = \sum_i (\underline{n} \wedge A)^i \partial_i$

(305) $$(\underline{n} \wedge A)^1 = -g^{-1/2} \sum_i A^i g_{i2}, \quad (\underline{n} \wedge A)^2 = g^{-1/2} \sum_i A^i g_{i1}.$$

Then using (300) we obtain (in the frame ∂_1, ∂_2):

(306) $$\overrightarrow{\mathrm{curl}}_\Gamma f = g^{-1/2}(-\partial_2 f, \partial_1 f).$$

In an orthogonal coordinate system where the Riemannian metric is

(307) $$ds^2 = g_1 (dx^1)^2 + g_2 (dx^2)^2,$$

and the Lebesgue measure on Γ is $d\Gamma = (g_1 g_2)^{1/2} dx^1 dx^2$, we obtain with $A = \sum A^i \partial_i$, $B = \sum B^i \partial_i$, and with $g_o = (g_1 g_2)^{-1/2}$:

(308) $$\begin{cases} \mathrm{grad}_\Gamma f = \sum_j \frac{1}{g_j} \frac{\partial f}{\partial x^j} \partial_j, & \mathrm{div}_\Gamma B = g_o \sum_j \frac{\partial}{\partial x^j}(B^j (g_1 g_2)^{1/2}) \\ \overrightarrow{\mathrm{curl}}_\Gamma f = g_o(-\partial_2 f\, \partial_1 + \partial_1 f\, \partial_2), & \mathrm{curl}_\Gamma A = g_o\left(\frac{\partial}{\partial x^1}(A^2 g_2) - \frac{\partial}{\partial x^2}(A^1 g_1)\right). \end{cases}$$

We also have for the Laplace-Beltrami operator:

(309) $$\Delta_\Gamma f = g_0 \sum_j \frac{\partial}{\partial x^j}\left(\frac{(g_1 g_2)^{1/2}}{g_j}\frac{\partial f}{\partial x^j}\right).$$

Application to Γ a cylindrical surface or a sphere. Then we have:

$g_\phi = \rho^2$, $g_z = 1$ for a cylinder, $g_\phi = r^2\sin^2\theta$, $g_\theta = r^2$ for a sphere.

Remark 20. Using both relations (289) and (308) in an admissible coordinate system in R^3 (with $g_n = g_3 = 1$), we obtain the expressions of the operators grad, curl, div, with respect to the tangential and normal differential operators

(310) $$\begin{cases} \text{i) } \operatorname{grad} f = \operatorname{grad}_\Gamma f + \frac{\partial f}{\partial n} n, \quad \operatorname{div} B = \operatorname{div}_\Gamma t_\Gamma B + \frac{\partial B^n}{\partial n} + 2R_m B^n, \\ \text{ii) } \operatorname{curl} A = \operatorname{curl}_\Gamma (t_\Gamma A)\, n + \frac{\partial}{\partial n}(n \underline{\wedge} A) - \overrightarrow{\operatorname{curl}}_\Gamma A^n + 2R_m\, n \underline{\wedge} A, \end{cases}$$

where n is identified with $\frac{\partial}{\partial n}$, where $t_\Gamma B$ is the projection of B on the tangent plane, $B^n = n.B$, and R_m is the mean curvature given with $g_0 = (g_1 g_2)^{-1/2}$ by:

$$2R_m = (g_1 g_2)^{-1/2}\frac{\partial}{\partial n}((g_1 g_2)^{1/2}) = g_0 \frac{\partial}{\partial n} g_0^{-1}.$$

We recall that A *is given by* $A = \sum A^i \partial_i$. Then the Laplace-Beltrami operator is:

(310)' $$\Delta f = \Delta_\Gamma f + \frac{\partial^2 f}{\partial n^2} + 2R_m \frac{\partial f}{\partial n}.$$

The proof relative to the div operator follows from formula:

(311) $$\operatorname{div} B = \operatorname{div}_\Gamma t_\Gamma B + (g_1 g_2)^{-1/2}\frac{\partial}{\partial n}(B^n (g_1 g_2)^{1/2}).$$

For the curl operator, we have:

(312) $$\begin{aligned} \operatorname{curl} A = g_0\left(\frac{\partial(A^2 g_2)}{\partial x^1} - \frac{\partial(A^1 g_1)}{\partial x^2}\right)\frac{\partial}{\partial n} &+ g_0\left(-\frac{\partial}{\partial n}(A^2 g_2)\partial_1 + \frac{\partial}{\partial n}(A^1 g_1)\partial_2\right) \\ &+ g_0\left(\frac{\partial A^n}{\partial x^2}\partial_1 - \frac{\partial A^n}{\partial x^1}\partial_2\right), \end{aligned}$$

and the following expression due to (305):

$$n \underline{\wedge} A = (g_1 g_2)^{-1/2}(-(A^2 g_2)\partial_1 + (A^1 g_1)\partial_2),$$

from which we easily deduce (310). Note that from (310):

(313) $$n.\operatorname{curl} A = \operatorname{curl}_\Gamma (t_\Gamma A).$$

We emphasize that *the decomposition of curl A given by* (310) *does depend on the basis, and is not "intrinsic"*: in other words the normal derivative in (310)ii) is not a "*true*" vector, i.e., is not the "covariant derivative". We will give the main formulas in cylindrical then spherical coordinates at the end of this Appendix.

⊗

7. MAXWELL EQUATIONS WITH DIFFERENTIAL FORMS

7.1 Maxwell equations with differential forms in R^3

We first claim that there is no unique definition for electromagnetic quantities. We can define them either as r-fields or as differential forms. We make the following choice.

The electric induction D, the electric field E and the current density J may be taken as time dependent even 1-fields on R^3 (or on open sets in R^3). We denote by ω^1_D, ω^1_E, ω^1_J the time dependent (even) 1-form corresponding to them, through the Riemannian metric g. The magnetic induction B and the magnetic field H may then be taken as time dependent odd 1-fields in R^3, and will be identified with time dependent (even) 2-forms ω^2_B, ω^2_H; thus in a Cartesian system, for all vector field A:

$$\omega^1_A = GA = \sum_i A^i dx^i; \quad \omega^2_A = {}_*\,\omega^1_A = A^1\, dx^2 \wedge dx^3 + A^2\, dx^3 \wedge dx^1 + A^3\, dx^1 \wedge dx^2.$$

Then the evolution Maxwell equations are with $\dot{\omega}_A = \dfrac{\partial \omega_A}{\partial t} = \omega_{\partial A/\partial t}$

$$\text{(314)} \qquad d\omega^1_E + \dot{\omega}^2_B = 0, \quad \delta\omega^2_H - \dot{\omega}^1_D = \omega^1_J, \quad (\text{or } d\omega^1_H - \dot{\omega}^2_D = \omega^2_J).$$

7.2. Maxwell equations with differential forms in R^4

We can also consider differential forms in R^4 (rather than in R^3 with time dependence) and it is better to do that for a deep understanding of the transformation laws of the electromagnetic field. But we have to take space R^4 equipped with the bilinear form ds^2 given in a Cartesian system by:

$$\text{(315)} \qquad g = ds^2 = \sum_{k=1,2,3} (dx^k)^2 - c^2 (dt)^2$$

(with c the velocity of light in free space). This metric is not positive, therefore not Riemannian; it is called a pseudo-Riemannian metric or more precisely a Lorentz metric. Space R^4 equipped with this metric is called Minkowski space. It allows us to define (as in the Riemannian case) an isomorphism, also denoted by G (or G_L), mapping r-fields onto r-forms, and a unique odd 4-form, denoted by μ_c or v^g, defined as in formula (14), so that:

$$\text{(316)} \qquad \mu_c = |\det g^{ij}|^{-1/2} \omega_1 \wedge \omega_2 \wedge \omega_3 \wedge \omega_4\,,$$

for (ω_j), j = 1, 2, 3, 4, a "basis" such that $(\omega_j, \omega_k) = g^*(\omega_j, \omega_k) = g^{jk}$.

For a Cartesian system of coordinates $\mu_c = v^g = c\, dx^1 \wedge dx^2 \wedge dx^3 \wedge dt$. Therefore μ_c satisfies with (316):

$$(317) \qquad (\mu_c, \mu_c)_g = \Sigma\, |\det g^{ij}|^{-1} \det g^*(\omega_j, \omega_k) = -1.$$

Then there also exists a Hodge transformation $*$ (which is an isomorphism from the r-forms space onto the $(4-r)$-forms space) so that

$$(318) \qquad \alpha \wedge *\beta \overset{\text{def}}{=} (\alpha, \beta)_{g_x} \mu_c, \qquad (\beta \wedge \alpha, \mu_c)_g = (\alpha, *\beta)_g,$$

thus with (21)'

$$*\beta = \beta \underset{g}{\lrcorner} \mu_c, \qquad *1 = \mu_c, \qquad *\mu_c = -1, \qquad **\alpha^r = (-1)^{r+1}\alpha^r.$$

Then the codifferential operator is defined by:

$$(319) \qquad \delta\omega^r = * d * \omega^r, \qquad \text{or} \quad *\delta\omega^r = (-1)^r d * \omega^r.$$

From 1-forms and 2-forms ω_E^1, ω_B^2, ω_D^1, ω_H^2 defined in R^3, we define the following 2-forms ω_F^2, $\tilde{\omega}_L^2$ in a Cartesian system of coordinates by:

$$(320) \qquad \omega_F^2 \overset{\text{def}}{=} \omega_E^1 \wedge dt + \omega_B^2, \qquad \tilde{\omega}_L^2 \overset{\text{def}}{=} \omega_H^1 \wedge dt - \omega_D^2.$$

In fact ω_F^2, $\tilde{\omega}_L^2$ *are fundamental entities, respectively even and odd forms,* and we have to define the electomagnetic field from them. From the current J and the charge ρ, we first define a field $j = (J, \rho)$ or a 1-form ω_j by:

$$(321) \qquad \omega_j = \omega_j^1 = Gj = G(\sum_k J^k \partial_k + \rho \frac{\partial}{\partial t}) = \sum_k J^k dx^k - c^2 \rho dt.$$

We can easily see that the 2-forms ω_F^2, $\tilde{\omega}_L^2$ correspond to the 2-fields (or to the skew symmetric tensors) F (even) and L (odd) defined by

$$(322) \quad F = G^{-1}\omega_F = \begin{pmatrix} (B) & -E/c^2 \\ E/c^2 & 0 \end{pmatrix}, \quad L = G^{-1}\tilde{\omega}_L^2 = \begin{pmatrix} (-D) & -H/c^2 \\ H/c^2 & 0 \end{pmatrix},$$

with

$$(B) = \begin{pmatrix} 0 & B^3 & -B^2 \\ -B^3 & 0 & B^1 \\ B^2 & -B^1 & 0 \end{pmatrix}, \qquad (D) = \begin{pmatrix} 0 & D^3 & -D^2 \\ -D^3 & 0 & D^1 \\ D^2 & -D^1 & 0 \end{pmatrix}.$$

Using the inner product (21), we obtain ω_E^1, ω_H^1 from ω_F^2 and $\tilde{\omega}_L^2$ (and then ω_B^2, ω_D^2) by

(323)
$$\left|\begin{array}{ll}\omega^1_E = i_{\partial/\partial t}\omega^2_F, & \omega^1_H = i_{\partial/\partial t}\tilde{\omega}^2_L,\\ \omega^2_B = \omega^2_F - \omega^1_E \wedge dt, & \omega^2_D = -\tilde{\omega}^2_L + \omega^1_H \wedge dt.\end{array}\right.$$

Then we can write the Maxwell equations in the differential form

(324)
$$d\omega^2_F = 0, \qquad d\tilde{\omega}^2_L = \frac{1}{c} i_j \mu_c = i_j \mu.$$

With the following (even) 2-form:

(325)
$$\omega^2_{\tilde{L}} = -c_* \tilde{\omega}^2_L = \omega^2_H + c^2 \omega^1_D \wedge dt, \quad \text{with } \tilde{L} = G^{-1}\omega^2_{\tilde{L}} = \begin{pmatrix} (H) & -D \\ D & 0 \end{pmatrix}$$

we can also write *Maxwell equations in the form*

(324)'
$$d\omega^2_F = 0, \qquad \delta\,\omega^2_{\tilde{L}} = \omega^1_j .$$

From these equations, using the usual properties, we immediately see that:
i) the current j must satisfy

$$\delta\omega^1_j = 0, \qquad \text{i.e.,} \quad \operatorname{div} J + \frac{\partial \rho}{\partial t} = 0;$$

ii) from Poincaré lemma there exists a vector field (called the vector potential) A so that $\omega^2_F = d\omega^1_A$ and ω^1_A is defined up to an exact form $\omega^1_{A'} = \omega^1_A + df$.

We say that the vector potential A is defined up to a *gauge transformation*. See also Abraham-Marsden-Ratiu [1] for some developments in the free space case.

7.3. Transformation laws. Lorentz and evolution transformations

Here we recall the action of mappings on differential forms and tangent fields, previously used in (63) for example. Let M and M' be two (continuously differentiable) manifolds and let u be a *diffeomorphism* from M to M' (i.e. u and its inverse are continuously differentiable). Let x' = u(x). The *pullback* of an r-form ω' on M' by u is an r-form on M, ω = u*(ω) defined by

(326)
$$(u^*\omega')_x(v_1,\dots,v_r) = \omega'_{x'}(Tu_x(v_1),\dots,Tu_x(v_r)),$$

where Tu_x is given in local charts by the matrix $(\frac{\partial x'^i}{\partial x^k})$ with x' = u(x).
For a vector field X' on N, we also define:

(327)
$$(u^*X')_x \stackrel{\text{def}}{=} (Tu_x)^{-1}X'_{u(x)}.$$

Pullback has the properties:

(328)
$$du^*\omega' = u^*d\omega', \quad u^*(\omega'_1 \wedge \omega'_2) = u^*\omega'_1 \wedge u^*\omega'_2, \quad u^*(i_{X'}\omega') = i_{u^*X'}u^*\omega'$$

with respect to exterior derivation, to exterior product and to inner product.

Application to the Electromagnetic Field and to the Maxwell equations

Let u be a diffeomorphism on R^4. The pullback of Maxwell equations (324), for an electromagnetic field given by F' and L', with a current j' is

$$(329) \qquad d\,u^*\omega^2_{F'} = 0, \qquad d\,u^*\,\tilde{\omega}^2_{L'} = \frac{1}{c} i_{u^*j'} u^*\mu_c \,.$$

7.3.1. Lorentz transformations

A Lorentz transformation u is a linear transformation preserving Lorentz metric, that is for all vector fields X, Y in R^4:

$$(330) \qquad g(u^*X, u^*Y) = g(X,Y), \quad \text{or also} \quad u^*g = g \text{ with } (326).$$

Then u commutes with the Hodge transformation and satisfies $u^*\mu_c = \mu_c$.
Thus we see that $u^*\omega^2_{F'}$ and $u^*\,\tilde{\omega}^2_{L'}$ satisfy Maxwell equations with the current u^*j'.

A *special Lorentz transformation* u with the velocity v in R^3 (the velocity of an observer S' with respect to the observer S, each observer having charts $u_S(N) = (x,t)$, $u_{S'}(N) = (x',t')$ of the physical system, which are exchanged by $u_{S'S} = u$) is defined by $(x',t') = u(x,t)$ with

$$(331) \qquad x'_v = \operatorname{ch}\theta\, x_v - \operatorname{sh}\theta\, ct = \beta(x_v - vt) \quad \text{and} \quad \pi^0_v x' = \pi^0_v x,$$
$$c\,t' = -\operatorname{sh}\theta\, x_v + \operatorname{ch}\theta\, ct = \beta\,(-(v/c)\,x_v + ct),$$

where $x_v = x.v/v$, $v = |v| < 1$, $\operatorname{th}\theta = v/c$, $\operatorname{ch}\theta = \beta = (1 - \frac{v^2}{c^2})^{-1/2}$, $\operatorname{sh}\theta = \beta v/c$, and π^0_v is the orthogonal projection on the plane orthogonal to v. Thus

$$(332) \quad x' = \frac{\beta}{v^2}(x.v - v^2 t)\,v + \pi^0_v x, \quad t' = \beta\,(t - x.v/c^2), \qquad \pi^0_v x = -\frac{1}{v^2}(x \,\underline{\wedge}\, v)\,\underline{\wedge}\, v.$$

Let E', B', H', D' (and thus F', L') be the electromagnetic field at (x', t'). Then at (x, t), E, B, H, D (and F, L) are obtained by the pullback of the 2-forms $\omega^2_{F'}$, $\tilde{\omega}^2_{L'}$ by u and from relations (323). Since $\omega^2_{B'} = \frac{1}{v}(dx'_v \wedge \omega^1_{B' \underline{\wedge} v}) + \frac{1}{v}\pi_v B'\,\omega^2_v$, we have:

$$\frac{1}{c}\omega^1_{E'} \wedge cdt' + \omega^2_{B'} = \frac{1}{c}\omega^1_{\pi^0_v E'} \wedge cdt' - \frac{1}{v}\omega^1_{B' \underline{\wedge} v} \wedge dx'_v + \frac{v}{c}\pi_v E'\, dx'_v \wedge cdt' + \frac{1}{v}\pi_v B'\,\omega^2_v$$

Then using (331), and since $dx'_v \wedge cdt'$ and ω^2_v are invariant, we obtain

$$(333) \qquad \pi^0_v E = \beta(\pi^0_v E' + B' \,\underline{\wedge}\, v), \qquad B \,\underline{\wedge}\, v = \beta\,(B' \,\underline{\wedge}\, v + \frac{v^2}{c^2}\pi^0_v E')$$

$$\text{and} \qquad \pi^0_v H = \beta(\pi^0_v H' - D' \,\underline{\wedge}\, v), \qquad D \,\underline{\wedge}\, v = \beta\,(D' \,\underline{\wedge}\, v - \frac{v^2}{c^2}\pi^0_v H').$$

Furthermore using (331), we have

$$(333)' \qquad J_v = \beta\,(J'_v + \rho' v), \qquad \rho c = \beta\,((v/c)\,J'_v + \rho' c),$$

and

(333)'' $\quad E'.v = E.v,\ \ B'.v = B.v,\ \ D'.v = D.v,\ \ H'.v = H.v,\ \ J'\underline{\wedge} v = J\underline{\wedge} v.$

Inverse formulas are obtained changing v into $-v$. *The transformation law of the electromagnetic field is given by*:

$$E' = \beta E + (1-\beta)(E.v)v/v^2 + \beta\, v\underline{\wedge} B, \qquad B' = \beta B + (1-\beta)(B.v)v/v^2 - \beta\, v\underline{\wedge} E/c^2$$

$$D' = \beta D + (1-\beta)(D.v)v/v^2 + \beta\, v\underline{\wedge} H/c^2, \qquad H' = \beta H + (1-\beta)(H.v)v/v^2 - \beta\, v\underline{\wedge} D$$

$$J' = J + (\beta - 1)(J.v)v/v^2 - \beta\rho v, \qquad \rho' = \beta(\rho - (J.v/c^2)).$$

We see that electric field and magnetic induction (resp. electric induction and magnetic field) are coupled through these transformations.

7.3.2. Transformation laws for solid motions and for pure Galilean motions

Here we consider transformations U in R^4 such that:

(334) $\quad U(x,t) = (x',t') = (u(x,t),t)$ with $u(.,t) = u_t$ an isometry in R^3;

u represents the evolution of a solid medium which is in the connected domain Ω at time 0 and in the domain $\Omega_t = u_t(\Omega)$ at time t. Since Ω is a connected set, u is naturally extended to the whole space R^3; u_t is given for all t by

(335) $\quad u_t x = x_t = R_t x + \tau_t$, with R_t a rotation and τ_t a translation in R^3.

The time derivative of (335) is:

(336) $\quad \dfrac{dx_t}{dt}\Big|_{x_t} = \xi_t\underline{\wedge} R_t x + v_t^{\tau} = \xi_t\underline{\wedge}(x_t - \tau_t) + v_t^{\tau} = \xi_t\underline{\wedge} x_t + v_t^{e},$

with $v_t^e = v_t^\tau - \xi_t\underline{\wedge}\tau_t$, v_t^τ being the time derivative of τ_t and ξ_t a vector in R^3.

If we differentiate $(R_t x, R_t y) = (x,y)$ with respect to time, we see that ξ_t is independent of x; ξ_t is the *angular velocity* (Arnold [1] p. 140). Then we define

(337) $\quad \xi_M^t x = \xi_t\underline{\wedge} x + v_t^e.$

Now (335) allows us to define *(generalized) Galilean transformations* on R^4 by U: $x' = u_t x$, $t' = t$. We assume that we have two observers S and S': S is linked to the solid medium, S' is at rest, each having charts $u_s(N) = (x,t)$, $u_{s'}(N) = (x',t')$ of the physical system, which are exchanged by $u_{s's} = U$.

Pure Galilean transformations are given by (with v the velocity of S with respect to S'):

(335)' $\quad x' = u_t x = x + vt,\ \ t' = t,\ \ \ v \in R^3$ (thus with $R_t = I$, $\tau_t = vt$).

Let E', B', H', D' (and thus F', L') be the electromagnetic field at (x', t') for S', and E, B... be the corresponding electromagnetic field at (x,t) for S. The pullback of the 2-form $\omega^2_{F'}$ by U is given by

(338) $$U^*\omega^2_{F'} = U^*(\omega^1_{E'} \wedge dt') + U^*\omega^2_{B'} = u_t^*\omega^1_{E'} \wedge dt' + U^*\omega^2_{B'}.$$

Since u_t is an isometry we have:

(339) $$u_t^*\omega^1_{E'} = \omega^1_{u_t^*E'}, \quad u_t^*\omega^2_{B'} = \omega^2_{u_t^*B'},$$

and also,

$$(U^*\omega^2_{B'})_{x,t} = B'^1(x_t,t)\left((R_t dx)^2 + \frac{\partial x_t^2}{\partial t}dt\right) \wedge \left((R_t dx)^3 + \frac{\partial x_t^3}{\partial t}dt\right) + \ldots,$$

and then developing, we have

(340) $$U^*\omega^2_{B'} = u_t^*\,\omega^2_{B'} - u_t^*\omega^1_{B' \underline{\wedge}\, \xi^t_M x'} \wedge dt'.$$

With (337), we finally obtain:

(341) $$U^*\omega^2_{F'} = \omega^1_{u_t^*(E' - B' \underline{\wedge}\, \xi^t_M x')} \wedge dt + u_t^*\omega^2_{B'}.$$

Similarly we have

(341)' $$U^*\tilde{\omega}^2_{L'} = \omega^1_{u_t^*(H' + D' \underline{\wedge}\, \xi^t_M x')} \wedge dt - u_t^*\omega^2_{D'}.$$

Thus we obtain the transformation law

(342) $$\left|\begin{array}{ll} E(x,t) = (u_t^*(E' - B' \underline{\wedge}\, \xi^t_M x'))_x & \text{and} \;\; B(x,t) = (u_t^*(B'))_x, \\ H(x,t) = (u_t^*(H' + D' \underline{\wedge}\, \xi^t_M x')_x & \text{and} \;\; D(x,t) = (u_t^*(D'))_x, \end{array}\right.$$

that is also

(342)' $$\left|\begin{array}{ll} E(x,t) = R_t^{-1}(E'(x_t,t) - B'(x_t,t) \underline{\wedge}\, \xi^t_M x_t), & B(x,t) = R_t^{-1}(B'(x_t,t)), \\ H(x,t) = R_t^{-1}(H'(x_t,t) + D'(x_t,t) \underline{\wedge}\, \xi^t_M x_t), & D(x,t) = R_t^{-1}(D'(x_t,t)). \end{array}\right.$$

The transformation laws of current and charge are given by $j = U^*j'$, thus by

(343) $$\begin{pmatrix} J \\ \rho \end{pmatrix}(x,t) = \begin{pmatrix} R_t^{-1} & -R_t^{-1}\xi^t_M x_t \\ 0 & 1 \end{pmatrix}\begin{pmatrix} J' \\ \rho' \end{pmatrix}(x_t,t),$$

thus also by

(343)' $$J(x,t) = R_t^{-1}(J'(x_t,t) - \xi^t_M x_t\, \rho'(x_t,t)), \quad \rho(x,t) = \rho'(x_t,t).$$

Application to the Galilean transformation (335)'. We have $\xi^t_M x = v$, thus

(344) $$\left|\begin{array}{ll} E(x,t) = E'(x+vt,t) - B'(x+vt,t) \underline{\wedge}\, v, & D(x,t) = D'(x+vt,t), \\ H(x,t) = H'(x+vt,t) + D'(x+vt,t) \underline{\wedge}\, v, & B(x,t) = B'(x+vt,t), \end{array}\right.$$

and

$$J(x,t) = J'(x+vt,t) - \rho'(x+vt,t)v, \qquad \rho(x,t) = \rho'(x+vt,t).$$

We verify that these transformation laws are in agreement with those relative to Lorentz transformations for small velocities (v/c<<1), changing v into $-v$.

CONCLUSION

We can define "proper electromagnetic field" in moving bodies as fields on the tangent space TM (identified with $R^3 \times R^3$) according to (from (344) for moving particles, or from (342) for solid medium):

$$(345) \quad \left| \begin{array}{ll} E(x,v) \stackrel{def}{=} E(x) + v \underline{\wedge} B(x), & B(x,v) \stackrel{def}{=} B(x), \\ H(x,v) \stackrel{def}{=} H(x) - v \underline{\wedge} D(x), & D(x,v) \stackrel{def}{=} D(x). \end{array} \right.$$

Then the force (called Lorentz force) on a particle with charge q and velocity v in an electromagnetic field is:

$$(346) \quad F = q\,E(x,v) = q\,(E + v \underline{\wedge} B).$$

Furthermore since u_t is an isometry for all t, μ_c is invariant by U, and thus (see (329)) the transformed electromagnetic field (by any transformation (334), therefore by any Galilean transformation) satisfies the Maxwell equations. We insist here on the fact that we have transformed equations (324) and not (324)', because U does not commute with the Hodge transformation in R^4. The necessity to use (324) rather than (324)' is due to the transformation law (328). The definition of E, B, D, H from the fundamental 2-forms according to (323) implies that they have different transformation properties with respect to time and space inversions.

From the two fundamental forms, it is also possible (see for example Jones [1] p. 112) to define the *Maxwell stress tensor* in $T^1{}_1(M)$, (T) by:

$$(347) \quad (T) = \begin{pmatrix} H \otimes B + E \otimes D - \frac{1}{2}(H.B + E.D) & B \underline{\wedge} D \\ -E \underline{\wedge} H / c^2 & \frac{1}{2}(H.B + E.D) \end{pmatrix}.$$

There is no universal agreement for such a tensor (see for example Eringen-Maugin [1] p. 62), but it is the key point of the relations of electromagnetism with mechanics, and to define forces in a solid medium.

Then conservative laws are obtained from Maxwell equations, either using the Maxwell tensor or associated differential forms: for example we have the relation giving the usual conservative law for the Poynting vector, see (51) chap.1

$$(348) \quad -d(\omega^1_E \wedge \omega^1_H) = (\omega^1_H \wedge \dot{\omega}^2_B + \omega^1_E \wedge \dot{\omega}^2_D) + \omega^1_E \wedge \omega^2_J.$$

7.3.3. Transformation of the constitutive relations

Now we assume that the electromagnetic field for the observer S' satisfies the constitutive relations:

(349) $D' = \varepsilon E', \quad B' = \mu H', \quad J' = \sigma E'.$

We can obtain the constitutive relations for the observer S either using Lorentz transformation laws (333) to (334) giving E, D, B, H as functions of the field "at rest", or using (342) for a given evolution of the body, and in particular using (344) for a translation motion at a velocity v.

i) *Lorentz transformation*. By projection on the velocity v, and on the orthogonal plane to v, we easily obtain from (333), (333)', (333)" the relations

(350)
$$\left| \begin{array}{l} D + v\underline{\wedge} H/c^2 = \varepsilon (E + v\underline{\wedge} B), \qquad B - v\underline{\wedge} E/c^2 = \mu (H - v\underline{\wedge} D), \\ \qquad J_v - v\rho = (\sigma/\beta)\, E_v, \qquad \pi_v^0 J = \sigma\beta(\pi_v^0 E + v\underline{\wedge} B) \end{array} \right.$$

thus $J - \rho v = \sigma\beta\, (E + v\underline{\wedge} B - (E.v)\, v/c^2).$

We can solve relations (350) with respect to B, D as functions of E, H. Then we obtain with $\eta = \dfrac{1 - \varepsilon\mu c^2}{1 - \varepsilon\mu v^2}$, $\pi_v^0 E$ (or $\pi_v^0 H$) = projection of E (or H) on an orthogonal plane to v

(351)
$$\left| \begin{array}{l} B = \mu\, (H - \eta \dfrac{v^2}{c^2} \pi_v^0 H) + \eta \dfrac{1}{c^2} v\underline{\wedge} E = \mu\, H + \eta \dfrac{1}{c^2} v\underline{\wedge}(E + \mu\, v\underline{\wedge} H), \\ D = \varepsilon\, (E - \eta \dfrac{v^2}{c^2} \pi_v^0 E) - \eta \dfrac{1}{c^2} v\underline{\wedge} H = \varepsilon\, H - \eta \dfrac{1}{c^2} v\underline{\wedge}(H - \varepsilon\, v\underline{\wedge} E)\,. \end{array} \right.$$

These relations are of chiral type (see (95) chap.1) but for small v/c, we have the same constitutive laws for D and B as in (349). Note that free space has always the same constitutive laws.

ii) *Galilean transformation*. Using inverse relations to (344), we have from (349):

(352) $B = \mu(H + v\underline{\wedge} D), \quad D = \varepsilon(E - v\underline{\wedge} B), \quad J + \rho v = \sigma\, (E - v\underline{\wedge} B).$

Solving these relations with respect to B, D as functions to E, H, we obtain with $\theta = \varepsilon\mu(1 - \varepsilon\mu v^2)^{-1}$:

(353)
$$\left| \begin{array}{l} B = \mu\, H + \theta\, v\underline{\wedge}\, (E - \mu\, v\underline{\wedge} H), \\ D = \varepsilon\, E - \theta\, v\underline{\wedge}\, (H + \varepsilon\, v\underline{\wedge} E). \end{array} \right.$$

These relations are also of chiral type.

7.3.4. Boundary transmission conditions for a solid moving body

We consider as in section 6.3.2 the evolution of a solid medium in the (bounded) connected set Ω (in R^3) at time 0 and in Ω_t at time t, with an evolution given by u as in (335), and thus $\Omega_t = u_t(\Omega)$. We assume that the boundary Γ of Ω is regular and given by equation $\phi(0,x) = 0$. Then the equation of the bôundary of Ω_t, Γ_t, is:

$$\phi(t,x) = \phi(0,u_t^{-1}x) = 0.$$

This is also the equation of the boundary $\tilde{\Gamma}$ of the domain $\tilde{\Omega} = \{(x,t) \text{ with } x \in \Omega_t\}$ in R^4. Then the (outward if $\phi < 0$ in Ω) unit normal n_t to Γ_t at x (resp. ν to $\tilde{\Gamma}$ at (x,t)) is:

$$(354)\quad n_t = \operatorname{grad}\phi/|\operatorname{grad}\phi|\,,\quad \nu = (\phi'_t, \operatorname{grad}\phi)/\left((\phi'_t)^2 + |\operatorname{grad}\phi|^2\right)^{1/2} \text{ with } \phi'_t = \frac{\partial\phi}{\partial t}.$$

With $v = \frac{dx}{dt}$, we have: $\phi'_t = -\,v.\operatorname{grad}\phi$. Thus the unit normals ν and n to $\tilde{\Gamma}$ and to Γ_t are related by:

$$(355)\quad \nu = \left(1 + (v.n)^2\right)^{-1/2} \cdot \frac{1}{|\operatorname{grad}\phi|}\{\phi'_t\,,\, \operatorname{grad}\phi\} = \left(1 + (v.n)^2\right)^{-1/2}.\{-\,v.n,\ n\}.$$

Then for all smooth function ϕ on R^4 we have with i = 1, 2, 3, from Stokes formula:

$$(356) \int_{\tilde{\Gamma}} \phi\,\nu^i\, d\tilde{\Gamma} = \int_{\tilde{\Omega}} \frac{\partial\phi}{\partial x^i}\, dxdt = \int_R dt \int_{\Omega_t} \frac{\partial\phi}{\partial x^i}\, dx = \int_R dt \int_{\Gamma_t} \phi n^i\, d\Gamma_t = \int_{\tilde{\Gamma}} \phi n^i d\Gamma_t dt,$$

thus we have with (355) for i = 1, 2, 3:

$$(357)\quad n^i\, d\Gamma_t dt = \nu^i\, d\tilde{\Gamma}\,,\quad d\Gamma_t\, dt = \left(1 + (v.n)^2\right)^{-1/2} d\tilde{\Gamma}\,,\qquad \nu^o\, d\tilde{\Gamma} = -\,v.n\, d\Gamma dt.$$

Then we consider an electromagnetic field E, B, D, H satisfying the Maxwell equations (1) of chap.1, or (324), in the differential form with F and L. If we have no concentration of currents nor charges at the boundary of the moving medium, and if the electromagnetic field is "regular" on each side of the boundary, what are the continuity relations of the electromagnetic field across the boundary ? We can answer this question in two different ways:

i) either using an elementary point of view calculating expressions such as:

$$\langle \operatorname{curl} E + \frac{\partial B}{\partial t}\,,\, \phi\rangle = \langle E,\, \operatorname{curl}\phi\rangle - \langle B,\, \frac{\partial\phi}{\partial t}\rangle\,,\qquad \forall\phi \in \mathcal{D}(R_x^3 x R_t),$$

then using the Stokes formula. Surface integral terms are of the following form:

$$\int_R dt \int_{\Gamma_t} -\,\underline{n \wedge E}\,\phi\, d\Gamma_t - \int_{\tilde{\Gamma}} \nu_o B\phi\, d\tilde{\Gamma} = \int_{\tilde{\Gamma}} (-\,\underline{n \wedge E} + v.n\, B)\,\phi\,(1 + (v.n)^2)^{-1/2}\, d\tilde{\Gamma}\,;$$

ii) or using "Schwartz" jump formula (94) for differential forms such as:

$$(358)\qquad \Sigma_\nu \wedge [\omega_F^2]_{\tilde{\Gamma}} = 0,$$

thus with (95) and (355) (here only Euclidean structure of R^4 matters):

(359) $(\omega_n^1 - n.v\,dt) \wedge [\omega_E^1 \wedge dt + \omega_B^2]_{\tilde{\Gamma}} = 0,$

that is:

(360) $[(\omega_n^1 \wedge \omega_E^1 - n.v\,\omega_B^2) \wedge dt]_{\tilde{\Gamma}} + [\omega_n^1 \wedge \omega_B^2]_{\tilde{\Gamma}} = 0,$

therefore, using (273):

(361) $[n\underline{\wedge} E]_{\tilde{\Gamma}} - n.v\,[B]_{\tilde{\Gamma}} = 0, \quad [n.B]_{\tilde{\Gamma}} = 0.$

From the relation:

(362) $n\underline{\wedge}(v\underline{\wedge}A) = (n.v)\,A - v\,(n.A),$

for all vectors A in R^3, we finally get:

(363) $n\underline{\wedge}[E - v\underline{\wedge}B]_{\tilde{\Gamma}} = 0, \quad [n.B]_{\tilde{\Gamma}} = 0.$

Similarly with the 2-form $\tilde{\omega}_L^2$ we have:

(363)' $n\underline{\wedge}[H + v\underline{\wedge}D]_{\tilde{\Gamma}} = 0, \quad [n.D]_{\tilde{\Gamma}} = 0.$

Summing up results, using relations (342),(342)', we have with $n' = R_t n$:

(364) $[n'\underline{\wedge}\,E'] = 0, \quad [n'.B'] = 0, \quad [n'\underline{\wedge}\,H'] = 0, \quad [n'.D'] = 0.$

They are the usual transmission conditions for an observer moving with the velocity of the solid medium.

The Hodge decomposition for a regular manifold M with boundary for r ≠ 0, n.

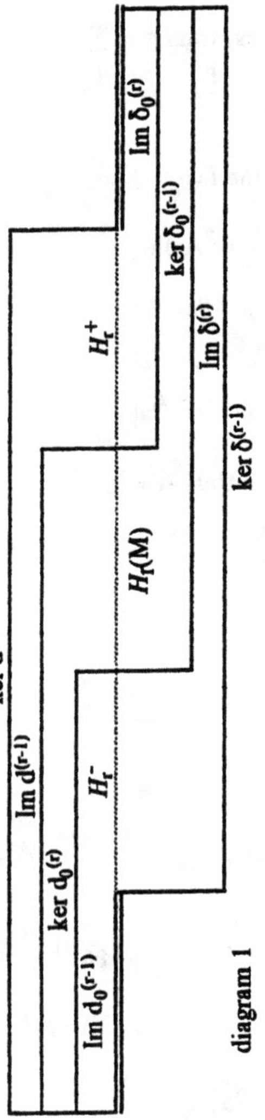

The Hodge decomposition for a regular manifold M without boundary

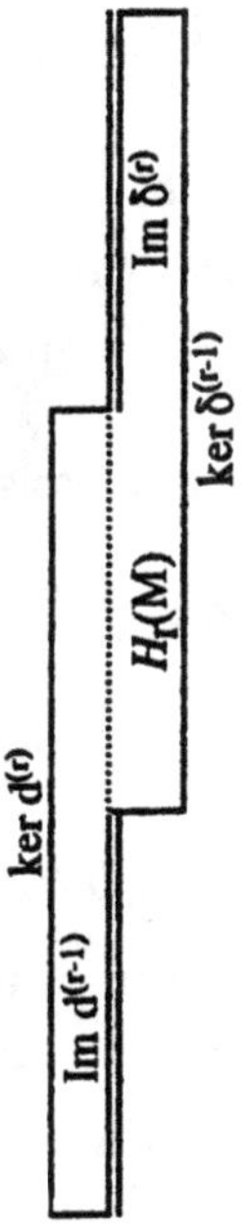

diagram 2

Table of the main differential operators in an orthogonal coordinate system.

Let (x^1, x^2, x^3) be a (direct) local coordinate system, with Riemannian metric:

$ds^2 = \sum g_j (dx^j)^2$. Let $g^o \overset{\text{def}}{=} (g_1 g_2 g_3)^{1/2}$.

Cylindrical system (ρ, ϕ, z) see (291); $g_\rho = 1$, $g_\phi = \rho^2$, $g_z = 1$; thus $g^o = \rho$.

Spherical system (r, θ, ϕ) see (292); $g_r = 1$, $g_\phi = r^2 \sin^2\theta$, $g_\theta = r^2$; thus $g^o = r^2 \sin\theta$.

1) Orthonormal basis (e_1, e_2, e_3) for vectors, thus $A = \sum A_j e_j$, with $e_j = (g_j)^{-1/2}\partial_j$

$$\text{grad}\, f = \sum (g_j)^{-1/2}\frac{\partial f}{\partial x_j} e_j \overset{\text{cyl}}{=} \frac{\partial f}{\partial \rho} e_\rho + \frac{1}{\rho}\frac{\partial f}{\partial \phi} e_\phi + \frac{\partial f}{\partial z} e_z \overset{\text{sph}}{=} \frac{\partial f}{\partial r} e_r + \frac{1}{r}\frac{\partial f}{\partial \theta} e_\theta + \frac{1}{r\sin\theta}\frac{\partial f}{\partial \phi} e_\phi .$$

$$\text{div}\, A = \frac{1}{g^o}\sum \frac{\partial}{\partial x_j}(g^o g_j^{-1/2} A_j), \quad \Delta f = \text{div grad}\, f = \frac{1}{g^o}\sum \frac{\partial}{\partial x_j}\left(\frac{g^o}{g_j}\frac{\partial}{\partial x_j} f\right)$$

$$\text{div}\, A \overset{\text{cyl}}{=} \frac{1}{\rho}\frac{\partial}{\partial \rho}(\rho A_\rho) + \frac{1}{\rho}\frac{\partial A_\phi}{\partial \phi} + \frac{\partial A_z}{\partial z} \overset{\text{sph}}{=} \frac{1}{r^2}\frac{\partial}{\partial r}(r^2 A_r) + \frac{1}{r\sin\theta}\left[\frac{\partial}{\partial \theta}(\sin\theta\, A_\theta) + \frac{\partial}{\partial \phi}(A_\phi)\right]$$

$$\text{curl}\, A = \frac{1}{g^o}\begin{vmatrix} g_1^{1/2} e_1, & g_2^{1/2} e_2, & g_3^{1/2} e_3 \\ \partial_1, & \partial_2, & \partial_3 \\ g_1^{1/2} A_1, & g_2^{1/2} A_2, & g_3^{1/2} A_3 \end{vmatrix} \overset{\text{cyl}}{=} \frac{1}{\rho}\begin{vmatrix} e_\rho, & \rho e_\phi, & e_z \\ \partial_\rho, & \partial_\phi, & \partial_z \\ A_\rho, & \rho A_\phi, & A_z \end{vmatrix} \overset{\text{sph}}{=} \frac{1}{r^2\sin\theta}\begin{vmatrix} e_r, & r e_\theta, & r\sin\theta\, e_\phi \\ \partial_r, & \partial_\theta, & \partial_\phi \\ A_r, & r A_\theta, & r\sin\theta\, A_\phi \end{vmatrix}.$$

2) Differential basis $(\partial_1, \partial_2, \partial_3)$ for vectors; thus $A = \sum A^j \partial_j$.

$$\text{grad}\, f = \sum \frac{1}{g_j}\frac{\partial f}{\partial x_j}\partial_j \overset{\text{cyl}}{=} \frac{\partial f}{\partial \rho}\partial_\rho + \frac{1}{\rho^2}\frac{\partial f}{\partial \phi}\partial_\phi + \frac{\partial f}{\partial z}\partial_z \overset{\text{sph}}{=} \frac{\partial f}{\partial r}\partial_r + \frac{1}{r^2}\frac{\partial f}{\partial \theta}\partial_\theta + \frac{1}{r^2\sin^2\theta}\frac{\partial f}{\partial \phi}\partial_\phi .$$

$$\text{div}\, A = \frac{1}{g^o}\sum \frac{\partial}{\partial x_j}(A_j g^o)$$

$$\text{div}\, A \overset{\text{cyl}}{=} \frac{1}{\rho}\frac{\partial}{\partial \rho}(\rho A_\rho) + \frac{\partial A_\phi}{\partial \phi} + \frac{\partial A_z}{\partial z} \overset{\text{sph}}{=} \frac{1}{r^2}\frac{\partial}{\partial r}(r^2 A_r) + \frac{1}{\sin\theta}\left[\frac{\partial}{\partial \theta}(\sin\theta\, A_\theta) + \sin\theta\frac{\partial A_\phi}{\partial \phi}\right]$$

$$\text{curl}\, A = \frac{1}{g^o}\begin{vmatrix} \partial_1, & \partial_2, & \partial_3 \\ \frac{\partial}{\partial x_1}, & \frac{\partial}{\partial x_2}, & \frac{\partial}{\partial x_3} \\ g_1 A_1, & g_2 A_2, & g_3 A_3 \end{vmatrix} \overset{\text{cyl}}{=} \frac{1}{\rho}\begin{vmatrix} \partial_\rho, & \partial_\phi, & \partial_z \\ \frac{\partial}{\partial \rho}, & \frac{\partial}{\partial \phi}, & \frac{\partial}{\partial z} \\ A_\rho, & \rho^2 A_\phi, & A_z \end{vmatrix} \overset{\text{sph}}{=} \frac{1}{r^2\sin\theta}\begin{vmatrix} \partial_r, & \partial_\theta, & \partial_\phi \\ \frac{\partial}{\partial r}, & \frac{\partial}{\partial \theta}, & \frac{\partial}{\partial \phi} \\ A_r, & r^2 A_\theta, & r^2\sin^2\theta\, A_\phi \end{vmatrix}$$

$$\Delta f \overset{\text{cyl}}{=} \frac{1}{\rho}\frac{\partial}{\partial \rho}\rho\frac{\partial f}{\partial \rho} + \frac{1}{\rho^2}\frac{\partial^2 f}{\partial \phi^2} + \frac{\partial^2 f}{\partial z^2} \overset{\text{sph}}{=} \frac{1}{r^2}\frac{\partial}{\partial r}r^2\frac{\partial f}{\partial r} + \frac{1}{r^2\sin\theta}\left[\frac{\partial}{\partial \theta}\sin\theta\frac{\partial f}{\partial \theta} + \frac{\partial}{\partial \phi}\frac{1}{\sin\theta}\frac{\partial f}{\partial \phi}\right].$$

3) Definition of the covariant derivative ∇_X by $\nabla_{gX}(fY) = g[f\, \nabla_X Y + X(f)Y]$,

with Γ^h_{ij} the Christoffel symbols, $\Gamma^1_{31} = \Gamma^1_{13} = \frac{1}{2g_1}\frac{\partial g_1}{\partial n}$, $\Gamma^2_{32} = \Gamma^2_{23} = \frac{1}{2g_2}\frac{\partial g_2}{\partial n}$, 0 otherwise,

with $x_3 = n$, and $\nabla_{\partial_i}(\partial_j) = \sum \Gamma^h_{ij}(x)\partial_h$. Then $\nabla_n B = \sum_{j=1,2}\frac{\partial B^j}{\partial n}\partial_j + \frac{1}{2}\sum_{j=1,2} B^j\frac{1}{g_j}\frac{\partial g_j}{\partial n}\partial_j$,

for $B = \sum_{j=1,2} B^j\partial_j$.

$u(x,t)$ in $R^3_x x R_t$	supp u on C_+	supp u on C_-	supp u on C
$\square u(x,t) = \delta(x,t)$	$\Phi_+(x,t) = Y(t)\dfrac{\delta(r-t)}{4\pi r}$, $r = \lvert x\rvert$	$\Phi_-(x,t) = Y(-t)\dfrac{\delta(r+t)}{4\pi r}$, $r = \lvert x\rvert$	$\Phi_0 = \frac{1}{2}(\Phi_+ + \Phi_-) = \frac{1}{4\pi}\delta(r^2 - t^2)$
$\mathcal{F}_x u(\xi,t) = v(x,t)$ $(\partial_t^2 - \xi^2)\, v = \delta(t)$	$Y(t)\dfrac{\sin(\lvert\xi\rvert t)}{\lvert\xi\rvert}$	$-Y(-t)\dfrac{\sin(\lvert\xi\rvert t)}{\lvert\xi\rvert}$	$\frac{1}{2}\,\mathrm{sign}\,t\,\dfrac{\sin(\lvert\xi\rvert t)}{\lvert\xi\rvert}$
$\mathcal{F}_t u(x,k) = w(x,k)$ $-(k^2 + \Delta)\, w = \delta(x)$	$\dfrac{e^{ikr}}{4\pi r}$	$\dfrac{e^{-ikr}}{4\pi r}$	$\dfrac{\cos kr}{4\pi r}$
$\mathcal{F}_{x,t} u(\xi,k) = z(\xi,k)$ $-(k^2 + \xi^2)\, z = 1$	$\mathrm{pv}\dfrac{1}{\xi^2 - k^2} + i\pi\,\delta(\xi^2 - k^2)\,\mathrm{sign}\,k$	$\mathrm{pv}\dfrac{1}{\xi^2 - k^2} - i\pi\,\delta(\xi^2 - k^2)\,\mathrm{sign}\,k$	$\mathrm{pv}\dfrac{1}{\xi^2 - k^2}$

supp u on C, $\square u = 0$
$R = \Phi_+ - \Phi_- = \frac{1}{2\pi}\,\mathrm{sign}\,t\,\delta(r^2 - t^2)$
$\mathcal{F}_x R(\xi,t) = \dfrac{\sin(\lvert\xi\rvert t)}{\lvert\xi\rvert}$
$\mathcal{F}_t R(x,k) = i\,\dfrac{\sin kr}{2\pi r}$
$\mathcal{F}_{x,t} R(\xi,k) = 2i\pi\,\delta(\xi^2 - k^2)\,\mathrm{sign}\,k$

C is the light cone, C^+ the future cone, C^- the past cone.

$$\delta(r^2 - t^2) = \frac{1}{2r}[\delta(r+t) + \delta(r-t)],\ t \neq 0,\ r = \lvert x\rvert .$$

References

ABRAHAM R., MARSDEN J., RATIU E. [1] Manifolds, Tensor Analysis, and Applications. Addison-Wesley (1983).

ABRAMOWITZ M., STEGUN I.A. [1] Handbook of Mathematical Functions. Dover (1972).

ADAMS R.A. [1] Sobolev Spaces. Academic Press (1975).

AGMON S., HÖRMANDER L. [1] Asymptotic properties of solutions of differential equations with simple characteristics. J. Anal. Math. 30, 1-38 (1976).

ARNOLD V. [1] Méthodes mathématiques de la mécanique classique. Mir (1974).

ARTOLA M., CESSENAT M. [1] Diffraction d'une onde électromagnétique par un obstacle borné à permittivité et perméabilité élevées. C. R. A. S. 314, Série I, 349-354 (1992).

[2] Sur les conditions de Leontovich; Les grands systèmes des sciences et de la technologie. ed J. Horowitz, Lions J. L. R.M.A. 28 Masson (1994)

[3] Problèmes d'homogénéisation diélectrique conducteur dans les composites en structure périodique. Rapports CEL-V/DS-EM 270/90, (1990), 438/90, (1990), 5/91, (1991), 65/91, (1991), 140/91 (1991).

AYDIN K., HIZAL A. [1] On the Completeness of the Spherical Vector Wave Functions. J. Math. Anal. Appl. 117, 428-440 (1986).

BAMBERGER A., HA DUONG T. [1] Formulation variationnelle espace-temps pour le calcul par potentiel retardé d'une onde acoustique par une surface libre. Rapport 107. C.M.A.P. Polytechnique (1984).

[2] Formulation variationnelle espace-temps pour le calcul par potentiel retardé d'une onde acoustique. Math. Mech. Appl. Sci. 8, 405-435, 598-608 (1986).

[3] Diffraction d'une onde acoustique par une paroi absorbante: nouvelles équations intégrales. Rapport 121. C.M.A.P. Polytechnique (1985).

BARDOS C. [1] Utilisation des notions de propagation de singularité pour le controle et la stabilisation des problèmes hyperboliques. Gazette des Mathématiciens 36, avril 1988.

BARDOS C., RAUCH J. [1] Maximal positive boundary value problems as limits of singular perturbation problems. Trans. A.M.S. 270, 2, 377-408 (1982).

BARDOS C., LEBEAU G., RAUCH J. [1] Gevrey regularity and distribution of the scattering frequencies. Invent. Math. 90, 77-114 (1987).

[2] Controle et stabilisation dans les problèmes hyperboliques. Appendix of the book: LIONS J.L. Controlabilité exacte des systèmes distribués, collection R.M.A. Masson (1988).

BENDALI A. [1] Approximation par éléments finis de surface de problèmes de diffraction des ondes électromagnétiques. Thèse de Doctorat (1984).

[2] Problèmes aux limites extérieur et intérieur pour le système de Maxwell en régime harmonique. Rapport 59. C.M.A.P. Polytechnique (ERA/CNRS 747) (1980).

BENSOUSSAN A., LIONS J.L., PAPANICOLAOU G., [1] Asymptotic Analysis for Periodic Structures. North-Holland (1978).

BETHUEL F., BREZIS H. [1] C.R.A.S. tome 310, Série I, 859-864 (1990).

BONNEMASON P., STUPFEL B. [1] Modeling high frequency scattering by axisymmetric perfectly or imperfectly conducting scatterers. Electromagnetics 13, 111-129 (1993).

BOSSAVIT A. [1] Un nouveau point de vue sur les éléments mixtes. MATAPLI, Bulletin n°20, SMAI 23-35 (Oct. 1989).

[2] Modèles variationnels et méthodes mixtes en électromagnétisme. (Computation of Eddy currents and Lorentz forces in a system of moving conductors). Ecole d'été CEA EDF INRIA d'analyse numérique (1988).

[3] Edge-elements for scattering problems. Bulletin de la direction des études et des recherches, série C, EDF 2, 17-33 (1989).

[4] Electromagnétisme en vue de la modélisation. Mathématiques et Applications. Springer (1993).

BOURBAKI N. [1] Eléments de Mathématique. Algèbre. Hermann (1970).

[2] Eléments de Mathématique. Variétés différentielles et analytiques. Fascicule de résultats, Paragraphe 8-15. Hermann (1971).

BREZIS H., CORON J.M. [1] Large Solutions for harmonic maps in two dimensions. Comm. Math. Phys. 92, 203-215 (1983).

BREZIS H. [1] Analyse fonctionnelle Théorie et applications. Masson (1983).

[2] Liquid crystals and energy estimates for S^2-valued maps, in: Theory and Applications of Liquid Crystals, IMA 5.

BROWN W.F. [1] Micromagnetics. 18 - Interscience Tracts on Physics and Astronomy. Wiley (1963).

BUTZER P.L., BERENS H. [1] Semi-groups of Operators and Approximation. Springer (1967).

CADILHAC M., PETIT R. [1] On the roles of physical intuition, computation and mathematical analysis in electromagnetic diffraction theory. Radio Science 22 n°7, 1247-1259 (Dec. 1987).

[2] On the diffraction problem in electromagnetic theory: a discussion based on concepts of functional analysis including an example of practical application. Huygens'Principle 1690-1990. Theory and Applications. Elsevier (1992).

CADILHAC M. [1] in PETIT [1].

CALDERON A. [1] The Multipole Expansion of Radiation Fields. J. Rational Mech. Anal. 3, 523-537 (1954).

CESSENAT M. [1] Problèmes de Helmholtz par méthode de séparation des variables en coordonnées polaires. Rapport CEL-V/DS-EM 048/89 (1989).

[2] Sur quelques opérateurs liés à l'équation de Helmholtz en coordonnées polaires, transformation H.K.L (Hankel-Kantorovitz-Lebedev). C.R.A.S. 309, Serie I, 25-30 (1989).

[3] Résolution des problèmes de Helmholtz par séparation des variables en coordonnées polaires. C.R.A.S. 309, Serie I, 105-109 (1989).

[4] Résolution des problèmes de Maxwell en regime harmonique par des méthodes intégrales. Rapport CEL-V/DS-EM 587 (1987).

[5] Scattering d'une onde incidente par un obstacle mince d'indice élevé. Problème stationnaire. loi d'induction de Faraday. Rapport CEL-V/DS-EM 262/94 (1994).

CHAZARAIN J., PIRIOU A. [1] Introduction à la Théorie des Equations aux Dérivées Partielles Linéaires. Gauthier Villars (1981).

COLTON D., KRESS R. [1] Dense sets and far field patterns in electromagnetic wave propagation. SIAM J. Math. Anal. 16, 5 (1985).

[2] Inverse Acoustic and Electromagnetic Scattering Theory. Applied Math. Scences (1993).

COSTABEL M. [1] A symmetric method for the coupling of finite elements and boundary elements 3d. PROC. MAFELAP VI. Academic Press (to appear).

[2] A remark on the regularity of solutions of Maxwell's equations on Lipschitz domains, Math. Methods Appl. Sci., 12, 365-368 (1990).

COSTABEL M., STEPHAN E. [1] A direct boundary integral equation method for transmission problems. J. Math. Anal. Appl. 106, 367-413 (1985).

COSTABEL M., WENDLAND W.L. [1] Strong ellipticity of boundary integral operators. J. für Math. Band 372 (1986).

DAUTRAY R., LIONS J.L. [1] Mathematical Analysis and Numerical Methods for Science and Technology. Vols.1-6. Springer-Masson (1988-1990).

Da SILVA PASSOS A. [1] Méthodes mathématiques du traitement du signal. Eyrolles (1989).

DERMENJIAN Y., GUILLOT J.C. [1] Théorie spectrale de la propagation des ondes acoustiques dans un milieu stratifié perturbé. J. Diff. Equations. 62, 357-409 (1986).

DESPRES B. [1] Domain Decomposition Method and the Helmholtz Problem. Mathematical and Numerical Aspects of Wave Propagation Phenomena. ed. Cohen G., Halpern L, Joly P. SIAM, Strasbourg (1991).

DIEUDONNE J. [1] Eléments d'analyse. Vols. 1-8 Gauthier Villars (1971).

DUNFORD N., SCHWARTZ J.T. [1] Linear Operators. Part II. Interscience (1964).

DUVAUT G., LIONS J.L. [1] Les inequations en mécanique et en physique. Dunod (1972).

de GENNES P.G. [1] The Physics of Liquid Crystals. Clarendon Press (1974).

ENGQUIST B., MAJDA A. [1] Absorbing boundary conditions for the numerical simulation of waves. Math. Comp. 31, n°139, 629-651 (1977).

[2] Radiation boundary conditions for acoustic and elastic wave calculations. Comm. Pure Appl. Math. 32, 313-357 (1979).

ERICKSEN J.L. [1] Hydrostatic theory of liquid crystals. Arch. Rat. Mech. Anal. 9, 371-378 (1962).

ERICKSEN J.L., KINDERLEHRER ed. [1] Theory and Applications of Liquid Crystals. Springer (1987).

ERINGEN A.C., MAUGIN G.A. [1] Electrodynamics of Continua I and II. Springer (1990).

FOURNET G. [1] Electromagnétisme à partir des équations locales. Masson (1979).

FRIEDRICHS K.O. [1] Symmetric positive linear differential equations. C.P.A.M. 11, 333-410 (1958).

GILBARG D., TRUDINGER N.S. [1] Elliptic Partial Differential Equations of Second Order. Springer (1983).

GILKEY P.B. [1] Invariance Theory, the Heat Equation and the Atiyah Singer Index Theorem. Publish or Perish (1974).

GIRAULT V., RAVIART P.A. [1] Finite Element Approximation of the Navier-Stokes Equations. Lecture Notes in Math. 749. Springer (1979).

[2] Finite Element Methods for the Navier Stokes equations Theory and Algorithms. Springer (1986).

GOVIND S., WILTON D.R., GLISSON A.W. [1] Scattering from Inhomogeneous Penetrable Bodies of Revolution. I.E.E. Transactions on Antennas and Propagation AP-32, 11 (Nov. 1984).

GOLDSTEIN C. [1] Scattering theory for elliptic differential operators in unbounded domains. J. Math. Anal. Appl., 723-745 (1974).

[2] Eigenfunction expansion associated with the Laplacian for certain domains with infinite boundaries, I-III Trans. Amer. Math. Soc.,(I) 135, 1-32 (1969), (II) 135, 33-50 (1969), (III) 143, 283-301 (1969).

GRAY G.A., KLEINMAN R.E. [1] The integral equation method in electromagnetic scattering. J. Math. Anal. Appl. 107, 455-477 (1985).

GRISVARD P. [1] Elliptic Problem in Nonsmooth Domains. Pitman (1985).

[2] Boundary Value Problems in Non-Smooth Domains. Univ of Maryland. Lecture note 19 (1980).

[3] Equations opérationnelles abstraites dans les espaces de Benach et problèmes aux limites dans des ouverts cylindriques. Annali della Scuola Normale Superior di Pisa, vol. XXI, Fasc.III, 307-347 (1967).

GUELFAND I.M., GRAEV M.I., VILENKIN N.Ja. [1] Les distributions. Vol. 5. Dunod (1970).

HARRINGTON R.F. [1] Time Harmonic Electromagnetic Fields. McGraw Hill (1961).

HAZARD C., LENOIR M. [1] On the solution of time-harmonic scattering problems for Maxwell's equations. (to appear).

HELIOT J.P. [1] Equations intégrales couplées pour la diffraction électromagnétique par un dielectrique ou par l'assemblage d'un conducteur et d'un dielectrique. Rapport CESTA/SI 613 (1992).

HELGASON S. [1] The Radon transform on Euclidean spaces, compact two-point homogeneous spaces and Grassmann manifolds. Acta Math. 113, 153-180 (1965).

HÖRMANDER L. [1] Linear Partial Differential Operators. Springer (1976).

[2] The Analysis of Linear Partial Differential Operators. Vols. 1-4 Springer (1983-1984).

[3] An Introduction to Complex Analysis in Several Variables. Van Nostrand (1966).

JACKSON J.D. [1] Classical Electrodynamics. Wiley (1975).

JOLY P., MERCIER B.[1] Une nouvelle condition transparente d'ordre 2 pour les équations de Maxwell en dimension 3. Rapport INRIA 1047 (1989).

JOLY P., ROBERTS J.E. [1] Approximation of the surface impedance for a stratified medium. Rapports INRIA 1365 (Jan.1991).

JONES D.S. [1] Acoustic and Electromagnetic Waves. Clarendon Press (1986).

[2] Methods in Electromagnetic Wave Propagation. Vols.1-2. Clarendon Press. (1987).

KATO T. [1] Perturbation Theory for Linear Operators. Springer (1976).

KLEINMAN R.E, MARTIN P.A. [1] On single integral equations for the transmission problem of acoustics. SIAM J. Appl. Math. 48 n°2 (April 1988).

KLEINMAN R.E., ROACH G.F. [1] Boundary Integral Equations for the three-Dimensional Helmholtz Equation. SIAM Review 16 n°2 (April 1974).

KLEINMAN R.E., WENDLAND W.L. [1] On Neumann's Method for the Exterior Neumann Problem for the Helmholtz Equation. J. Math. Anal. Appl. 57, n°1 (Jan. 1977).

KLEMAN M. [1] Points Lignes Parois dans les fluides anisotropes et les solides cristallins. Vol. 1 et 2. Les Editions de Physique (1977)

KNAUFF W., KRESS R. [1] On the exterior boundary value problem for the time harmonic Maxwell equations. J. Math. Anal. Appl. 72, 215-235 (1979).

KOBAYASHI S., NOMIZU K. [1] Foundations of differential geometry. Vols. 1-2. Wiley (1963).

KRASNOSEL'SKII M.A., POKROVSKII A.V. [1] Systems with Hysteresis. Springer (1980).

KRESS R. [1] On the Existence of a Solution to a Singular Integral Equation in Electromagnetic Reflection. J. Math. Anal. Appl. 77, 555-566 (1980).

[2] On the Boundary Operator in Electromagnetic Scattering. Proceedings of the Royal Society of Edinburgh 103 A, 91-98 (1986).

LAKHTARIA A., VARADAN V.K, VARADAN V.V. [1] Time Harmonic Electromagnetic Fields in Chiral Media. Lecture Notes in Physics 335. Springer (1989).

LANDAU L., LIFCHITZ E. [1] Electrodynamique des milieux continus. Mir (1969).

LANG S. [1] Introduction aux Variétés Differentiables. Dunod (1967).

LASIECKA I., TRIGGIANI R. [1] Sharp Regularity Theory for Second Order Hyperbolic Equations of Neumann Type. Ann. Mate. Pura Appl. 157, 285-367 (1990).
LAX P.D., PHILLIPS R.S. [1] Scattering Theory. Academic Press (1967).
[2] Scattering Theory for the acoustic equation in an even number of space dimension. Indiana Univ. Math. J. 22, 101-134 (1972).
[3] Local boundary conditions for dissipative symmetric linear differential operator. C.P.A.M. 13, 427-455 (1960).
LESLIE F.M. [1] Some constitutive equations for liquid crystals. Arch. Rational Mech. Anal. 28, 265-283 (1968).
LIONS J.L. [1] Perturbations Singulières dans les Problèmes aux Limites et en Contrôle Optimal. Lecture Notes in Mathematics 323. Springer (1973).
[2] Problèmes aux limites dans les équations aux dérivées partielles. Presses de l'université de Montréal (1965).
[3] Contrôlabilité exacte Perturbations et Stabilisation de Systemes distribués. Vols. 1-2 Collection R.M.A. Masson (1988).
[4] Equations Différentielles Opérationnelles et Problèmes aux Limites. Springer (1961).
LIONS J.L., MAGENES E. [1] Problèmes aux Limites non Homogènes et Applications. Vols 1-3. Dunod (1968-1970).
MALLIAVIN P. [1] Géometrie différentielle intrinsèque. Hermann (1972).
MARSDEN J. [1] Applications of Global Analysis in Mathematical Physics. Publish or Perish (1974).
MASMOUDI M. [1] Résolution numérique de problèmes exterieurs. Thèse de spécialité en Mathématique (1979).
MAYERGOYZ I.D., FRIEDMAN G. [1] Generalized Preisach Model of Hysteresis. IEEE Transactions on Magnetics 24 n°1 (Jan. 1988).
MAYERGOYZ I.D. [1] Mathematical Models of Hysteresis. Ecole d'été CEA-INRIA-EDF. Analyse numérique (1986).
[2] Mathematical Models of Hysteresis. Phys. Rev. Lett. 56 n°15 (April 1986).
MAYSTRE D., VINCENT P. [1] Diffraction d'une onde électromagnétique plane par un objet cylindrique non infiniment conducteur de section arbitraire. Optics Commun 5, 327-330 (1972).
[2] in PETIT [1].
METHEE P.D. [1] Sur les distributions invariantes dans le groupe des rotations de Lorentz. Comments Math. Helv. 28, 225-269 (1954).
[2] L'équation des ondes avec second membre invariant. Comments Math. Helv. 32, 153-164 (1957).
[3] Transformées de Fourier de distributions invariantes liées à la résolution de l'équation des ondes. Colloque international du CNRS Nancy 145-163 (1956).
MEYER.Y. [1] Ondelettes Vol. 1 Ondelettes et opérateurs. Hermann (1990).
MORAWETZ C. [1] The decay of solutions of the exterior initial-boundary value problem for the wave equation, CPAM 14, 561-568 (1961).

[2] The limiting amplitude principle CPAM 15, 349-361 (1962).
MORAWETZ C., RALSTON J., STRAUSS W. [1] Decay of the solution of the wave equation outside a non trapping obstacle. Comm. Pure Appl. Math. 30, 447-508 (1977).
MORREY C.B. [1] Multiple Integrals in the Calculus of Variations. Springer (1966).
MÜLLER C. [1] Foundations of the Mathematical Theory of Electromagnetic Waves. Springer (1969).
MURAT F., TARTAR L. [1] Calcul des Variations et Homogénéisation. Les Méthodes de l'homogénéisation: théorie et applications en physique. 323-369. Ecole d'été d'analyse numérique. CEA-EDF-INRIA Eyrolles (1985).
NECAS J. [1] Les méthodes directes en théorie des équations elliptiques. Masson (1967).
NEDELEC J.C., STARLING F. [1] Integral Equation Methods in a Quasi Periodic Diffraction Problem for the time Harmonic Maxwell's Equations. Rapport 179. C.M.A.P. Polytechnique (April 1988).
NEDELEC J.C. [1] Cours de l'école d'été d'analyse numérique CEA-EDF-INRIA (1978).
PAQUET L. [1] Problèmes mixtes pour le système de Maxwell. Annales Faculté des Sciences de Toulouse. IV, 103-141 (1982).
PAZY A. [1] Semigroups of Linear Operators and Applications to Partial Differential Equations. Applied Math. Sciences 44, Springer (1983).
PEETRE J. [1] Another Approach to Elliptic Boundary Problems. Comm. Pure Appl. Math. 14, 711-731 (1961).
PETIAU G. [1] La théorie des fonctions de Bessel. CNRS (1955).
PETIT R. [1] Electromagnetic Theory of Gratings. Topics in Currents Physics. Springer (1980).
[2] Ondes electromagnétiques en radioélectricité et en optique. Masson (1989).
PETIT R., CADILHAC M. [1] Electromagnetic Theory of Gratings: some advances and some comments on the use of the operator formalism. J. Opt. Soc. Am. A.7, n°9, 1666-1674 (Sept.1990).
PUJOLS A. [1] Time dependent Integral Method for Maxwell Equations. Rapport CESTA/SI A.P.6589 (1991).
[2] Equations intégrales Espace-Temps pour le système de Maxwell. Application au calcul de la Surface Equivalente Radar. Thèse de Doctorat de Math. Appl. (1991).
RALSTON J. [1] Solution of the wave equation with localised energy. Comm. Pure Appl. Math. 22, 807-823 (1969).
RAMM A.G. [1] Scattering by Obstacles. Reidel Publ.Comp. (1955).
RAUCH J., TAYLOR M. [1] Exponential Decay to hyperbolic equations in bounded Domains. Indiana University Math Journal 24 n°1, 7-86 (1974).
RICHTMYER R.D. [1] Principles of Advanced Mathematical Physics. Vol.1 Springer (1978).

ROBERTS J.E., THOMAS J.M. [1] Mixed and Hybrid Methods pp.523-639 Handbook of Numerical Analysis, Vol. II Ed. Ciarlet P.G., Lions J.L., Elsevier (1991).
ROUBINE E.A. [1] Antennes. Masson (1986).
SADOSKY C. [1] Interpolation of Operators and Singular Integrals. Dekker (1979).
SANCHEZ HUBERT J., SANCHEZ PALENCIA E. [1] Vibration and Coupling of Continuous Systems Asymptotic Methods. Springer (1989).
SANCHEZ PALENCIA E. [1] Non-Homogeneous Media and Vibration Theory. Lecture Notes in Physics 127, Springer (1980).
SAUT J.C., SCHEURER B. [1] Unique Continuation for Some Evolution Equations. J. Diff. Equations, 66 n°1 (Jan. 1987).
SCHMIDT E. G. [1] Spectral and scattering theory for Maxwell's equations in an exterior domain. Arch. Rational Mech. Anal. 28, 284-322 (1967-1968).
SCHULENBERGER J. [1] The Debye potential. A scalar factorization for Maxwell's equations. J. Math. Anal. Appl. 63, 502-520 (1978).
SCHWARTZ L. [1] Théorie des Distributions. Hermann (1966).
STEIN E.M., WEISS G. [1] Introduction to Fourier Analysis on Euclidean Spaces. Princeton (1971).
TARTAR L. [1] Topics in Nonlinear Analysis. Publications mathématiques d'Orsay.
[2] Nonlinear Partial Differential Equations Using Compactness Methods. Univ. of Wisconsin Madison (1976).
TAYLOR M. E. [1] Pseudodifferential operators. Princeton (1981).
TREVES F. [1] Topological Vector Spaces. Distributions and Kernel. Academic Press (1967).
[2] Basic Linear Partial Differential Equations. Academic Press (1975).
TRIEBEL H. [1] Interpolation Theory: Function Spaces Differential Operators. North Holland (1978).
von WESTENHOLTZ C. [1] Differential Forms in Mathematical Physics. North Holland (1981).
VISINTIN A. [1] On Landau-Lifshitz Equations for Ferromagnetism. Japan J. Appl. Math. 2, 69-84 (1985).
WEBER C. [1] A local compactness theorem for Maxwell's equations, Math. Meth. Appl. Sci., 2 (1980), pp. 12-25.
WESTON V.H. [1] Theory of Absorbers in Scattering. IEEE Transactions on Antennas and Propagation, 578-584 (Sept. 1963).
WILCOX C.H. [1] Scattering Theory for the d'Alembert Equation in Exterior Domains. Lecture Notes in Math. 442 Springer (1975).
ZIEMIAN B. [1] On G-Invariant Distributions. J. Diff. Equations 35, 66-86 (1980).

Index

Notations

D, E, B, H electromagnetic field
J, ρ electric current, electric charge
ω the angular frequency, ν the frequency
ε_0, μ_0 permittivity and permeability of free space
c the velocity of light in free space
ε, μ permittivity and permeability of the medium,
ε', μ' their real part, ε'', μ'' their imaginary part, σ conductivity of the medium
k the wavenumber, $k^2 = \omega^2\varepsilon\mu$, λ the wavelength, $\lambda = 2\pi/k$ in free space
$Z = \omega\mu/k = k/\omega\varepsilon$ the impedance of the medium, $Y = 1/Z$ its admittance
C_+ the future (or forward) light cone
Φ the (outgoing) elementary solution

Ω an open set in R^n, Γ the boundary of Ω, n a unit normal to Γ
$\gamma_n v = n.v|_\Gamma$ the normal trace of the field v on Γ
$\pi_\Gamma v$ or $t_\Gamma v$ its tangential trace
$\gamma_\tau v = n \wedge v|_\Gamma$ for Ω in R^3 the vector product of n with v at the boundary
$[v]_\Gamma$ the jump of v across Γ, $[v]_\Gamma = v|_{\Gamma_i} - v|_{\Gamma_e}$, with n oriented from Γ_i to Γ_e

Differential operators:
grad, div (in R^n), curl in R^3
grad_Γ, div_Γ, curl_Γ, $\overrightarrow{\text{curl}}_\Gamma$ on a surface Γ (in R^3)
Δ the Laplacian
$\Box = \frac{1}{c^2}\frac{\partial^2}{\partial t^2} - \Delta$ the d'Alembertian

Integral operators: (for Helmholtz) *L*, *P*, 3.1.3. (67), L, K, J, R 3.1.3 (70)
(for Maxwell) L_m, P_m, 3.1.3 (95), T, R 3.1.3 (100)
Riesz operators R_g, R_c, R_g^*, R_c^* 5.1 (96), (96)'
Calderon projectors P_i, P_e, (and S) 3.1.3 (72) and (104), G_i, G_e, their graphs
Pseudodifferential operators: C_i, C_e, *the Calderon operators*

Transformations:
F Fourier transformation, *FL* Fourier-Laplace transformation, *L* Laplace transformation, *R*, R^d, $R^d{}_2$, R^w Radon transformations

Specific notations for the Appendix:
$g = ds^2$ the Riemannian metric, G its associated isomorphism
ω a r-form, $d\omega$ its exterior derivative, $\delta\omega$ its (exterior) coderivative
$d_o = d_m$, d_M, $\delta_o = \delta_m$, δ_M the minimal or maximal realizations of d, δ A.3(111)
Δ^+, Δ^-, Δ^D, Δ^N, Laplacian operators
C_r^+, C_r^- Calderon or capacity operator A.3 (221)

Regular functions $D(\Omega)$ 2.1, $D(\bar{\Omega})$ 2.1, (2), C^∞ regular functions
Distributions $D'(\Omega)$, $S'(R^n)$ tempered distributions, $E'(R^n)$ with compact support

$D'_+(X)$, $L_+(R_t,X)$, $L_+^0(R_t,X)$, X a Banach space

Lipschitz functions $C^{k,\alpha}(\Omega)$, $C^{k,\alpha}(\bar{\Omega})$, $k \in N$, $0<\alpha\leq 1$

L^2 **functions** $L^2(\Omega)$, $L^2(\Gamma)$, $L_t^2(\Gamma)$ 2.1, $L^2_{loc}(\bar{\Omega})$ 2.13 (179) 3.1 (57)
Usual Sobolev spaces $H^s(\Omega)$, $H^s(\Gamma)$, $s \in R$, Sobolev spaces on Ω or Γ, 2.1
$H^1(\Omega)$, $H_0^1(\Omega)$, $H^{1/2}(\Gamma)$, $H^{-1/2}(\Gamma)$; $H(\Delta,\Omega)$, $H^1(\Delta,\Omega)$ 2.1 (3)
Some less usual Sobolev spaces $H^1_{loc}(\bar{\Omega})$ 3.1 (57), $H^{1/2}_{oo}(\Gamma_1)$ 2.7 (67)
$H^{1/2}_{oo}(\Gamma_1)'$ 2.7 (68), (68)', $H^{1/2}_{oo}(\tilde{\Gamma}_0)$ 2.7 (72), $H^{-1/2}_{oo}(\tilde{\Gamma}_0)$ 2.7 (72)
and also $H_k(R^2)$ 5.1 (104), $\tilde{H}_k(R^2)$ 5.1 (105), $H_k^{1/2}(R^2)$, $H_k^{-1/2}(R^2)$ 5.1 (35)
Beppo Levi spaces $W^1(R^n)$, $W^1(\Omega)$, $W_0^1(\Omega)$
"Sobolev" spaces with time : $e^{\gamma t}H^s_\gamma(\Omega x R)$, $e^{\gamma t}H^s_\gamma(\Gamma x R)$, $D_\gamma(B)$
Spaces for div and curl :
$H(div,\Omega)$, $H_o(div,\Omega)$, $H(curl,\Omega)$, $H_o(curl,\Omega)$, 2.1 (5), (6), (7), (8)
$H^{-1/2}(div,\Gamma)$, $H^{-1/2}(div,\Gamma)$ trace spaces 2.4 (36)

$H^s(\bar{\Omega}')$, $H_{loc}(div,\bar{\Omega}')$, $H_{loc}(curl,\bar{\Omega}')$, 2.2 (27),(37) Ω' unbounded open set

Some less standard spaces on curl and div
$H^{-s}(div,\Omega)$, 2.7 (79), $H^{-s}(div,\Gamma_1)$ 2.4 (42), 2.7 (83), $H^{-s}(curl,\Gamma_1)$ 2.7 (84)
$H_0^{-s}(div,\Gamma_1)$ 2.7 (85), $H_0^{-s}(div,\Gamma_1)$ 2.7 (85), $X(div,\Gamma_1)$ 2.7 (86), $X(curl,\Gamma_1)$ 2.7 (87)
$H^s(curl,\Omega)$ 2.10 (128)
$H^1_{\Gamma_o}(\Omega)$ 2.9 (106), $H_{\Gamma_o}(curl,\Omega)$ 2.9 (106), $H_{\Gamma_o}(div,\Omega)$ 2.9 (106)
cohomology spaces $H^1(\Omega)$, $H^2(\Omega)$ 2.8 (104), $H(\Gamma)$ 2.9
$J(\Omega)$, $\tilde{J}(\Omega)$, $J_\tau^1(\Omega)$, $J_\nu^1(\Omega)$ 2.9 (111), $W^i(\Omega)$ Whitney spaces 2.14

Spaces for quasiperiodic functions $D_{K,L}$, $D'_{K,L}$, 5.2.1 (172) $H^s_{K,L}(R^2)$, $H^s_{K,L}(curl,P)$
Spaces for the Appendix Let M be a Riemannian manifold (with boundary Γ)

Spaces of regular r-forms $D_r(M)$, $\underline{D}_r(M)$, $D_r(\bar{M})$

Spaces of generalized r-forms $D'_r(M)$, $\underline{D}'_r(M)$, $S'_r(R^n)$

Spaces of L^2 r-forms $L^2_r(M)$

Spaces of Sobolev r-forms $H^s_r(M)$, $\underline{H}^s_r(M)$, $H^+_r(M)$, $H^-_r(M)$, $H^1_{ro}(M)$
$V_r^+ = H^1_{rt}(M)$, $V_r^- = H^1_{rn}(M)$

Spaces of r-forms **based on the exterior differential:**
$H_r(d,M)$, $H_r(\delta,M)$, $H_r(d,\delta,M)$, $H^s_r(d,M)$, $H^s_r(\delta,M)$, $H_{r,o}(d,M)$, $H_{r,o}(\delta,M)$
$H_{rn}(d,\delta,M) = H_{ro}(d,M) \cap H_r(\delta,M)$, $H_{rt}(d,\delta,M) = H_{ro}(\delta,M) \cap H_r(d,M)$.

Series on Advances in Mathematics for Applied Sciences

Aims and Scope

This Series reports on new developments in mathematical research relating to methods, qualitative and numerical analysis, mathematical modeling in the applied and the technological sciences. Contributions related to constitutive theories, fluid dynamics, kinetic and transport theories, solid mechanics, system theory and mathematical methods for the applications are welcomed.

This Series includes books, lecture notes, proceedings, collections of research papers. Monograph collections on specialized topics of current interest are particularly encouraged. Both the proceedings and monograph collections will generally be edited by a Guest editor.

High quality, novelty of the content and potential for the applications to modern problems in applied science will be the guidelines for the selection of the content of this series.

Instructions for Authors

Submission of proposals should be addressed to the editors-in-charge or to any member of the editorial board. In the latter, the authors should also notify the proposal to one of the editors-in-charge. Acceptance of books and lecture notes will generally be based on the description of the general content and scope of the book or lecture notes as well as on sample of the parts judged to be more significantly by the authors.

Acceptance of proceedings will be based on relevance of the topics and of the lecturers contributing to the volume.

Acceptance of monograph collections will be based on relevance of the subject and of the authors contributing to the volume.

Authors are urged, in order to avoid re-typing, not to begin the final preparation of the text until they received the publisher's guidelines. They will receive from World Scientific the instructions for preparing camera-ready manuscript.

SERIES ON ADVANCES IN MATHEMATICS FOR APPLIED SCIENCES

Vol. 17 The Fokker-Planck Equation for Stochastic Dynamical Systems and Its Explicit Steady State Solution
by C. Soize

Vol. 18 Calculus of Variation, Homogenization and Continuum Mechanics
eds. G. Bouchitté et al.

Vol. 19 A Concise Guide to Semigroups and Evolution Equations
by A. Belleni-Morante

Vol. 20 Global Controllability and Stabilization of Nonlinear Systems
by S. Nikitin

Vol. 21 High Accuracy Non-Centered Compact Difference Schemes for Fluid Dynamics Applications
by A. I. Tolstykh

Vol. 22 Advances in Kinetic Theory and Computing: Selected Papers
ed. B. Perthame

Vol. 23 Waves and Stability in Continuous Media
eds. S. Rionero and T. Ruggeri

Vol. 24 Impulsive Differential Equations with a Small Parameter
by D. Bainov and V. Covachev

Vol. 25 Mathematical Models and Methods of Localized Interaction Theory
by A. I. Bunimovich and A. V. Dubinskii

Vol. 26 Recent Advances in Elasticity, Viscoelasticity and Inelasticity
ed. K. R. Rajagopal

Vol. 27 Nonstandard Methods for Stochastic Fluid Mechanics
by M. Capinski and N. J. Cutland

Vol. 28 Impulsive Differential Equations: Asymptotic Properties of the Solutions
by D. Bainov and P. Simeonov

Vol. 29 The Method of Maximum Entropy
by H. Gzyl

Vol. 30 Lectures on Probability and Second Order Random Fields
by D. B. Hernández

Vol. 31 Parallel and Distributed Signal and Image Integration Problems
eds. R. N. Madan et al.

Vol. 32 On the Way to Understanding The Time Phenomenon: The Constructions of Time in Natural Science – Part 1. Interdisciplinary Time Studies
ed. A. P. Levich

SERIES ON ADVANCES IN MATHEMATICS FOR APPLIED SCIENCES

Vol. 33 Lecture Notes on the Mathematical Theory of the Boltzmann Equation
ed. N. Bellomo

Vol. 34 Singularly Perturbed Evolution Equations with Applications to Kinetic Theory
by J. R. Mika and J. Banasiak

Vol. 35 Mechanics of Mixtures
by K. R. Rajagopal and L. Tao

Vol. 36 Dynamical Mechanical Systems Under Random Impulses
by R. Iwankiewicz

Vol. 37 Oscillations in Planar Dynamic Systems
by R. E. Mickens

Vol. 38 Mathematical Problems in Elasticity
ed. R. Russo

Vol. 39 On the Way to Understanding the Time Phenomenon: The Constructions of Time in Natural Science — Part 2. The "Active" Properties of Time According to N. A. Kozyrev
ed. A. P. Levich

Vol. 40 Advanced Mathematical Tools in Metrology II
eds. P. Ciarlini et al.

Vol. 41 Mathematical Methods in Electromagnetism
Linear Theory and Applications
by M. Cessenat

www.ingramcontent.com/pod-product-compliance
Lightning Source LLC
Chambersburg PA
CBHW050453190326
41458CB00005B/1268
* 9 7 8 9 8 1 0 2 2 4 6 7 7 *